Isotope Geochemistry

Isotope Geochemistry

The Origin and Formation of Manganese Rocks and Ores

Vladimir Kuleshov

Edited by

J. Barry Maynard

AMSTERDAM • BOSTON • HEIDELBERG • LONDON • NEW YORK • OXFORD
PARIS • SAN DIEGO • SAN FRANCISCO • SINGAPORE • SYDNEY • TOKYO

Elsevier
Radarweg 29, PO Box 211, 1000 AE Amsterdam, Netherlands
The Boulevard, Langford Lane, Kidlington, Oxford OX5 1GB, United Kingdom
50 Hampshire Street, 5th Floor, Cambridge, MA 02139, United States

Library of Congress Cataloging-in-Publication Data
A catalog record for this book is available from the Library of Congress

British Library Cataloguing-in-Publication Data
A catalogue record for this book is available from the British Library

ISBN: 978-0-12-803165-0

For information on all Elsevier publications
visit our website at https://www.elsevier.com/

Publisher: Candice Janco
Acquisition Editor: Amy Shapiro
Editorial Project Manager: Tasha Frank
Production Project Manager: Vijayaraj Purushothaman
Cover Designer: Greg Harris

Typeset by SPi Global, India

Contents

Foreword

Questions about the geochemistry of manganese, the regularities of the distribution of manganese deposits, and the composition of manganese ores and conditions of their formation have been treated in an extensive scientific literature, consisting of over 5000 titles. Among them are commonly known works by V.I. Vernadskii, A.E. Fersman, A.G. Betekhtin, N.S. Shatskii, N.M. Strakhov, and S. Roy. Substantial contributions to the explanation of the nature of manganese-ore deposits have been provided by the research of I.M. Varentsov, J.B. Maynard, K.F. Park, J. Ostwald, B. Bolton, F. Veber, N. Beukes, J. Gutzmer, G.S. Dzotsenidze, D.G. Sapozhnikov, E.A. Sokolova, L.E. Schterenberg, and many other Russian and international researchers.

Despite the accumulation of a vast array of data on the geology of manganese deposits and particularly pertaining to the chemical composition of manganese rocks and ores, many questions of manganese ore-genesis remain only partially answered. This is the case, first and foremost, with genetic models of the formation of the principal industrial types of manganese ores that are contained in such giant deposits and manganese-ore basins as the Kalahari (Republic of South Africa), groups of Oligocene deposits of the Paratethys (Ukraine, Georgia, Kazakhstan, and Bulgaria), the northern Urals (Russia), the Gulf of Carpentaria (Groote Eylandt, Arnhem Land, and elsewhere in Australia), and others.

The structure of any model of ore genesis, including manganese ores, is predicated upon the presence of a logically complete and factually justified conceptual basis. That is, the model should account for such important questions as the sources of ore and non-ore components, the conditions of formation (*exogenous conditions*: climate, paleogeography, type of paleo-water body, physicochemical conditions; *endogenous conditions*: temperature, pH, Eh, pressure), as well as the ore-formation process's evolution over time (in part for concrete deposits, in whole for the history of the establishment of the Earth's lithosphere). Naturally, the conditions will vary for different industrial ore types.

To date the question also remains open as to the principal regularities of the evolution of the processes of accumulation of manganese in rocks of the lithosphere over the entire course of the Earth's formation. The formation of manganese rocks and ores has occurred unevenly over the course of geological history; this has been recorded in epochs and periods of manganese accumulation and is contingent upon the predominance of a given mechanism (model) of manganese ore-genesis.

In the present work, by means of generalizing the data available from the literature and particularly factual material, an attempt has been made to briefly clarify certain particularities in the genetic aspect of the formation of the manganese deposits themselves as much as the principal regularities of manganese ore-genesis in the history of the geological development of rocks of the lithosphere.

The principal types of manganese ores of deposits under development are oxides and carbonates. The oxides present the greater practical interest; however, the principal reserves of manganese—with the exception of the braunite-lutite of the deposits of the Kalahari manganese-ore field (Republic of South Africa)—are contained mainly within carbonate rocks. Therefore, the study of carbonates is undoubtedly of great practical significance.

Isotope research constitutes one of geology's most informative high-precision methods. Data on the isotopic composition of the carbon and oxygen found in manganese carbonates in many cases allow the identification of the principal regularities of the genesis of these carbonates and enable a more refined understanding of many aspects of the formation of the manganese deposits themselves. Established regularities of the distribution of isotopic composition are likewise useful in prospecting for new manganese deposits.

Although by now the geology and material composition of many known and industrially developed manganese deposits have been studied in detail and a massive base of isotope data has been gathered, systematic isotope research in manganese ores for all purposes remains to be conducted. It is precisely this "gap" in the scientific literature that the author of the present work aims to fill.

This monograph provides a generalization of the isotope data for a representative collection of natural manganese carbonates, collected from the modern sediments of lakes, seas, and oceans as well as directly from known manganese deposits in countries of the former USSR (Russia, Ukraine, Georgia, and Kazakhstan) and in other countries (Australia, Republic of South Africa, Ghana, Gabon, Brazil, and elsewhere). This work characterizes in detail the principal genetic types of manganese rocks and ores and the particularities of their formation. The obtained isotope data allow for determining the genetic classification of manganese deposits. It has been established that the manganese-ore-forming process in the sedimentary basins does not reach its completion in the early diagenetic stage, but continues into later diagenesis and subsequently in the stage of catagenesis (epigenesis).

The particularities, illustrated here, of the accumulation of manganese in the stratisphere, conditioned by the predominance of a given mechanism (model) of manganese ore-genesis, have allowed a delineation of the principal epochs and periods of manganese accumulation in the history of the Earth's development.

The conducting of the isotope research and the writing of the present monograph were accomplished with the participation and constant support of my colleagues. Over the course of an extended period of studying manganese deposits, discussion of the isotope data took place with the direct participation of my instructor and head of the Laboratory of Geochemistry of Isotopes and Geochronology of the Geological Institute of the Russian Academy of Sciences (GIN RAS), Professor V.I. Vinogradov and my colleague, doctor of Geological-Mineralogical Sciences, B.G. Pokrovsky, to whom the author is eternally grateful.

The author expresses deep appreciation to those colleagues and specialists in the field of the geology and geochemistry of the deposits of manganese ores and manganese-bearing sediments, who for the purposes of isotope research kindly lent their personal collections: A.I. Brusnitsyn (SPbGU), E.V. Starikovaia (SPbGU), A.G. Rozanov (IO RAS), V.N. Sval'nov (IO RAS), L.E. Schterenberg (GIN RAS), E.A. Sokolovaia (GIN RAS), and Zh.V. Dombrovskaia (IGEM RAS). The author thanks A.F. Bych, Iu.V. Mirtov, S.M. Mirtova (ZapSibGU, Novokuznetsk), and B.A. Gornostai (PGO "Arkhangel'skgeology," Nar'ian-Mar) for their helpful assistance in the selection of mineral material during the fieldwork.

Invaluable support in understanding the geology of the supergiant deposit of the Kalahari manganese-ore field was provided by Johannesburg University (Republic of South Africa) Professors N. Beukes and J. Gutzmer, to whom the author expresses sincere gratitude.

The author is grateful for his mentor in the field of the geochemistry of manganese deposits, a tireless reviewer of practically all scientific publications on the isotope geochemistry of manganese deposits and principal researcher at the Geological Institute of the Russian Academy of Sciences, doctor of Geological-Mineralogical Sciences, I.M. Varentsov.

Finally, author extends special gratitude to the editor of this monograph, professor of Department of Geology, University of Cincinnati, J. Barry Maynard. His comments, notes, and addition of new data on geology and geochemistry of world's manganese deposits allowed us to produce a significantly improved English edition.

Editor's Preface

MANGANESE AND ITS ROLE IN GEOCHEMISTRY

Manganese is the 10th most abundant element in the Earth's crust. Most of its industrial use is in steel making with a much lesser amount going into the production of batteries. It is very similar to iron in its chemical properties. Both are commonly found in +2 and +3 valences with high spin states for the 3d electrons and with similar ionic radii. Mn and Fe^{2+} ions have radii 0.83 and 0.78 Å, while the 3+ ions have 0.70 and 0.65 Å (Li, 2000, Table I-4). Accordingly, manganese is commonly found substituted in small amounts in iron minerals. Manganese, however, also has access to a higher valence state, +4, which gives rise to a plethora of complex manganese oxide minerals that do not have Fe counterparts. By contrast, Mn sulfides are quite rare compared to their Fe cousins. The net result is a tendency, in sedimentary systems with a large redox gradient, to partition iron into the more reducing parts of the system as the sulfide, whereas manganese will move toward areas of higher oxidation potential and tends to precipitate when it encounters mildly oxidizing conditions.

It follows from the above that manganese has geochemical significance in its own right, both as an abundant constituent of the Earth's crust and as a critical industrial metal. It has additional significance in two ways: first, its oxides are highly effective adsorbents for other metals (especially Cu, Pb, Zn plus Ba) so these minerals carry a record of the composition of fluids they have been exposed to. Second, its various redox states provide a useful window into the history of oxidation levels at the Earth's surface.

Among the most effective ways to probe the mechanisms of action of manganese in Earth surface environments is to study the behavior of the stable isotopes of manganese minerals. Therefore, the appearance of a new book with many new details on the isotope geochemistry of the manganese deposits in the former Soviet Union and elsewhere was very welcome when the Russian edition of this book came out in Dec. 2013 (Kuleshov, 2013). To bring this information to a wider audience, we present the translation of the original Russian text with some updates.

By way of introducing the subject, I present a few preliminary observations. An examination of the distribution of manganese among the various reservoirs that make up the Earth reveals much about how the element behaves in geochemical cycles. Table P1 compares manganese and iron in some common rock reservoirs and in some key rock types and types of natural waters. The geochemistry of manganese closely resembles that of iron, but iron has such a greater crustal abundance that it normally swamps out any manganese present. Therefore an understanding of manganese behavior, especially when it comes to the formation of ore deposits, entails an understanding of how manganese and iron differ.

Note the similarity of Mn/Fe ratios in all solid reservoirs. Therefore, ordinary sedimentary processes will not separate manganese from iron. Seawater has higher Mn/Fe ratios, and in the open ocean surface waters are somewhat enriched in Mn compared with deep waters. Note, however, the very strong enrichment of Black Sea deep water in Mn, which suggests an important role for anoxic basins in the genesis of Mn deposits.

Manganese ores are far from uniformly distributed in time and space. The Early Proterozoic of South Africa saw the formation of the world's largest endowment of manganese. It is followed in size by a much younger array of deposits ringing the present-day Black Sea that formed in the Oligocene. Table P2 shows production and reserves of manganese by country, as compiled by the USGS (2016).

Table P1 Distribution of Mn and Fe in Various Reservoirs of the Earth

Reservoir	^{25}Mn	^{26}Fe	Mn/Fe
Carbonaceous chondrite C1 (ppm)	1990	190,400	0.010
Upper mantle (ppm)	1000	64,000	0.016
Average basalt (ppm)	1550	83,000	0.019
Average granite (ppm)	390	21,100	0.018
Upper continental crust (ppm)	600	35,000	0.017
Sedimentary rocks			
Mudstone (ppm)	728	41,800	0.017
Sandstone (ppm)	852	34,500	0.025
Limestone (ppm)	418	9500	0.044
Average sediment (ppm)	852	35,800	0.024
Seawater			
Atlantic shallow (μg/L)	104.4	111.7	0.93
Atlantic deep (μg/L)	98.9	390.9	0.25
Pacific shallow (μg/L)	104.4	11.2	9.34
Pacific deep (μg/L)	43.9	111.7	0.39
Black Sea shallow (μg/L)	0.56	0.293	1.91
Black Sea deep (μg/L)	333	4.11	81.0

"Based on data in Lin Yuan-Hui's 'Compendium of Geochemistry'."

Table P2 World Mine Production and Estimated Reserves (Thousands of Metric Tons)

	Mine Production			Reserves	Host Rock Ages	
	2013	2014	2015	2015	Ma	
South Africa	4300	5200	6200	200,000	2250	
Ukraine	300	422	390	140,000	28	
Australia	2980	3050	2900	91,000	95	
India	920	945	950	52,000	3200	770
Brazil	1120	1040	1000	50,000	2100	700
China	3000	3000	3000	44,000	367	230
Gabon	1970	1860	1800	22,000	2143	
Ghana	533	418	390	13,000	2050	
Kazakhstan	390	385	390	5000	28	
Mexico	212	236	240	5000	156	
Malaysia	430	378	400	NA	unknown	unknown
Burma	157	98	100	NA	unknown	unknown
Others	597	740	740	Small		
World	16,900	17,800	18,000	620,000		

NA, not available.
Source: http://minerals.er.usgs.gov/minerals/pubs/commodity/manganese/index.html.

Table P3 The Dominant Mn Minerals in Commercial Deposits

Mineral	Formula	% Of Deposits Where Dominant
Rhodochrosite	$MnCO_3$	32
Braunite	$Mn^{2+}Mn_6^{3+}SiO_{12}$	23
Cryptomelane	$K(Mn^{4+},Mn^{2+})_8O_{16}$	9
Manganite	$MnO(OH)$	8
Pyrolusite	MnO_2	5
Hausmannite	$Mn^{2+}Mn_2^{3+}O_4$	3
Romanechite	$(Ba,H_2O)_2(Mn^{4+},Mn^{3+})_5O_{10}$	4
Amorphous oxides	–	2
Kutnohorite	$Ca(Mn,Mg,Fe^{2+})(CO_3)_2$	2
Mn calcite	$(Ca,Mn)CO_3$	2
Todorokite	$(Na,Ca,K)_2(Mn^{4+},Mn^{3+})_6O_{12}\cdot 3–4.5(H_2O)$	2
Others (oxides)		8

Modified from Maynard, J.B., 2010. The chemistry of manganese ores through time: a signal of increasing diversity of earth-surface environments. Econ. Geol. 105 (3) 535–552.

Note the dominance of the South African deposits and their growing production numbers. Note also that the Oligocene deposits occupy a strong second place. Therefore, large manganese deposits are not confined to the Precambrian, as are those of iron. Whatever process leads to the formation large manganese deposits, it is not one that requires an oxygen-free atmosphere or any other sort of extreme geochemical conditions. It therefore behooves us to seek analogs in the modern for mechanisms of manganese ore genesis.

I mentioned that there is a large array of manganese minerals known, numbering in the hundreds. Only a few, however, make up the dominant ore minerals in commercial-scale ore deposits (Table P3). The prominence of the carbonates is noteworthy. It suggests again that reducing conditions are key in the genesis of manganese ores, and also provide a tool, because carbonates carry two isotope signals, one from carbon and one from oxygen. Of the two, carbon is the more stable in the face of later changes and so tends to reflect original sedimentary conditions, whereas oxygen exchanges more readily with fluids with which it comes into contact. Therefore, oxygen reflects more the later behavior of the system.

Finally, we should say a word about the fantastic mineral endowment present in the Kalahari deposits. Not only is this the world's greatest repository of manganese, it also contains a treasure trove of mineralogical specimens. The world of manganese mineralogy revealed has been documented in a pair of beautifully prepared books: "The Manganese Adventure: the South African Manganese Fields" by Cairncross et al. (1997) and "The Kalahari Manganese Field—The Adventure Continues." 2013 by Cairncross and Beukes (2013) (see Maynard, 2013). These books show us both the external beauty of these minerals and the internal beauty of an understanding of manganese geochemistry.

ACKNOWLEDGMENTS

The cost of the translation of this volume was underwritten by Upstream Resources, LLC; by the Nic Beukes Publication Research Fund, Geology Department, University of Johannesburg; and by the Jenks Economic Geology Fund of the University of Cincinnati. The author and editor express our sincere thanks for this help in bringing this project to fruition.

CHAPTER

MANGANESE ROCKS AND ORES

1

Manganese is a silvery-white, brittle metal, possessing a density of 7.2–7.46 g/cm^3, a hardness of 5–6 (Mohs scale), and a melting temperature of 1244°C. Manganese is a transition metal, belonging to the group of siderophiles (a geochemical class after V.M. Goldschmidt) and occupying the 25th place (atomic number) of the VII group of the 4th period in D.I. Mendeleev's periodic table; it has an atomic weight equal to 55. Among its atoms are known one stable isotope—^{55}Mn—and 11 radioactive isotopes—from ^{49}Mn to ^{58}Mn (Lavrukhin and Iurkin, 1974). The mean content of manganese in the Earth's crust constitutes approximately 0.1% (by weight) (*Kratkii spravochnik*…, 1970).

The basic electronic configuration of manganese is $1s^22s^22p^63s^23p^63d^54s^2$. Its ions can have up to 10 oxidation states (Salli, 1959; Emsli, 1993), of which in the conditions of the Earth's crust are realized only four—Mn^{2+} (d^5), Mn^{3+} (d^4), Mn^{4+} (d^3), and Mn^{7+} (d^0). Of these four, only the 2^+ and 4^+ occur stably in natural waters. Mn^{3+} does occur in the solid state, as in minerals such as manganite (see Post, 1999 for a review of the structures of manganese oxides), where it is stabilized by crystal field effects. This stabilization energy applies even more strongly to manganese in the 4^+ valence state but not to Mn^{2+}. The extra energy component causes Mn^{4+} in octahedral positions in minerals to be strongly favored over Mn^{2+} in solution (Crerar et al., 1980, p. 296). Moreover, minerals that contain Mn^{2+}, such as rhodochrosite, tend to be light colored, whereas Mn^{4+} minerals tend to be dark, often black, which arises because of the splitting of d orbitals of Mn^{4+} in the imposed crystal field, which gives rise to excited states with the same spin multiplicity and enhanced ability to absorb light photons.

The principal consumer of manganese (>90%) is the metallurgical industry, where it is used predominantly in the form of alloys with iron, "ferromanganese," and silicon, "silicomanganese," as well as in the form of metallic manganese (95–99% Mn), applied in the deoxidation and desulfurization of iron, in the formation of liquid slag, and in the alloying of steel (from 1–2% to 12–14% Mn). In a comparatively small quantity, manganese is used in the production of alloys with nonferrous metals such as copper, aluminum, and nickel, for example in the production of manganin, bronzes, and brasses. Only 5–10% of the metal is consumed in electrical systems for the production of dry-cell batteries and in the chemical industry, in ceramic and glass production, and in the agricultural sector for additives in mineral fertilizers and in feed for livestock.

In nature, manganese is composed of various *manganese rocks* (synonym: *manganoliths*)—a class of sedimentary rocks (understood in the broad sense of those that formed as a result of the processes of the full cycle of sedimentogenesis: from the physical and chemical destruction of the parent rocks of the terrain to the transformation of sediments into sedimentary-rock basins up to the stage of catagenesis, inclusively), consisting predominantly (>50%) of manganese minerals (Mn content—15–20% and greater); commonly it is used in the literature as a synonym for manganese ore.

Isotope Geochemistry. http://dx.doi.org/10.1016/B978-0-12-803165-0.00001-X

There are distinguished, depending on the manganese content, manganese-containing (5–15%) and manganiferous (up to 5% Mn) subsurface rocks (carbonates, jasperoids, cement of conglomerate rocks, etc.).

Manganese rock by composition is represented by two subgroups of the chemical and biochemical group of sedimentary rocks (Frolov, 1964a,b)—*carbonate and oxide (manganolites)*. Manganolites represent sedimentary rocks and ores, the predominant component of which are oxides and hydroxides of manganese (Geologicheskii slovar', 1973. T.1, p. 411), which in natural conditions are found in the form of layer deposits, lenses, concretions, and weathering crusts (in terms of manganese and manganese-containing rocks) (Kuleshov, 2011a).

Because of the variety of oxidation states, there are a large number of manganese minerals known. The Webmineral site lists 190 minerals with Mn contents of 25% or greater. Among these, however, only a relative handful—30 or so, predominantly the oxides, hydroxides, and carbonates—dominate the phases in commercial ores.

The predominant minerals of the oxides and hydroxides of manganese are represented by several groups:

1. those with high valence of manganese: modification: pyrolusite, ramsdellite; the group of nsutite — $Mn_{1-x}^{4+}Mn_{x}^{2+}O_{2-2x}\cdot(OH)_{2x}$; the group (Ba, Na, K, Pb)$Mn_8O_{16}\cdot xH_2O$ (hollandite, coronadite, cryptomelane, and manjiroite); psilomelane (or romanechite) [(Ba, K, Mn, Co)$_2Mn_5O_{10}\cdot xH_2O$]; the group of burnesite (Ca, Na)(Mn^{2+}, Mn^{4+})$_7O_{14}\cdot 3H_2O$; todorokite (Na, Ca, K, Mn^{2+})(Mn^{4+}, Mn^{2+}, Mg)$_6O_{12}\cdot 3H_2O$;ranciéite (Ca, Mn^{2+}) $Mn_4^{4+}\cdot nH_2O$, and the group of hydroxides of manganese—MnOOH (groutite, feitknechtite, manganite, crednerite, quenselite, and janggunite);
2. those with lower valence of manganese: braunite $3Mn_2O_3\cdot MnSiO_3$; and bixbyite $(Mn, Fe)_2O_3$; and
3. minerals of the isomorphic system Fe_3O_4-Mn_3O_4: jacobsite, hausmannite, and vredenburgite.

Mineral carbonates of manganese are less valuable than raw manganese ore and are used primarily in the capacity of flux material in ferromanganese smelting. Manganese carbonates are represented by minerals of the isomorphic series: rhodochrosite-calcium rhodochrosite—manganocalcite—manganiferous calcite [$MnCO_3$-$(Mn_m,Ca_n)CO_3$)], and oligonite-manganosiderite-kutnohorite—$(Mn_m,Fe_n,Mg_k)CO_3$.

In nature, manganese-containing rocks of metamorphic genesis are also widely distributed. These are formed as a rule as a result of the metamorphism of initially sedimentary manganese and manganese-containing rocks. In them, manganese is a component of metamorphic minerals—silicates (silicates of manganese, manganese-containing garnets, amphiboles, and pyroxenes), carbonate rocks that vary in terms of composition and degree of metamorphism, and "manganized" schists and phyllites. In many cases, in the weathering crusts on these rocks are formed large deposits of rich oxide ores of manganese.

A significant group is formed by manganese-containing rocks of hydrothermal genesis, composing ore-bearing bodies (as a rule—veins) with low contents of manganese and insignificant reserves. In vein rocks, manganese is embedded within its own minerals—oxides, carbonates, silicates, and sulfides—as frequently as it occurs as a component of various minerals. In the oxides of manganese of hydrothermal veins of deep genesis, other elements are also commonly present—Pb, Ba, Zn, Ag, etc. Spatially and genetically hypogenous minerals of manganese are commonly connected with barite, fluorite, calcite, sulfides of nonferrous metals, and gold-silver mineralization. In the industrial regard, manganese-containing rocks of this type, as a rule, do not present interest.

Potentially present in the mineral assemblage of manganese rocks are detrital quartz, oxides and hydroxides of iron, clayey minerals, zeolites, and rarely phosphates and sulfides.

The predominant types of manganese ores, according to Betekhtin (1946), are braunite-hausmannite, psilomelane-pyrolusite and psilomelane-vernadite, quartz-pyrolusite, rhodochrosite, and opal- and chlorite-rhodochrosite.

In terms of reserves, deposits of manganese can be divided into unique (greater than 100 million metric tons of metal—only South Africa and the Ukraine), very large (50 million tons), large (5 million tons), and small (less than 5 million tons), based on the U.S. Geological Survey annual estimates of worldwide production and reserves (see Editor's Preface).

In terms of the dimensions and form of ore bodies, the variability of thickness, internal structure, and quality of the ores, deposits of manganese (sections of large deposits for development by independent enterprises) correspond to the first, second, and third groups of complexity of the "Classification of reserves of deposits and inferred resources of solid commercial minerals" (GKZ, 1997).

To the first group belong deposits of simple geological structure with ore bodies, represented by fairly large horizontal sheet or low-inclined deposits with consistent thickness, even distribution of manganese, and regular intervals of different types of ores; they are embedded in terrigenous-carbonate formations of oceanic genesis (eg, the Nikopol and Bolshoi Tokmak deposits, Ukraine) (Table 1.1).

To the second group belong deposits likewise connected with terrigenous-carbonate rocks of oceanic genesis, but of more complex geological structure, represented by fairly large, moderately pitching sheet deposits with inconsistent thickness, uneven distribution of manganese, complex and irregular

Table 1.1 World Mine Production and Estimated Reserves (Thousands of Metric Tons)

	Mine Production			Reserves
Mine	**2013**	**2014**	**2015**	**2015**
South Africa	4300	5200	6200	200,000
China	3000	3000	3000	44,000
Australia	2980	3050	2900	91,000
Gabon	1970	1860	1800	22,000
Brazil	1120	1040	1000	50,000
India	920	945	950	52,000
Others	597	740	740	Small
Malaysia	430	378	400	NA
Ghana	533	418	390	13,000
Kazakhstan	390	385	390	5000
Ukraine	300	422	390	140,000
Mexico	212	236	240	5000
Burma	157	98	100	NA
World	16,900	17,800	18,000	620,000

NA, *not available.*
http://minerals.er.usgs.gov/minerals/pubs/commodity/manganese/index.html

combination of different types of ores, and the presence of barren interbeds (eg, Chiatura deposit, Georgia; the northern Ural group of deposits, Russia; as well as certain volcanogenic- (hydrothermal) sedimentary and metamorphic deposits with large and medium sheet deposits of complex structure and inconsistent thickness, with uneven distribution of manganese and irregular intervals of different types of ores).

To the third group belong numerous supergenic deposits with fine lensoid and nodular deposits, with uneven mineralization and complex morphology, as well as deposits of other industrial types with fine sheet and lensoid deposits of complex structure, with inconsistent thickness and conditions of occurrence, uneven distribution of manganese, and irregular intervals of different types of ores, with numerous interbeds and inclusions of barren rocks (eg, Iuzhno-Khingansk and Mazul'sk deposits).

It follows to note that in the present work, the term "manganese ore" is used in the scientific understanding. In the strict sense this term has an economic connotation: "ore – a natural mineral raw material, containing metals or their compounds in a quantity and in a form suitable for their industrial use… They are distinguished as naturally rich ores, or poor ores requiring enrichment…" (Gornaia entsiklopediia, 1989, t. 4, p. 412), and "ore – a mineral material, from which it is technologically possible and economically viable to extract by bulk method metals or minerals for their use in the national economy…" (Geologicheskii slovar', 1973, t. 2, p. 193).

Many of the principal manganese ore deposits and prospective deposits accounted for in the national register of mineral reserves of Russia are in fact subeconomic with prevailing metallurgical and mining technologies. Examples include the northern Urals group of deposits: Tin'inskoe, Loz'vinskoe, Iurkinskoe, Berezovskoe, etc.), of the Komi Republic (Parnokskoe), of the Evreiskaya Autonomous Oblast', of Irkutsk (Nikolaevskoe), and of other regions (Potkonen, 2001).

The majority of cases, especially for carbonates and oxides of iron-manganese ores, are unprofitable and cannot be considered ores. In metallurgy today, the majority of these "ores" strictly defined are used in the capacity of a flux or in the capacity of an additive to high-quality, predominantly imported (Kazakhstan, Australia, Republic of South Africa) ores, in order to cut the production cost of the finished product.

CHAPTER

MANGANESE CARBONATES IN MODERN SEDIMENTS

2

An understanding of the genesis of ancient carbonate, oxide, and oxide-carbonate-manganese rocks and ores requires a conceptualization of the conditions of the generation of manganese-containing deposits in modern sedimentation basins—oceans, seas, and lakes.

As is known, the process of modern ferromanganese rock and ore formation is fairly widespread. It occurs as much on the bed of the global ocean as within the boundaries of marginal and inland water bodies (seas and lakes). The accumulation of Fe-Mn rocks and ores is manifest principally in the form of various oxide-ferromanganese crusts and concretion nodules, as well as metal-bearing sediments of the zones of subaqueous discharge of hydrothermal systems. In the majority of cases, they are of diagenetic origin and are characterized by mineralogical-geochemical characteristics reflective of the specific conditions of their formation (type of water body, location, source of ore material, etc.).

The accumulation of manganese in the form of carbonates—authigenic carbonates of complex composition—is likewise a fairly widespread phenomenon in the Pleistocene-Holocene sediments of lakes, seas, and oceanic areas of near-continental lithogenesis. Characteristic for these is a variable content of Mn, Ca, Mg, and Fe. It is proposed (Logvinenko et al., 1972) that conditions favorable to the generation of such kind of manganese carbonate (rhodochrosite) are created in the sediments during the process of diagenesis with low contents of organic matter. Such conditions are observed in the transitional zone of the ocean: from the littoral, where sediments are strongly reduced, toward the pelagic, where sediments are oxidized.

At present, there have been detailed studies of the mineralogy and geochemistry of the manganese-bearing sediments and ferromanganese crusts and nodules; the principal patterns of their genesis and distribution have been established alike within the boundaries of the offshore areas of the global ocean and in the marginal and inland seas and lakes. The results of these investigations have been thoroughly expostulated in an extensive scientific literature, and their analysis has long moved beyond the scope of the questions addressed in the present work.

Undoubtedly, the elucidation of the conditions of the formation of manganese carbonates is crucial for an understanding of the processes and sources of material necessary for the genesis of deposits of terrestrial manganese. This is due mainly to the fact that the primary manganese-containing rocks ("proto-ore") of high-quality manganese ores of many of the world's developed deposits have

Isotope Geochemistry. http://dx.doi.org/10.1016/B978-0-12-803165-0.00002-1

been manganese carbonates embedded in the sequences of rocks of sedimentary and volcanogenic-sedimentary genesis. Highly important in this regard, as will be shown below, is isotope research—the study of the isotopic composition of carbon and oxygen. In the present work, primary attention is devoted to precisely these questions.

It follows to note an important point: the circumstance that, among the known deposits of manganese within continents, rocks analogous to the oceanic iron-containing nodules have not yet been discovered. This fact represents one of the characteristic particularities of the evolution of manganese-ore genesis in the formation history and evolution of the Earth's lithosphere, and evidently is contingent upon the particularities of the development of the oceans and consolidated blocks (of the lithosphere) as tectonic structures.

2.1 MANGANESE CARBONATES IN OPEN OCEANIC SEDIMENTS

At present, the most fully studied in terms of isotope geochemistry are the manganese carbonates in the sediments of the Pacific Ocean. As a result the work of the research vessel *Glomar Challenger* in the offshore Peruvian littoral of the Pacific Ocean (the Guatemala basin), manganese-containing carbonates were discovered in the cores of several wells. Isotope research conducted by Coleman et al. (1982) and by our research (Sval'nov and Kuleshov, 1994), as well as the incidental isotope data of other authors (Pedersen and Price, 1982) for Mn carbonates of this region, indicates a substantial distinction between the isotopic composition of its carbon and oxygen in comparison by open oceanic sedimentary carbonates (organogenous) which are in the isotope equilibrium with the bicarbonate of oceanic water (DIC—dissolved inorganic carbon). The isotopic composition of manganese carbonates from other regions of the global ocean practically has not been studied, with the exception of isolated, fragmentary data (Morad and Al-Aasm, 1997; Meister et al., 2009).

2.1.1 ISOTOPIC COMPOSITION AND GENESIS OF CALCIUM RHODOCHROSITE IN THE SEDIMENTS OF THE GUATEMALA DEPRESSION (PANAMA BASIN, PACIFIC OCEAN)

2.1.1.1 Materials and methods of research

On the 41st voyage of the research vessel *Dmitrii Mendeleev*, a detailed study of ferromanganese concretions and their host sediments was conducted (Rozanov, 1989). One of the study areas was the Guatemala depression of the Pacific Ocean. The collection of sediments (dredging samples; core length up to 412 cm) was carried out at water depths of 3490–3740 m (Fig. 2.1; Sval'nov and Kuleshov, 1994).

Nodules and scattered fine aggregates of crystals of authigenic manganese carbonate were detected in the sediments at the stations 3850, 3884, and 3899 (Fig. 2.2). For the study of the isotopic composition of carbon and oxygen, nodules of manganese carbonate were used, as well as the shells of planktonic foraminifera extracted from the nodules and the host sediments.

The results of mineralogical research of the examined samples indicate that the poorly soluble carbonates are represented by manganese carbonate—calcium rhodochrosite.

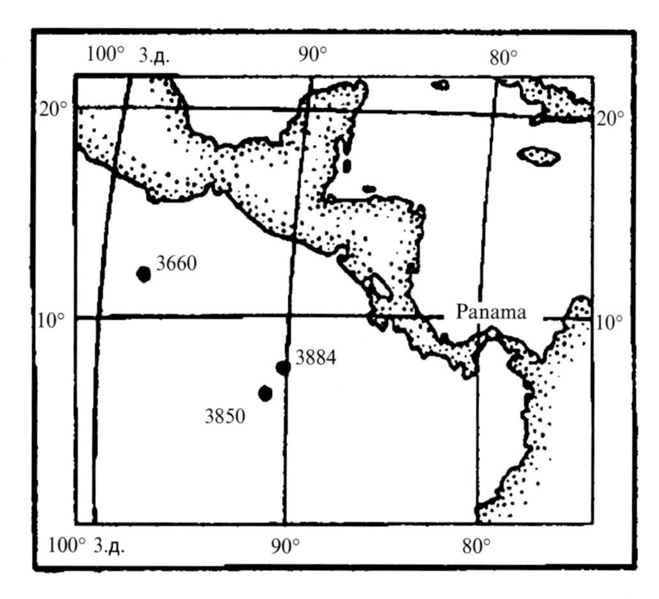

FIG. 2.1

Location of studied holes in the Guatemala depression (Panama Basin, Pacific Ocean) (Свальнов and Кулешов, 1994).

2.1.1.2 Composition of sediments

At station 3899 (12°16.3′N, 97°06.0′W; depth 4095 m), the thickness of the uncovered quaternary sequence is equal to 382 cm. The core generally is represented by grayish-green and greenish-gray pelitic, clayey-diatom oozes, rich in radiolarians.

Nodules of manganese carbonate with dimensions up to 5 cm have been detected within the sediments of 295–300 cm. One of the nodules gradually transitions into a disintegration of ferromanganese concretion, which during diagenesis is stripped of practically all manganese. The interior part of the nodules is dense and gray in color; the exterior wall is friable and brownish-gray. In the nodules occur rare remnants of planktonic foraminifera, radiolarians, and diatoms. Directly above the horizon with the nodules (interval of 293–295 cm), fine aggregates of manganese carbonate compose around 3% of the area of the thin rock section and below (the horizon 300–302 cm), they compose up to 45%.

The clayey-diatom oozes of the core are characterized by a fairly stable chemical composition (Table 2.1). The content of detrital organic carbon fluctuates from 0.7% to 3.0%, declining downward along the sequence. However, at depths of 42–47 and 301–306 cm are observed relative maximums (respectively 3.0% and 1.6%). The carbonate content of the sediments in the conversion to $CaCO_3$ is less than 1% and only directly below the horizon with the nodules it reaches 8.3%. Precisely here was detected the peak of manganese against the backdrop of consistently lower concentrations, with a certain enrichment of ooze by vanadium.

In the behavior of other analyzed elements anomalies were not detected. Their vertical distribution in the sequences is adequately explained by diagenetic conditions and the interchange of the pore waters with the sediment (Sval'nov and Kuleshov, 1994).

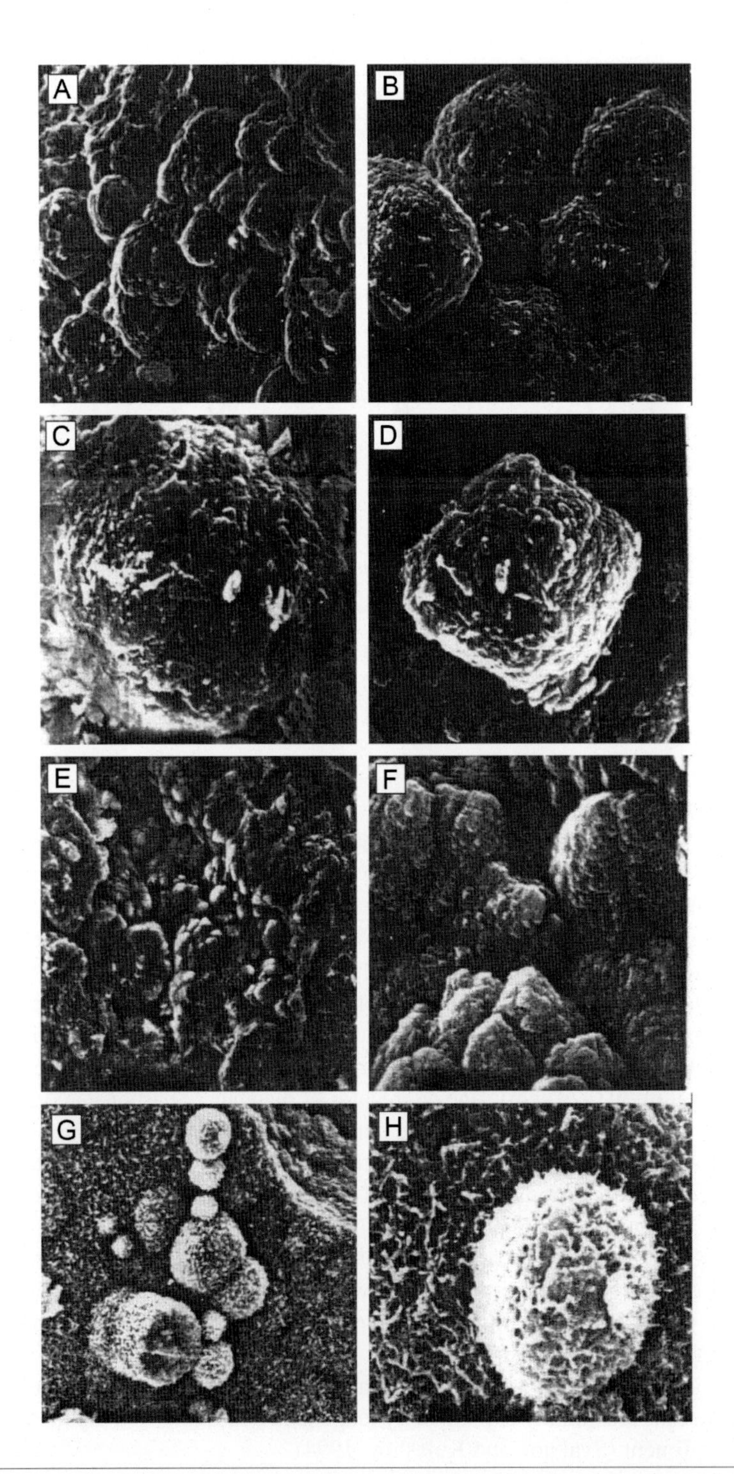

FIG. 2.2

Electron-microscopic photographs of calcium rhodochrosite: (A–F) concretions of spheroidal rhodochrosite (A—mag. 500, B—mag. 1500, C—mag. 2000, F—mag. 2500); (G) authigenic clay minerals at the contact of rhodochrosite and decomposing ferromanganese concretion, mag. 400; (H) the same, mag. 2000.

Table 2.1 Chemical Composition of the Principal Types of Studied Sediments of the Guatemala Basin (Pacific Ocean) (% Dry Weight) (Sval'nov and Kuleshov, 1994)[a]

Number of Station (Coordinates)	Horizon (cm)	Type of Sediment[b]	SiO_2	TiO_2	Al_2O_3	Fe_2O_3	MnO	MgO	CaO	K_2O	LOI	Total
3850 (6°37.5′N, 93°21.0′W)	10–14	T	67.6	0.32	12.4	3.16	0.97	1.50	1.06	3.38	7.2	97.6
	30–35	Tlo	46.9	0.46	10.2	4.91	1.54	1.97	11.2	2.04	18.3	97.5
	130–135	Crlo	47.5	0.52	9.3	6.08	1.02	3.08	8.22	1.36	20.5	97.6
	340–345	CR	50.0	0.56	10.1	6.19	2.80	3.53	5.18	1.47	17.8	97.6
3884 (7°27.0′N, 92°43.8′W)	0–5	RD	54.4	0.62	10.5	4.83	3.94	3.32	1.48	1.43	17.1	97.6
	12–16		48.3	0.56	9.7	5.27	2.27	3.22	6.25	1.35	20.6	97.5
	25–30		53.2	0.53	9.5	6.18	0.55	4.07	3.19	1.46	18.8	97.5
	55–60		54.7	0.49	9.5	5.03	0.86	4.14	2.74	1.67	18.4	97.6
	87–93		56.9	0.45	10.2	4.80	0.40	3.23	3.44	2.14	16.0	97.6
	119–120	CRd	55.5	0.62	11.0	6.20	2.04	3.53	1.37	1.63	15.7	97.6
	125–130	CR	59.1	0.57	10.6	8.08	0.42	2.96	1.38	1.85	12.5	97.5
	285–290		56.1	0.58	10.6	5.80	0.42	3.03	2.28	1.65	17.0	97.5
3899 (12°16.3′N, 97°06.0′W)	0–5	CDr	55.0	0.60	12.5	6.30	0.25	3.38	1.11	1.55	16.7	97.4
	25–30		54.0	0.74	14.0	6.11	0.13	3.25	1.81	1.94	15.5	97.5
	105–110		56.4	0.78	12.7	5.92	0.19	3.67	1.43	1.99	14.5	97.6
	260–265		56.2	0.66	13.1	6.34	0.61	3.61	1.25	1.80	14.0	97.6
	290–295		54.0	0.79	15.0	6.63	0.58	3.59	1.84	2.01	13.0	97.4
	340–345		56.3	0.91	14.3	6.00	0.23	2.83	1.40	2.10	13.4	97.5
	377–382		57.1	0.63	13.2	6.82	0.26	3.70	1.03	2.11	12.7	97.6

[a] *Analyst T.G. Kuz'mina (IO RAN).*
[b] *LOI, loss on ignition; T, tephra; Tlo, the same, weakly limy; Crlo, clayey-radiolarian, low limy; oozes: CR, clayey-radiolarian; CRd, the same, enriched by diatoms; RD, radiolarian-diatom; CDr, clayey-diatom, enriched by radiolarians.*

At station 3884 (7°27.0′N, 92°43.8′W; depth 3585 m) was studied a core of sediments with a length of 340 cm. In the 249–255 cm interval among the clayey-lime ooze, rich in radiolarians, were found yellowish-gray compressed nodules of manganese carbonate with dimensions up to 1.5 cm. The latter contain up to 15% radiolarians, 3% diatoms, and detritus of isolated planktonic foraminifera. Fine aggregates of authigenic carbonate have been detected also at depths of 48–50, 60–62, and 119–121 cm (respectively 20%, 2%, and 3% of the area of the thin section). The content of detritus of planktonic foraminifera in the host oozes does not exceed 1%.

Regarding chemical composition, sediments at station 3884 are somewhat depleted relative to sediments of station 3899 in organic matter and aluminum, but rich in manganese (see Table 2.1). The carbonate content of clayey-radiolarian and radiolarian-diatom oozes commonly consists of less than 1% $CaCO_3$. The content of detrital organic carbon fluctuates within the range of 0.25–1.64%. A maximum was detected at the depth of 30–35 cm. Comparatively elevated concentrations of manganese are associated with the sediments with accumulations of authigenic carbonate, and its maximum (3.66%) has been noted in the layer 0–5 cm. Clayey-radiolarian oozes (interval 119–121 cm), as well as the host ferromanganese concretion, are rich in cobalt and copper.

At station 3850 (6°37.5′N, 93°21.0′W; depth 3660 m) the thickness of the uncovered quaternary cross section is equal to 376 cm. The core generally is represented in varying degree by oxidized pelitic clayey-radiolarian oozes, occasionally weakly limy, tuffitic, or rich in diatoms. The lower ash bed includes a ferromanganese concretion with diameter of approximately 3–4 cm. Still another concretion (with dimensions along the long axis up to 7 cm) was found among the weakly limy clayey-radiolarian oozes at a depth of 56–63 cm. In the 330–340 cm interval light-brown clayey-radiolarian oozes include yellowish-white nodules of manganese carbonate with dimensions along the long axis up to 3 cm. Scattered fine aggregates of authigenic carbonate have been detected as well at a depth of 146–148 cm, where they compose 3–5% of the area of the thin section. Their host clayey-radiolarian ooze contains rounded bodies of hydroxides of manganese (up to 7%), detritus of skeletons of planktonic foraminifera (5–7%), and coccoliths (1–2%). In nodules of manganese carbonate are found isolated radiolarians and approximately 2% of detritus of planktonic foraminifera.

The content of the detrital organic carbon fluctuates within the range 0.20–0.66%, gently decreasing downward along the sequence. The carbonate content of the sediments is extremely variable and in places increases from <1% to 40% $CaCO_3$. The opposite tendency is traced in the distribution of manganese. On the whole, relative to stations 3884 and 3899 the sediments at station 3850 are poor in organic matter, but rich in manganese and carbonate of calcium.

Thus, accumulations of authigenic manganese carbonate in the examined cores are observed in various types of sediments with contents of organic carbon of 0.3–1.6%, with lower carbonate content ($CaCO_3$: from <1% and up to 8.3%) and a manganese concentration of 0.4–4.5%.

2.1.1.3 Composition of manganese carbonate

Authigenic Mn-carbonate is represented by the spheroid aggregates with N_o 1.736 and N_e 1.545. Under the scanning electron microscope, growths of spheroidal isolations of manganese carbonate have been traced (Fig. 2.2A–F). At station 3899 at its contact with the decayed ferromanganese concretion under the scanning electron microscope have been detected isolations of authigenic clayey minerals that occasionally inherit, probably, the form of biogenic detritus (Fig. 2.2G and H).

X-ray analysis has shown that the researched samples are composed of rhodochrosite. The upward bias of the basal reflection (2.899–2.1912 Å) provides evidence of the isomorphic replacement of parts of manganese ions by ions of calcium and magnesium in rhodochrosite. The purest rhodochrosite was detected at station 3850 (Fig. 2.3). The presence of rhodochrosite is supported by the results of

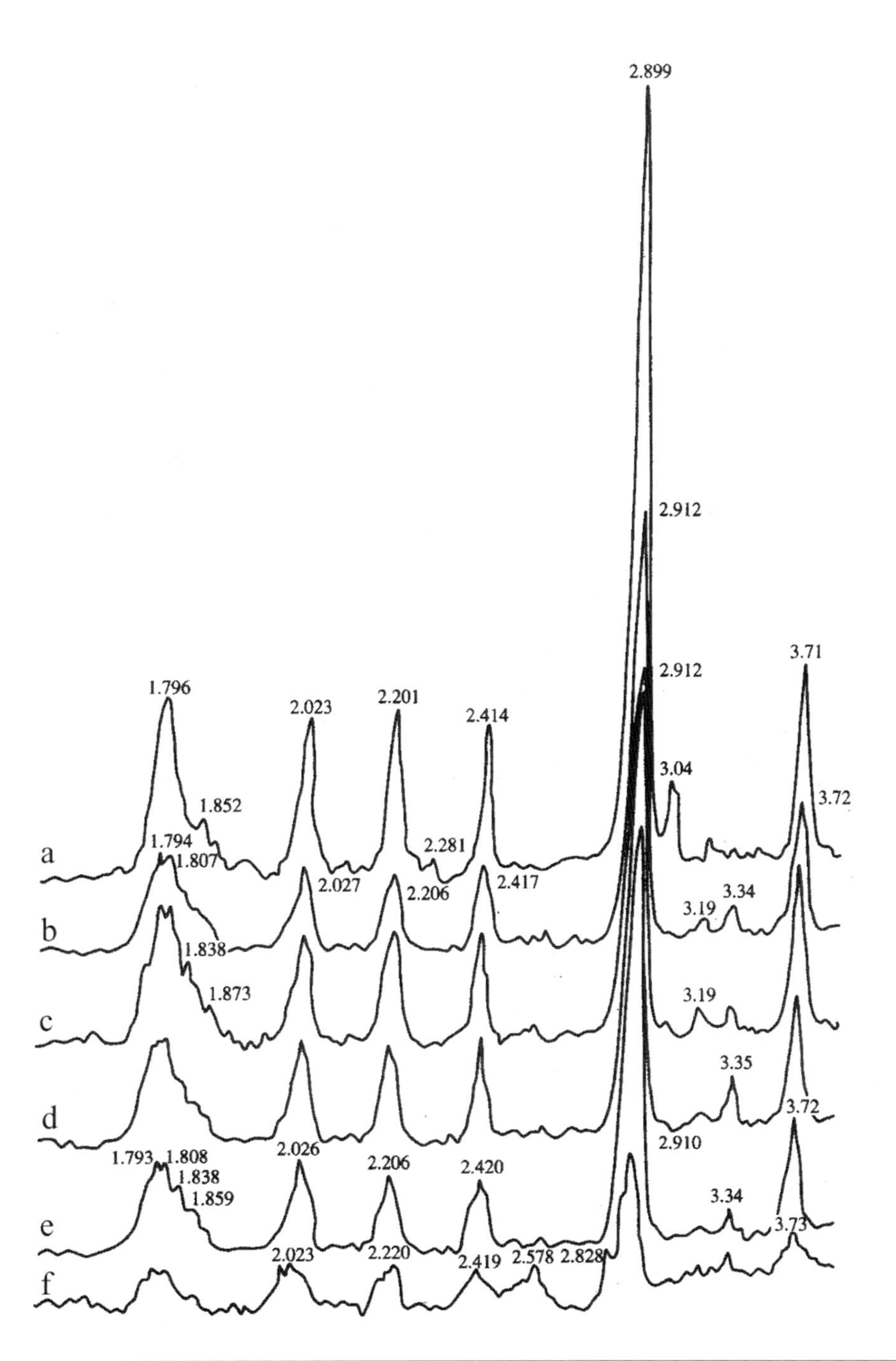

FIG. 2.3

X-ray diffractograms of calcium rhodochrosite: (a) yellowish-white nodule, station 3850, 330–340-cm horizon; (b) yellowish-gray flattened nodules, station 3884, 249–255-cm horizon; (c–f) station 3899, 295–300-cm horizon; (c—yellowish-gray lose nodule, d—nodule's brownish-gray loose outer shell, e—nodule's gray, dense inner part, f—contact of rhodochrosite and ferromanganese concretion).

the study of the chemical composition of the authigenic nodules (see Table 2.1). After normalizing to 100% carbonate components, the mineral formulas (Sval'nov and Kuleshov, 1994) are

$(Mn_{60.7}Ca_{32.5}Mg_{6.5}Fe_{0.3})CO_3$—friable nodule;
$(Mn_{60.4}Ca_{32.7}Mg_{6.6}Fe_{0.3})CO_3$—friable external wall of the nodule; and
$(Mn_{64.6}Ca_{32.4}Mg_{2.8}Fe_{0.2})CO_3$—dense interior part of the nodule.

Thus, in terms of the complex of indicators in the sediments of the Guatemala depression, authigenic carbonate is represented by calcium rhodochrosite with a prominent impurity of magnesium. A similar composition for rhodochrosite was discovered in this depression previously (Lynn and Bonnati, 1965)—$(Mn_{50–80}Ca_{20–50})CO_3$. In well 503 (south of the Guatemala depression) calcium rhodochrosite forms numerous nodules in the reduced sediments of the late Miocene-Holocene (Coleman et al., 1982). These authors propose that rhodochrosite was formed near the surface of the bed as a result of the interplay of bicarbonate of the benthic and pore waters with divalent manganese of very high concentration.

2.1.1.4 Isotopic data

The overwhelming majority of available isotope data for modern carbonate-manganese deposits belong to lacustrine or shallow marine sediments. Particularly oceanic manganese carbonates are rather weakly studied with regard to their isotopic composition, a shortcoming resulting from the insufficient degree of surveying thus far carried out in those regions of the global ocean most likely to contain such carbonates.

The Guatemala depression of the Pacific Ocean, as is known, is one of the known regions of the global ocean to feature wide development of manganese carbonates in modern sediments. The isotope data for oceanic sediments that are available in the literature (Coleman et al., 1982; Pedersen and Price, 1982) refer to samples of carbonates taken in precisely this region. The former work (Coleman et al., 1982) provides a detailed study of the material and isotopic composition of carbon and oxygen in 17 samples of carbonates rich in manganese that were uncovered at wellsites 503A and 503B (south of the Guatemala depression). The range of measured variations in isotopic composition in these samples ranged from −3.8‰ to −1.2‰ for $\delta^{13}C$ and from 4.06‰ to 5.99‰ (relative to the standard PDB) for $\delta^{18}O$. The isotope data previously obtained by this book's author (Sval'nov and Kuleshov, 1994) turned out to be analogous (Table 2.2, Fig. 2.4): the $\delta^{13}C$ values vary within the range of −2.5‰ to −1.1‰, and $\delta^{18}O$ values within the interval 34.9–37.1‰ (relative to the standard SMOW).

The data cited above are substantially distinct from the isotope data particular to lake and sea manganese carbonates in that the $\delta^{13}C$ values are higher in the former. At present, there is no means for comparing the isotopic composition of the studied rhodochrosites with analogous accumulations from other parts of the offshore areas of the global ocean. However, as will be shown below, rhodochrosites from the sediments of the Guatemala depression are substantially distinct in terms of their isotope characteristics from analogous accumulations in other types of water bodies in that the former have higher contents of the heavy isotope ^{13}C. This is evidently connected with the specific conditions of the formation of manganese carbonates in the sediments of this region of the global ocean.

It has been established that practically all of the studied carbonate-manganese accumulations (fossil as well as modern sea and lake; Kuleshov, 1999, 2011a,b) are characterized by low $\delta^{13}C$ values. This is due to the fact that Mn carbonates in sedimentary series are formed in a stage of diagenesis with active participation by isotopically light carbon dioxide. The latter forms during the oxidation of organic matter in the environment of the sediments themselves. For this reason, it can be proposed that the generation of Mn carbonates in sediments of the Guatemala depression during diagenesis has occurred

Table 2.2 Isotopic Composition of Carbon and Oxygen of Sediments of the Guatemala Basin

Number of Analysis	Number of Sample	Location of Sample Collection and Its Characteristics	$\delta^{13}C$, ‰ (PDB)	$\delta^{18}O$, ‰ (SMOW)
2850	3884-250	Station 3884, >249–253 cm		
		Yellowish-gray carbonate nodules:		
		Calcite	−1.9	35.2
		Rhodochrosite	−1.6	35.5
2851	3899-1	Station 3899, >295–300 cm		
		Brownish-gray friable carbonate nodules:		
		Calcite	−2.6	35.7
		Rhodochrosite	−2.3	34.9
2852	3899-2	Station 3899, >295–300 cm		
		Brownish-gray friable clast of carbonate nodule:		
		Calcite	−2.6	35.3
		Rhodochrosite	−2.3	36.9
2853	3899-3	Station 3899, >295–300 cm		
		Gray dense internal part of carbonate nodule:		
		Calcite	−2.3	35.4
		Rhodochrosite	−1.9	36.6
2854	3899-295	Station 3899, >295–300 cm		
		Yellowish-gray carbonate nodule with biogenic remnants:		
		Calcite	−3.0	32.1
		Rhodochrosite	−2.5	36.0
2855	3899-301–306	Station 3899, >301–306 cm. Greenish dark-gray clayey-diatom ooze enriched by radiolarians:		
		Rhodochrosite	−1.9	37.1
2856	3859-335	Station 3850, >330–340 cm		
		Yellowish-white carbonate nodules:		
		Calcite	−0	33.3
		Rhodochrosite	−1.1	36.3
2857	3899	Station 3899, >295–300 cm. Sediment washed out of the nodules:		
		Rhodochrosite	−2.3	–
2858	3899	Station 3899, >295–300 cm. Carbonate nodules:		
		Calcite	−0.9	–
		Rhodochrosite	−1.0	–
2859	3899	Station 3899, >300–302 cm. Sediment washed out by water:		
		Rhodochrosite	−2.0	–

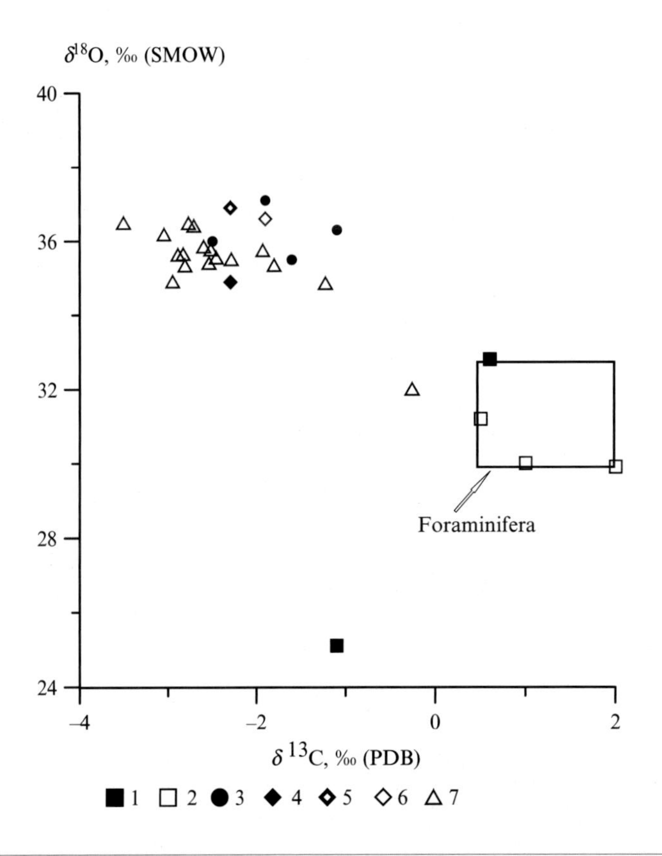

FIG. 2.4

$\delta^{13}C$ vs. $\delta^{18}O$ in carbonate nodules of Guatemala depression (Panama Basin, Pacific Ocean). 1—Foraminifera from nodules, 2—foraminifera from enclosing sediments, 3—nodules from stations 3884, 3895, and 3899; 4–6—nodule from station 3899, horizon 295–300 cm: 4—loose nodule, 5—outer shell, 6—inner part; 7—data (Coleman et al., 1982).

in several other, fairly specific conditions distinguished by the insignificance of the role of CO_2 of organic origin in this process. The formation of manganese carbonates occurred principally as a result of the dissolution and redeposition of biogenic calcium carbonate of the sediments (evidently, shells of foraminifera and limy skeletal remnants of other organisms).

For certain samples available to us, it was possible to study different genetic varieties of manganese carbonates contained within a single sample. Thus, the isotopic composition of the carbon of the rhodochrosite from the host sediment and from carbonate nodule of the same interval of core sample 3899 turned out to be identical: $\delta^{13}C$ varies within the range −2.5‰ to −2.0‰ (samples 2851, 2852, 2857, 2859). Furthermore, for certain nodules, their inner part is distinguished by a higher values of $\delta^{13}C$ (−1.9‰, an. 2853) in comparison with that of their outer part (−2.3‰, an. 2852). Analogous results have been obtained in an earlier work (Coleman et al., 1982), which found that for two samples out of the five studied, there was a gradual increase from the periphery toward the center in the heaviness of the isotopic composition of the carbon.

On the whole, the isotopic composition of carbonate-manganese matter is not dependent on its form of occurrence in the sediment (ie, nodule, dispersion, etc.). The observed insignificant variation of the values of $\delta^{13}C$ and $\delta^{18}O$, within different samples as much as within the boundaries of a single nodule (as was proposed in the work; Coleman et al., 1982), was to all appearances connected with local conditions (ie, permeability of the sediment, physical-chemical heterogeneity, etc.) of the generation of Mn carbonates.

By way of comparison, the isotopic composition of carbon and oxygen in the foraminifera collected from the host sediment as well as from the carbonate-manganese nodules themselves was studied (Table 2.3). Here was manifest a constant difference in the isotopic composition of foraminifera and of manganese carbonates from nodules. The former are characterized on the whole by higher $\delta^{13}C$ values (approximately 2‰) in relation to that of the nodules. This difference is evidently due to the different isotopic composition of the original bicarbonate and by the temperature of the generation of carbonate matter. As is well known, bicarbonate of the upper horizons of the hydrosphere is characterized as a rule by higher $\delta^{13}C$ values—a consequence of the higher bioproductivity of the near-surface horizons of the water column. Therefore, carbonate from foraminifera, which form from bicarbonate of surface waters, should have heavier $\delta^{13}C$ values.

Thus, it can be stated that authigenic manganese carbonate in the sediments of the Guatemala depression (Pacific Ocean) is represented by calcium rhodochrosite and was formed mainly as a result of the dissolution and redeposition of limy shells of foraminifera and skeletal fossils of other organisms. Furthermore, the isotopically light carbon dioxide that formed as a result of the oxidation of C_{org} plays a sharply subordinate role in this process. One source of manganese of the studied nodules was the host sediments, and occasionally also Fe-Mn concretions, which supplied Mn as it was remobilized during the process of diagenesis.

A principal feature of the formation of Mn carbonates in the sediments of the Guatemala depression is the low geochemical activity of organic matter in this process, which resulted in the relatively high values of $\delta^{13}C$ in the studied nodules.

Table 2.3 Isotopic Composition of Carbon and Oxygen of Foraminifera From the Sediments of the Guatemala Basin

Number of Analysis	Number of Sample	Location of Gathering the Sample	$\delta^{13}C$, ‰ (PDB)	$\delta^{18}O$, ‰ (SMOW)
Foraminifera (>0.25 mm) from the sediment				
3350	3850	Station 3850, >330–340 cm	2.0	29.9
3351	3884	Station 3884, >240–253 cm	1.0	30.0
3352	3899	Station 3899, >295–300 cm	0.5	31.2
Foraminifera from nodules				
3353	3850	Station 3850, >335 cm	0.6	32.8
3354	3884	Station 3884, >250 cm	−1.1	25.1

2.1.2 MANGANESE CARBONATES IN THE SEDIMENTS OF THE CENTRAL AMERICAN TROUGH (EL GORDO UPLIFT, PACIFIC OCEAN)

Shterenberg et al. (1992) studied the sediments of the ocean slope of the Central American trough, retrieved at station 82105 (18°08.3′N, 104°43.2′W, depth 4300 m) during the eighth voyage of the research vessel *Akademik Nikolai Strakhov* in 1989.

The general length of a core sample of the sediments comprised 457 cm. In the uppermost parts of the cross section (0–5 cm) is located silty ooze with an impurity of tuffogenic material. Within the interval 5–15 cm, the ooze is silty and tuffaceous with frequent burrows. At depths of 15–57 cm, the ooze is likewise silty and dark-brown with an impurity of tuffaceous material. Lower in all of the core sample of the sediments is observed silty gray ooze with inclusions that are irregular in form and interlayers of silty oozes that contain in their composition an impurity of tuffaceous material. In the lowermost parts of the cross section (300–457 cm) of the core sample of the sediments of this station, as well as those of neighboring stations (82108, 82101, and 82096), sulfides of iron, represented principally by pyrite, have been discovered in relatively large quantities.

Enumerated in Table 2.4 are the contents of the macroelements, as well as of CO_2 and C_{org}, in sediments of four horizons: 15–25, 70–80, 240–250, and 400–430 cm.

The highest contents of MnO have been established in the sediments of the 15–25 cm horizons, where in a fraction >0.1 mm have been discovered oxides and carbonates of manganese. Downward in the sediments occurs an increase in the quantity of Fe^{2+}. The study of the fraction >0.1 mm has allowed the establishment of morphologically different textures of manganese carbonates that are distinguished by their chemical and mineralogical composition.

1. Comparatively thin (fractions of a millimeter), crust-type textures, distributed most frequently in the internal walls of oxide pipes. The thickness of these crusts is not constant and varies over short distances. Between the walls of pipes composed of oxides of manganese and carbonate

Table 2.4 Content of Elements in Sediments of Station 82105 (Central American Trough, Pacific Ocean), Mass % (Sval'nov and Kuleshov, 1994)

Interval of Gathering of Samples (cm)	SiO_2	TiO_2	Al_2O_3	Fe_2O_3	FeO	CaO
15–25	45.8	0.80	17.6	6.89	0.14	3.04
70–80	48.9	0.74	17.5	7.41	0.17	2.38
240–250	48.0	0.90	18.2	7.30	0.66	2.03
400–430	50.1	1.00	17.3	6.61	0.73	2.67
Interval of Gathering of Samples (cm)	**MgO**	**MnO**	**Na_2O**	**K_2O**	**CO_2**	**C_{org}**
15–25	2.96	1.77	4.00	1.92	0.65	1.26
70–80	2.85	1.36	3.29	2.06	1.09	1.00
240–250	2.94	0.76	4.00	2.29	2.10	1.09
400–430	2.86	1.11	3.77	2.15	None	0.88

crusts are commonly observed thin, almost filmy, glossy black laminae. Semiquantitative spectral analysis of this glossy film has shown that it is composed of X-ray-amorphous silicon dioxide and hydroxides of manganese.

2. Compressed, small in magnitude (rarely reaching 0.1 mm), commonly pale-colored textures, characterized by flattened upper surfaces. These represent sediments that have been cemented by manganese carbonates; from the outer surface they are likewise partially oxidized.
3. Lenses and inclusions among the oxide pipes, strongly manifest in observations of a section of the latter. These carbonate textures are frequently distributed according to the transverse structure of the pipe itself.
4. Carbonate-manganese textures in the form of pale-colored (pinkish-white and cream-white) inclusions among gray ooze covering from the external side of the walls of the oxide pipes.
5. Fine, elongated (fractions of a millimeter) carbonate forms with an outer surface with botryoidal appearance. Diffractograms of certain samples are shown in Fig. 2.5. They provide evidence that the studied samples represent multicomponent textures with varying primary reflections: from 2.89 to 2.93 Å. Reflections of 2.96–2.97 could provide evidence of the presence of manganocalcite (Vasil'ev, Vasil'eva, 1980).

The results of the analyses and their evaluation permit the delineation of three principal groups of manganese carbonates of complex composition, distinguished from each other by the contents of the primary and secondary cations: $(Mn_{53}Ca_{41}Mg_6)CO_3$, $(Mn_{58}Ca_{39}Mg_3)CO_3$, and $(Mn_{61}Ca_{37}Mg_2)CO_3$.

The chemical-mineralogical composition of carbonates of manganese found in the upper part of the sediments of the ocean slope of the Central American trough, retrieved at station 82105, is fairly similar to the composition of manganese carbonates formed under diagenesis in the hemipelagic zone of a range of other regions of the Pacific Ocean. The participation of diagenetic processes in the formation of manganese carbonates at this stage is also supported by the study of ooze waters removed from sediments at a range of horizons (Shterenberg et al., 1992).

The quantity of manganese in ooze waters does not remain constant, but rather regularly changes in a downward direction. In the upper parts of the sequence, where micrograined forms of manganese carbonates of complex composition have been established, the Mn^{2+} content in the ooze water constitutes 2 mg/L, and in the lower parts, all of 0.1 mg/L. We note that in the upper part of the sequence of the sediments, the ooze water has an increased content of calcium, which participates together with manganese in the formation of carbonates of complex composition.

HCO_3^- content regularly increases with depth (from 0.317 g/L in the depth interval 10–25 cm to 0.7681 g/L at the depth of 420–430 cm), while the magnitudes of pH in the interstitial waters decrease: from 7.83 (depth 10–25 cm) to 7.75 (depth 420–430 cm).

For the purposes of isotope research, a sample of sediments was selected that had the greatest contents of manganese (depth 15–25 cm), here embedded in rhodochrosite. The isotopic composition for this sample constitutes (‰): $\delta^{13}C_{PDB} = -6.01$ and $\delta^{18}O_{SMOW} = 35.0$.

The obtained isotope data are in agreement with analogous data for known sedimentary-diagenetic accumulations of manganese carbonate from lakes and seas (Kuleshov and Chistiakova, 1979; Kuleshov and Shterenberg, 1988; Suess, 1979), and from certain regions of the Pacific Ocean (Zheng et al., 1986); the data provide evidence for the active participation of organic matter in the process of formation of manganese carbonates under diagenesis.

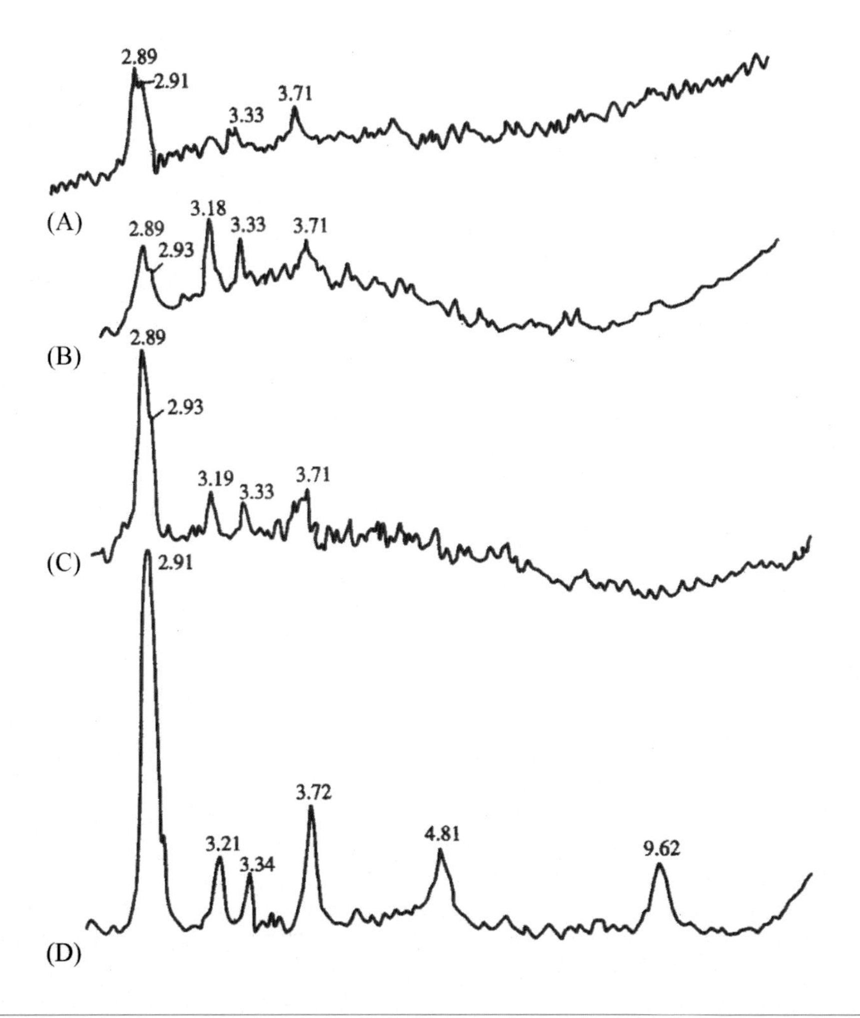

FIG. 2.5

X-ray diffractograms of manganese carbonates from sediments of the Central American Trough (El-Gardo elevation, Pacific Ocean). (a, b) Manganese carbonate crusts; (c) small carbonate nodules in ooze; (d) manganese oxide-carbonate hard matter.

2.2 MANGANESE CARBONATES IN MODERN MARINE SEDIMENTS

Manganese-bearing sediments within the boundaries of modern marine water bodies are distributed as widely as they are in the oceanic areas; they are represented principally by oxide structures (ie, nodules, crusts, scattered material in oozes) of ferromanganese composition. The presence of manganese carbonates in marine sediments is exceptionally limited and localized predominantly in northern seas (Baltic, White, Kara, and others) (Gorshkova, 1931, 1967; and others).

2.2.1 MANGANESE CARBONATES IN THE SEDIMENTS OF THE BALTIC SEA

The Baltic Sea is one of a small number of regions of the offshore areas of the global ocean where the processes of manganese carbonate formation have developed in the modern sediments. It is a very shallow sea, typically only 60–80 m deep, but contains several areas of substantially deeper water (Bornholm, Carlsö, Farö, Gotland and Landsort Deeps) which have limited circulation and consequently oxygen-free (stagnated). But two deepest, the Landsort Deep (450 m) and the Gotland Deep (250 m), are characterized by manganese-bearing sediments predominantly in the form of carbonates (Geologiia..., 1976; Lepland and Stevens, 1998; Neumann et al., 2002; Fonselius and Valderrama, 2003 http://helcom.fi/baltic-sea-trends/environment-fact-sheets/hydrography/water-exchange-between-the-baltic-sea-and-the-north-sea-and-conditions-in-the-deep-basins/).

From the moment of the first finds of Mn-containing carbonates in the deepwater sediments of the Baltic (Gorshkova, 1960, 1967) up to the present, the mineralogy, geochemistry, and other aspects of their formation have been studied in detail. However, isotope research of authigenic Mn carbonates remains insufficient.

Thus, Suess (1979) has studied the isotopic composition only of carbon in a sample of rhodochrosite from the ooze deposits of the Landsort Deep. The obtained $\delta^{13}C$ value turned out to be rather low (−13‰), providing him the basis to conclude, drawing likewise from isotope data on other carbonates from Baltic sediments, that the formation of Mn carbonates was diagenetic, with the active participation in this process by carbon dioxide of organic origin. The latter was formed as a result of the microbial decay of organic matter.

2.2.1.1 Factual material

The isotopic composition of carbon and oxygen has been studied in the carbonate component of the sediments along the sequence of the core sample of the oozes, retrieved by a direct-flow core sampler from a depth of 243 m at station 3173, situated in the central part of the Gotland Basin (57°20.9′N, 20°18.8′E) (Kuleshov and Rozanov, 1998; Figs. 2.6 and 2.7).

The Gotland deep is located between 57° and 58°N and 20°E, between the Swedish Island Gotland and the west coast of Latvia and spread about 260–280 km in meridional and 70–80 km in latitudinal directions. The water depth on an average is 175–200 m (max 250 m). The water column is characterized by a chemical stratification (halocline) with saline (bottom water) and surface desalted water due to exchange of Baltic Sea water body and Northern Sea waters. In the present Gotland Deep, a permanent halocline at 70–80 m depth separates the dense bottom water with a salinity of 11–14‰ from surface water with a salinity of −7‰. Below 125–150 m, the water is periodically anoxic (Lepland and Stevens, 1998; Emeis and Struck, 1998; Winterhalter, 2001).

Holocene deposits of the studied station are represented by terrigenous pelitic green-gray and black-gray oozes, which at a depth of 168 cm from the surface of the sediment are deposited on the Late Glacial blue-gray clays. The sediments are strongly reduced (Eh = −200 mV). They are rich in organic matter (C_{org} = 9.9 − 5.79%) and contain free hydrogen sulfide, which arises in the underlying stratum of Late Glacial clays, causing as well their high degree of reduction (Eh = −280 mV).

The magnitudes of pH according to the available data are located within the range of 7.4–7.7. In the surface layer of the sediment, these magnitudes occasionally drop to 7.2. The underlying Late Glacial clays are characterized by lower magnitudes: around 7.0–7.1.

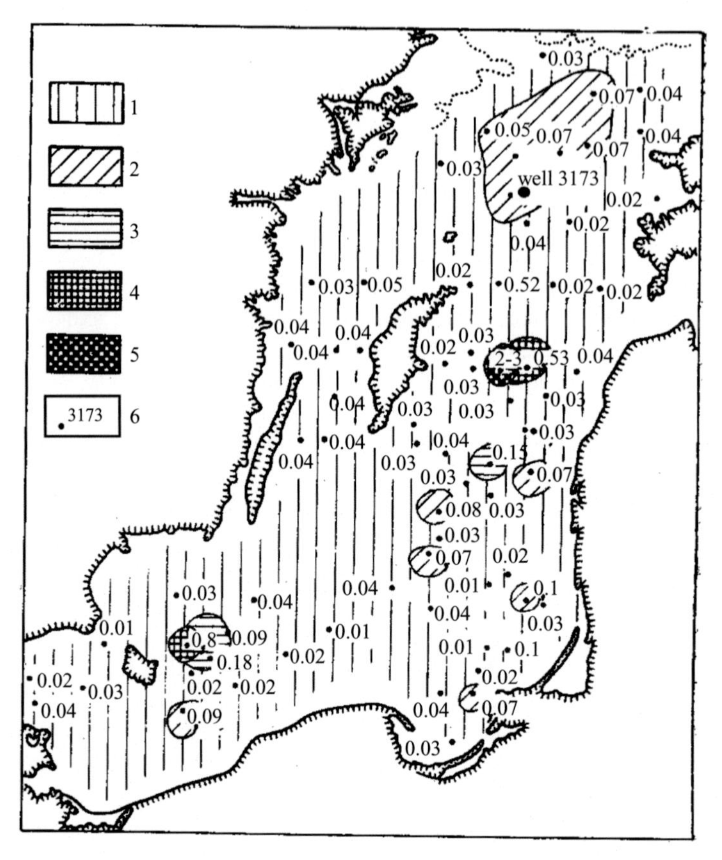

FIG. 2.6

Location of station 3173 and manganese contents (%) in upper layer of Baltic Sea deposits (Горшкова, 1967). 1—<0.05; 2—0.05 to 0.1; 3—0.1 to 0.2; 4—0.5 to 1.0; 5—2 to 3; 6—location of station.

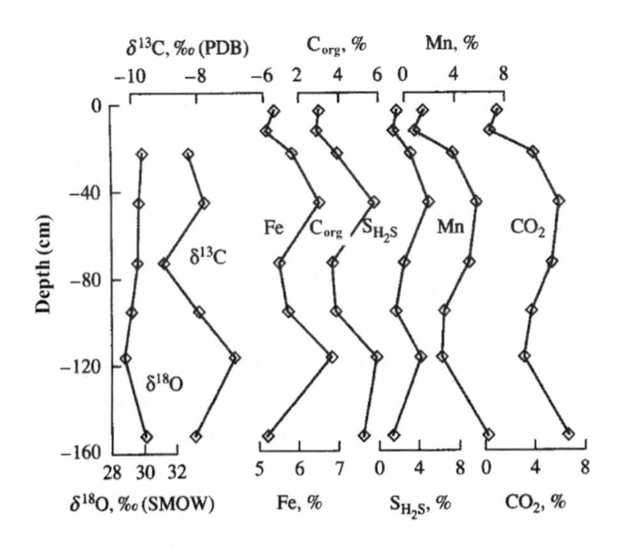

FIG. 2.7

Distribution of $\delta^{13}C$ and $\delta^{18}O$ in manganese carbonates and chemical composition of mud, station 3173 (Baltic Sea) (Кулешов and Розанов, 1998).

2.2.1.2 Composition of authigenic carbonates

Authigenic carbonates in the sediments of the deepwater basins of the Baltic Sea have a complex composition. X-ray research of a powder mount has enabled the establishment of the presence of carbonates corresponding to the isomorphic mineral series from calcium and dolomite to rhodochrosite and ankerite. According to the calculations of Manheim (1961), the composition of the carbonate he studied in dark-green ooze (Gotland deep, station S-33, depth 188 m) corresponds to the formula: $(Mn_{72}Ca_{16}Mg_{12})CO_3$. In his calculations Manheim did not account for iron, which he considered to have precipitated during a separate phase (sulfides), although he did not exclude the possibility of an impurity of this element in the carbonates. In the opinion of Blazhchishina (1976), the formula arrived at by Manheim is not entirely precise, insofar as the content of pyrite and siderite in the sediments is insignificant, while neither of these minerals compensate for the total content of iron in the sample.

A similar composition of authigenic manganese-containing carbonate was obtained by Suess (1979) for the deepwater sediments of the Landsort Deep: $Mn_{0.85}Ca_{0.10}Mg_{0.05}$. According to the generalization of Calvert and Price (1977), in Baltic sediments the composition of mixed carbonate varies within wider boundaries: from $Me_{0.6}Mn_{0.4}CO_3$ to $Me_{0.1}Mn_{0.9}CO_3$, where Me represents the divalent cations Ca, Mg, and Fe.

A slightly different formula of a mixed carbonate, which does account for iron, has been provided by Hartman: $(Mn_{56.8}Ca_{25.46}Mg_{9.72}Fe_{8.02})CO_3$ (Hartman, 1964).

The composition of the authigenic carbonates in our studied core sample of the sediments of station 3173 likewise represents an isomorphic mixture of complex mineral composition. Thus, according to the data of Emel'ianov et al. (1986), in the core sample of the cited station, with the exception of samples with decreased content of manganese and carbonate carbon, occurs precisely those mixed carbonates close to the composition: $Me_{0.3}Mn_{0.7}CO_3$ and $Me_{0.2}Mn_{0.8}CO_3$, where Me represents the cations Ca^{2+}, Mg^{2+}, and Fe^{2+}. In earlier investigations, diffractometric analysis has shown that beginning from a depth of around 1 m from the surface of the sediment, carbonate of complex composition gives way to weakly crystallized rhodochrosite (Emel'ianov and Pustel'nikov, 1975).

X-ray investigations (analyst Pokrovskaia, Geology Institute, Russian Academy of Sciences) of the powder mounts of dry sediment taken from two intervals of the core sample, which are characterized by the highest content of Mn and CO_2—samples 2122 (interval 40–51 cm) and 2126 (interval 150–155 cm) (Table 2.5)—have revealed the absence of the reflexes particular to carbonate minerals in the X-ray of sample 2122 . At the same time, in sample 2126 distinctly the reflex corresponding to the basal spacing 2.886 Å and the less intensive reflex corresponding to the basal spacings (Å): 2.409, 2.194, 2.021, etc. were uncovered. Formally, such a selection of reflexes could provide evidence for the presence of dolomite in the sample. But, taking into account the high content of Mn in the sample, it is more likely that we are dealing with rhodochrosite with an insignificant impurity of calcite and dolomite.

Thus, the results of the preceding investigations together with our data provide evidence that in the sediments of the deep-water basins of the Baltic Sea, authigenic carbonates are present mainly in the form of a complex isomorphic mixture of minerals ranging from calcium and dolomite to rhodochrosite and siderite. Moreover, in the upper layers of the sediment, carbonate matter is X-ray-amorphous, while only toward the lower parts of the Holocene deposits its decrystallization has been observed (to all appearances, not a full decrystallization).

Table 2.5 Content of Iron, Manganese, Sulfur, and Organic Matter (C_{org}) and Isotopic Composition of Carbon and Oxygen of Carbonates in Oozes of Well 3173, Baltic Sea

Characteristics													
N (Sample)	Interval (cm)	Sample	Nature	Color	S_{H_2S} (%)	S_{pyr} (%)	Fe_{tot} (%)	Fe_{pyr} (%)	Mn_{tot} (%)	CO_2 (%)	C_{org} (%)	$\delta^{13}C$, ‰ (PDB)	$\delta^{18}O$, ‰ (SMOW)
2019	0–5	Ooze	Terrigenous, pelitic	Dark greenish-gray	1.68	1.15	5.38	1.00	1.44	1.00	2.94	–	–
2020	10–15	Same			1.35	0.94	5.20	0.82	0.73	0.33	2.85	–	–
2121	20–25	Sapropel ooze, micro-laminated	Terrigenous, pelitic	Dark-gray	3.07	2.59	5.80	2.25	3.68	3.85	3.84	−8.3	29.9
2122	40–51	Same			4.86	4.44	6.50	3.86	5.60	5.94	5.70	−7.8	29.7
2123	70–76	Same			2.39	2.02	5.50	1.76	4.96	5.28	3.60	−9.1	29.6
2124	90–100	Ooze	Terrigenous, pelitic	Green-gray	1.55	1.27	5.70	1.10	2.95	3.63	3.81	−8.0	29.2
2125	113–120	Ooze, micro-granular		Green-gray	4.05	3.69	6.80	3.21	2.80	3.08	5.79	−6.9	28.8
2126	150–155	Same			1.38	1.18	5.20	1.03	6.54	6.60	5.19	−8.1	30.1

– indicates below detection.

2.2.1.3 The isotope data and their discussion

The carbonates we studied in the sediments of the Gotland Basin are characterized by insignificant variations in the isotopic composition of carbon ($\delta^{13}C = -9.1‰$ to $-6.9‰$) and oxygen ($\delta^{18}O = 28.8‰$ to $30.1‰$) (see Table 2.5). In comparison with normal sedimentary marine carbonates, the studied samples are essentially rich in light isotopes of carbon ^{12}C. On the whole, such isotopic characteristics are particular to carbonates of diagenetic origin. The richness in light isotopes of the carbonates is caused by the participation in their formation of isotopically light carbon dioxide of biogenic origin, which forms as a result of the oxidation of organic matter in conditions of diagenesis.

Analogous isotope data for rhodochrosite from the upper layer of the modern sediments of the Gotland Basin of the Baltic have been obtained by Neumann et al. (2002).

According to the results of Suess (1979), as well as those of Force and Cannon (1988), the light isotopic composition of these carbonates and their form of occurrence in the sediment, which are pseudomorphs of carbonates (bacterial), provide evidence that their formation was a result of the replacement of the carbonate substrata in the interstitial water in the presence of decayed (destroyed) organic matter.

Earlier, Lein et al. (1986) studied the isotopic composition of organic and mineral carbon in Baltic Sea sediments. According to their obtained data, in the sediments of the deep-water basins the values of $\delta^{13}C$ of the total carbon dioxide, representing the product of the decay of carbonic matter in 3% hydrogen chloride (HCl), vary from $-21.6‰$ to $-8.1‰$, and in the oozes of the Gotland Basin from $-18.1‰$ to $-8.1‰$. These authors arrive at the result that the source of isotopically light microbial carbon dioxide in the sediment is the anaerobic microbiological decay of organic matter; they explain the observed variations of $\delta^{13}C$ values as the result of the different degree of participation in the process of diagenetic carbonate formation by carbon of different origins: microbial CO_2, sedimentary carbonates, and organic matter of benthic sediments. Proposed by them as well was a method of calculating the share of participation by isotopically light microbial carbon dioxide in the composition of ooze carbonates.

Lower ratios of isotopic composition of organic carbon from sediments have been obtained (Sohlenius et al., 1996; Lepland and Stevens, 1998): $-30‰$ to $-24‰$.

As has been noted, the isotopic composition of oxygen in the carbonates of our study on the whole is rather consistent along the sequence. Only at a depth of 100–120 cm from the surface of the sediment was noted an insignificant lightening of the isotopic composition of oxygen ($\delta^{18}O$—to 28.8‰).

The distribution of the ratios of the isotopic composition of carbon along the core section of the core sample, in contrast to that of oxygen, is nonuniform. The observed divergences (peaks) of $\delta^{13}C$ values on the whole are correlated to the chemical composition of the sediment (Fig. 2.7). As can be seen, a direct correlation is observed with the contents of Fe_{total}, C_{org}, and less distinctly with that of S_{pyrite}, while there is an inverse relationship with thc contents of Mn and CO_2. Moreover, as has already been noted numerous times previously (Emelyanov et al., 1982; Emel'ianov et al., 1986), the concentration of manganese is neatly correlated to the content of CO_2 (Fig. 2.8A). The noted correlational relationship of the isotopic composition of carbon to the chemical composition of the ooze sediment could provide evidence on the complex and mutually dependent geochemical redistribution of chemical material in the zone of diagenesis of ooze sediment, which in a general approximation can be regarded as a closed geochemical system with regard to the examined elements.

2.2.1.4 Genesis of carbonates

The authigenic origin of the carbonate component in Holocene oozes of the deep-water basins of the Baltic Sea is not a cause for any doubt among researchers today. Evidence for this is found in the

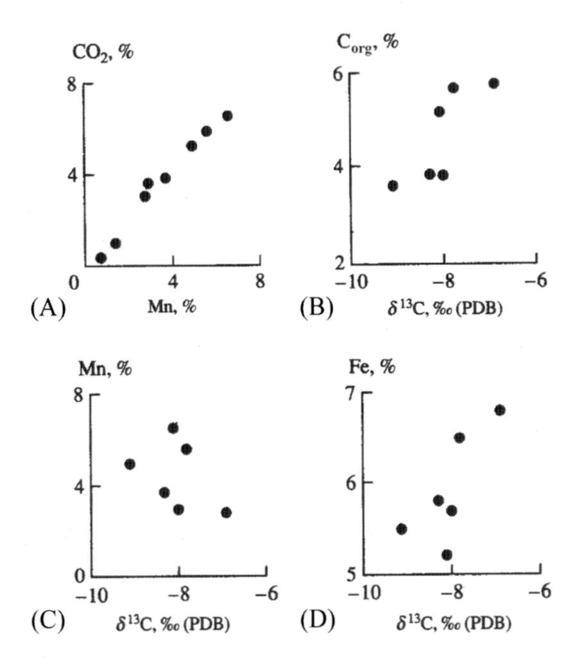

FIG. 2.8

Relationship between CO_2 and Mn (A), carbon isotopic composition and C_{org} contents (B), Mn (C) and Fe (D) in Baltic Sea sediments (station 3173) (Кулешов and Розанов, 1998).

isotope data available in the literature (Suess, 1979; Leni et al., 1986; Sohlenius et al., 1996; Böttcher and Huckriede, 1997; Neumann et al., 2002) as well as from our own data on carbon and oxygen. Nevertheless, our obtained patterns of variation for the $\delta^{13}C$ and $\delta^{18}O$ values along the sequence in carbonates of the Holocene oozes and the established correlational relationship of the isotopic and chemical composition of the sediments allow more detailed elucidation of certain aspects of the conditions of carbonate formation.

In this way, the high $\delta^{18}O$ values and their insignificant variations along the sequence provide evidence for the high stability of the isotope-geochemical system of hydrosphere sediments. It can be supposed that despite the rather complicated history of development of the Baltic in the period following the Baltic Ice Lake, the hydrological regime remained fairly stable in terms of its isotope geochemistry. There might have been expected a more significant variation of $\delta^{18}O$ values, insofar as the history of development of the Baltic after the Subarctic period includes periods of sharp decrease of the water area of the sea and of significant reduction in the inflow of marine (ocean) waters and, naturally, an increase of the relative share of fresh water in the basin (the stage of the Echeneis and Ancylus seas), as well as the stages of intense exchange by the water mass with the Global Ocean. The latter is connected with the intense inflow (transgression) into the water area of the Paleobaltic by saline marine waters (stage of the seas Yoldia, Ancylus, Litorina, and particularly the Post-Litorina and the modern Baltic) (Gudelis, 1976; Sohlenius et al., 1996; Fonselius and Valderrama, 2003; Nausch et al., 2015).

Consequently, changes in the degree of the isolation of the Baltic hydrosphere from the global ocean led to changes in the salinity and chemical composition of its waters and, naturally, to changes in the relative quantities of fresh water. All of this likely had an impact on the isotopic composition of the oxygen of the Baltic hydrosphere and ultimately on the composition of the deposited carbonates. Taking into account the obtained isotope data, it can be supposed that in the deepwater basins of the Baltic Sea, evidently owing to stagnation phenomena, these processes did not lead to significant changes in the isotopic composition of the oxygen of the benthic waters.

Another, no less probable cause of the observed fluctuations of the $\delta^{18}O$ values in the carbonates within the interval of 100–120 cm could be the general warming of the climate (up to 5 °C) at the time of the deposition of these oozes.

Variations in the isotopic composition of the carbon in the oozes along the sequence are caused mainly by the different degree of participation in their composition by isotopically heavy carbon of sedimentary (carbonate) origin and isotopically light carbon of organic (oxidized C_{org}) origin. This is caused by the different initial saturation of the sediment of carbonate shell material and the quantity of the C_{org} oozes. To go by the mean $\delta^{13}C$ value of the organic matter in the modern sediments of the Baltic Sea, which is equal to −26‰ to −23‰ (Lein et al., 1976; Lepland et al., 1998), 30–40% of the carbon dioxide of the studied carbonates has a microbial origin.

The correlational relationship between the contents of C_{org} and the $\delta^{13}C$ value (Fig. 2.8B) indicates that the isotopic composition of the carbon of the carbonates is lightened by a decrease in content of organic matter. This could provide evidence on the different shares of participation by oxidized organic matter in the composition of authigenic carbonates. With a high content of organic matter in the sediment, only a small portion of it (at the examined station 26.1–44.1, in the mean 36.7% of C_{org}; Emel'ianov, 1986) is consumed during microbial oxidizing-reduction processes.

In this way, the more oxidized the C_{org} the less of it remains in the sediment and, consequently, the more of it is fixed in the composition of authigenic carbonates and the lighter the isotopic composition of the carbon in them.

The principal anaerobic oxidizers of organic matter in the sediments of the examined station are sulfates of ooze waters (in mean 77% CO_2, formed anaerobically; Emel'ianov et al., 1986), but with the presence of elevated quantities of manganese its contribution to the oxidation of organic matter becomes highly pronounced (up to 30% CO_2). This is seen clearly in the graph of the relationship of $\delta^{13}C$ values to the contents of Mn (Fig. 2.8C): the higher the contents of manganese, the lighter the isotopic composition of the carbon and, consequently, the greater the share of participation by microbial carbon dioxide in the process of the formation of the carbonates. Additionally, it is evident that the carbonates thus forming accumulate CO_2 independently from the source of oxidation (sulfates, oxides of Mn^{4+}, or Fe^{3+}).

The correspondence of the isotopic composition of carbon to the contents of iron indicates an inverse relationship: the higher the concentration of Fe, the higher $\delta^{13}C$ values (Fig. 2.8D). An analogous, but less distinct relationship can be noted for the correspondence of S_{pyrite} to $\delta^{13}C$. The noted relationship reflects the conditions of sedimentation in the relationship of Fe, and primarily in the relationship of organic matter, an elevation in the content of which activates the processes of sulfate reduction (ie, pyrite) (Volkov, 1984).

2.2.1.5 Sources of manganese

The process of manganese carbonate formation in the sediments of the Baltic Sea, in comparison with the ancient basins of the known Oligocene manganese deposits, for example Nikopol' or Chiatura, developed completely otherwise. If for Mn carbonates of the indicated deposits, as will be shown below, the observed relationship of contents of MnO and the isotopic composition of carbon can provide evidence on the line of mixing of carbonates of sedimentary origin and the isotopically light CO_2 supplied to the zone of carbonate formation (as a rule, diagenesis and catagenesis) with the catagenic dissolutions from the zones of petroleum and gas generation (Kuleshov and Dombrovskaia, 1997a,b), then in the Baltic sediments the isotopic composition of carbon dioxide is formed in the course of the oxidizing-reduction diagenetic processes as a result of the chemical redistribution of matter in the mass of the sediments themselves.

Without argument, one principal, evident source of manganese in the waters and sediments of the Baltic Sea is continental runoff. Detailed studies of the forms of the supply of Mn by rivers and its balance in the Baltic basin at present have been published in fundamental works by Gorshkova, Varentsov, Emel'ianov, and a range of Russian and international researchers.

Nevertheless, the supply of manganese and organic matter to the zones of the deepwater depressions of the Baltic Sea by means of expelled petroleum waters, in the opinion of the author of the present volume (Kuleshov and Rozanov, 1998), cannot be fully excluded. Evidence of this can be provided above all by the numerous finds of petroleum deposits in the offshore areas and by the geological structure of the Baltic bed (Levin and Fel'dman, 1970). For example, in the geological cross section across the Gotland Basin (Geologiia..., 1976) it is apparent that the modern oozes here are deposited on top of Ordovician, Silurian, and Devonian deposits, which, with the exception of the Ordovician, are petroleum-bearing.

In the tectonic regard, the region of the studied station belongs to the Gotland horst of the Gotha elevated zone. From the west and the east, it is bounded by faults of submeridional strike that are manifested in the sedimentary cover in the form of flexures (Geologiia..., 1976). The central part of the Gotland block is intersected by deep faults, and the northern part is bounded by a series of faults that have been established on the basis of gravimetric measurements in the sea (Golub and Sidorov, 1971; Effendiev, 1967) and geological measurements on the island of Gotland. The faults form a single tectonic zone that begins in Sweden and stretches from the north-west to the south-east until the eastern coast, where it links with the Liepāja-Saldus and the Mažeikiai-Kuršėnai tectonic zones of the Baltic Syneclise (Levin, Fel'dman, 1970; Beckholmen and Tirén, 2009, 2010a,b).

Within the boundaries of the bed of the Gotland depression have been noted numerous zones of faults, which are clearly visible in the geological section (Geologiia..., 1976, p. 74). It can be supposed that precisely along these zones of tectonic activity the embedded waters of the petroleum-bearing strata of the Lower Paleozoic could have been discharged. Additionally, the embedded discharge of such waters could also have occurred from Devonian petroleum-bearing deposits within the boundaries of the eastern rim of the depression.

An impulse for the movement of the embedded waters of the Paleozoic deposits might have issued from the most recent tectonic movement of the Boreal and the subsequent Post-Glacial period of development of the Baltic. The primary feature of tectonic activity of this period in the eastern Baltic region is the preponderance on the part of peculiarly tectonic movements on the constantly waning glacio-isostatic background. At that time in the Baltic region separate districts of sinking and raising (relative and absolute) were taking shape, which was a result of the autonomous movements of separate tectonic structures.

It follows to note that widespread seeps of thermogenic methane that are related to faults of pre-Cambrian crystalline basement (Söderberg and Flodén, 1992), on our opinion, do not have connection with the source of manganese of Baltic Sea sediments.

2.2.2 FERROMANGANESE CONCRETIONS IN THE SEDIMENTS OF ONEGA BAY (WHITE SEA, RUSSIA)

In the offshore areas of Onega Bay, ferromanganese concretions are distributed quite irregularly. In the main part of the bay they have not been found (despite the rather detailed testing using a bottom grab and ground pipes). Here are distributed only thin Fe-Mn-fouling films in the pebbles. At separate stations pebbles were retrieved with Fe-Mn-borders and isolated microconcretions. Concretions particularly are concentrated in a small territory within the boundaries of the axial depression and the Onega shore near the village Liamtsa (Fig. 2.9) (Kuleshov and Chistiakova, 1989). Here have been uncovered continuous fields composed of concretions of spherical, discoid, and pellet forms. Also occurring are Fe-Mn borders on pebbles and microconcretions.

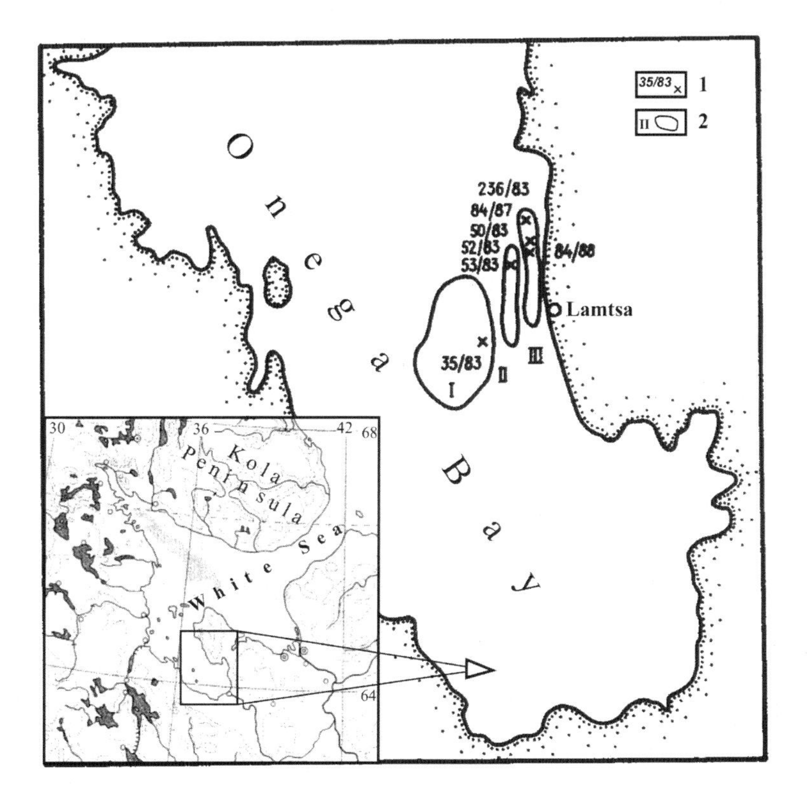

FIG. 2.9

Location of Fe-Mn nodules and crusts in Onega Bay (White Sea). 1—Number of sample and its location, 2—areas of: I—Fe-Mn-fouling films and crusts on pebbles, II—spheroidal (predominantly) nodules of axial depression, III—concretions and nodules spheroidal and flattened form.

All concretions possess a core surrounded by concentrically zoned layers of intergrowths. Textures have been found that provide evidence for breaks in the growth of the concretions as well as for the erosion of the concretions in defined stages. At certain spots in the concretions, a secondary redistribution of ferromanganese material is observed along the burrows that disrupt the initial concentric zonality. Fairly frequently microtextures are observed that provide evidence of incorporation and embedding of shells of foraminifera, balanus, and rarely of other organisms inside the concretions during the process of their growth. This circumstance is particularly important for the interpretation of the data on the isotopic composition of the carbon and oxygen in the carbonate of the concretions.

Mineral composition of Fe-Mn nodules is fairly diverse. Here have been identified the following minerals: manganese—vernadite, birnessite, buserite I, asbolane-buserite, todorokite; iron—feroxyhyte. The principal mineral phase of the concretions is represented by vernadite and feroxyhyte (Chistiakova, 1987).

As a rule, in the peripheral parts of the concretions vernadite predominates. Present as well is buserite I or asbolane-buserite, and in the internal parts birnessite is manifest. This is in full agreement with the general trend of diagenetic transformations of manganese minerals, the final product of which is birnessite (Drits et al., 1985).

The study of the chemical composition enables the identification of variations among concretional nodules of various morphological types. Above all, there are sharp distinctions between Fe-Mn borders on pebbles and those on concretions. This difference is manifest in the content as well as the ratio of Fe/Mn, the content of alumino-silicate impurity, and the presence of C_{org}. Hence, the greatest content

Table 2.6 Chemical Composition of Ferromanganese Nodules of Onega Bay (White Sea) (Kuleshov and Chistiakova, 1980)

Components	Fe-Mn-Fringe on Pebble (Sample 35/83)	Discoid Concretion (Sample 236/83)	Spheroid Concretion (Sample 50/83)
SiO_2	41.4	23.6	30.6
TiO_2	0.24	0.23	0.29
Al_2O_3	6.38	4.72	5.96
Fe_2O_3 (total)	17.7	22.0	15.8
MnO (total)	8.32	14.4	16.9
MgO	1.66	2.14	2.14
CaO	2.14	2.06	2.04
P_2O_5	0.82	1.28	0.72
Na_2O	2.30	2.32	2.36
BaO	0.00	0.12	0.11
SrO	0.02	0.02	0.02
NiO_2	0.12	0.12	0.14
H_2O^+	5.64	7.53	7.08
H_2O^-	7.58	12.5	8.48
CO_2	1.74	1.55	1.54
C_{org}	0.19	0.33	None

of alumino-silicate impurity is noted in pebbles, in which the content of manganese oxides constitutes 8.3%, and the manganese modulus (Fe/Mn) in mean is equal to 0.42.

Concretions of compressed form are characterized by a close manganese modulus (0.48) with a general high content of Mn and Fe (Table 2.6). As in the pebbles, C_{org} is present in such concretions.

Spherical concretions are characterized by a higher (12.21%) content of Mn. The manganese modulus here is equal to 1.12%. C_{org} is absent in concretions of this type.

In all concretions are observed contents of certain lesser elements (Ni, Co, Mo) that are elevated in comparison with their Clark. Despite their contents varying within wide boundaries, they are on the whole to be found at an order of magnitude lower than those in the pelagic concretions of the global ocean.

The content of CO_2 in the specimens is low, as a rule, usually not exceeding 1–1.5%. Forms of CO_2 entry into the composition of the matter of the concretions have not been established. In diffractograms peaks of carbonate minerals have not been observed. Evidently, CO_2 enters the composition of X-ray-amorphous compounds of manganese and iron carbonates.

2.2.2.1 Isotope data

The results of the study of the isotopic composition of the carbon and oxygen of the studied Fe-Mn concretions are reported in Table 2.7 and indicated in the graph of Fig. 2.10. From these results it follows that the $\delta^{13}C$ values are characterized by fairly low magnitudes and vary within the interval from −23.5‰ to −17.3‰. The isotopic composition of the oxygen is likewise variable. The $\delta^{18}O$ values fluctuate within

Table 2.7 Isotopic Composition of Carbon and Oxygen of the Carbonate Component of Ferromanganese Nodules of Onega Bay (White Sea)

No. Analysis	No. Sample	Morphology	C_{org}, %	CO_2, %	$\delta^{13}C$, ‰ (PDB)	$\delta^{18}O$, ‰ (SMOW)
2132	35/83	Fe-Mn-fringe on pebble	0.19	1.74	−18.8	17.9
2133	50/83	Spheroidal concretion	None	1.54	−18.5	8.8
2137	50/83		Not determined	Not determined	−23.5	3.2
2139	52/83		–	–	−24.1	3.3
2140	52/83		–	–	−23.0	3.2
2135	53/83	Spheroidal concretion	–	–	−23.0	4.2
2141	53/83		–	–	−17.3	11.2
2143	236/83	Discoid concretion	–	–	−22.5	3.4
2144	236/83		–	–	−23.4	2.5
2134	236/83		None	1.55	−17.9	10.8
2136	84/87	Spheroidal concretion	Not determined	Not determined	−21.4	7.6
2142	84/88		–	–	−21.8	3.9

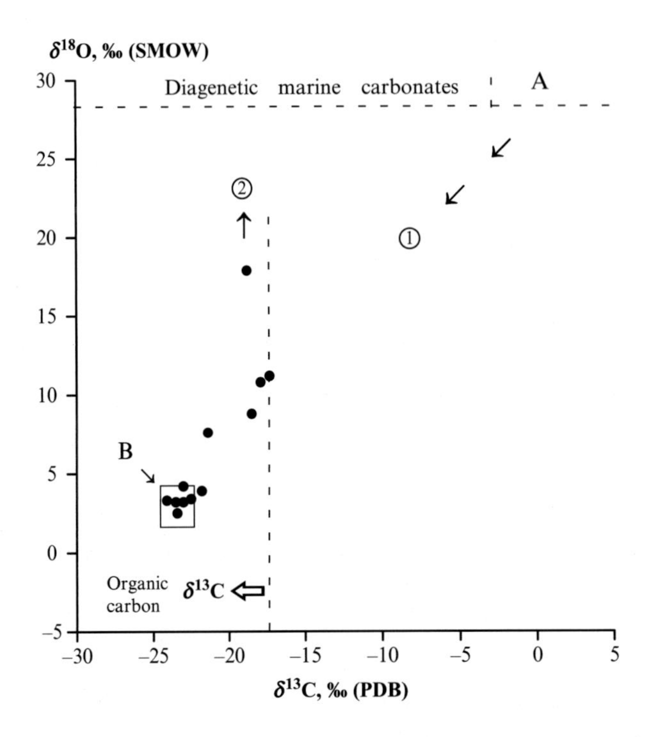

FIG. 2.10

$\delta^{13}C$ vs. $\delta^{18}O$ in Fe-Mn nodules in Onega Bay (White Sea) (Кулешов and Чистякова, 1989). A—Areas of nonequilibrium diagenetic carbonates. B—carbonate in oolites of Fe and Mn oxides. ① and ②—see text.

the interval from 2.5‰ to 17.9‰. An analogous isotopic composition, as indicated in one work (Zheng et al., 1986), characterizes the ferromanganese concretions of the northern part of the Pacific Ocean.

The low $\delta^{13}C$ values discovered in the studied samples provide evidence that the carbonate carbon of the concretions is not situated in isotopic equilibrium with the dissolved bicarbonate of marine water ($\delta^{13}C = -2$‰ to 2‰). Such light values of $\delta^{13}C$ are commonly particular to carbon of organic origin. For this reason, it can be confirmed that carbon of organic origin predominated in the generation of the carbonate constitution of Fe-Mn concretions.

The $\delta^{18}O$ values likewise provide evidence for the disequilibrium of carbonate carbon with marine water. The former are substantially enriched in the light isotope ^{16}O by comparison with normal sedimentary marine carbonates (in isotopic equilibrium with DIC of marine water). Such low $\delta^{18}O$ values (up to 2–4‰) are potentially a consequence of the oxidation of organic carbon to CO_2 by oxides of manganese and iron within the Fe-Mn nodules themselves.

Thus, the data on the isotopic composition of the carbon and oxygen permit the conclusion that the carbonate constitution of Fe-Mn nodules of Onega Bay of the White Sea has a diagenetic origin and is formed evidently at the stage of early diagenesis as a result of the oxidation of organic matter by oxides of Mn and Fe (up to the carbonate form). The values of $\delta^{13}C$ for this carbonate should approach −25‰ to −22‰, and for $\delta^{18}O$ should approach 2–6‰. In the graph of Fig. 2.10, this carbonate matter occupies section "A."

A linear relationship has been established in the distribution of the isotopic composition of the carbon and oxygen of the studied samples. This is due primarily to the insignificant impurity in the composition of the concretions of a (sedimentary) carbonate that is stable with the marine water. Direct evidence of this is also provided, as noted above, by the finds of fossils of carbonate shells of foraminifera and other organisms embedded in the oxide-ferromanganese matter of the concretions. However, the addition of a stable carbonate is insignificant; in certain samples, it is practically imperceptible.

The observed linear relationship could potentially complicate the occurrence of certain samples outside the line of mixing. This is understood inasmuch as the process of diagenesis also occurs at the present time. This gives rise to certain cases of concretions found with unstable isotopic equilibrium; for example, in sample 35/83 we find a residual content of C_{org}, which provides evidence of an incomplete process of diagenetic oxidation of organic matter. The displacement of the samples from the line of equilibrium along the $\delta^{18}O$ axis could be caused by the oxygen isotope exchange of authigenic (diagenetic) carbonate with marine water.

The established distribution pattern of $\delta^{13}C$ and $\delta^{18}O$ in the carbonate component of diagenetic ferromanganese nodules of Onega Bay is not unique; as will be shown below, it also corresponds to the character of the distribution of the isotopic composition of the carbon and oxygen in the modern Fe-Mn-diagenetic accumulations of Lake Punnus-Yarvi (Karelian Isthmus).

2.3 MANGANESE CARBONATES IN LAKE SEDIMENTS (ON THE EXAMPLE OF THE KARELIAN LAKES)

Ferromanganese and carbonate-manganese concretions and nodules are widely distributed in the lakes of the humid zone of North America (Canada, northwest USA) and Europe (Finland, Karelia) and are known also in the lakes of Belarus and the southern Urals (Shterenberg et al., 1966; Robbins and Callender, 1975; Shniukov et al., 1976; Zhukhovitskaia, 1986; Asikainen and Wehrle, 2007; and others). However, their maximum distribution is known in the Baltic Shield and the adjoining parts of the Russian Plate. Here especially are singled out the lakes of Karelia, specifically Lake Punnus-Yarvi (Lake Krasnoe). The sediments of this lake serve as a case study of modern lacustrine manganese-ore formation (Krotov, 1950a,b; Zavalishin, 1951; Sokolova, 1961; Semenovich, 1958; Strakhov et al., 1968; and others); for this reason, we have taken the ferromanganese accumulations of this lake as the object of our isotope research.

2.3.1 FERROMANGANESE CONCRETIONS IN LAKE PUNNUS-YARVI

Lake Punnus-Yarvi is distinguished among the lakes of the Karelian Isthmus sharply by the most developed process of modern ferromanganese sediment, which is manifested in the form of oxides of Fe-Mn and carbonate-Mn nodules in the upper layers of the Holocene sediments. This circumstance has increased its interest for researchers of diverse disciplines.

The author conducted a study of the isotopic composition of the carbon and oxygen of carbonate minerals and carbon components that participate in the structure of Fe-Mn nodules. The author proposes that the identification of their distribution patterns could permit a more reliable interpretation of the isotope data for manganese ores of the largest deposits found in the strata of terrigenous and terrigeno-carbonate rocks that are known within the boundaries of Russia, Ukraine, Georgia, Kazakhstan, Mexico, China, and other countries.

2.3.1.1 The object of the study and characteristics of the samples

Lake Punnus-Yarvi is situated in the central part of the Karelian Isthmus. It extends from the north-west to the south-east for approximately 9.6km, and its width at its greatest reaches 2.8km. The depth of the water body reaches 14m. The lakebed is characterized by steep slopes and a fairly level bottom. It is classified as a low-mineralization lake and is practically not distinguished in this regard from other lakes of Karelia and the Kola Peninsula. The Ca content in its water varies from 7.1 to 9.1mg/L; Mg content from 1.9 to 2.4mg/L; HCO_3^- content from 25 to 29mg/L; SO_4^{2-} content from 10.5 to 16.5mg/L; Cl content from 1.3 to 4.7mg/L; and SiO_2 from 3 to 8mg/L. The level of dissolved organic matter in water is comparatively low (4.5–10.8mg/L) (Kuleshov and Shterenberg, 1988).

The replenishing of Lake Punnus-Yarvi with iron, manganese, and a range of other elements, as well as that of other water bodies of the Karelian Isthmus, Karelia, and the Kola Peninsula, is conducted primarily by means of surface runoff and groundwater. Mobilization of these elements occurs in the soil of podzolic and stagnant areas. According to Zavalishin (1951), the content of iron in the peaty podzols does not exceed 3–4mg/L and, as a rule, is found in the form not of minerals, but of metal-organic compounds. Divalent iron in free mineral form appears in small quantities only periodically (in spring) and in short duration.

The Suontaka-Yoki River, being the main artery (60.1% of the general inflow of the lake), according to Semenovich (1958), supplies from 0.5 to 3.5mg/L of iron. The quantity of this river's supply of manganese is markedly lower (up to 0.85mg/L).

The sediments of Lake Punnus-Yarvi can be divided into three groups: modern, transitional, and ancient. The modern deposits of the littoral, up to depths of approximately 2–4m, are represented by small pebbles and sands of various degrees of granularity. All clastic material of this zone is represented primarily by the erosion products of granitoids and more rarely by fragments of Cambrian sandstones. Sands, as seen in Fig. 2.11, reach their greatest distribution in the northern part of the lake, where they form wide fields reaching 50–70m. Downward along the slope of the bed of the water body they become increasingly fine-grained. The mineral grains and fragments of rocks are poorly rounded.

In the central part of the lake, dark-greenish-gray viscous oozes with impurities of silty (50–53%) and sandy (up to 10%) material are widely developed. Among the oozes have been established microzones and bands of black color containing somewhat elevated quantities of sulfides of iron, primarily hydrotroilite, which rapidly oxidize in air. The thickness of the oozes is not constant, reaching a maximum up to 16m in the central part of the water body. Between the near-shore sands and the deepwater oozes, there lie transitional deposits comprising light-colored silty sands and significant impurity of sandy-silty material.

The ancient deposits of Lake Punnus-Yarvi are light-gray banded glacial-lacustrine clays. In Table 2.8 are reported the mean contents of Mn, Fe, and C_{org} in the modern and ancient deposits of Lake Punnus-Yarvi. Judging by these analyses, all the modern sediments to some degree are rich in the elements mentioned above. Their contents do not remain constant and regularly increase in the direction from the near-shore toward the more deep-water deposits.

The ferromanganese ores of Lake Punnus-Yarvi and other lakes of Karelia and the Kola Peninsula can be conditionally divided into two different types (Shterenberg et al., 1970). The first type comprises "substantially diagenetic" ores, which in Lake Punnus-Yarvi include Fe-Mn nodules that are situated

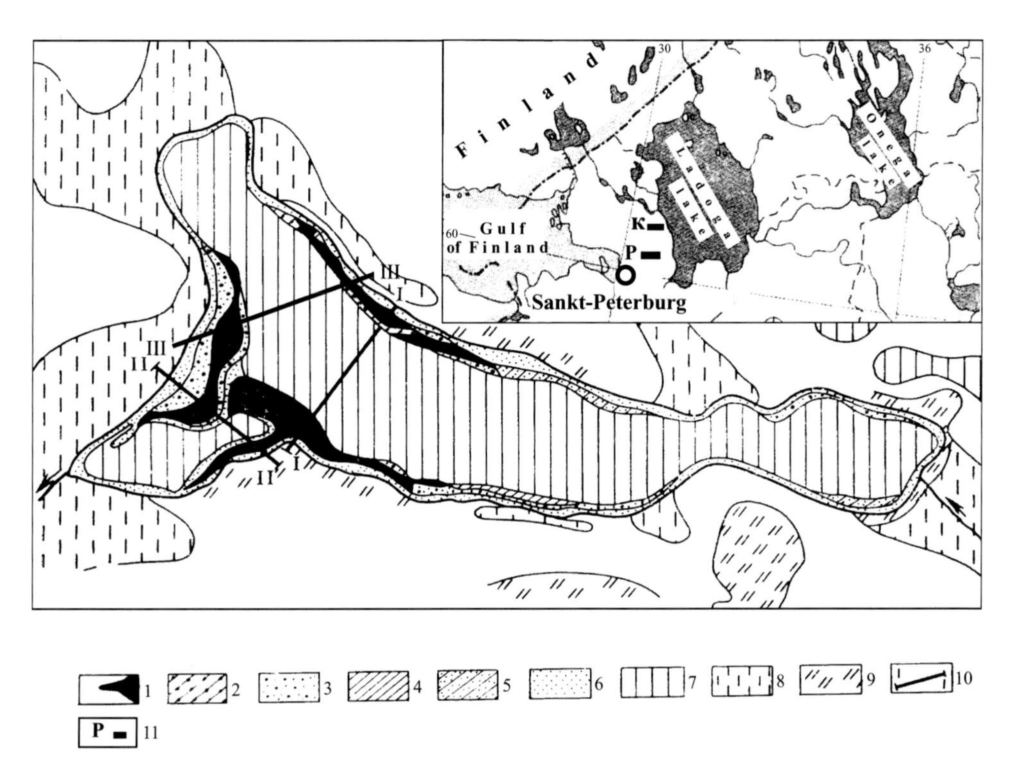

FIG. 2.11

Sediments and ores distribution in Punnus-Yarvi Lake (according to Семенович, 1958) and location of sampling profiles on isotope analysis (Кулешов, Штеренберг, 1988). 1—Fe-Mn concretions of ore fields; 2—same, in aleurite sediments; 3—same, in sands; 4—lacustrine varved clay; 5—aleurolites; 6—sands; 7—mud; 8—swamp (bog, moor?) without hydrous ferric oxide; 9—peat bog; 10—location profiles of sampling; 11—location of studied lakes: P—Punnus-Yarvi, К—Konchozero.

Table 2.8 Mean Contents of Mn, Fe, and $C_{(org)}$ in Sediments of Lake Punnus-Yarvi, % (Semenovich, 1958)

Deposit	Mn	Fe	Mn:Fe	C_{org}
Modern				
Sands	0.11	1.50	0.063	0.49
Sands, silty	0.21	3.89	0.053	2.11
Oozes, silty-clayey	0.44	7.37	0.06	7.3
Ancient				
Clays, varved	0.07	3.66	0.016	0.19

primarily on the surface of the principal ore field (see Fig. 2.11). The role of bog waters in replenishing this water body (as with other lakes of this territory) is minimal.

The ores near the cape Mikhkiur-Niemi are mainly spherical in form, 1–2.5 cm in diameter, but sometimes reach 4–4.5 cm, and rarely larger. Their surface is coarsely textured and even somewhat hummocky. The color is black, luster is dull. A zonal-concentric structure is characteristic for these ores (Fig. 2.12).

The composition and structure of "substantially diagenetic" ores do not remain static and regularly varies along the littoral as much as along the vertical. From the central part of the ore deposit in the direction toward the deep-water zone of the lake the dimensions of ore oolites decrease, up to full disappearance. They become more ferruginous with elevated contents of phosphorus and minor contents of manganese and barium.

The dimensions of the ore oolites are greatest in the central part of the ore field, where they are characterized by elevated contents of manganese and barium, with minimal quantities of iron and phosphorus.

Along the direction toward the shoreline, in the zone of near-pinching out, the ores lose their spherical form to acquire an increasingly compressed form, in the marginal shore area approaching thin incrustations and films on the rocks and pebbles. There is a simultaneous increase in this direction in their iron content and a decrease in manganese.

The composition and form of the manganese enrichments along the separately taken sections do not remain constant. In the upper parts of the sequences, the concretions have the greatest dimensions and contain more manganese and barium, and in the lower parts of the sequences are found the finer varieties, insignificantly enriched in iron and phosphorus.

Information remains scant as to the mineral composition of the "substantially diagenetic" Fe-Mn ores. Basic to the composition of the ores are X-ray-amorphous hydroxides of manganese and iron of "wad" psilolemane type, which are classified at present by many researchers as vernadite and hydrogoethite. In the Fe-Mn concretions of Lake Punnus-Yarvi, Chukhrov et al. (1989) have established hollandite, cryptomelane, vernadite, birnessite, feroxyhyte, and others.

Ores of the second type are conditionally termed "dominantly sedimentary." They are formed due to the effects of the processes of bog formation and are intensively developed in catchment areas confined as a rule to the shallow-water zones of the lake. These manganese enrichments have smaller dimensions than do the first type varieties. Their surface is smooth; their structure is homogenous and solid. Ores of the "substantially sedimentary" type are rich in iron and phosphorus with extremely small quantities of manganese and barium. A principal role is played in their composition by such minerals as hydrogoethite, vivianite, and hydroferrichlorite.

Apart from these extreme types of Fe-Mn accumulations, an intermediate type has also been established in the lakes of Karelia and the Kola Peninsula.

In Fe-Mn-oxide concretions of the first type, from the zone of the ore deposit's far-pinching out toward its near-pinching out—that is, toward the shoreline—there is a marked growth in the Mn and Ba contents. Moreover, in the most developed in thickness zone of the ore horizon is observed the maximum of the contents of these elements. The barium contents according to our data approach 1% (Table 2.9; Kuleshov, Shterenberg, 1988). However, with X-ray (diffractometric) analysis of the samples of Fe-Mn concretions and of the various fractions isolated from them by means of heavy liquid separation, no barium minerals could be detected.

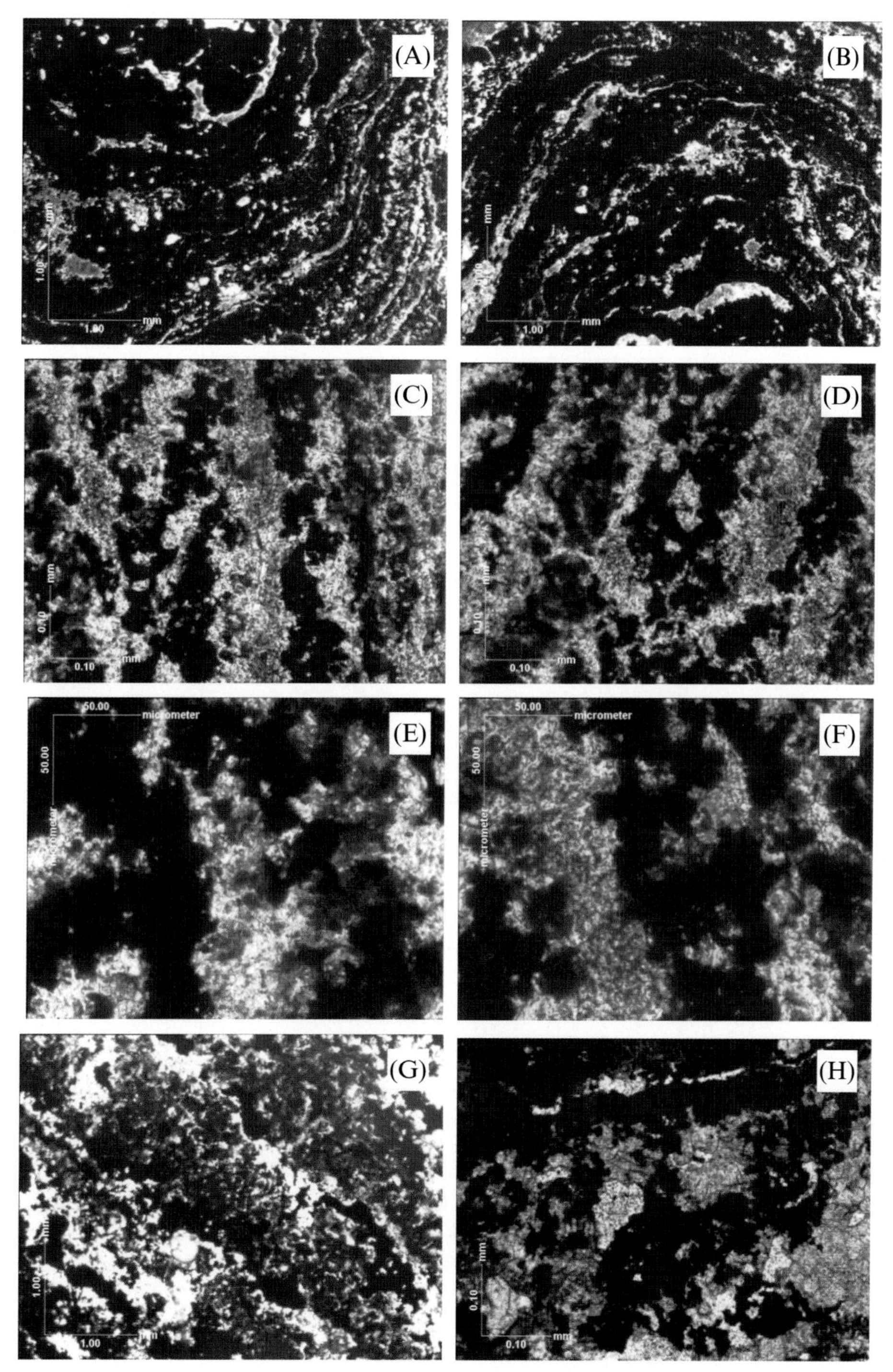

FIG. 2.12

Structure and composition of carbonate-oxide oolites of Punnus-Yarvi Lake (photomicrographs of thin sections at different magnifications). (A–E) Sample 1847, by miscellaneous increase; clearly visible concentric zonal structure (A–D: parallel nicols, E and F: crossed nicols); (G and H) sample 1844, crossed nicols.

Table 2.9 Chemical Composition of Ferromanganese Nodules of Lake Punnus-Yarvi (Karelian Peninsula)

Analysis No	SiO_2	TiO_2	Al_2O_3	Fe_2O_3	MnO_{total}	MgO	CaO	P_2O_5	Na_2O	K_2O	BaO	SrO	ZnO	CO_2	C_{org}	MnO	MnO_2	Total
1831	7.04	0.02	1.02	12.5	48.8	0.22	1.38	1.06	0.13	0.23	0.26	–	0.06	9.65	0.58	23.5	29.1	86.7
1832	17.9	0.04	0.94	26.7	23.3	0.2	1.88	1.1	0.11	0.11	1.02	0.02	0.08	11.5	0.85	18.6	6.3	87
1834	10.4	0.18	2.12	22	38.8	0.24	0.86	0.66	0.32	0.59	0.7	0.02	0.12	0.37	–	5.63	40.1	84.3
1835	8.36	0.08	1.92	14.7	47.3	0.36	1.34	0.64	0.33	0.64	0.08	0.02	0.1	0.97	–	7.19	49.1	85.7
1836	5.74	0.04	1.34	20.8	41	0.22	1.32	1.02	0.19	0.35	–	–	0.12	3.2	0.17	7.34	41.9	83.7
1837	7.24	0.04	1.28	27.3	37.2	0.16	0.98	1.34	0.13	0.43	–	0.04	0.1	2.2	0.34	6.9	36.4	84.3
1838	12.6	0.1	1.62	45.6	10.6	0.36	1.32	2.74	0.18	0.13	1.02	–	0.1	3.05	0.97	6.27	5.5	86.5
1839	22.7	0.12	2.9	50	2.84	0.26	0.94	2.68	0.34	0.58	0.18	–	0.12	0.35	1.29	3.02	0.28	85.8
1843	24.5	0.04	1.18	38.4	7.84	0.3	1.16	1.58	0.12	0.24	0.54	–	0.12	2.22	1.31	5.82	2.61	80.2
1844	8.16	0.02	0.98	33.8	28.8	0.14	0.76	1.58	0.09	0.28	0.78	0.02	0.12	2	0.66	7.9	26.2	83.5
1845	6.56	–	1.16	28.7	34.2	0.16	1.14	1.46	0.15	0.36	0.18	0.2	0.12	4.55	0.52	12.7	27	84.8
1846	6.68	0.04	1.3	24.7	38.5	0.18	1.04	1.34	0.17	0.41	–	0.02	0.14	3.47	0.6	10.3	34.6	85
1847	4.38	0.02	1	21.8	45.1	0.24	0.88	1	0.14	0.46	–	0.04	0.14	1.9	0.35	8.91	44.3	85.6
1848	4.5	0.02	0.74	21.7	44.2	0.18	0.98	0.92	0.42	0.46	–	0.04	–	1.5	0.19	5.73	47.8	85.2
1849	3.38	0.02	0.82	18.6	48.8	0.14	0.46	0.74	0.08	0.5	–	0.04	0.08	1.25	0.2	8.05	49.7	84

Analyses conducted by M.A. Stepanets (GIN AN USSR). –, content undetected.

The magnitude of the oxidation-reduction potential of the ore zone of Lake Punnus-Yarvi, according to the data of Strakhov et al. (1968) and the data of Sokolova (1962), ranges from +550 to +600 mV, while the pH fluctuates within the range of 6.5–6.9. The magnitude of pH in water above the ooze is closed to 7, and Eh varies from +355 to +440 mV.

The principal ore field of Lake Punnus-Yarvi is represented by substantially diagenetic accumulations. It stretches from the cape Miukhkiur-Niemi in the north-western direction to the opposite shore of the lake in an almost continuous band; a narrow band of barren sediments has been identified only in the most deepwater part of the bay (see Fig. 2.11). On the north-eastern shore of the lake, as is the case in the other shallow areas of Lake Punnus-Yarvi, the ores are deposited in more narrow zones among sandy deposits. Within the boundaries of the principal ore layer, the ore layer reaches a thickness of 0.20–0.22 m at its greatest.

The CO_2 content in the sediments in almost all areas of the lake, according to the data of Strakhov et al. (1968), is highly negligible and as a rule does not exceed a few tenths of a percent. However, in the area of Fe-Mn ores where measured, CO_2 content is sharply elevated, approaching 2.5–3.5%. Here the relationship of CO_2 to the ore facies is clearly visible.

The elevated quantities of CO_2 in the ore zone are connected with the presence of carbonates of manganese and iron in their composition. This connection is seen in the graphs illustrating the relationship among Mn, Ca, and CO_2 in the sediments and ores of Lake Punnus-Yarvi (Figs. 2.13 and 2.14).

Elevated quantities of organic matter in the lakes occur within the shallow zone, composed of sandy sediments, in which Fe-Mn-concretional ores form. According to the data of Strakhov et al. (1968),

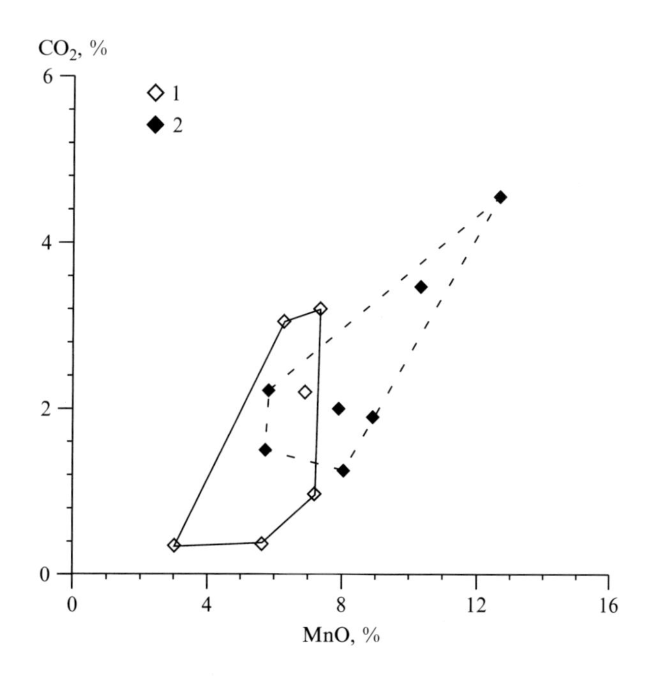

FIG. 2.13

MnO vs. CO_2 (wt.%) in nodules of profile II-II′, Punnus-Yarvi Lake. 1—Eastern part of profile; 2—western part of profile.

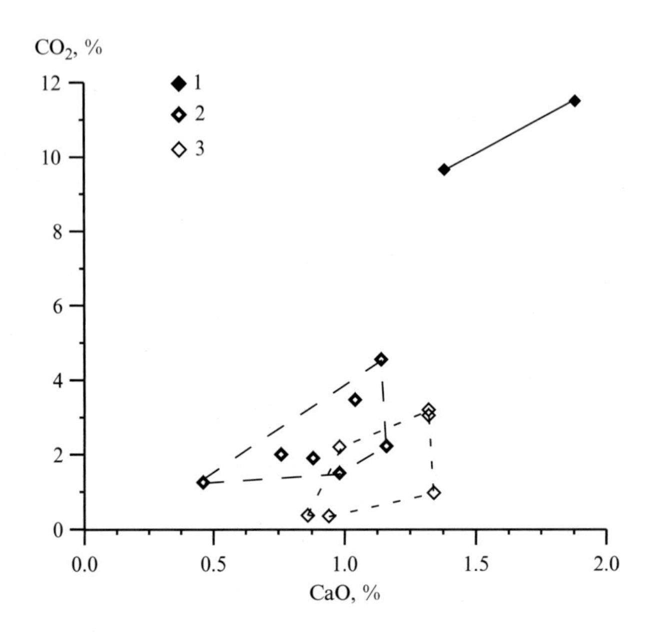

FIG. 2.14

CaO vs. CO_2 (wt.%) in nodules of Punnus-Yarvi Lake. 1—Southern part of profile I-I′; 2—eastern part of profile II-II′; 3—western part of profile II-II′.

the C_{org} content in the ore zone on the whole is low in comparison with that in the fine-grained ooze deposits of the deep zone. The C_{org} content in Fe-Mn-concretional ores of the first type ranges from insignificant (around 0.05%) to 1.45%, at a mean of around 0.75%.

Thus, the ore nodules of Lake Punnus-Yarvi on the whole are enriched by organic matter. According to Sokolovaia-Dubinina and Deriugina (1967a,b), the enrichment of Fe-Mn nodules by organic matter is connected with the accumulation of remnants of the plants (reeds and cane) in the zone of manganese, which engenders favorable conditions for the development of microorganisms that restore manganese. The remains of diatoms in certain cases are visible under a scanning electron microscope (Fig. 2.15).

Ferromanganese concretions at Lake Punnus-Yarvi were collected by a boat by means of a stratometer in three profiles: I-I′, II-II′, and III-III′ (see Fig. 2.11). Each profile was conducted at a large number of stations, which allows for detailed study of the particularities of the distribution of Fe-Mn ores in terms of area and dimensions. An object to the most complete study were concretions only in profile II-II′, in which the contents of Mn^{2+}, Mn^{4+}, Fe_{total}, C_{org}, and other elements were determined in addition to the isotopic composition of carbon and oxygen (see Table 2.8). The conditions of the deposition of the studied Fe-Mn concretions are indicated in the cross sections in these profiles (Figs. 2.16–2.18). They were set along the core of the stratometer.

Our data in general agree with the results of previous research (Strakhov et al., 1968; Shterenberg et al., 1970). In the eastern part of the profile II-II′ (Fig. 2.17), with increasing distance from the shoreline, there is an increase in Fe and C_{org} while the Mn content (both Mn^{2+} and Mn^{4+}) decreases. The content of CO_2 in the manganese enrichments is distributed unevenly. The maximum concentrations of carbon dioxide are noted inside the principal ore field, somewhat migrating into the near-shore zone.

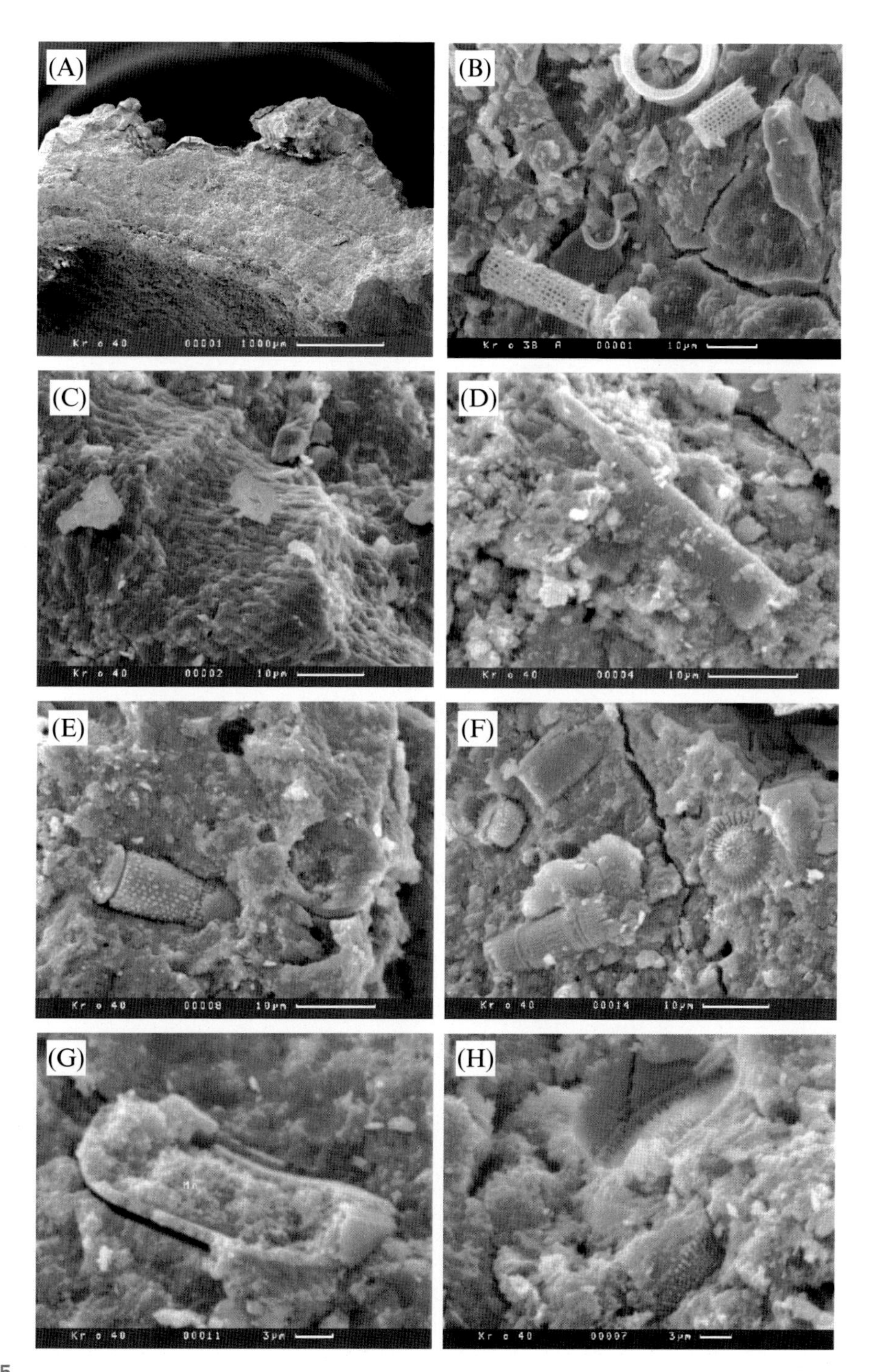

FIG. 2.15

Structure and composition of carbonate-oxide oolites of Punnus-Yarvi Lake (scanning electron microscope photomicrographs, photo Zhegallo and Shkol'nik). (A) General view; (B–H) remains of diatomic algae in ore matter.

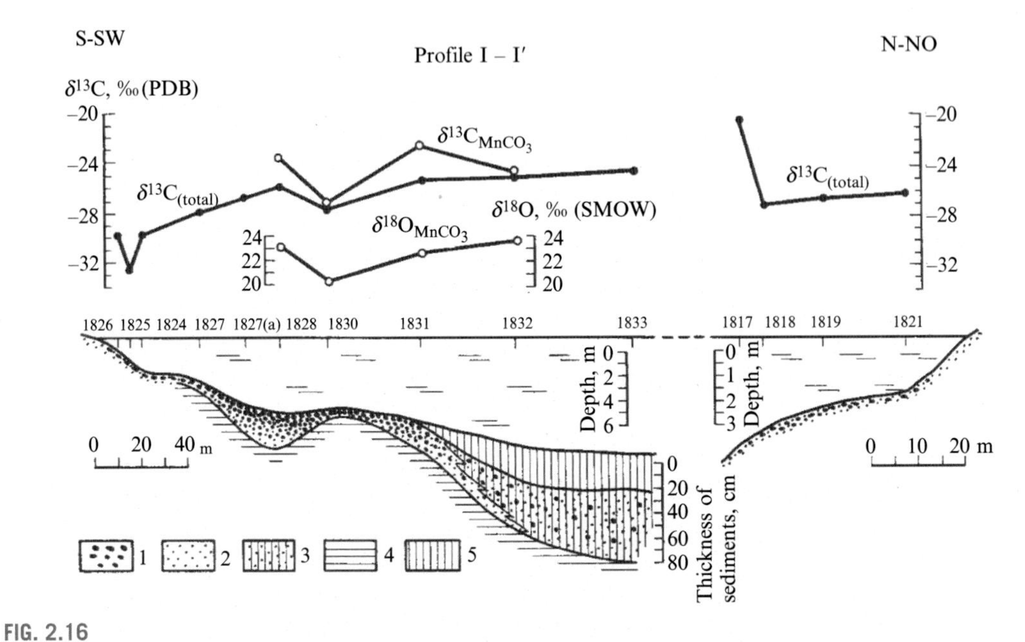

FIG. 2.16

$\delta^{13}C$ vs. $\delta^{18}O$ in ore nodules of profile I-I′, Punnus-Yarvi Lake. 1—Fe-Mn nodules; 2—sand; 3—ooze sand; 4—clay; 5—ooze.

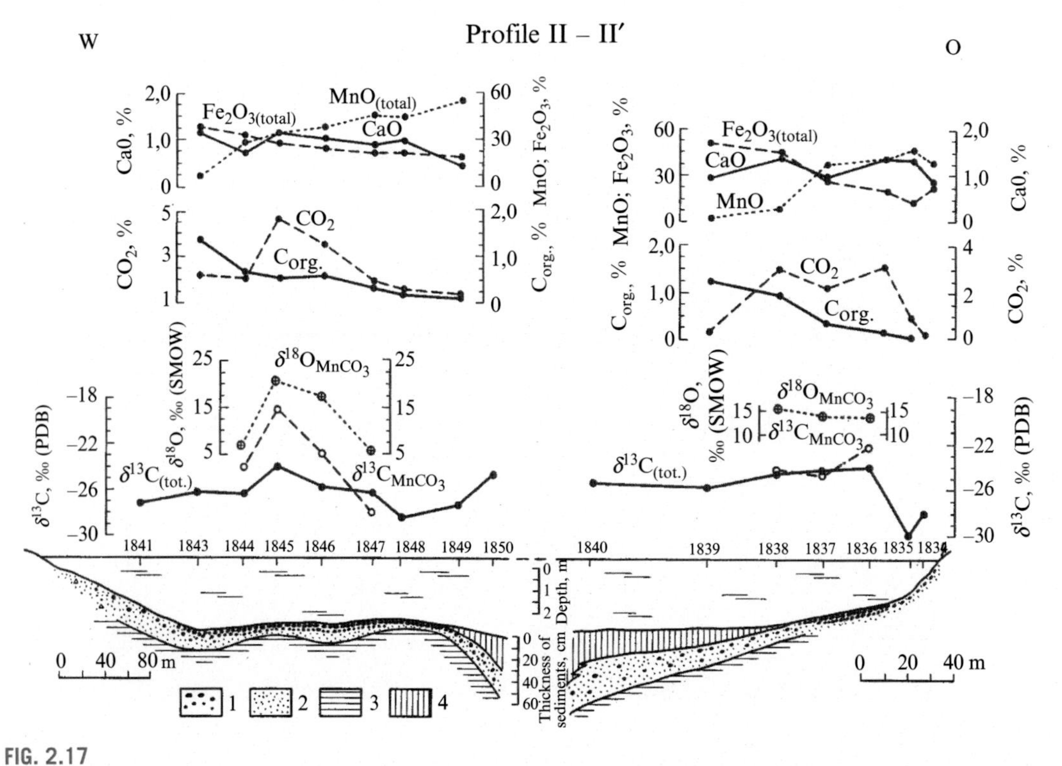

FIG. 2.17

$\delta^{13}C$ vs. $\delta^{18}O$ in ore nodules of profile II-II′, Punnus-Yarvi Lake. 1—Fe-Mn nodules; 2—sand; 3—clay; 4—ooze.

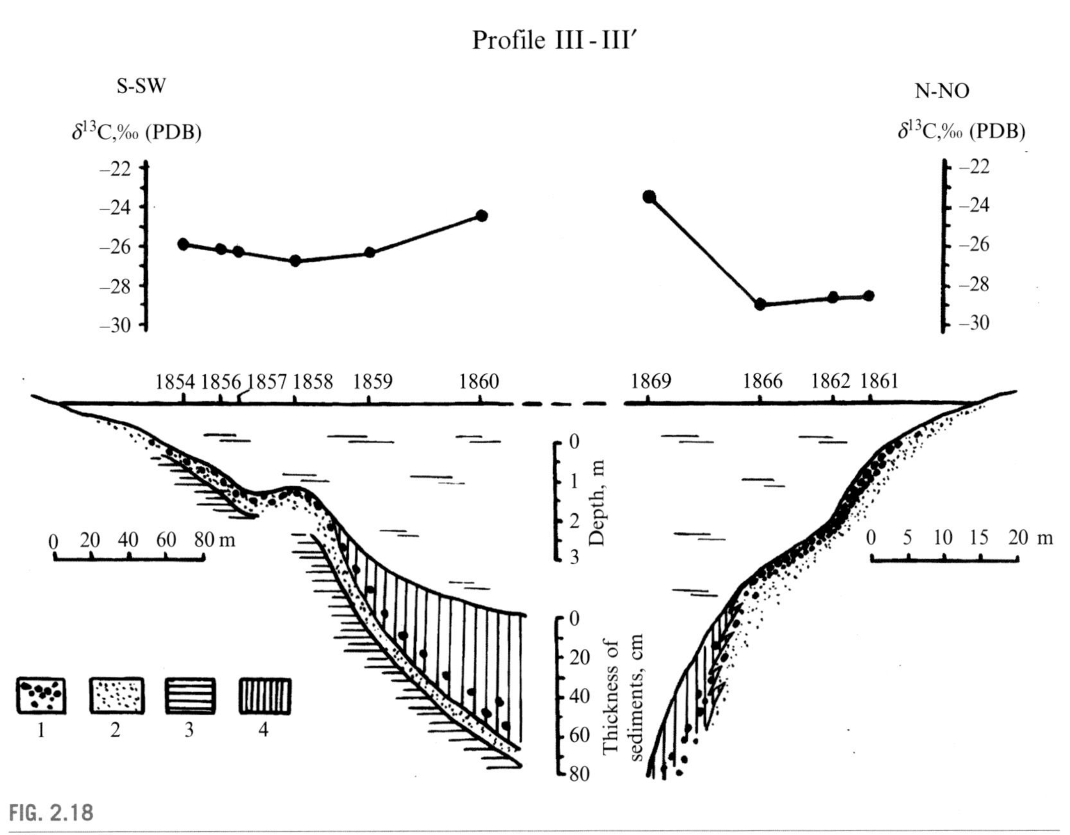

FIG. 2.18

$\delta^{13}C$ vs. $\delta^{18}O$ in ore nodules of profile III-III′, Punnus-Yarvi Lake. 1—Fe-Mn nodules; 2—sand; 3—clay; 4—ooze.

In different parts of the ore field, the ratios of the principal chemical components in the ore nodules can be different and are determined by local conditions of ore formation. Thus, for example, for ores of the western and eastern parts of our studied profile II-II′ were noted certain differences in the ratios of CaO and MnO (Fig. 2.19). Moreover, for CaO they are somewhat higher in the eastern part of the profile. The distribution of the ratios of MnO_{tot} and C_{org} likewise differs markedly in these parts of the profile (Fig. 2.20). At the same time, it follows to note that the character of the ratios of Fe_2O_3-MnO_{tot} in the cross section of the ore field in the western and eastern parts of the profile remains approximately identical (Fig. 2.21). This could provide evidence for a single mechanism of formation of ore components within the boundaries of the studied profile II-II′ and evidently within the ore field as a whole.

In the western part of the studied profile, an inverse relationship is noted in the distribution of Fe, Mn, and C_{org}: contents of C_{org} and Mn increase on the open part of the lake, while iron content, conversely, decreases. In the opinions of Strakhov et al. (1968) and of Shterenberg (1979), this is connected most likely with the predominance here of substantially sedimentary ores that formed under the influence of bog waters.

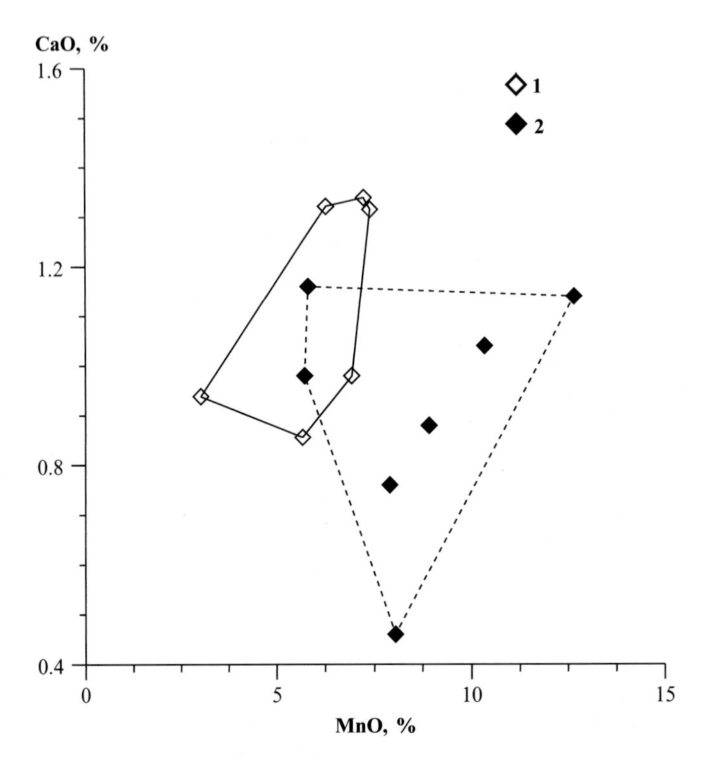

FIG. 2.19

CaO vs. MnO (wt.%) in nodules of profile II-II′, Punnus-Yarvi Lake. 1—Eastern part of profile; 2—western part of profile.

According to Shterenberg et al. (1966), the basic mineral in the carbonate phase of Fe-Mn ores of Lake Punnus-Yarvi, which achieve 7–8% from the gross of their compositions, is substantially calcium rhodochrosite with insignificant isomorphic impurities of $CaCO_3$ and $FeCO_3$. Sokolova-Dubinina and Deriugina (1967a,b) likewise have shown that the basic carbonate-manganese mineral of Fe-Mn concretions of Lake Punnus-Yarvi is rhodochrosite.

This result is corroborated by our research (Kuleshov and Shterenberg, 1988). In Fig. 2.22 are shown the results of the X-ray (diffractometric) analysis of two samples of Fe-Mn nodules that were collected within the boundaries of the principal ore field. These data provide evidence of the wide spectra of carbonate minerals that participate in ore formation—from siderite ($FeCO_3$), the principal interplanar distance of which is equal to 2.78 Å; to siderite with small isomorphic impurities of $CaCO_3$ and $MnCO_3$; to rhodochrosite ($MnCO_3$) with $d=2.79$ Å (the principal carbonate phase in ores); and further to the series of intermediate ore formation between rhodochrosite and calcite, which we classify as calcium rhodochrosite with interplanar distances of 2.91, 2.92, 2.95, and 3.01 Å. The series of carbonate minerals in the studied nodules is capped by calcite ($CaCO_3$), the interplanar distance of which is equal to 3.03 Å.

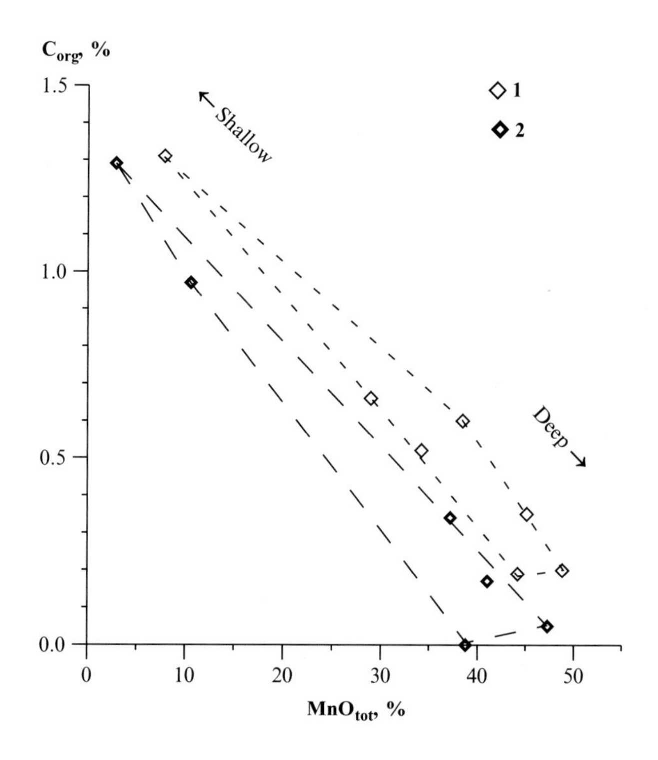

FIG. 2.20

C_{org} vs. MnO_{tot} in Fe-Mn nodules of profile II-II′, Punnus-Yarvi Lake. 1—Eastern part of profile; 2—western part of profile.

The results of the conducted methodological research (Kuleshov, 2013) indicate that the isotopic composition of the carbon in the CO_2 that was isolated by various methods as a rule is the same, except for a few cases where the carbonate carbon is enriched insignificantly by heavy isotopes. This allows us to regard the carbon of CO_2 in the various carbonate minerals as behaving in the same way, isotopically.

2.3.1.2 Isotope data and their discussion

The obtained isotope data for ferromanganese nodules of Lake Punnus-Yarvi (collected in profiles I-I′, II-II′, and III-III′) are represented in Table 2.10 and are shown in the corresponding graphs (see Figs. 2.16–2.18 and 2.23).

The range of variations of $\delta^{13}C$ (PDB) for the carbonate component of these accumulations varies from −28.2‰ to −19.2‰, and that of $\delta^{18}O$ (SMOW) from 6.0‰ to 23.7‰. The isotopic composition of the carbon in CO_2 of the total sample ($CO_{2\,total} = CO_2$ from carbonate + CO_2 from Mn-Fe oxide matter by heating) varies from −32.5‰ to −20.4‰, and that of $\delta^{18}O$ from 0.9‰ to 17.7‰, although the primary quantity of the latter is confined to the interval 9–15‰. (Data on the isotopic composition of the oxygen of the $CO_{2\,total}$ are not included in the tables, insofar as they do not bear a direct genetic factor.)

From the cited data it is apparent that all $\delta^{13}C$ values are negative, indicating that carbon is of organic origin. They are much lower in comparison with lacustrine carbonates deposited in isotopic

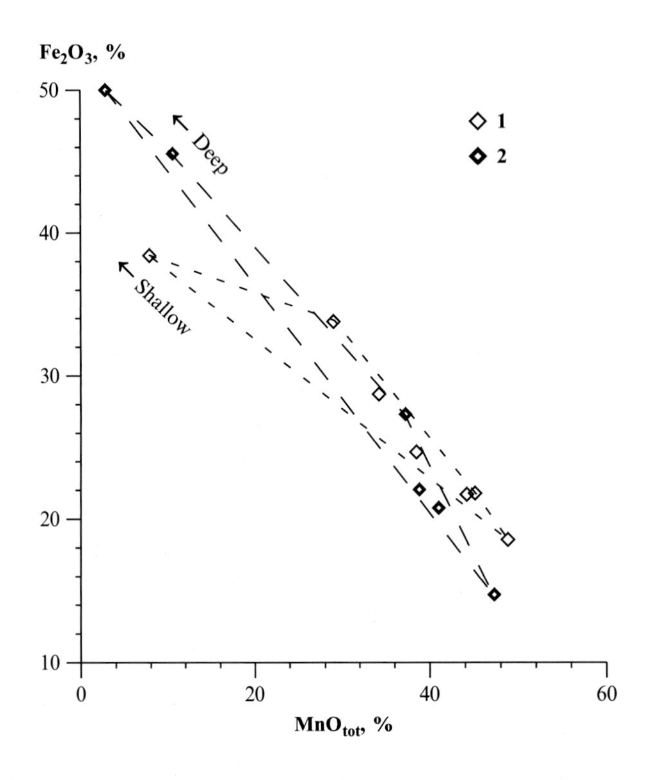

FIG. 2.21

Fe_2O_3 vs. MnO_{tot} in Fe-Mn nodules of profile II-II′, Punnus-Yarvi Lake. 1—Eastern part of profile; 2—western part of profile.

equilibrium with the bicarbonate of the lake water. Mollusk shells serve as an example of the latter. Thus, the values of $\delta^{13}C$ for two shells of a bivalve mollusk (freshwater mussel) from the given lake turned out to be equal to −9.5‰ and −8.5‰, while those of $\delta^{18}O$ were 21.5‰ and 21.6‰, respectively. Such high $\delta^{13}C$ and $\delta^{18}O$ values (ie, close to the composition of the mollusk shells) characterize carbonate substance in sediments of the lakes of the Baltic countries (Martmaa et al., 1982) and other lakes of humid districts (Poliakov et al., 1982).

Consequently, the oxidized organic matter of sediments of the lake constitutes a source of carbon dioxide in the studied Fe-Mn ores. Moreover, insofar as the values of $\delta^{13}C$ for the total sample and the carbonate component in the same sample are analogous (Table 2.10), we are justified in saying that the oxidation of organic matter within the ore nodules occurred without a notable isotopic shift in the carbon. Isotopically, carbonates and residual C_{org} appear to be "conservative" and do not equilibrate (Bottinga, 1969).

The variation of isotopic composition of oxygen in the carbonate component of the studied samples spans a rather wide range of $\delta^{18}O$ values particular to the oxygen of carbonates, on the one hand, isotopically balanced with oxygen of the lake water, and on the other, enriched by light isotopes displaced in composition to the oxygen of the oxides of Fe and Mn. Moreover, as is apparent from the graph in Fig. 2.23, in certain cases (profile II-II′, western part) is noticed a linear correlation in the distribution of the isotopic composition of carbon and oxygen.

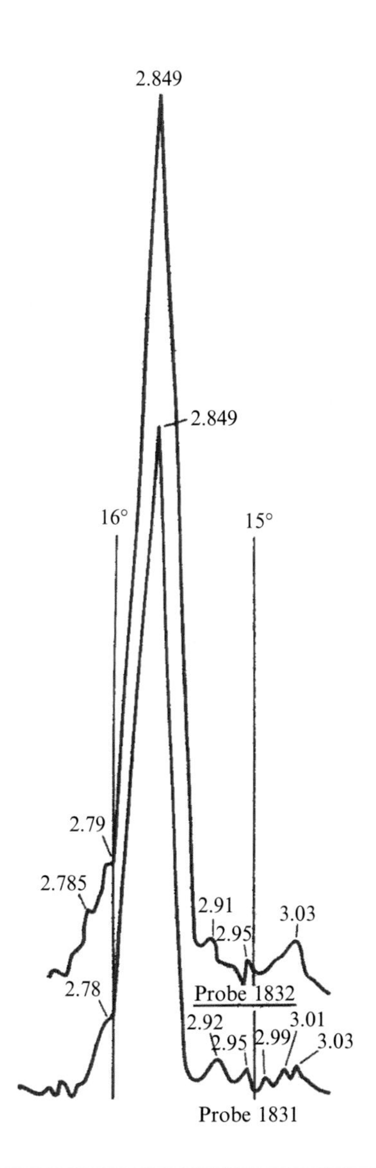

FIG. 2.22

X-ray diffractograms of carbonate-oxide manganese ores, Punnus-Yarvi Lake (Кулешов and Штеренберг, 1988).

The observed correlation most likely represents a line of mixing of the material from various sources with different initial isotopic composition. The probability that this is caused by the processes of fractionation of isotopes simultaneously for carbon and for oxygen in the process of diagenesis (or an experiment) is extremely low, insofar as such a correlation is noted as much for the gas phase of the sample of total carbon dioxide (CO_{2total}) as it is for the carbonate component of the ferromanganese nodules.

Table 2.10 Isotopic Composition of Carbon and Oxygen of Ferromanganese Ore Oolites of the Lakes of Karelia

No Sample	Characteristics of Sample	$\delta^{13}C$, ‰ (PDB)	$\delta^{18}O$, ‰ (SMOW)
Lake Punnus-Yarvi			
Profile I-I′			
(A) Northern part			
1817	Ore oolites in sand and fine gravel		
	L=55 m. H=4.25 m. Total carbon	−20.4	
1818	The same. L=45 m. H=3.70 m. Total carbon	−27.3	
1819	The same. L=30.0 m. H=2.7 m. Total carbon	−25.2	
	Rhodochrosite	−26.7	19.1
1821	The same. L=25.0 m. H=2.1 m. Total carbon	−26.2	
(B) Southern part			
1824	Sand and fine gravel with large and rare oolites. L=20 m. H=2.75 m		
	Total carbon	−29.5	
1825	The same. L=15 m. H=1.9 m. Total carbon	−32.5	
	–, L=10 m. H=0.9 m. Total carbon	−29.7	
1826	Solid crust of cemented oolites		
1827	L=45 m. H=3.5 m. Total carbon	−27.7	
1827a	The same. L=65 m. H=5.15 m. Total carbon	−26.5	
1828	–, L=80 m. H=5.95 m. Total carbon	−25.1	
	Rhodochrosite	−25.1	23.2
1830	–, L=100 m. H=4.85 m. Total carbon	−27.6	
	Rhodochrosite	−27.4	20.4
1831	The same. L=140 m. H=6.9 m. Total carbon	−25.2	
	Rhodochrosite	−22.8	22.5
1832	Scattered oolites in sandstone ooze		
	L=180 m. H=8.65 m. Total carbon	−25.0	
	Calcite	−25.0	21.8
	Rhodochrosite	−24.6	23.7
1833	The same, L=230 m. H=9.25 m. Total carbon	−24.5	
Profile II-II′			
(A) Eastern part			
1834	Fine gravel, cemented by ore to the crust		
	L=7 m. H=1.0 m. Total carbon	−28.2	
1835	The same. L=12 m. H=1.6 m. Total carbon	−30.1	
	Valve of mussel	−7.6	22.5
1836	Dense layer of oolites. L=30 m. H=2.2 m		
	Total carbon	−24.1	
	Rhodochrosite	−22.1	14.1

Table 2.10 Isotopic Composition of Carbon and Oxygen of Ferromanganese Ore Oolites of the Lakes of Karelia—cont'd

No Sample	Characteristics of Sample	$\delta^{13}C$, ‰ (PDB)	$\delta^{18}O$, ‰ (SMOW)
1837	Dense layer of oolites with sand		
	$L=50$ m. $H=2.5$ m. Total carbon	−24.3	
	Rhodochrosite	−24.6	13.9
1838	Sand with ore oolites. $L=75$ m. $H=2.9$ m		
	Total carbon	−24.8	
	Rhodochrosite	−24.7	15.7
1839	Sand with rare ore oolites		
	$L=100$ m. $H=3.1$ m. Total carbon	−25.9	
1840	Sand with rare fine ore oolites		
	$L=150$ m. $H=3.15$ m. Total carbon	−25.4	
(B) Western part			
1841	Sand with ore oolites. $L=70$ m. $H=2.15$ m		
	Total carbon	−27.2	
1843	The same. $L=120$ m. $H=3.2$ m. Total carbon	−26.4	
1844	Dense layer of oolites. $L=160$ m. $H=3.05$ m		
	Total carbon	−28.4	
	Rhodochrosite	−24.1	6.7
1845	The same. $L=190$ m. $H=2.85$ m. Total carbon	−24.1	
	Rhodochrosite	−19.2	20.6
1846	The same. $L=230$ m. $H=2.85$ m. Total carbon	−25.9	
	Rhodochrosite	−23.0	17.5
	Sample of water: Bottom	−11.6	−11.6
	Surface	−11.6	−11.6
1847	Dense layer of oolites. $L=275$ m. $H=2.8$ m		
	Total carbon	−28.2	
	Rhodochrosite	−26.4	6.0
1848	The same. $L=300$ m. $H=2.75$ m. Total carbon	−28.3	
	Oxides ($SiO_2=4.50\%$)		−4.8
1849	Ore oolites with sand. $L=350$ m. $H=3.1$ m		
	Total carbon	−27.5	
	Oxides ($SiO_2=3.38\%$)		−2.3
1850	Sandstone ooze with scattered oolites		
	$L=380$ m. $H=3.35$ m. Total carbon	−24.7	
Profile III-III′			
(A) Southern part			
1853	$L=2.5$ m. $H=1.05$ m. Valve of mussel	−8.5	21.5
1854	Sand with ore oolites. $L=65$ m		
	$H=1.35$ m. Total carbon	−25.8	
1855	Valve of mussel	−9.5	21.6

(Continued)

Table 2.10 Isotopic Composition of Carbon and Oxygen of Ferromanganese Ore Oolites of the Lakes of Karelia—cont'd

No Sample	Characteristics of Sample	$\delta^{13}C$, ‰ (PDB)	$\delta^{18}O$, ‰ (SMOW)
1856	The same. $L=85$ m. $H=1.75$ m. Total carbon	−26.1	
1857	The same. $L=100$ m. $H=2.25$ m. Total carbon	−26.3	
1858	The same. $L=130$ m. $H=2.05$ m. Total carbon	−26.8	
1859	The same. $L=170$ m. $H=4.0$ m. Total carbon	−26.3	
1860	Scattered oolites in sandy ooze		
	$L=230$ m. $H=5.25$ m. Total carbon	−24.4	
(C) Northern part			
1861	Layer of dense oolites with impurity of sand		
	$L=15$ m. $H=1.45$ m. Total carbon	−28.5	
1862	Crust of cemented oolites with fine gravel and sand. $L=20$ m. $H=2.95$ m		
	Total carbon	−28.7	
	Sample of water: Bottom		−11.4
	Surface		−11.5
1866	The same. $L=30$ m. $H=4.2$ m. Total carbon	−29.0	
1869	Sandstone ooze with rare ore oolites. $L=45$ m. $H=9.3$ m		
	Total carbon	−23.4	
Lake Konchozero			
1804	Sand with rare ore oolites		
	$L=75$ m. $H=3.8$ m. Total carbon	−23.5	
1805	The same. $L=85$ m. $H=5.5$ m. Total carbon	−25.4	
1806	Sandstone clay with abundant inclusions of oolites. $L=100$ m. $H=5.8$ m		
	Oolites of the upper part of the core ($M=20$ cm)		
	Rhodochrosite	−25.5	7.7
	Oolites of the lower part of the core ($M=20$ cm)		
	Rhodochrosite	−24.8	7.3
	Sample of water		−9.0
1808	The same. $L=140$ m. $H=7.5$ m		
	Total carbon	−25.4	
Lake Urosozero			
1800	Unsorted sand with fine reddish brown ore nodules		
	$L=60$ m. $H=1.6$ m.		
	Total carbon	−26.5	
1801	The same. $L=30$ m. $H=1.2$ m		
	Total carbon	−27.5	
1802	The same. $L=17$ m. $H=0.8$ m. Total carbon	−26.0	
	Sample of water		−11.6

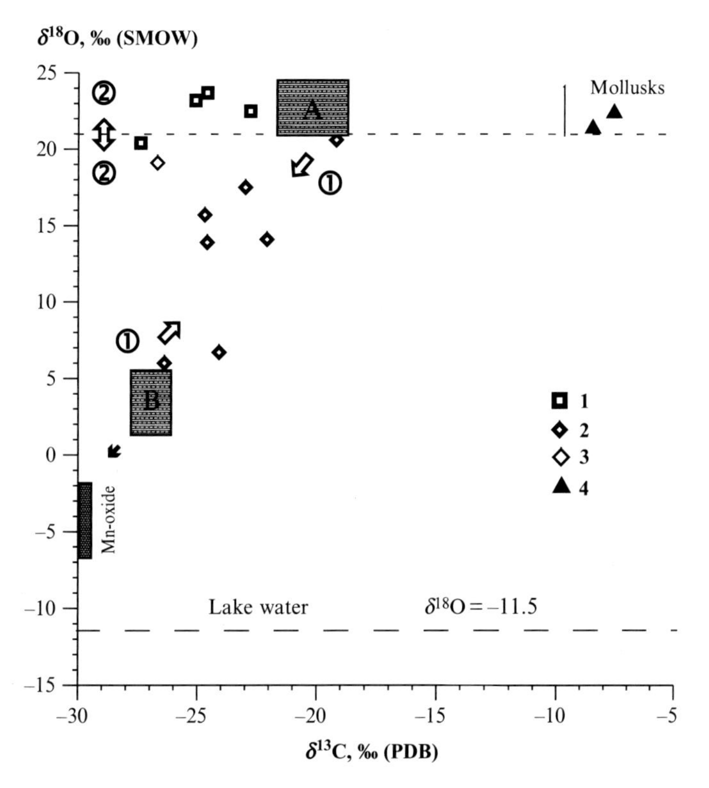

FIG. 2.23

$\delta^{13}C$ vs. $\delta^{18}O$ in carbonate-oxide nodules of Punnus-Yarvi Lake. A—Equilibrium diagenetic lacustrine carbonates; B—nonequilibrium diagenetic lacustrine carbonates. 1—Southern part of profile I-I′; 2—western part of profile II-II′; 3—nodule of Konchozero Lake; 4—mollusks.

Based on the interpretation of the line of mixing, it can be supposed that the initial sources of CO_2 should be characterized by isotopic ratios of carbon and oxygen particular to the values of the end-members within the boundaries of the obtained $\delta^{13}C$ and $\delta^{18}O$ values (see Fig. 2.23). As is evident, one source should be found in the region of relatively negative values for the isotopic composition of carbon and oxygen (region "A"); its $\delta^{13}C$ values should be close to −23‰ to −19‰, and $\delta^{18}O$ values close to 21.5‰—that is, they should be balanced (or close to balanced) with the bicarbonate of the ooze waters of the lake. This source can be provisionally identified as the carbon dioxide of balanced lacustrine diagenetic carbonates (bearing in mind the equilibrium in carbon from lake waters).

It follows to note that we do not see a perceptible addition of sedimentary carbonates balanced with the lake water (the isotopic ratios correspond to the values of the mollusk shells). This once more provides evidence to the effect that the carbonate matter in ore nodules has a diagenetic origin.

Another source of CO_2 is more enriched in light isotopes of carbon and oxygen (region "B"). Its values of $\delta^{13}C$ lie within the boundaries from −25‰ and lower, while its values of $\delta^{18}O$ approach those of the oxygen of oxides of Mn and Fe (−7‰ to −2‰; see Table 2.10). In natural compounds such a light isotopic composition of carbon as a rule is characteristic of certain groups of organic matter (Kodina and Galimov, 1984).

It can be supposed that the oxidation of organic compounds by the oxygen of oxides of iron and manganese in our samples is also a principal source of isotopically light carbon. Unfortunately, at present we lack data on the composition of the organic matter in the ores themselves. However, greenish oozes from the central, most deep-water part of the lake, according to analyses (analyst L.F. Ivanova, Geology Institute, USSR Academy of Sciences), in a general content of 5.62% organic matter contain 2.50% humic acids, 2.28% kerogen, and 0.84% bituminoids.

Of note is the circumstance that the carbon isotopic composition of the total sample ($CO_{2total} = CO_2$ from carbonate + CO_2 from Mn-Fe oxide matter by heating) within the boundaries of the studied profiles in the near-shore zones is characterized by a sharply expressed decrease of the $\delta^{13}C$ values in comparison with those of the deeper water sections. It can be supposed that this bears a general character and is one of the characteristic features of the distribution of $\delta^{13}C$ values within the ore fields.

To all appearances, the noted enrichment by light isotope ^{12}C of the ore matter in the near-shore zones is contingent upon the initial composition of the organic matter itself (the CO_2 contents in ores of these zones are minimal). Presently we cannot draw any straightforward conclusions regarding the origin and source of organic matter of these zones. It is known from isotopic studies of Swedish lakes that most CO_2 is derived from oxidation of terrestrial material rather than from primary productivity within the lake (Jonsson et al., 2001a,b). However, to answer this question correctly for the Karelian lakes, we should first determine the chemical and isotopic composition of all forms of organic matter distributed within the boundaries of the lake and in the ores themselves, as well as the variation of the isotopic composition of the organic matter in the course of its diagenetic transformation. Such an investigation has so far not been undertaken.

The isotopic composition of the carbon dioxide of source "B," as a rule, is enriched in the light isotope ^{16}O (up to $\delta^{18}O = 6‰$), which is much lower in relation to that of the carbonate in isotope equilibrium with the lake water. This is noted principally for samples of profile II-II′, while simultaneously the carbonate component of the ores of profile I-I′ as a rule is in equilibrium with the oxygen. This could provide evidence for the intensity of the processes of oxidation of organic matter inside ore nodules leading to the reduction of oxide forms of manganese and iron and their conversion into the carbonate phases.

The interpretation of the line of mixing provides an understanding of the observed correlation in the studied profiles I-I′ and II-II′ between the distribution of the isotopic composition of carbon and oxygen with the content of CO_2. This is clearly seen in the graph of Fig. 2.24. As a rule, the greater the carbonate component in a sample, the higher the $\delta^{13}C$ and $\delta^{18}O$ values.

The isotopic composition of the carbon dioxide of the ore nodules becomes heavier in profile with the distance from the shore toward the profundal zone. This is caused primarily by an increase in the quantity of isotope equilibrium carbonates in the sample. In this same direction, the ore nodules become oxide-carbonate and, in certain cases, even carbonate.

In the near-shore zones of the ore field, the carbonate component is the lowest, and here on the whole are noted the heaver values of $\delta^{13}C$ and $\delta^{18}O$. This is due to an increase in the relative share of isotopically unbalanced (nonequilibrium with DIC of lake water) diagenetic carbonates (area "B," see Fig. 2.23) in the composition of the carbonate component of the nodules.

The presence of isotopically unbalanced carbonates in the ore nodules enables us to consider the occurrence of the oxidation of organic matter up to CO_2 within the ore matter under diagenesis (Strakhov, 1960; Varentsov et al., 1973, 1977). The latter could combine with Mn^{2+} (more rarely Fe^{3+}) and form

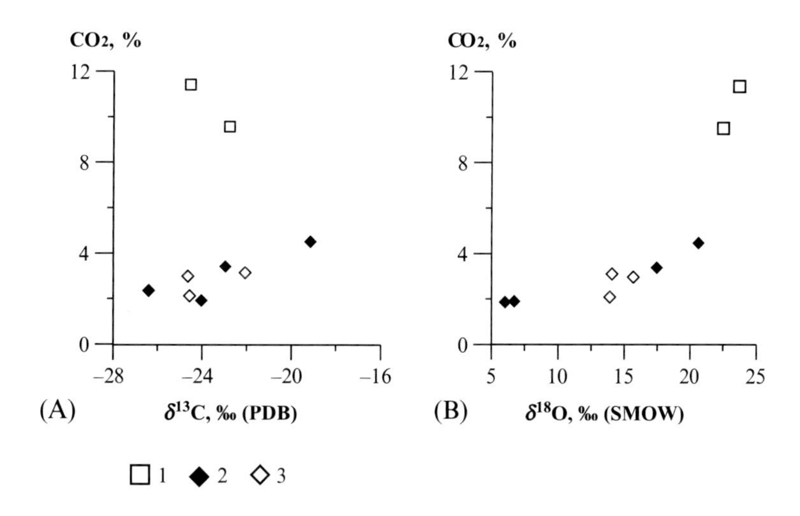

FIG. 2.24

$\delta^{13}C$ (A) and $\delta^{18}O$ (B) vs. CO_2 (%) in nodules of Punnus-Yarvi Lake. 1—Southern part of profile I-I′; 2—western part of profile II-II′; 3—eastern part of profile II-II′.

carbonate minerals. For this reason, we are justified in expecting to find within the carbonate parts of the ore nodules an entire "spectrum" of isotope ratios that vary from those particular to diagenetic isotopically balanced carbonates to those intrinsic to isotopically unbalanced carbonates. In the graph of Fig. 2.23 this direction is indicated by arrow ①.

Also in support of the cited result is an observed correlation with regard to the contents of the manganese oxides and the contents of the organic matter (see Fig. 2.20). Higher concentrations of MnO_{tot} have likewise been noted in ore nodules with minimal C_{org} contents.

On the whole is noted a tendency toward an increase in the heaviness of the isotopic composition of the carbon and oxygen with an increase of the CO_2 and Mn^{2+} contents (Figs. 2.24 and 2.25). This could provide evidence for the isotopically unbalanced diagenetic carbonate having a lighter value or becoming isotopically balanced over time as a result of the isotope-exchange processes with oxygen of the lake water.

The Mn^{2+} and CaO in the nodules are associated primarily with the carbonate phase. Moreover, carbonates of manganese predominate, as reflected in the correlation between the CO_2 contents and the MnO contents (Fig. 2.13).

At the same time, with the formation of carbonates from CO_2 due to the oxidation of organic matter inside the ore nodules, another direction is noted in the distribution of $\delta^{18}O$ values: the formed carbonates becoming balanced to the oxygen with the lake water. Furthermore, the isotopic composition of the carbon in them remains as before—that is, isotopically light—while the isotopic composition of the oxygen acquires higher values. In the graph of Fig. 2.23, this direction is indicated by arrow ②.

The obtained isotope data support the results of previous research on the diagenetic formation of ore nodules (and in part the formation of their carbonate component). We generally adhere to the interpretations of the geochemistry and behavior of matter during diagenesis that have been fairly well elucidated in the fundamental works of Strakhov (1960, 1965, 1976) (Strakhov et al., 1968) and those of a range of his successors.

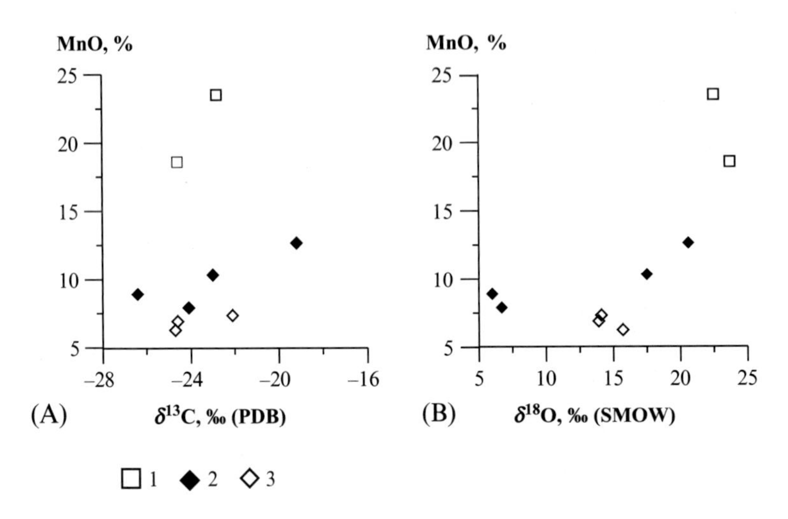

FIG. 2.25

$\delta^{13}C$ (A) and $\delta^{18}O$ (B) vs. MnO (%) in nodules of Punnus-Yarvi Lake. 1—Southern part of profile I-I′; 2—western part of profile II-II′; 3—eastern part of profile II-II′.

2.3.2 FERROMANGANESE CONCRETIONS IN LAKE KONCHOZERO

Lake Konchozero belongs to the Konchozero group of lakes—Konchozero, Ukshozero, Surgubskoe, Munozero, Pertozero, and others—which are linked by tributary streams. They belong to the Shuia River basin and share similar water compositions. The catchment surface is composed mainly of mafic crystalline rocks (diabases, gabbro-diabases, etc.). The composition of the detrital minerals in the sediments of these lakes is similar to the ferromanganese nodules that form within those sediments.

Lake Konchozero has the form of a narrow, long water body, extending from the north-west to the south-east at 22.5 km (Fig. 2.26). The width at its greatest reaches 3 km; the greatest depth is 19.5 m. The water inflow into the lake is supplied primarily from the northerly Lake Pertozero, inasmuch as tributaries are small in number and low in volume. The runoff descends through the Kosalminskii channel into Lake Ukshozero.

In the structure of the lakebed have been identified two deep depressions, which are the result of the fault zones stretching along the western and eastern shores of the lake (Shterenberg, 1979). Noted as well is a series of smaller depressions extending in the same direction, which make more complex the relief of the bottom of the water body.

The chemical composition of the water of Lake Konchozero has been studied in detail by Shterenberg (1979). Divalent iron has not been detected in the surface water. Likewise, there is little Fe^{3+} in the surface and near-bottom water. Insignificant quantities of iron have been detected in dissolved metal-organic complexes. Manganese likewise has not been detected in the surface and near-bottom water. Al contents constitute 0.04–0.08 mg/L. There is very little silicon in the water: in the surface layers it is 0.8 mg/L, and in the near-bottom layers it increases insignificantly to 1.1–1.2 mg/L. The content of CO_2 in the near-bottom water is somewhat elevated in comparison with that of the surface water. The pH of the water constitutes 6.9–7.5. The content of bicarbonate anion, being a primary component of the water, constitutes 36.8–42.9 mg/L. The quantity of phosphorus usually does not exceed 0.0022 mg/L.

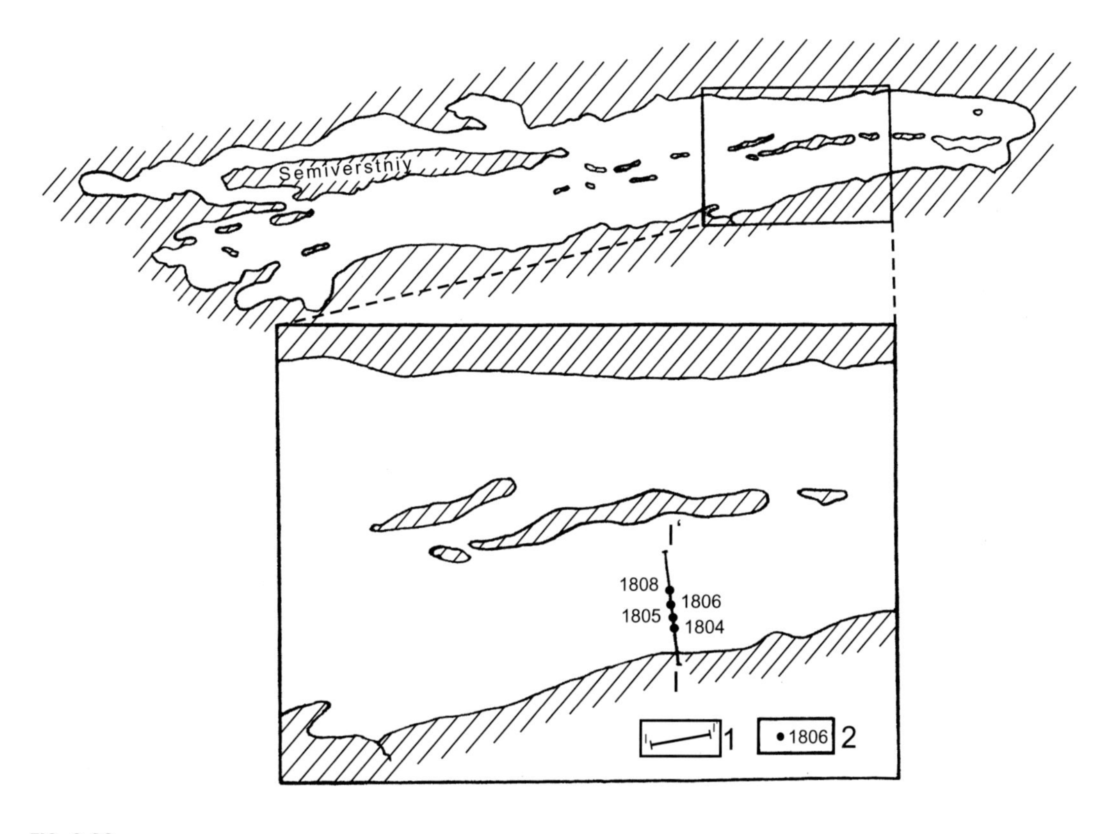

FIG. 2.26

Location of sampling profiles on isotope analysis on Konchozero Lake; (Кулешов, Штеренберг, 1988). 1—Location of sampling profiles, 2—number of observing stations (corresponds to the number of analysis samples).

The sediments of the lake are represented primarily by detrital rocks. The shallow-water near-shore zone (to a depth of a few meters) is filled to a significant degree by large boulders, pebbles, and gravel. These are covered by a thin (1–2 cm) layer of silty-sandy ooze of light-gray color. Coarse-grained sediments (boulder, gravel, and sand) are on the whole not widely distributed; they form in plan a narrow band of deposits bordering the shoreline. Toward the profundal zone, sands give way to silts that are commonly strongly clayey. The most deep-water parts of the lake are occupied by grayish-bluish and greenish clayey-silty oozes.

2.3.2.1 Ferromanganese enrichments

Ferromanganese enrichments are localized in Lake Konchozero at depths of 2.5–8.0 m, more rarely at greater or lesser depths. Commonly they are associated with silts or silty sands; however, there are cases of large nodules (dimensions up to 2.0–2.5 cm in diameter) situated among light-gray oozes on the borders with the banded lacustrine-glacial clays, occasionally overlapping the clays. The nodules mainly have spherical, rarely somewhat flattened form. In certain cases, a decrease is noted in their dimensions from the surface toward the depths of the sediments.

The thicknesses of the ore zone vary within a relatively wide range: from tenths of a millimeter (a relatively thin rusty-brown film on the surface of the sediments) up to 8–10 cm (accumulations of Fe-Mn concretions that are spherical in form). The size of the ore nodules in profile from station to station likewise does not remain constant and regularly changes as in other lakes of Karelia and the Kola Peninsula (Shterenberg et al., 1968, 1970). In the near-shore zone of the upslope-margin of the ore-bearing sediments among the sands have been discovered fine nodules of compressed form, rarely reaching 0.5–0.8 cm in diameter. It is here that ferromanganese crusts with a thickness up to 3–5 mm have been discovered, representing terrigenous-detrital grains cemented by hydroxides of Fe and Mn. In a range of cases, pebbles and small-sized boulders have been found that bear Fe-Mn crusts on their surface.

At depths of 5–6 m (more rarely of 7 m), where ooze-silty sediments are developed, the spherical ore nodules have their greatest diameter (2–2.5 cm). Toward the bottom of the lakebed, in the zone of the fringe zone of the deep-water side, the quantity and size of the nodules gradually decrease up to full disappearance. In these cases at depths of approximately 8–11 m in the upper part of the ooze, deposits can be observed only as thin (tenths of a millimeter) red-brown film.

Among the oozes of the deepwater zone within the boundaries of the southern part of Lake Konchozero, no ferromanganese concretions have been uncovered. Here in the oozes Shterenberg has established very fine, irregularly rounded structures of vivianite, oxidized on the surface up to β-kertschenite, thereby acquiring a dark-blue color. Similar phosphate-ferruginous minerals—$Fe_3(PO_4)_2$—have been established in a ranges of lakes of the Karelian Isthmus by Semenovich (1953, 1958) and Shterenberg et al. (1968).

It must likewise be noted that as in many other water bodies, subsurface ore nodules have been established in Lake Konchozero. These are deposited under a layer of ooze in the horizon of fine-grained sand with a thickness of 5 cm and underlie light-gray heavy clay.

The ore nodules have a rusty dark-brown color and are friable and porous; they are spherical in form with dimensions up to 1.5 cm in diameter. Their internal structure is close to conchoidal-concentric (Fig. 2.27). The ore components are represented primarily by hydroxides of iron and manganese. Rarely noted is the presence of thinly scattered carbonate matter—manganese carbonates (calcium rhodochrosite).

Judging by the contents of Fe, Mn, C_{org}, and CO_2 (Table 2.11; Shterenberg et al., 1979) along with their mean magnitudes in the various types of sediments and ores, in Lake Konchozero, as in other Karelian water bodies, there are two types of distribution of elements: Clark and ore. With the Clark (scattered) type of distribution of their element, contents increase in the direction from the more coarse-grained varieties (of sands) toward the oozes situated in the deepwater parts of the lake. Moreover, a certain—albeit in this case minor—shift has been observed in the Mn/Fe ratios in the direction toward the more deepwater sediments. Conversely, during the ore-formation process, the contents of Mn, Fe, P, and C_{org} increase from the profundal zone toward the littoral—more precisely, toward the zone of ore formation. In this same direction likewise increases the magnitude of the manganese modulus, approaching in mean up to 0.1.

The behavior of Mn, Fe, C_{org}, and CO_2 in profile I-I′ (see Fig. 2.28) has been examined previously (Shterenberg et al., 1979). In that work, Shterenberg noted a contrasting distribution of magnitudes of the manganese modulus in the oozes of the profundal zone and the Fe-Mn concretions. In all cases, this

FIG. 2.27

Structure of ferromanganese-oxide nodules of Punnus-Yarvi Lake (scanning electron microscope photomicrographs, photo E.A. Zhegallo and E.L. Shkol'nik). (A, B) General view (by miscellaneous increase); (C, D) visible microbial structure; (D) visible remains of diatomic algae.

Table 2.11 Content of Mn, Fe, C_{org}, and CO_2 in Ores and Sediments of Lake Konchozero (Profile II-II′) (Shterenberg et al., 1979)

No Station	Characteristics of Sediment	Mn %	Fe %	C_{org} %	CO_2 %	$\frac{Mn}{Fe}$
15	Sand, various granularity	0.08	2.45	0.38	0.10	0.032
	Ore	0.08	7.25	0.18	0.14	0.011
14	Ooze, hosting the ore	0.16	4.85	–	–	0.035
	Ore	6.85	8.92	0.46	1.10	0.76
13	Ooze, light-gray	0.10	6.17	4.18	0.14	0.015
7	Ooze, hosting the ore	0.10	3.38	–	None	0.031
8	The same	0.48	5.73	1.91	None	0.083
	Ore	2.11	16.6	0.75	0.12	0.12
12a	Ooze, hosting the ore	0.08	3.04	6.00	None	0.026
10	The same	0.15	5.59	2.14	0.10	0.027
	Ore	1.46	15.2	2.52	0.80	0.09

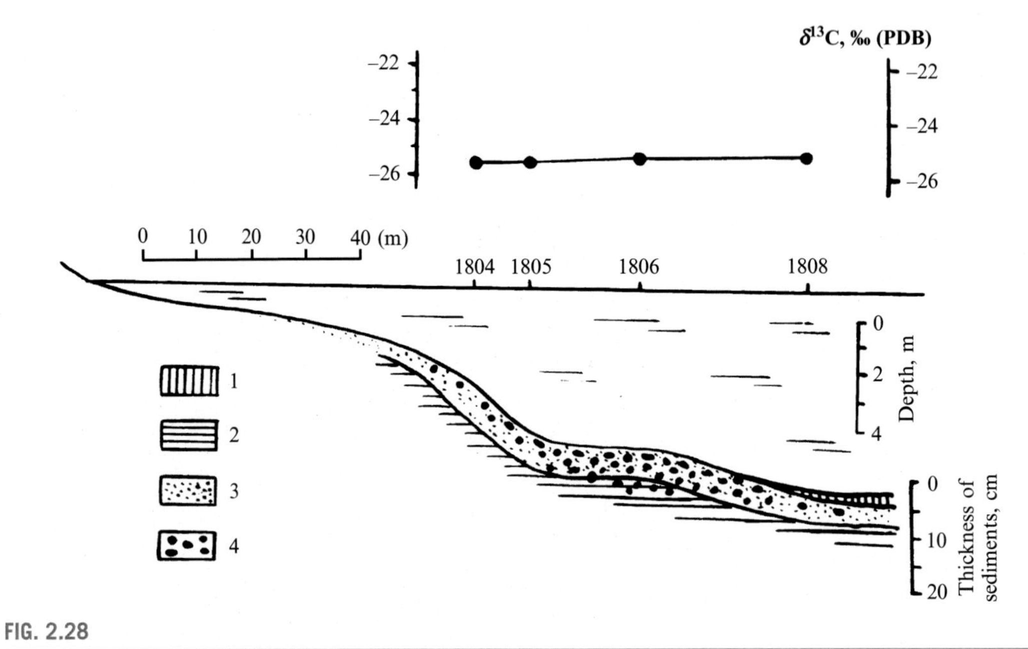

FIG. 2.28

$\delta^{13}C$ vs. $\delta^{18}O$ in carbonate nodules of Konchozero Lake. 1—Mud; 2—clay; 3—sand; 4—ore oolites.

ratio is elevated toward the shallow-water sections, which provides evidence for a shift of the manganese contents relative to that of iron in the direction toward the littoral.

The mechanism of seepage of Mn, Fe, and other components in the ooze waters along the bed has been indicated previously by Strakhov (1965). The possibility of a migration of Mn and Fe (bicarbonate forms and complex organic compounds) is indicated as well by the data obtained by Shterenberg (1979) in the analysis of ooze waters processed in a centrifuge. It turned out that the ooze waters collected from relatively deepwater sections of the bottom contained up to 8.0 mg/L of common iron and up to 4 mg/L of manganese. At the same time, ooze water expelled from the near-shore sandy-gravel sediments (Mn content 0.1 mg/L; Fe not detected) is practically indistinguishable from the content of these elements in natural water.

All ore bodies gravitate toward the relatively shallow-water environment or the sublittoral zones and are associated primarily with fine sand to silt-sized sediments. Ore horizons are localized not only close by the shoreline, but also near islands, on the slopes of subaqueous rises, and the like.

The formation of Fe-Mn ores of spherical and other forms is undoubtedly connected with the diagenetic processes during which iron and manganese are fed from the deeper horizons of the ooze into their upper parts, and not connected with their settling from the natural water—even less so from the surface water.

The results of the isotope research confirm the results regarding the formation of Fe-Mn nodules under diagenesis due to the redistribution of the matter of the sediment itself.

2.3.2.2 Isotope data and their discussion

The studied samples were collected in profile from the western shore of the lake (in the vicinity of the village Tsarevichi) toward the profundal zone in the direction of the island Dlinnyi (it corresponds to profile I-I′ in the work of Shterenberg, 1979). Oolites from the core of pipes of four stations were analyzed for isotopic composition (Fig. 2.26). The isotopic composition of the carbon of the general sample was confined within a narrow range: $\delta^{13}C$ varies from 25.5‰ to 23.5‰ (Table 2.10). The carbonate component was studied for oolites from the upper part of the sequence of sediments of station 1806. Their $\delta^{13}C$ value turned out to be equal to 18.5‰, and $\delta^{18}O$ value was equal to 9.2‰. Radiographic study of the powder mount of this sample did not indicate the presence of carbonate minerals; therefore, it can be concluded that the carbonate-manganese (possibly also ferruginous) compounds in the given sample were X-ray amorphous.

A distinctive particularity of the oolites (nodules) of station 1806 is that they are not limited only to the upper layer of sand, but are also present in the underlying light-gray clays. The isotopic composition of the common carbon in the oolites from sands—that is, from the upper part of the sequence—and from the underlying clays turned out to be identical (−25.5‰ and −24.8‰, respectively).

Thus, in distinction from Lake Punnus-Yarvi, within the boundaries of our studied profile in Lake Konchozero in the direction from the littoral to the profundal zone as well as along the cross section of sediments, we found no regular variations in the isotopic composition of the common carbon in the studied oolites.

The carbonate matter in the oolites of station 1806 in the ratio of oxygen isotope composition is isotopically unbalanced (isotope disequilibrium) with the water of the natural and ooze waters and is classified to the type of isotopic disequilibrium diagenetic carbonates. As has already been noted, such carbonates are found frequently in the nodules of Lake Punnus-Yarvi.

For comparison, we studied nodules of iron oxides of modern sediments of Lake Urosozero (Karelia). Here in the near-shore zone (at a distance from the shore of 10–15 to 50–60 m) of the southern part of the lake at depths of 1–2 m are widely developed oolites of iron oxides, which in places form a fairly dense crust with a thickness of several centimeters. The isotopic composition of their common carbon is identical and constitutes from −27.5‰ to −26.0‰ (see Table 5.3)—that is, it is characterized by those same magnitudes as the carbon of nodules of lakes Punnus-Yarvi and Konchozero.

Thus, the obtained magnitudes of isotopic composition provide evidence that the carbon of ferromanganese and iron oxide nodules of lakes Konchozero and Urosozero, as well as Lake Punnus-Yarvi, is of organic origin. Particularly sedimentary carbon—carbon balanced with the dissolved bicarbonate of the lake water—was not detected in the studied samples.

2.4 ISOTOPIC-GEOCHEMICAL REGULARITIES OF THE FORMATION OF MANGANESE CARBONATES IN MODERN SEDIMENTS

In order to arrive at an explanation of the regularities of the formation of manganese carbonates in modern sediments and their isotopic composition, we studied (as shown above) a wide circle of carbonate-manganese enrichments, formed in diverse environments of sedimentogenesis: oceanic, near-shore marine, and lacustrine. In selecting the research objective, we sought to exclude those regions in which the influence of modern hydrothermal processes leave a notable "impression" on the course (process) of normal sedimentation. Therefore, in the interpretation of geochemical and isotope data we have excluded from our discussion the influence of hydrothermal (and volcanogenic) processes in the formation of the isotopic composition of authigenic manganese carbonates.

The isotope data obtained by us together with the data available in the literature entirely confirm the existing theory of a diagenetic origin of manganese carbonates in lake as well as in sea and ocean sediments. The observed magnitudes of the isotopic composition of carbon and oxygen and the regularities of their distribution within the sediment permit a more detailed explication of certain aspects of their formation. Thus, in Fig. 2.29 in the $\delta^{13}C$—$\delta^{18}O$ coordinates have been supplied all isotope data

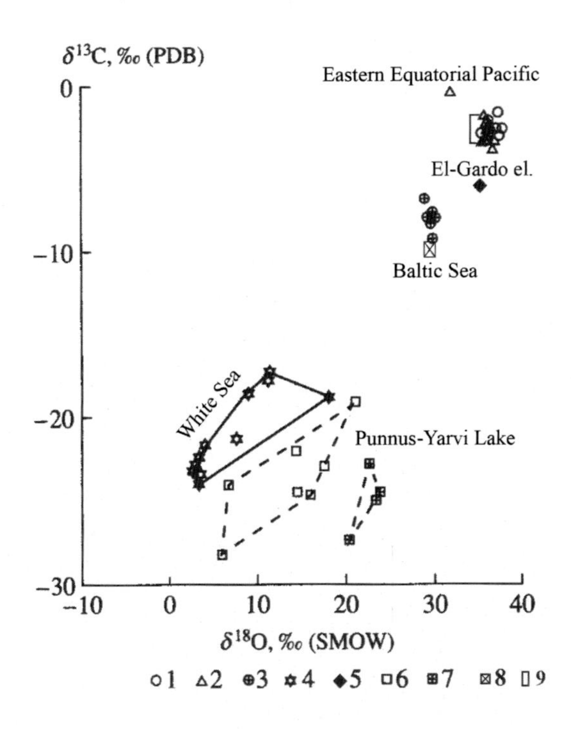

FIG. 2.29

Distribution of $\delta^{13}C$ and $\delta^{18}O$ values in manganese carbonates from recent sediments. 1—Guatemala depression (Panama Basin, Pacific Ocean); 2—Guatemala depression (Panama Basin, Pacific Ocean), data from Coleman, M., Fleet, A., Dobson, P., 1982. Preliminary studies of manganese-rich carbonate nodules from leg 68, suite 503, Eastern equatorial Pacific. Initial Reports of the DSDP, vol. 68. U.S. Govt. Printing office, Washington, DC, pp. 481–489; 3—Gotland Deep, Baltic Sea; 4—Onega Bay (White Sea); 5—El Gardo Rise, Middle American trench, Pacific Ocean; 6–7—Punnus-Yarvi Lake, Karjala Isthmus; 6—profile II-II′, western part; 7—profile I-I′, southern part; 8—Gotland Deep, Baltic Sea, data from Neumann, Th., Heizer, U., Leosson, M.A., Kersten, M., 2002; 9—Eastern Equatorial Pacific [Meister et al., 2009]. Early diagenetic process during Mn-carbonate formation: evidence from the isotopic composition of authigenic Ca-rhodochrosites of the Baltic Sea. Geochim. Cosmochim. Acta 66(5), 867–879.

considered in the present work; it is apparent that the Mn carbonates in the sediments of various types of water bodies occupy various positions in these coordinates: in the lower left, with the lowest values of $\delta^{13}C$ and $\delta^{18}O$, are the lacustrine and carbonate of oxide Fe-Mn concretions of Onega Bay (White Sea), and in the upper right, with the most heavy isotopic composition, are the rhodochrosites of the sediments of the Guatemala depression (Pacific Ocean). Mn carbonates of the deep-water sediments of the Baltic Sea occupy an intermediate position.

Such a distribution of isotopic composition is determined by a single determining factor: the degree of participation by organic matter (more precisely, by carbon dioxides of microbial origin that formed within the sediment during the process of the oxidation of organic matter under diagenesis) in the process of the formation of diagenetic manganese carbonates. In the final account, this is determined by the initial contents of C_{org} in the sediment. The $\delta^{13}C$ values could provide evidence for the relative share of carbon dioxide of organic or sedimentary origin in the composition of the studied carbonates. Thus, we observe the maximum contents of carbon dioxide formed due to the oxidation of C_{org} in lacustrine carbonates and the carbonate component of Fe-Mn concretions, whereas in Mn carbonates of deep oceanic sediments the content of organic-derived carbon dioxide is minimal.

The mechanism of the influence by organic matter on the process of manganese accumulation under diagenesis and its assumed forms have been examined in detail in the foundational works of Strakhov and his successors (Strakhov et al., 1968; Stravinskaya, 1980; Martynova, 2014).

The conclusion that a determining role is played by organic matter in the formation of diagenetic manganese carbonates is strongly confirmed by the observed regularity in the distribution of $\delta^{13}C$ values and MnO content in the studied samples (Fig. 2.30). A general regularity has been established: the higher the concentrations of manganese in sediments, the lighter the isotopic composition of the carbon—that is, the greater the quantity of CO_2 of microbial origin contained in the carbonate-manganese matter. This pattern, as shown below, is also observed in carbonate-manganese ores such as those of Chiatura and Kvirila (Kuleshov and Dombrovskaia, 1997a,b), Nikopol (Kuleshov and Dombrovskaia, 1988), Mangyshlak (Kuleshov and Dombrovskaia, 1990a,b), Usa (Kuleshov and Bych, 2002), and a range of other deposits.

Thus, the available isotope data provides evidence that manganese carbonates in modern sediments were formed under diagenesis with some degree of participation by the carbon of reduced organic matter. Sedimentary manganese carbonates—that is, those entering the sediments as a result of direct precipitation within the water column and consequently isotopically equilibrium with the dissolved bicarbonate of the sedimentary water body—to date have not been detected in marine systems. There is one interesting example recently discovered in Fayetteville Green Lake in New York, where rhodochrosite is forming in the water column with isotopic signature similar to the water (Havig et al., 2015). It is not yet known, however, how significant this production of manganese carbonate is compared to formation within the sediment.

This conclusion is in agreement with the data of Kravtsov (1998) in terms of the Gotland depression of the Baltic Sea; the natural water of the depression is strongly undersaturated in manganese carbonate (up to three orders of molar magnitude), which precludes the likelihood of the precipitation of manganese carbonates from these waters.

From these observations, it follows that some modification is needed in the upwelling model of formation of sedimentary deposits of manganese, as proposed by Sapozhnikov (1967). The precipitation of Mn into the sediment from deep waters rich in manganese that enter the near-shore zones

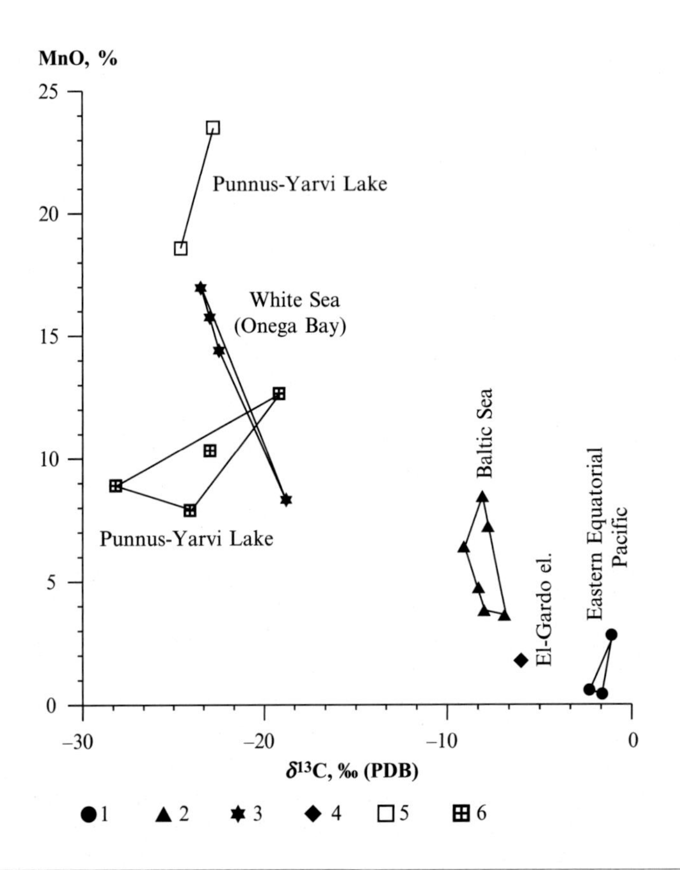

FIG. 2.30

Relationship between $\delta^{13}C$ values and MnO content in manganese carbonates from recent sediments. 1—Guatemala depression (Panama Basin, Pacific Ocean); 2—Gotland Deep, Baltic Sea; 3—Onega Bay (White Sea); 4—El Gardo Rise, Middle American trench, Pacific Ocean; 5–6—Punnus-Yarvi Lake, Karjala Isthmus; 6—profile II-II′, western part.

(upwelling) is apparently possible only in oxide form. Mn carbonates, judging by the available isotope data for modern deposits together with the results of the isotope research of carbonates of manganese ores of a range of deposits (Kuleshov and Dombrovskaia, 1993), are not formed by this method.

Special attention is due to the carbonates in the oxide Fe-Mn concretions and nodules. To all appearances, these carbonates represent a special type of diagenetic Mn carbonates, which were formed as a result of the oxidation of organic matter within the concretions or nodules evidently by the oxygen of manganese oxides. Their $\delta^{13}C$ values correspond to the isotopic composition of the initial organic matter, and the $\delta^{18}O$ values are determined by the oxygen of the same oxides.

In the isotope regard, this type of carbonate matter falls into two groups. The first represents the most "young" carbonate, which is being formed in the present time within concretions. The isotopic composition of its oxygen is determined by the composition of the oxygen of the oxides and is characterized by the lowest values of $\delta^{18}O$ (3–5‰ for concretions of Onega Bay; 6‰ or less for Lake Punnus-Yarvi).

With time, in process of the isotope-oxygen exchange with the water of the basin (sea or lake), the isotopic composition of the oxygen of these carbonates becomes heavier and reaches isotopic equilibrium with the oxygen of the water of the basin. In this case, they are not distinguished in terms of their isotopic characteristics from the diagenetic carbonates of ooze sediment ($\delta^{18}O$ = 28–30‰). These carbonates form the second group.

In the actual observations, we see an entire transitional series of carbonates from the first to the second group for samples of Lake Punnus-Yarvi (see Fig. 2.23) as much as for those of Onega Bay (see Fig. 2.10).

In this way, comparing the isotope data of diagenetic carbonates from modern sediments with the available analogous data for Oligocene and more ancient deposits of manganese, it can be concluded that carbonate-manganese ores of the deposits in terms of isotopic characteristics correspond most closely to carbonates from the sediments of inland seas. With regard to the accumulation of diagenetic Mn ores in the form of carbonates, particularly oceanic sediments are evidently of little prospect.

Lacustrine manganese enrichment presents a purely scientific interest for us—it provides an opportunity for detailed observation of various aspects of modern ferromanganese ore-genesis. In the practical regard, this type of manganese carbonate ore likewise is of little prospect.

The discussion of the question of the source of Mn exceeds the scope of the present work. Evidently, continental ablation is basic to all types of water bodies. However, in certain regions of the global ocean, with the active manifestations of hydrothermal activity and volcanism, we can observe a significant contribution of the ore material to the sea and ocean sediments (eg, in the Red Sea).

At the same time, with the consideration of one of the actively developed modern hypotheses on an elisional source of Mn for certain large deposits (Paragenezis…, 1990; Pavlov, Dombrovskaia, 1993; Kuleshov, Dombrovskaia 1997a,b), the probability cannot be excluded that an inflow of manganese into the sedimentary basin is produced by elisional waters (eg, in the Baltic Sea).

The isotope data examined above are pertaining to authigenic manganese (and manganese-containing) carbonates from modern sediments of various types of water bodies (ocean, sea, lake). The regularities of their distribution permit the following conclusions: (a) authigenic carbonates of manganese under diagenesis are generated with active participation by carbon dioxide of microbial origin formed as a result of the oxidation of organic matter contained in the sediments. The isotope data fully support the interpretation of N.M. Strakhov and his successors regarding the substantial role played by organic matter in the processes of diagenesis, including manganese carbonate ores; (b) to date, particularly sedimentary manganese carbonates that were deposited in isotopically equilibrium conditions with the bicarbonate of marine (and lacustrine) water by chemogenic (or biogenic) means have not been discovered. All studied manganese carbonates, whether from modern sediments or ancient deposits (continental deposits of manganese ores), have a diagenetic origin; (c) in the isotope regard, diagenetic manganese carbonates in modern sediments are represented by two types: isotopically balanced and isotopically unbalanced (isotope equilibrium in the oxygen from marine water). The latter are formed within concretions, crusts, and other morphological types of oxide accumulations with the oxidation of the carbon of organic matter by the oxygen of oxides of manganese and iron. To that type can be classified also carbonates of iron that are formed by the same method; (d) a correlation has been established between the isotopic composition of the carbon and the manganese content in the sediments, which provides evidence for a determining role played by the processes of the oxidation of organic matter in authigenic carbonates enriched by manganese. The details of the mechanism of such enrichment, along the geological and geochemical conditions of this process, require further study; (e) the formation of authigenic manganese-containing carbonates in the sediments of the global ocean and in the epicontinental bodies of water occurs under principally different conditions. Characteristic for the latter is the active participation in their formation by isotopically light carbon of oxidized organic matter.

CHAPTER

GENETIC TYPES, CLASSIFICATIONS, AND MODELS OF MANGANESE-ORE FORMATION

3

3.1 GENETIC TYPES AND CLASSIFICATION OF MANGANESE-ORE DEPOSITS

The scientific literature presents various classifications of deposits of manganese. In general plan, four genetic groups of deposits have been identified (see the works of A.G. Betekhtin, N.P. Kheraskov, N.S. Shatskii, S. Park, S. Roy, I.M. Varentsov, V.P. Rakhmanov, V.K. Chaikovskii, and others): (1) sedimentary (particularly sedimentary and volcanogenic-sedimentary), (2) magmatogenic (hydrothermal and contact-metasomatic), (3) metamorphosed (regional and contact metamorphism of sedimentary and magmatogenic ore accumulations), and (4) deposits of the weathering crust (residual, infiltrational, etc.). In the practical regard, these groups are far from equivalent.

The overwhelming majority of global stocks of manganese ores are contained in deposits confined to sedimentary basins. Depending on the composition and sediments and the source of manganese, they can be subdivided into strictly sedimentary and volcanogenic-sedimentary deposits.

There is likewise a typology of manganese deposits on a formational basis—manganese-bearing facies as defined parageneses of ores and sedimentary (including volcanogenic-sedimentary) rocks. For example, making use of data on a range of foreign countries, Varentsov and Rakhmanov (1974) have identified for deposits of Russia and the CIS countries the following manganese-bearing formations.

1. *Quartz-glauconite sandstone-clayey*. This formation is developed predominantly in tectonically stable regions—in platforms, slopes of crystalline shields, Phanerozoic fold belts, and the like. Associated with it are the known Oligocene deposits of Ukraine (Nikopol', Bol'shoe-Tokmak), Georgia (Chiatura), Kazakhstan (Mangyshlak), the Upper Pliocene deposits of the North Caucasus (Laba), the Paleocene deposits of the eastern slope of the northern Urals (Polunochnoe, Ivdel'), and others.
2. *Carbonate formations of geosynclines and platforms*. These include the Lower Cambrian deposits of Kuznetsk Alatau (Usinskoe), the Middle Permian pre-Urals (Ulu Teliak), the Archean Baikal area (Sagan-Zaba), and others.
3. *Carbonate-silicon formations*. Characteristic deposits: Upper Devonian central Kazakhstan (Karazhal'), Silurian Uzbekistan (Takhta-Karacha, Zeravshan Ridge), and others.
4. The group of volcanogenic-sedimentary formations of mio- and eugeosinlynes; includes two formations: spilite-keratophyre-silicon (Devonian deposits of the Primagnitogorsk synclinorium of the Urals) and porphyro-silicon (Durnov deposit of the Lower-Middle Cambrian, Salair).
5. *Manganese ferruginous-siliceous formation (jaspilite)*. A typical representative is the group of the Upper Proterozoic deposits of Brazil, Australia, Lesser Hinggan, and elsewhere.

Isotope Geochemistry. http://dx.doi.org/10.1016/B978-0-12-803165-0.00003-3

It should be accepted that the typology of deposits on the basis of manganese-bearing formations for many geological objects does not bear a genetic connotation. For example, the group of carbonate formations of geosynclines and platforms encompasses deposits of manganese formed under entirely different conditions and have different sources of manganese: carbonates of evaporate formations (Ulutelyak, Burshtyn), limestones of detrital off-reef facies (Usa), or carbonate strata of the Upper Archean with interbeds of manganese-containing rocks (Sagan-Zaba, Khoshchevato). This systematization likewise does not account for the series of ore-occurrences and minor deposits of the Karsk depression of the Komi Republic (Russia), represented by concretions of manganese carbonates; the same applies to the deposits of rich oxide ores of karst (eg, the Woodie-Woodie deposit, Australia), and the group of deposits of the Postmasburg region (Republic of South Africa).

Notwithstanding the detailed study of many deposits, including those under industrial development such as Mamatwan (Republic of South Africa), Groote Eylandt (Australia), Moanda (Gabon), Nikopol' and Bol'she-Tokmak (Ukraine), Chiatura (Georgia), and a range of other known deposits, many aspects of their formation to date remain examined only incompletely. This is due above all to the diversity of possible sources of manganese and the variety of conditions of formation of specific manganese ores (Fig. 3.1).

Most frequently, an exogenic source of manganese is involved (eg, continental runoff) (see the works of A.G. Betekhtin, V.I. Griaznov, and Iu.I. Selin, N.M. Strakhov, G.A. Avaliani, N. Beukes, J. Gutzmer, and many others). However, calculations of the quantity of manganese necessary for the formation of certain deposits (eg, giant deposits such as the Kalahari manganese-ore field, as well as Nikopol', Chiatura, Mangyshlak, and other deposits) suggest that the breakdown of rocks of the catchment area is not sufficient to supply the needed amounts of manganese. This has compelled many researchers to turn to a deep (volcanogenic, hydrothermal, etc.) source of manganese (the works of

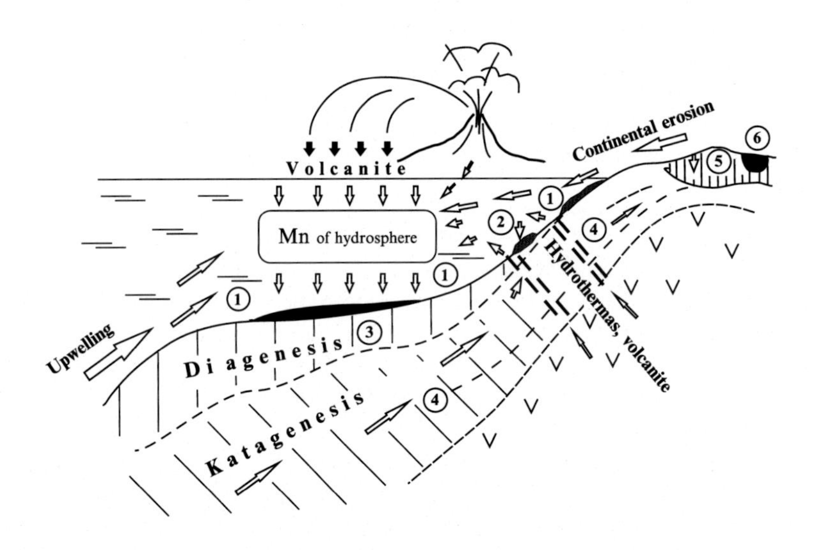

FIG. 3.1

Principal scheme of the main sources of manganese ore-formation. 1—Sedimentary; 2—hydrothermal (volcanic, exhalative)-sedimentary; 3—sedimentary-diagenetic; 4—katagenetic (metasomatic); 5—supergene, crust of weathering; and 6—supergene, ore bodies of karst cavities.

G.S. Dzotsenidze, A.I. Makharadze, G.A. Machabeli, M.M. Mstislavskii, I.P. Druzhinin, and other Russian and international researchers).

The most universal and least theoretically vulnerable has turned out to be the conceptualization of upwelling as the mechanism for the entry of manganese from the deep hydrosulfuric zone of marine (and oceanic) basins into the coastal parts of water bodies (the works of D.G. Sapozhnikov, N. Beukes, and others).

A fundamentally different point of view on the genesis of manganese deposits associated with the strata of sedimentary rocks is focused on the catagenetic (late-diagenetic) processes. The possibility of catagenetic manganese ore-formation in connection with the Mangyshlak and Laba deposits has been expressed by V.N. Kholodov, V.I. Dvorov, and E.A. Sokolova. In this case, petroleum waters contain large quantities of carbon dioxide, methane, hydrogen sulfide, and other aggressive components. Ore elements contained in such kind of waters could have been leached from underlying host rocks and transferred to the surface by large masses (Pavlov, 1989; Paragenesis et al., 1990; Pavlov and Dombrovskaya, 1993).

Thus, the existence of differing, at times diametrically opposite points of view on the origin of deposits of manganese ores is a testament to the complexity of the geological objects under study and, apparently, to the diversity of mechanisms, to the sources of (ore as well as non-ore) matter, and to the conditions of the formation of manganese and manganese-containing rocks.

Essential information regarding the conditions of formation and origin (source) of ore and non-ore components of manganese ores has been provided, in our view, by isotope research. Most informative with regard to manganese carbonates are the data on the isotopic composition of the carbon and oxygen included in the anion group CO_3^{2-}.

Reliable isotope criteria are available and characteristic for given geological processes. For example, sedimentary, diagenetic, hydrothermal, and catagenetic carbonates are characterized by specific particularities of the isotopic composition of carbon and oxygen ($\delta^{13}C$, $\delta^{18}O$).

Complementary (direct and circumstantial) information on the conditions of formation and source of matter of manganese ores can be comprised of data on the isotopic composition of the oxygen of manganese oxides, the isotopic composition of sulfur (of sulfates and sulfides, included in the composition of ores), as well as data on the isotopic composition of strontium. However, in the present scientific literature such information, unfortunately, is extremely scanty.

Undoubtedly, the predominant quantity of manganese rocks and ores are closely connected in the genetic regard with deposits bearing an initially *sedimentary nature*—that is, deposits that were formed in aquatic basins of sedimentation and are represented by (a) *biogenic-sedimentary* (*oolites*, *pisolites*, etc.), *diagenetic*, *and catagenic* (*late-diagenetic*) *origin* (having an exogenic source of ore components: weathering crusts, the washout of rocks of the feeding landmass, subaqueous leaching, or a redistribution of the ore matter in the basin of sedimentation, etc.) and (b) *volcanogenic-sedimentary* (an endogenic source of ore components: hydrothermals, submarine exhalations, etc.) rocks.

As a result of subsequent processes of transformation in the areas of regional metamorphism and in zones of contact metamorphism are formed *metamorphosed* and *metamorphic* manganese and manganese-containing rocks and ores.

In areas with humid and primarily tropical equatorial climate, various types of *hypergenic* deposits of rich manganese-oxide ores are formed in the weathering crusts.

Magmatogenic deposits, having a magmatic (deep, magmatogenic, or mantle) source of manganese, occur infrequently and do not present industrial interest.

Likewise, *contact-metasomatic* and *hydrothermal postmagmatic* types of ore-occurrences and minor deposits can be identified. Common sources of manganese in this case are manganese-containing rocks of volcanic origin as well as surrounding manganese-containing sedimentary rocks. Deposits and ore-occurrences of this type of ores comprise negligible stocks and as a rule lack practical significance.

It follows to note that regardless of the appeal of a wide circle of processes of manganese ore-formation (a detailed survey is provided in the work of Silaev, 2008), the formation of a given type of ores and deposits is contingent upon the type of sedimentary-rock basin and the regime of its tectonic development, which determined the accumulation of corresponding sediments (terrigenous-sedimentary, volcanogenic-sedimentary, various types of carbonate rocks, etc.).

The conditions of manganese accumulation in a sedimentary basins are determined by the geological development of the lithosphere in the geological history of the Earth (the existence and breakup of the proto-continents Paleopangaea, Rodinia, Gondwana, and their "splinters"), climatic conditions (periods of glaciations and aridizations), and biotic events (periods of the flourishing of the species composition of fauna and, accordingly, the general bioproductivity of the paleo-water bodies and the die-off and burial of organisms). All these processes are mutually connected with manganese ore-genesis and are reflected in the formation of manganese deposits.

In the mineralogical aspect, manganese rocks and ores are composed of oxides and carbonates, and more rarely by manganese silicates of various geneses. Sedimentary manganese oxides are represented as a rule by oolites, pisolites, and similar textures of coastal shallow-water zones of paleobasins; there are no doubts as to their initial sedimentary (apparently, biogenic-sedimentary) origin.

An insignificant part of similar oolite-like formations was formed in the sediment at the stage of early diagenesis, and subsequently in the environment of already lithified sediment and in the zone of hypergenesis. The geological, geochemical, paleogeographical, and other aspects of the formation of oxide oolite and pisolite manganese ores have been studied in sufficient detail on the examples of large deposits such as Groote Eylandt (Australia), Chiatura (Georgia), Nikopol' (Ukraine), Moanda (Gabon), and others and are well elucidated in the scientific literature.

Colossal resources of manganese in the sedimentary basins—the stratosphere (the Eastern Paratethys basin, the northern Urals manganese basin, the Jurassic Huayacocotla Basin of Mexico, and elsewhere)—are contained in carbonate rocks composed predominantly of minerals of the isomorphic series: manganoan calcite-manganocalcite-rhodochrosite-kutnohorite. Naturally, their sedimentary origin by means of precipitation from the water mass of the basin was assumed and long drew no objections. However, more recently obtained isotope data for a range of deposits globally have enabled detailed explanation of certain aspects of the conditions of the formation and the source of their constituents.

Significant resources of manganese (associated paragenetically, possibly also genetically) are contained in ferromanganese rocks of the banded iron formations (BIF)—banded ferruginous quartzites or layered hematite siliceous shales, represented by the necessary member of the complex of crystalline rocks and by practically all ancient consolidated shields of Australia, Africa, America, China, and Eurasia. Also interbedded with these rocks are manganese lutites of the Kalahari manganese-ore basin (Republic of South Africa), which constitute more than half of the manganese-ore potential of the deposits of the Earth's landmass.

These banded ferruginous quartzites are characterized as a rule by insignificant primary concentrations of manganese and do not constitute rich manganese, unless enriched by weathering. The overwhelming majority of deposits currently under development (South America, Africa, Australia, and elsewhere) belong to the zone of their hypergenic weathering and are developed primarily in the zone of equatorial tropical laterites.

3.2 MODEL EXAMPLES OF THE FORMATION OF MANGANESE DEPOSITS

Primary concentrations of manganese (sometimes reaching economic levels) in rocks and ores coincide exclusively with sedimentary or volcanogenic-sedimentary rocks (in many cases metamorphosed), which were formed in basins of sedimentation of various types and are represented by carbonates as well as by oxides and hydroxides of manganese (commonly in isomorphic mixture with iron). Subsequent processes of transformation of the initial manganese-containing rock, particularly during hypergenesis, as a rule lead to the formation of rich manganese-oxide ores.

The most important in the industrial and consequently in the scientific regard are without question the largest deposits, for instance the supergiant deposit of the Kalahari manganese-ore field (Republic of South Africa), the Eastern Paratethys manganese basin (Nikopol', Chiatura, Varna, Mangyshlak, and elsewhere), the Francevillian basin (Gabon), the Carpentaria basin (Groote Eylandt, Arnhem Land, and elsewhere in Australia), the North Urals manganese-ore basin (Polunochnoe, Ivdel', Berezovskoe, south Berezovskoe, Novo-Berezovskoe, Tyn'inskoe, and others), and a range of other known deposits of sedimentary-rock basins of the Earth's landmass.

Significant global reserves of manganese are contained in a range of small deposits of eugeosynclinal and miogeosynclinal basins (Primagnitogorsk, San Francisco, Kazakhstan, and elsewhere).

Of interest in the genetic regard, as models of formation of manganese deposits, are known large carbonate-associated deposits such as the Usa (Kuznetsk Alatau, Russia), as well as a range of deposits genetically associated with rocks of evaporite origin (Ulutelyak, Russia; Burshtyn, Ukraine; several deposits in Morocco; and elsewhere).

Of great practical significance are the rich manganese ores of lateritic and humid weathering crusts developed on manganese-containing rocks of various geneses—sedimentary, volcanogenic-sedimentary, metamorphic, and the like (deposits of African countries, Australia, Brazil, Russia, India, and elsewhere). The accumulation of manganese ores in this case is commonly accompanied by processes of karst development (Postmasburg region, Republic of South Africa; deposits of the Woodie-Woodie region, Western Australia; and elsewhere).

3.2.1 SEDIMENTARY-DIAGENETIC DEPOSITS

Some of the largest basins of sedimentation of manganese rocks are without question the basins of the Eastern Paratethys. Here are known such giant deposits as Nikopol' and Bol'she-Tokmak (South Ukrainian basin), as well as a range of large deposits of Georgia—Chiatura and Kvirila (West Georgia basin); Bulgaria—Obrochishte and others (Varna depression); Kazakhstan—Mangyshlak (South Mangyshlak trough); Turkey—Binkılıç and others (Thracian depression); and a range of small deposits and occurrences in Hungary and Slovakia (Pannonian depression) and the North Caucasus (Laba) (Fig. 3.2). Collectively, they make the time of Early Oligocene the largest manganese-forming episode of the Phanerozoic (Maynard, 2010, Fig. 3).

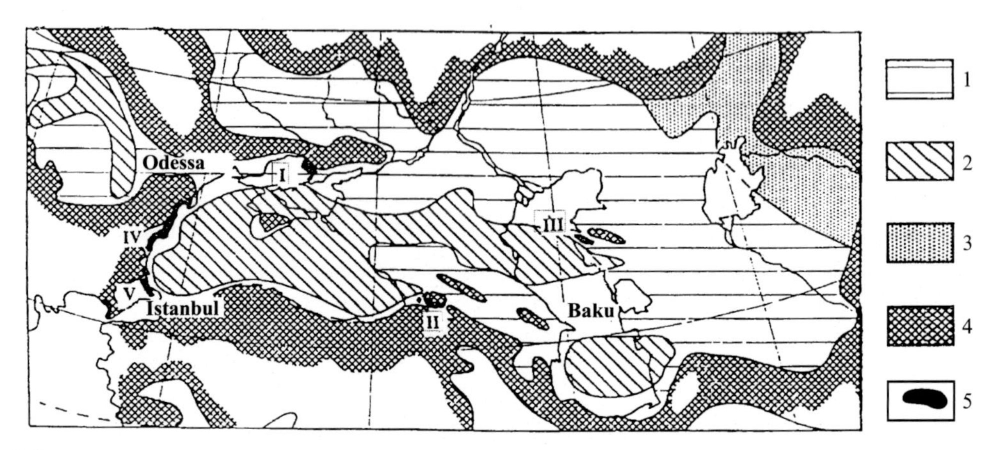

FIG. 3.2

Sketch map showing location of Mn-ore deposits within the Early Oligocene basins of the Eastern Paratetis (Varentsov, 2002). 1—Shelf regions, 2—bathyal deep basins, 3—periodically flooded coastal part of the land, 4—land, and 5—Mn-ore deposits: I—South Ukraine (Nikopol, Bol'shoy Tokmak, and others); II—Georgia (Chiatura, Kvirila, and others); III—Mangyshlak; IV—North-eastern Bulgaria (Obrochishte, and others); and V—North-western Turkey (the Trace basin, and others).

Formed in analogous paleogeographic, climatic, and facies environments are the deposits of the North Urals manganese basin and a range of sedimentary-diagenetic deposits of Europe, China, North America, and elsewhere.

Serving as a classic example of this type of deposit are the Nikopol' and Bol'she-Tokmak deposits, which contain over 1 billion tons of manganese.

3.2.1.1 Nikopol' manganese-ore basin

In the geological regard, the Nikopol' manganese-ore basin is one of the most studied and most fully elucidated in the scientific literature (Betekhtin, 1946; Griaznov and Selin, 1959; Griaznov, 1960, 1967; Nikopol'skii et al., 1964; Varentsov, 1963, 1964; Varentsov et al., 1967; Strakhov, 1964; Strakhov et al., 1968; etc.). For this reason, we limit ourselves to a brief survey of the geological situation and material composition of the carbonate ores.

In the tectonic regard, the examined basin is situated in the north-eastern part of the Black Sea tectonic depression, at its joint with the Ukrainian crystalline shield in the north and the Azov crystalline massif in the east (Fig. 3.3). Manganese ores of this basin are associated with Oligocene coastal-marine deposits found in the stratigraphic sequence of the Black Sea depression. All currently separate ore-bearing areas of the examined territory are parts of a single previously existing Oligocene marine basin and in structural, stratigraphic, and genetic plans are of a single type.

Geological setting and types of manganese ores

Participating in the geological structure of the Nikopol' manganese-ore basin are rocks of two structural stages: (1) the crystalline basement, represented by Precambrian rocks, and (2) the sedimentary cover, composed of rocks of Cretaceous, Paleogene, Neogene, and Quaternary age.

Sedimentary deposits of the Mesozoic and Cenozoic in correlation with the tilt of the basement surface are gradually pitched in a south-westerly direction toward the Black Sea depression. In this

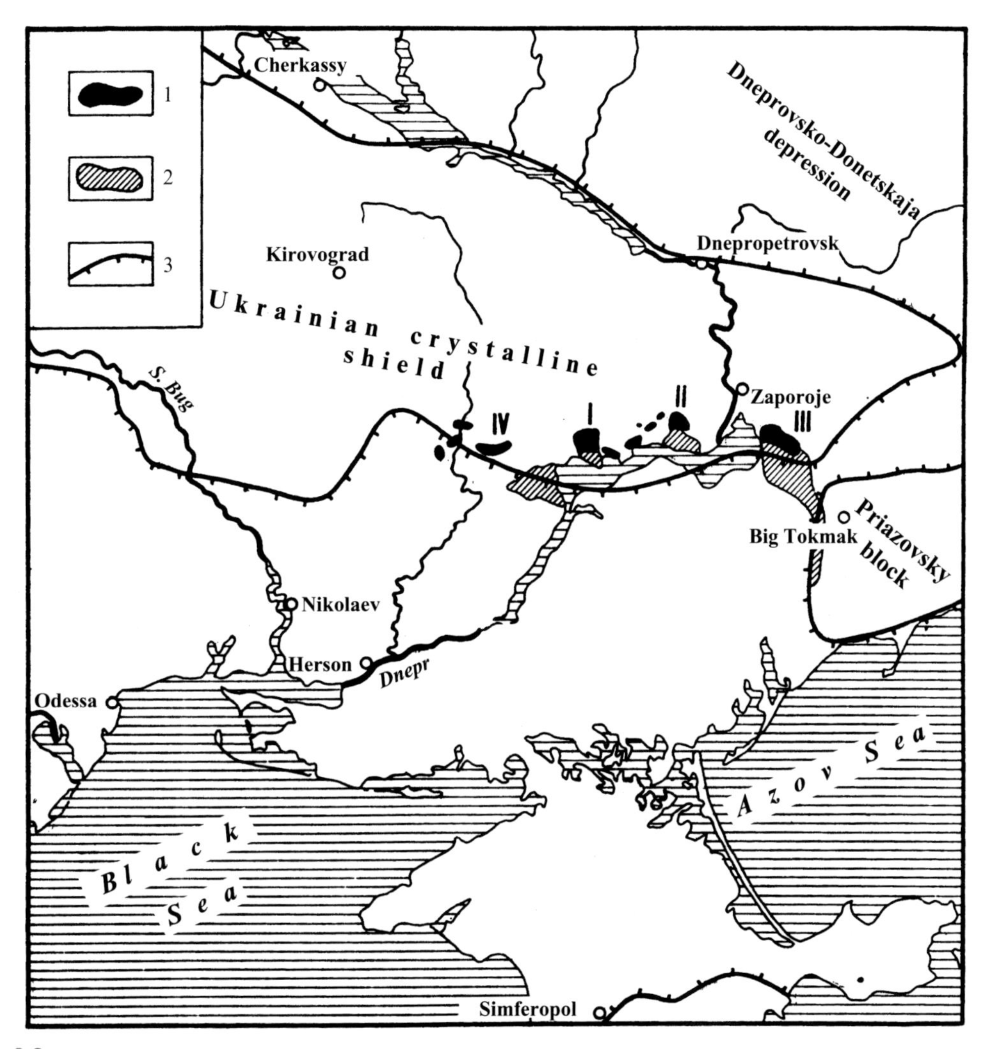

FIG. 3.3

Structural localization of Mn deposits in the Nikopol ore basin (Селин, 1984). 1—Oxide ores, 2—carbonate ores, and 3—boundary of structural regions: I—West ore field; II—East ore field; III—Bol'shoy Tokmak deposit; and IV—Inguletsk ore field.

same direction, there is an increase in the thickness and the completeness of the sequence, as well as a replacement of the continental facies by a shallow-water shore facies, and subsequently by a more deep-water marine facies.

Oligocene deposits in the area of the basin onlap upper Cretaceous-Eocene; where these deposits are absent, they lie on rocks of the crystalline basement or on their weathering crust.

Oligocene deposits are ore-containing and are composed of three sharply distinct members or rocks (from the bottom upwards): (1) subore sands, (2) manganese ores, and (3) supraore clays.

The ore horizon of the Nikopol' basin in the form of a single bed with a thickness from several centimeters up to 4.5 m stretches from the west to the east with interruptions for a distance of approximately 250 km. It represents a strata of sandstone-clay rocks containing concretional nodules of manganese carbonates and an impurity of friable manganese ores. The size of the concretions varies from several millimeters up to several dozens of centimeters. Concretions, frequently intergrown, form in the host rocks lensoidal bodies with a length up to several hundred meters (the so-called dense or slab ores).

Manganese ores are composed either by oxides (oxide type) or by carbonates (carbonate type), or jointly by oxides and carbonates (oxide-carbonate type).

At the Nikopol' deposits an observed mineral zonality is manifested in the replacement of oxide ores by carbonate ores in the direction from the shore of the Oligocene sea into the depth of the basin, as well as from top to bottom along the sequence of the ore body. Zones of oxide development of mixed (oxide-carbonate) and carbonate ores (Fig. 3.4A–C) have been identified. A typical geological cross section of the deposit is visible in the example of the Grushevskii-Basanskii area (Fig. 3.5).

In the Nikopol' manganese-ore basin, the predominant form of concentration of ore matter is the carbonate form. Carbonate ores constitute a large proportion of the ore deposits in the Western and Eastern areas; particularly large reserves are concentrated in the Bol'she-Tokmak deposit (Fig. 3.3).

Carbonate ore matter is represented by two types. The first type is friable ores. Here carbonate matter is scattered in the host sandstone-siltstone-clayey mass in the form of a thinly dispersed impurity and separate enriched nests. Its distribution is uneven; coagulations, nests, and interbeds are observed.

The second type is solid ("rocky") ores; they form nodules (lumps) and lenses, and rarely small beds. They are embedded among host sandstone-siltstone-clayey rocks or friable ores. Their form is irregular, commonly flattened along the vertical, and extended in the horizontal direction.

Bedded-lensoidal bodies do not occupy a definite place in the sequence of the ore body. They frequently form interbeds with a thickness of 0.1–0.25 and very rarely up to 0.6–1.5 m among nodular ores. In the sequence of the ore body can be found one or two such interbeds, partitioned by the friable mass. They are distributed unevenly about the area in the form of lenses with a length of 10–20, more rarely 150–300 m.

Among the solid ("rocky") carbonate ores of the Nikopol' basin two subtypes can be delineated. In the first, ore lumps are composed of light-gray carbonate and have massive, finely porous, or finely cellular textures. Ores of this subtype are distributed in the area of development of oxide-carbonate ores in the Western and Eastern ore-bearing areas and are classified as the poor.

The second subtype are richer carbonate ores characterized by an impurity of manganite. They are widely distributed in the area of development of carbonate ores in the Eastern area and the Bol'she-Tokmak deposit.

Dense ("rocky") ores of the second subtype commonly include concentrically layered textural elements, differentiated by size: ooliths, pisoliths, and concretions. Most widely distributed are pisoliths and small concretions (structures with dimensions from 2 to 25 mm). The ores are commonly rare-pisolitic. Pisoliths in them are separated by carbonate cement and in thin section do not appear to touch one another, so they have "floating" contacts with neighboring grains (Fig. 3.6).

Concentrically layered structures are commonly represented by powder-like earthy material of brown color, consisting of a mixture of manganite and hydrosilicates of iron and manganese (see Fig. 3.6C and D). Certain sections are partially or completely replaced by later large-spherulite carbonates (see Fig. 3.6E and F).

Concentrically layered oxide-hydrosilicate inclusions were formed earlier in relation to the mass cemented by their carbonates. Primary carbonate matter is dense, cryptocrystalline, and dark-colored. Their

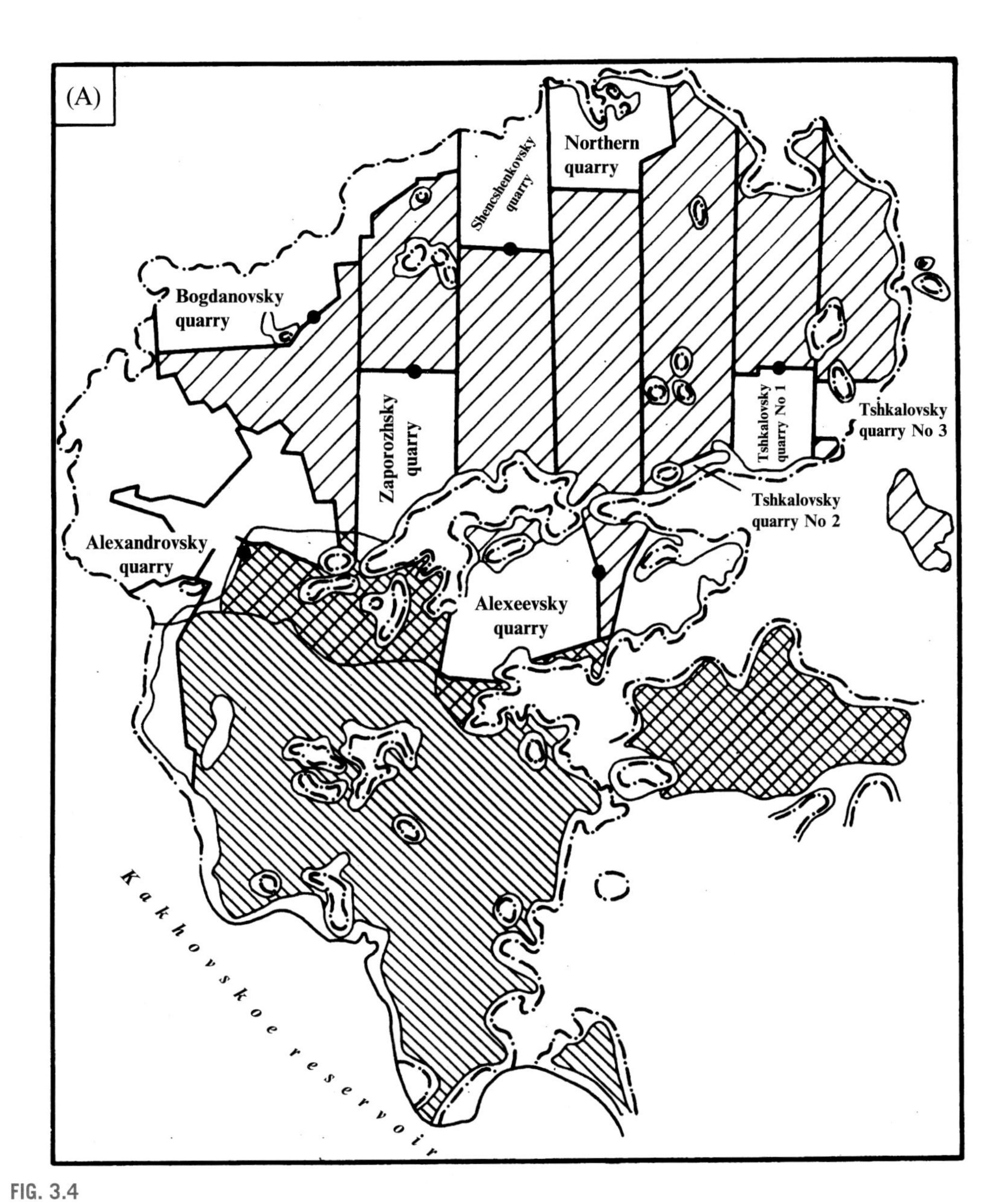

FIG. 3.4

Plans of ore bodies in the Nikopol manganese basin with location of studied sections. (A) Western ore area.

(Continued)

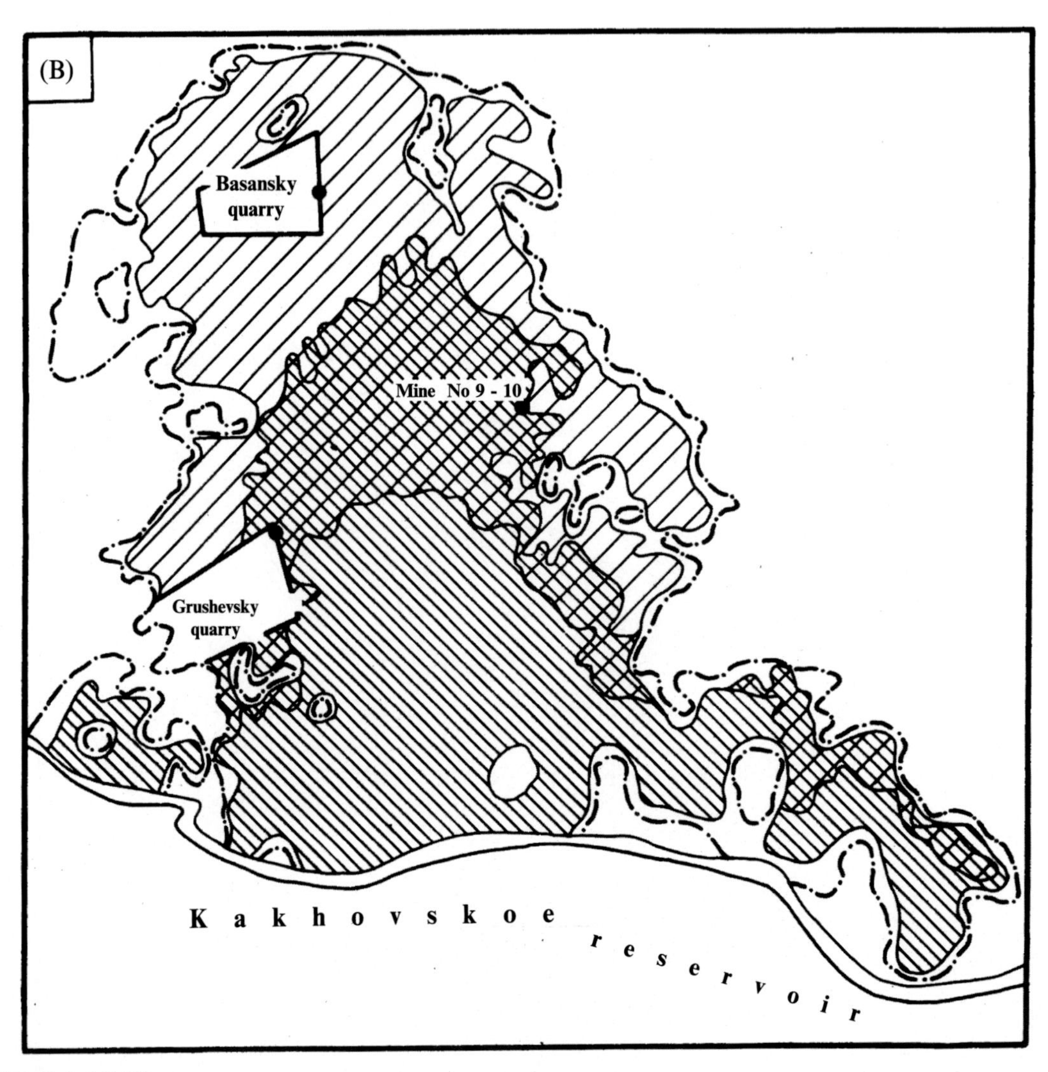

FIG. 3.4, CONT'D

(B) Eastern ore area.

(Continued)

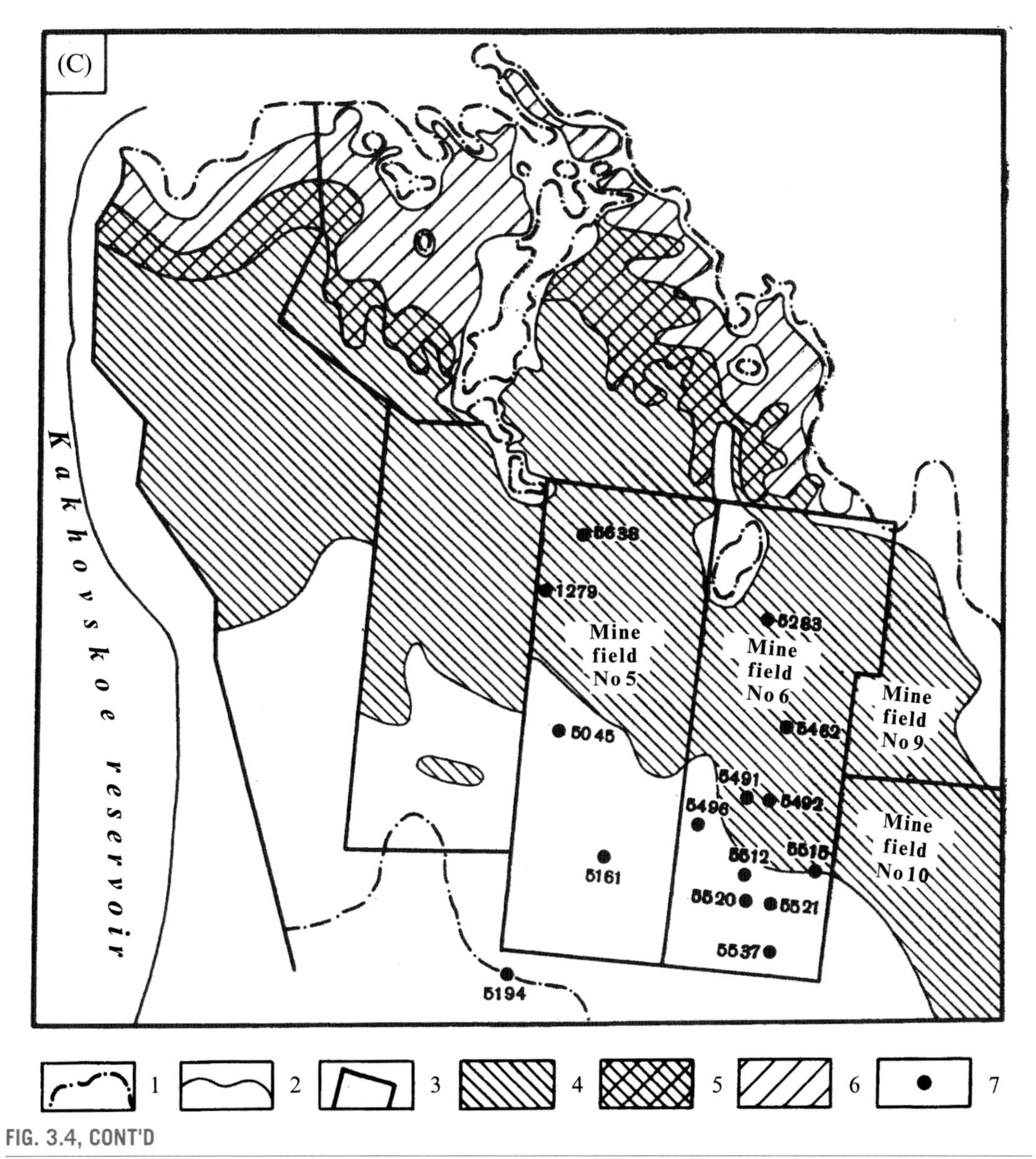

FIG. 3.4, CONT'D

(C) Bol'shoy Tokmak deposit: 1—boundary of ore bodies; 2—isoline of thickness 0.75 m ore body; 3—boundaries of quarries and mine fields; 4–6—areas of different types manganese ores: 4—carbonate, 5—oxide-carbonate, 6—oxide; and 7—location of studied sections.

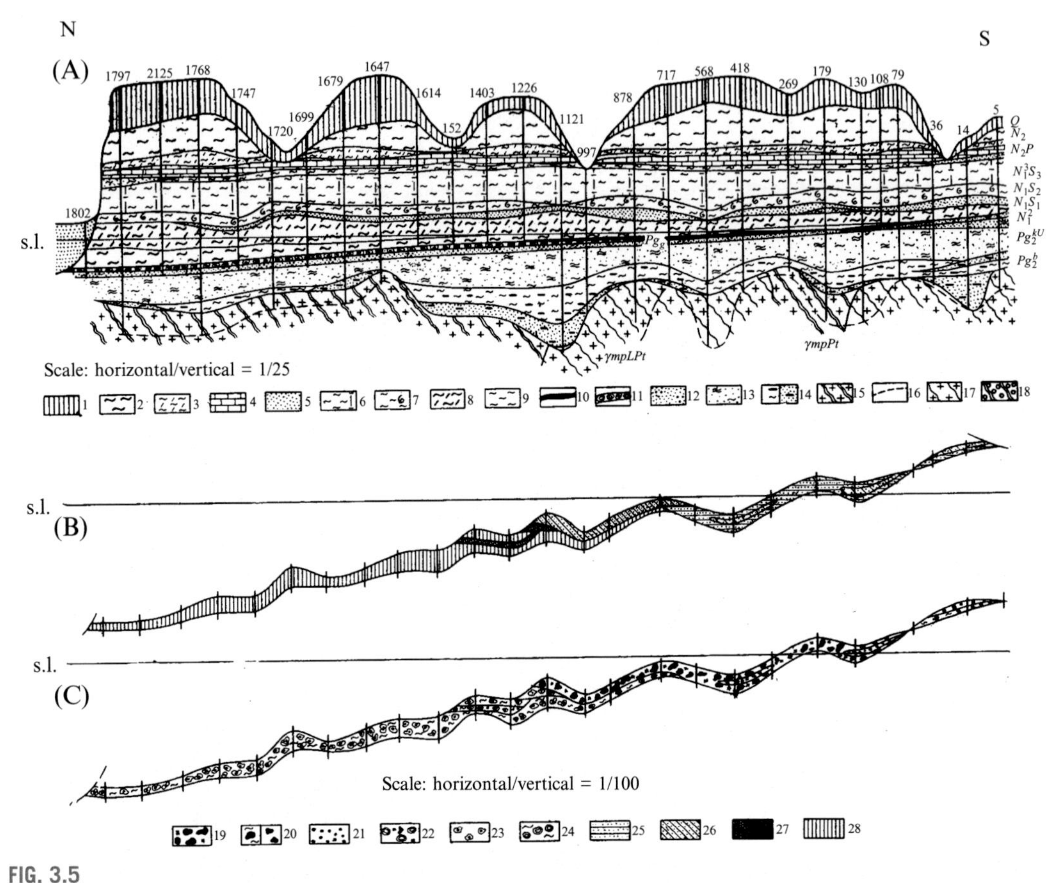

FIG. 3.5

Mineralogical zoning and ore body structure of Nikopol deposit, Grushevsko-Basanskaya section (Страхов и др., 1968). (A) Geological cross section; (B) Mineralogical composition of ore horizon; (C) Textures of ore horizon. 1—loam; 2—red-brown clay; 3—dark-gray clay; 4—shell limestone; 5—quartz sand; 6—calcareous clay; 7—dark-gray clay with detritus; 8—clay interbedded with marl; 9—gray-green clay; 10—oxide-manganese ore; 11—carbonate-manganese ore; 12—quartz-glauconite sand; 13—aleurite; 14—carbonaceous clay and sand; 15—manganite; 16—boundary of the weathering crust in crystalline rocks; 17—plagiogranite and migmatite; 18—pebbles; 19—lumpy-nodular oxide-manganese ore in soot-clay mass; 20—lumpy-nodular oxide ore in clay; 21—concretion and oolitic ore in the soot-clay mass; 22—mixed oxide-carbonate ore; 23—nodular carbonate ore with manganite pizolites; 24—nodules of carbonate-manganese ore in clay; 25—psilomelane ore; 26—manganite-psilomelane ore; 27—manganocalcite-manganite ore; 28—manganocalcite ore.

decarbonation and bleaching occurred subsequently. In microsections are noted traces of a strong leaching of the cement and inclusions with development of various voids (pores, caverns, cells, fractures) and recrystallized carbonate material along the walls of the voids. Here are also observed micro-cross-veins and microlenses. The voids are frequently filled by secondary, strongly recrystallized carbonates.

"Rocky" carbonate ores (Table 3.1) are composed up to 70–90% by carbonates of Mn and Ca; impurities constitute 10–30%. In friable ores, the content of impurities increases up to 40%. Impurities

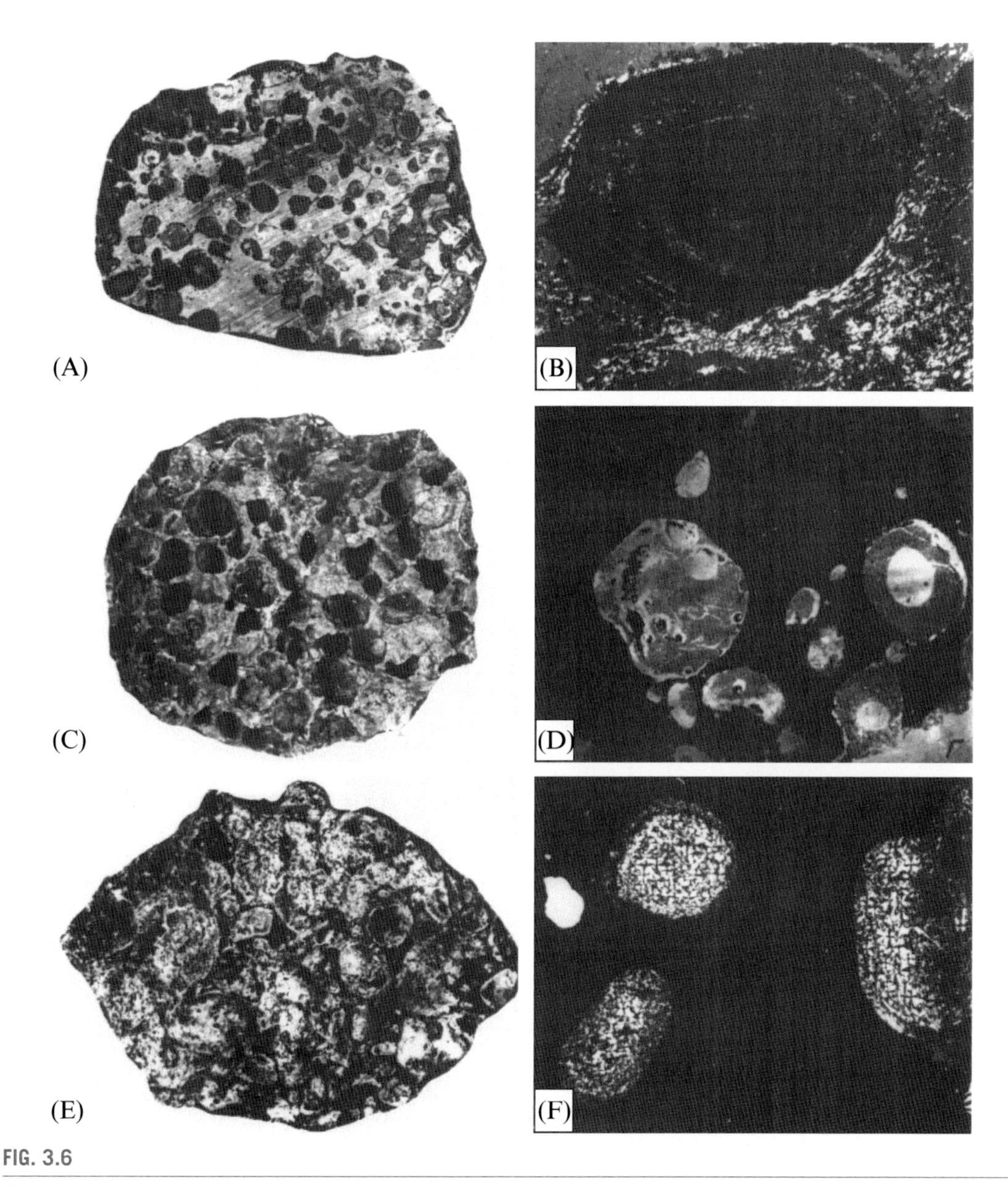

FIG. 3.6

Structure and texture features of manganese carbonate ores of Nikopol basin (photos by Zh.V. Dombrovskaja). (A) Pizolitic carbonate ore with manganese hydrosilicatic pizolites, lump of ore No. 721, decrease 0.8; (B) pisolite of the hydro-silicate Mn, partially replaced with Mn carbonate, transparent thin section No. 720, magnification 9.6, without the analyzer; (C) cellular-pisolitic carbonate ore, lump of ore No. 746, decrease 0.8; and (D) pizolitic cells in carbonate ores, partially filled with a siliceous material, transparent thin section No. 651, magnification 9.6, with the analyzer; (E) pisolitic carbonate ores with carbonate pisolite, lump of ore No. 624, decrease 0.8; (F) carbonate pisolite with spherulitic structure of the carbonate ore, transparent thin section No. 733, magnification 14.5, with the analyzer.

Table 3.1 Chemical Composition (in Mass%) of Carbonate-Manganese Ores of the Nikopol' Manganese-Ore Basin and Data on its Conversion Into Carbonate Matter (Compiled by Zh.V. Dombrovskaia)

	SiO_2	Al_2O_3	Fe_2O_3	Fe_2O_3	FeO	MnO_2	MnO	MgO	CaO	Na_2O
Number of Samples	**Total**									
Aleksandrovskii quarry										
600	17.7	3.40	2.08	–	–	13.5	16.7	4.36	12.6	0.79
1619	6.44	1.78	0.98	–	–	6.30	37.1	3.06	8.80	0.22
1622	4.90	1.38	1.54	–	–	1.71	38.7	2.60	12.1	0.18
1623	11.5	2.32	3.88	–	–	1.69	25.6	3.06	17.2	0.18
1624	12.5	1.86	9.4	–	–	0.87	27.7	1.58	14.0	0.14
594	15.1	2.27	1.6	–	–	0.61	31.7	2.30	13.5	0.19
1625	7.52	1.70	2.8	–	–	1.14	36.7	1.76	12.9	0.16
Alekseevskii quarry										
560	13.7	2.71	–	0.57	1.05	None	16.6	3.17	26.4	0.2
Bogdanovskii quarry										
574	11.3	1.58	1.17	–	–	10.2	22.8	3.10	15.0	0.35
576	17.1	1.98	13.4	–	–	1.65	21.8	3.10	10.3	0.29
Grushevskii quarry										
609	9.31	1.68	5.69	–	–	12.1	27.7	3.38	10.0	0.34
Bol'she-Tokmak deposit										
Well 5512, depth 93.50–95.65 m										
1628	7.30	2.06	1.54	–	–	1.60	29.9	2.38	18.1	0.12
1629	7.84	1.72	2.48	–	–	1.38	29.8	1.84	19.2	0.14
1630	4.78	1.66	2.20	–	–	0.97	27.2	1.98	24.0	0.12
1631	6.36	2.12	4.08	–	–	0.98	44.6	1.64	5.60	0.12
1632	4.94	1.68	2.72	–	–	4.60	39.2	1.48	8.80	0.10
Well 5545, depth 106.90 m										
1633	6.74	2.14	2.88	–	–	1.85	29.7	1.62	18.3	0.18
Well 1279, depth 48.20–50.70 m										
726									13.1	0.17
723									4.40	0.08
723									8.60	0.10
720									7.10	0.11
1628	7.3	2.06	1.54	–	–	1.60	29.9	2.38		
1629	7.84	1.72	2.48	–	–	1.38	29.8	1.84		
1630	4.78	1.66	2.20	–	–	0.97	27.2	1.98		
1631	6.36	2.12	4.08	–	–	0.98	44.6	1.64		
1632	4.94	1.68	2.72	–	–	4.60	39.2	1.48		
Well 5521, depth 106.0–108.0 m										
737	5.78	1.99	–	0.51	2.53	4.60	36.6	1.59	14.0	0.18
734	5.22	1.81	–	3.19	2.9		30.7	2.39	16.3	0.12
732	8.99	1.89	–	8.04	1.88		23.2	1.78	20.0	0.15
730	12.3	2.17	–	2.95	2.49		24.5	2.09	18.7	0.14
Well 5491, depth 93.40–95.10 m										
659	4.5	1.95	–	0.37	2.21	4.60	38.4	2.05	12.2	0.13
658	6.24	1.89	–	None	2.16		34.6	2.06	15.1	0.20
657	5.31	1.60	–	0.67	2.37		33.2	2.05	16.7	0.14

K_2O	CO_2	H_2O^+	H_2O^-	$MnCO_3$	$CaCO_3$	$MgCO_3$	$FeCO_3$	Sum of Insoluble Carbonates	
0.86	22.2	3.48	1.97	27.1	22.5	3.57	None	53.2	46.8
0.54	32.1	Not detected	Not detected	60.1	15.7	4.26		80.0	20.0
0.52	36.6			62.7	21.6	5.43	0.3	90.1	9.9
0.60	32.3			41.5	30.8	5.36	None	77.6	22.4
0.48	28.6			44.9	25.1	0.6		70.6	29.4
0.56	29.0	1.34	1.14	48.2	24.1	None		72.3	27.8
0.38	34.1	Not detected	Not detected	59.2	23.2	2.67		85.0	15.0
0.34	32.3	1.33	0.56	26.3	47.1	2.56	None	76.5	23.5
0.54	30.1	1.65	1.07	36.9	26.7	6.51	2.16	72.3	27.7
0.50	24.6	2.52	2.25	35.4	18.4	5.71	None	59.5	41.5
0.59	23	2.93	2.17	39.1	17.8	None	None	56.9	43.1
0.42	35.4	Not measured	Not measured	32.4	4.85	None	85.8	14.3	
0.36	34.2			48.2	34.3	1.05		83.6	16.4
0.38	37.7			44.1	43	3.53		90.6	9.40
0.46	34.3			72.3	10.1	3.44	1.00	86.8	13.2
0.40	33.4			63.5	15.8	3.11	1.50	85.9	14.1
0.36	34.1	Not measured	Not measured	48.2	32.7	2.25	None	83.1	16.9
0.27	31.9	0.92	1.50	56.5	23.3	None	None	79.8	20.2
0.10	34.3	0.79	1.68	80.6	7.82			88.4	11.6
0.17	34.4	0.83	1.60	72.5	15.3			87.8	12.2
0.26	32.9	1.60	1.23	69.4	12.7	1.37		83.4	16.6
0.3	32.3	1.65	0.94	59.6	24.9	None	None	80.6	19.5
0.3	32.6	1.74	1.17	49.8	29.1	1.30		80.2	19.8
0.33	30.2	2.09	1.28	37.6	35.7	0.19		73.6	26.5
0.35	30.4	1.47	1.16	39.7	33.5	0.97		74.1	25.9
0.27	34.2	1.09	0.98	62.3	21.9	1.37	None	85.5	14.5
0.33	33.6	1.26	0.98	56.1	26.9	0.46		83.4	16.6
0.29	34.1	1.70	0.72	53.8	29.9	0.65		84.3	15.7

are represented principally by grains of quartz, more rarely feldspars and glauconite; small quantities of clayey minerals and sulfides are present. Carbonate ores occasionally contain oxide minerals of manganese and iron.

The carbonate portion of the ore represents a mixture of carbonates of alternating composition of the $CaCO_3$-$MnCO_3$ isomorphic series. In all samples, with the exception of separate friable varieties, $MnCO_3$ predominates; it is present in quantities of 35–70%. Content of $CaCO_3$ constitutes 10–45%, that of $MgCO_3$ 0–8%. $FeCO_3$ is present up to 2% in isolated samples.

The carbonate mass is heterogeneous in mineral composition, as is distinctly manifest in the heating curves (Fig. 3.7). Carbonate samples commonly provide a broad endothermic effect in the 500–900°C range; that is, within the interval between the endothermic effects of rhodochrosite (500–600°C) and calcite (800–900°C).

In Fig. 3.8 are recorded the X-rays of the powder mounts of two samples. The set of peaks in sample 1643 provides evidence for the presence in the sample of two minerals. The first mineral, characterized by the reflections (*d*, Å): 1.777; 1.834; 2.014; 2.189; 2.866; 3.76, is evidently rhodochrosite with an insignificant impurity of Ca and Mg, which caused the somewhat elevated values of the basic reflection. The second mineral in this sample, with the basic reflection (*d*, Å) 2.952, is evidently kutnohorite of the Ca-Mg isomorphic series. In sample 609 (*d*, Å: 1.775; 1.833; 2.006; 2.180; 2.396; 2.861; 3.67) is present only one mineral—calciorhodochrosite.

Present in the X-rays are peaks caused by the presence of quartz in the samples.

Thus, the results of X-ray, thermal, and optical investigations indicate that solid (earlier) carbonate material is represented predominantly by a manganese mineral of a complex isomorphic series—from calciorhodochrosite to manganocalcite and kutnohorite with contents of $MnCO_3$ molecules reaching 76.3%. Noted as well is the presence of calcite.

Friable (secondary) carbonate is composed of a mixture of rhodochrosite and calcite; observed as well are traces of calciorhodochrosite. Carbonate material filling the voids is represented by rhodochrosite or calcite. This serves as evidence that the alteration of carbonate ores has led to the decomposition of primary Ca-Mn-carbonate molecules and the deposition from the solution of pure carbonates of Ca and Mn.

The samples for determining the isotopic composition of carbonate-manganese ores were collected from various ore-bearing areas of the Nikopol' basin. In the Western ore-bearing area, the samples were collected in the Aleksandrovskii, Alekseevskii, and Bogdanovskii quarries situated in the zones of development of oxide-carbonate and oxide ores; in the Eastern area, in the Grushevskii quarry and mines No. 9 and 10 situated in the zone of development of oxide-carbonate ores; and at the Bol'she-Tokmak deposit from the numerous wells of mine fields No. 5, 6, 9, and 10 situated in the zone of development of carbonate ores (see Fig. 3.4). As examples, we provide a description of two cross sections of the ore bed in the Aleksandrovskii quarry and along well 5512 of the Bol'she-Tokmak deposit.

In the Aleksandrovskii quarry, the ore bed is represented by (from bottom to top): carbonate (0.7 m), oxide-carbonate (0.8–0.9 m), and oxide (0.3 m) ores (Fig. 3.9). Carbonate ores are represented by a friable mass including rocky (the so-called "lump") nodules and lensoidal interbeds of "dense" ore with a thickness of 0.10–0.15 m. The friable mass is composed of a mixture of carbonate and clayey material and contains a significant quantity of terrigenous material, represented mainly by small clasts of quartz. Present in small quantities are glauconite and feldspars. Lumps and interbeds are represented by carbonate material of gray and dark-gray color with a large quantity of cells and pores, having a spherical form. Occasionally, the cells are filled by clayey material.

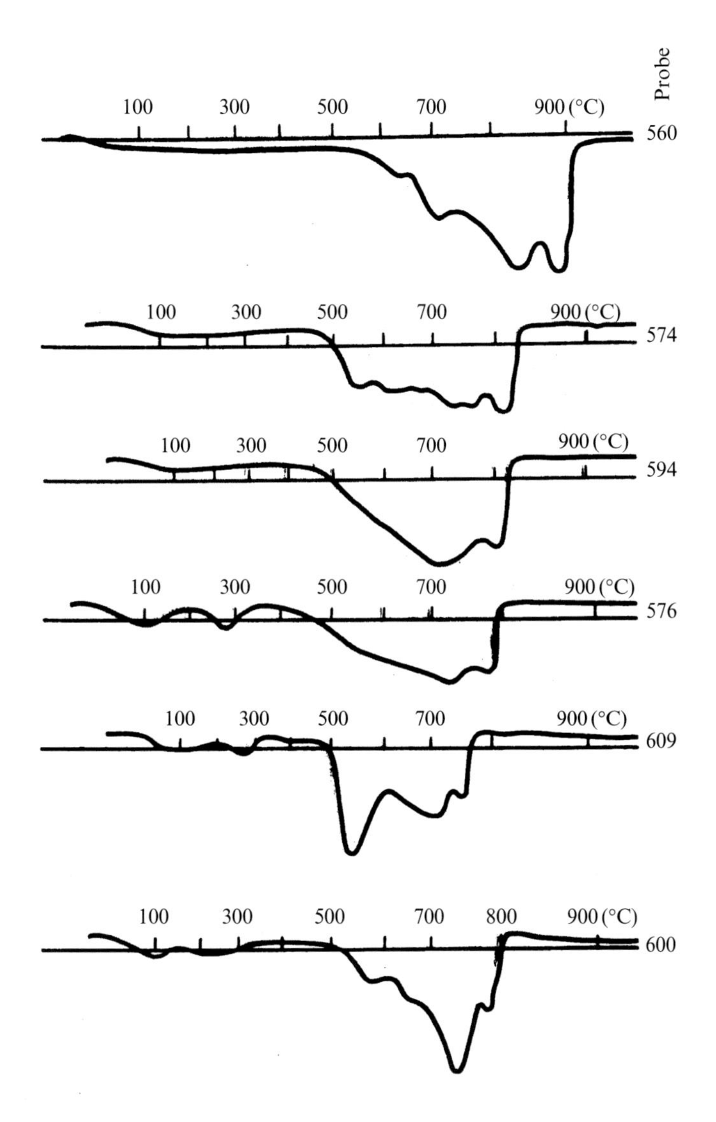

FIG. 3.7

Thermogram curves of carbonate ores from Nikopol basin.

The carbonate material of the nodules contains a smaller quantity of terrigenous impurity than does the hosting friable cement. According to the data calculated from chemical analyses (see Table 3.1), the carbonate ores are composed of $MnCO_3$—41.49–62.7%; $CaCO_3$—15.70–30.79%; and impurities—9.92–29.41%. In the heat curves (see Fig. 8.6, samples 594 and 600), there is a broad endothermic effect in the 500–820°C range, which is particular to a mixture of minerals of the calcite-rhodochrosite group. According to X-ray analysis data, the dense carbonate from sample 595 is composed of calciorhodochrosite with content of $MnCO_3$=76.3%, and the friable carbonate is composed of a mixture of that carbonate with calcite. C_{org} was not detected in the ores.

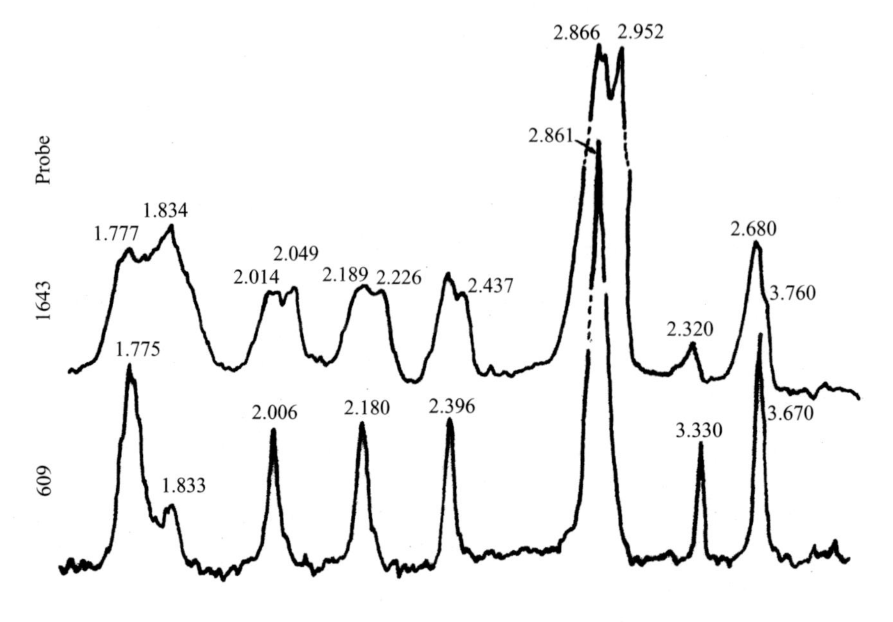

FIG. 3.8

X-ray diffractograms of carbonate-manganese ores, Nikopol deposit.

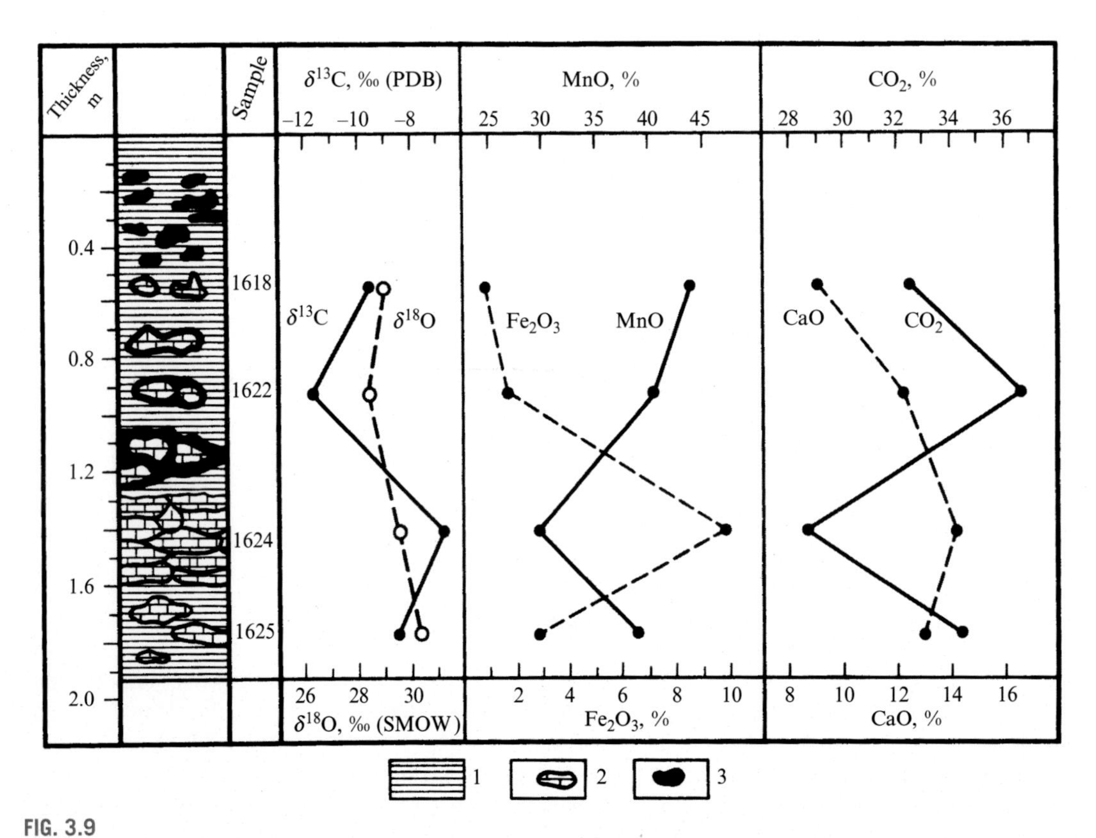

FIG. 3.9

Isotopic and chemical composition of carbonate-manganese ores from the sequence of the Alexandrovsky quarry.

The ore horizon of well 5512 with a thickness of 2.3 m is represented by a sandstone-clayey strata including carbonate ores in the form of nodules and nests (Fig. 3.10). In the basement of the strata predominate sands that are replaced upwards along the sequence by clays. Ore nodules of manganese carbonate are superimposed in the sandy as well as clayey parts of the sequence within the interval of depths from 93.5 to 95.8 m. The host clayey-carbonate mass contains a large quantity of grains of quartz and glauconite.

Isotope data and their discussion

All of the obtained isotope data are recorded in Table 3.2 and displayed in Fig. 3.11 [similar values have been obtained by Hein and Bolton (1992)]. As can be seen, carbonate ores of various areas in their isotopic composition do not differ and share the same isotopic particularities. This fact provides evidence of a single mechanism and of analogous conditions of manganese carbonate ore-formation within the boundaries of the examined territory.

Against the background of a relatively narrow range of variations in the isotopic composition of oxygen ($\delta^{18}O = 26.4–31.8‰$), the values of $\delta^{13}C$ in the studied rocks vary within a rather broad interval (−24.6‰ to −4.9‰). Such a broad spectrum of isotope ratios of carbon is characteristic not only for separate ore fields, but can be seen even within the boundaries of a single section (see Figs. 3.9 and 3.10). We have also identified two types of dense carbonate ores that do not differ in terms of isotopic composition, but rather in terms of their external features. For this reason, in the histogram (Fig. 3.11) and graph (Fig. 3.12) they have not been separated from each other and are indicated by the same sign.

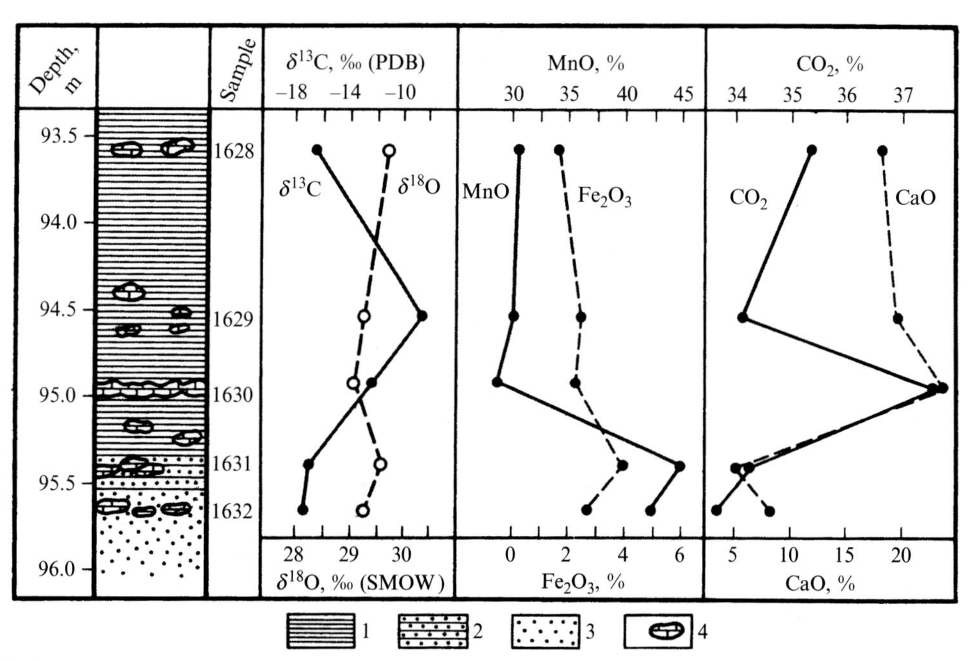

FIG. 3.10

Isotopic and chemical composition of carbonate-manganese ores from the borehole 5512.

Table 3.2 Isotopic Composition of Carbon and Oxygen of Carbonate-Manganese Ores of the Nikopol' Manganese-Ore Basin

Number of Analysis	Number of Sample	Characteristics and Location of Gathering of Sample	$\delta^{13}C$, ‰ PDB	$\delta^{18}O$, ‰ SMOW
1	2	3	4	5
Aleksandrovskii quarry				
2170	593	Dark-gray, dense	−9.3	27.6
2171	594	Gray, porous	−7	28.6
2172	595	Dark-gray, dense, cellular	−10.3	30.5
2173	596	Dark-gray, dense	−11.4	28
2174	597	Dense, cellular, ocherized	−10.1	28.5
2230	920p	Dark-gray, dense	−8.2	30.4
2231	920r	Light-gray, friable	−10	26.3
2232	921	Gray, dense, fine-cellular	−7.2	29.2
2233	922	Dark-gray, dense, cellular-cavernous	−7.9	30.8
1619	1813-1c	Oxide-carbonate	−9.5	29
1622	1814		−11.8	28.4
1624	1815b	Dense, carbonate	−6.8	29.5
1625	1816		−8.5	30.3
1935	1915		−6	30.3
Alekseevskii quarry				
1926	1906/9	Cavernous, carbonate	−5.7	28.6
2151	560	Carbonate, weakly dense	−12.6	28.4
2152	563		−11.2	27.2
Bogdanovskii quarry				
2169	576	Light-gray, pisolitic, friable	−7.9	28.6
1930	1910/20	Dense, carbonate	−8.8	28.4
1931	1911/21		−10.8	27.2
Eastern ore-bearing area				
Grushevskii quarry				
2235	956	Dark-gray, dense, pisolitic	−9.1	30.8
2236	957		−9.7	28.3
2237	959		−9.6	30.5
2238	971	Oversanded, carbonate	−10.9	29.2
1969	955k	Pisolite, strongly leached	−10.3	31
1967	958k	Pisolite, oxide-carbonate	−9.5	30.7
1968	958k	Cement	−10.1	30.4
1936	1916/32	Carbonate	−7.9	28.3
1969	955k	Pink carbonate from oxide-carbonate ore	−10.3	31

Table 3.2 Isotopic Composition of Carbon and Oxygen of Carbonate-Manganese Ores of the Nikopol' Manganese-Ore Basin—Cont'd

Number of Analysis	Number of Sample	Characteristics and Location of Gathering of Sample	$\delta^{13}C$, ‰ PDB	$\delta^{18}O$, ‰ SMOW
1	2	3	4	5
2264	955k	Pisolite, strongly leached	−10	27.9
2162	960	Dark-gray, dense	−16.1	27.2
2163	961	Fine-porous, dark-gray	−12.8	29.7
2164	962	Dark-gray, cellular	−12	28
2165	963	Dark-gray, fine-cellular	−11.4	30
2166	968	Dark-gray, cellular	−7.3	30.2
2167	969	Dark-gray, fine-dense carbonate nodules	−12.5	29.9
2153	605	Brownish-gray, pisolitic, carbonate	−9.3	28.5
2155	609		−13.8	28.2
Shaft 9–10				
2168	549	Gray rock with brown pisoliths	−14.4	28
2159	623	Brownish-gray, friable, carbonate	−11.7	28.9
2157	624	Coarse-pisolitic, rock	−9.1	28.9
Bol'she-Tokmak ore-bearing area				
Well 5512				
1628	1818	Lump; 93.5 m	−8.6	28.7
1630	1820	Cavernous-cellular; 95.0 m	−12.2	28.2
1629	1819	Cavernous-cellular; 94.6 m	−8.6	28.7
1631	1821	Lump; 95.4 m	−16.6	29.3
1632	1822	Lump, cavernous; 95.65 m	−17.3	28.6
Well 5545				
1633	1823	Lump; 106.9 m	−7.6	29
Well 5537				
1634	1824	Oolitic; 115.5 m	−9.8	29.5
1635	1825	Lump; 115.8 m	−7.9	27.8
Well 5491				
1638	1827	Lump; 93.45 m	−9.5	28.6
1639	1828	Lump; 93.8 m	−7.2	29.5
1640	1829	Lump; 94.3 m	−13.3	29.6
1641	1830	Massive; 94.8 m	−15.3	29.4

Continued

Table 3.2 Isotopic Composition of Carbon and Oxygen of Carbonate-Manganese Ores of the Nikopol' Manganese-Ore Basin—Cont'd

Number of Analysis	Number of Sample	Characteristics and Location of Gathering of Sample	$\delta^{13}C$, ‰ PDB	$\delta^{18}O$, ‰ SMOW
1	2	3	4	5
Well 1279				
2158	720	Pisolitic, rock,	−14.5	26
2193	722	Pisolite; 50.6 m	−24.4	26.2
1883	723p	Brown-gray, friable carbonate pisoliths in rock carbonate ore; 49.6 m	−12.1	25.8
2159	723	Rock, pisolitic cemented mass; 49.6 m	−17	27.8
2194	724	Gray, dense, pisolitic, concretional; 49.4 m	−15.5	26.2
2195	725	Dark-gray, dense, cavernous; 48.8 m	−13.8	30.4
2196	726	Dense with gray pisoliths; 48.4 m	−8.6	30.7
Well 6300				
1973	864k	Cemented pink carbonate of coarse-grained sandstone; 48.4 m	−13.7	27.4
Well 6179				
1974	893k	Secondary pink carbonate from vugs in oxide ore; 125.0 m	−14.8	26.4
Well 4496				
2176	699	Friable, sandstone; 93.4 m	−13.4	26.8
2177	700p	Dark-gray, dense, pisolitic-cellular; 93.3 m	−8.5	30.6
2178	700r	Pisolitic, weakly compacted; 93.3 m	−13	25.8
2179	701p	Pisolite; 93.2 m	−17.3	24.6
2180	701r	Light-gray, friable (nest in clay); 93.2 m	−17.1	25.6
2181	702p	Dark-gray, dense; 92.6 m	−9.2	31.1
2182	702r	Gray, compacted; 92.6 m	−15.7	28.8
2183	703p	Dark-gray, dense, pisolitic; 92.3 m	−15.8	28.8
2184	703r	Gray, nodules in clay; 92.3 m	−15.2	29.8
2185	705	Gray, friable, nest in clay; 91.2 m	−4.9	29.1
Well 5161				
2186	708	Dark-gray, rock; cuts in clay; 96.6.m	−15	27.1
2187	709r	Gray, fine nodules in clay; 96.8 m	−7.1	30.4
2188	710	Dark-gray, dense; 97.0 m	−14.4	28
2189	711	Dark-gray, dense, rare-cellular; 97.2 m	−12.6	30.4
2190	712	Dark-gray siltstone; 97.5 m	−2.5	5

Table 3.2 Isotopic Composition of Carbon and Oxygen of Carbonate-Manganese Ores of the Nikopol' Manganese-Ore Basin—Cont'd

Number of Analysis	Number of Sample	Characteristics and Location of Gathering of Sample	$\delta^{13}C$, ‰ PDB	$\delta^{18}O$, ‰ SMOW
1	2	3	4	5
Well 6422				
2161	919	Dark-gray, dense; 115.5 m	−18	29
Well 5722				
2267	976k	Pisolite fine, light-gray, dense; 110.7 m	−8.1	25.4
2239	976	Dark-gray rock, pisolitic, oversanded; 110.7 m	−15.5	27.9
Well 5194				
2197	727	Dark-gray, oversanded, cellular; 119.8 m	−18.9	29.9
Well 5521				
2198	730	Gray, dense, pisolitic-cellular; 107.8 m	−13.5	29.3
2199	731	Gray, friable; 107.5 m	−12.6	25.9
2200	732	Dark-gray, pisolitic-concretional; 107.3 m	−11.7	30.7
2201	733	Dark-gray, fine-cellular; 107.1 m	−10.8	30.1
2202	734	Dark-gray, concretional; 106.7 m	−14.4	28.2
2204	737	Dark-gray, cellular; 106.0 m	−8.4	30.2
2205	738	Gray, friable; 105.8 m	−11.5	27.3
Well 5283				
2207	746	Dark-gray, rock, pisolitic-cellular; 92.5 m	−10.8	29.5
Well 5462				
2208	749p	Dark-gray, dense, with pisoliths;62.0 m	−8.7	0.8
2209	947r	Gray, friable; 62.0 m	−15.9	27.3
Well 6522				
2210	815	Gray-brown, dense; 107.6 m	−8.7	29.2
2211	816	Gray, dense, strongly oversanded; 107.8 m	−14.5	29.3
2212	817	Gray, rock, cellular; 108.2 m	−8.3	29.8

Continued

Table 3.2 Isotopic Composition of Carbon and Oxygen of Carbonate-Manganese Ores of the Nikopol' Manganese-Ore Basin—Cont'd

Number of Analysis	Number of Sample	Characteristics and Location of Gathering of Sample	$\delta^{13}C$, ‰ PDB	$\delta^{18}O$, ‰ SMOW
1	2	3	4	5
Well 6088				
2213	824	Gray, rock, with pisoliths; 91.4 m	−17.6	31.8
2214	825	Brown-gray, rock, with brown pisoliths; 91.5 m	−15.7	27.8
2215	826	Gray, rock, oolitic; 91.6 m	−9.9	27.9
2216	827	Gray, dense, with pisolith; 91.8 m	−15.1	28.3
Well 6049				
2217	837	Gray, friable pisolite; 110.0 m	−14.8	28.7
Well 6076				
2218	841	Gray, rock, oolitic; 100.9 m	−12.8	27.7
Well 6468				
2219	847	Gray, rock, fine-cellular; 115.2 m	−10.3	30.6
2220	848	Gray, rock, rare-pisolitic, concretional; 115.6 m	−10	28.9
2260	848k	Pisolite dark-gray, leached; 115.6 m	−14.9	26.8
2221	849		−16.6	26.5
2222	850	Dark-gray, rock, with rare pisolith, leached; 116.8 m	−14.5	25.9
Well 9-V				
2223	877	Dark-gray, dense, fine-oolitic; 100.5 m	−9.8	30.1
2224	878p	Dark-gray, dense, cavernous;101.0 m	−9.8	29.8
2226	879	Dark-gray, dense, cavernous; 102.0 m	−13.7	30.6
2227	880	Brownish-gray, earthy; 102.5 m	−11.2	31
2228	881	Dark-gray, dense, cellular; 103.0 m	−13.9	27.8
2261	881k	Pisolite dark-gray, gray; 103.0 m	−13.6	25.6
2229	882	Gray, friable, with oolites; 103.5 m	−14	26.6
2262	882k	pisoliths, friable; 103.5 m	−15	25.8
2160	883	Brownish-gray, cellular; 104.0 m	−14.1	25.3
2263	885k	Pisolite dark-gray, friable; 102.0 m	−17.8	27.9
2240	980	Dark-gray, dense, cavernous;103.0 m	−11.6	29.4
2241	981	Dark-gray, dense, pisolitic-cellular; 103.4 m	−10.1	29.5

Table 3.2 Isotopic Composition of Carbon and Oxygen of Carbonate-Manganese Ores of the Nikopol' Manganese-Ore Basin—Cont'd

Number of Analysis	Number of Sample	Characteristics and Location of Gathering of Sample	$\delta^{13}C$, ‰ PDB	$\delta^{18}O$, ‰ SMOW
1	2	3	4	5
2242	982	Dark-gray, dense, cavernous, 103.8 m	−8.3	29.5
2243	983	Dark-gray, dense, cavernous, 104.2 m	−11	29.4
2244	984	Gray, friable, cellular; 104.6 m	−12.4	26.5
2245	985	Gray, compacted, oolitic; 104.8 m	−11.8	28.7
2246	986	Gray, compacted, oolitic; 105.0 m	−10.9	28.8
2247	987	Dark-gray, dense, cavernous;105.2 m	−13.6	31
Well 6516				
2248	1000	Gray, rock, oolitic;118.9 m	−13	31.1
2249	1001	Gray, rock, pisolitic-cellular; 119.0 m	−8.1	31.4
2250	1002	Gray, earthy; 119.1 m	−24.1	29.4
2251	1003	Gray, earthy; 119.2 m	−13.3	28
Well 6042				
2252	845	Sandstone with pink rhodochrosite cement; 124.45 m	−13.8	25.9
Well 6456				
2253	869	Gray, oolitic, weakly compacted; 125.4 m	−14.3	27.6
2254	870	Gray, rock, pisolitic; 125.65 m	−13.1	29.2
2255	871	Gray, friable, oolitic; 125.9 m	−10.5	28.5
Well 5045				
2257a	713	Gray, dense, pisolitic; section of dark color, calcite; 45.0 m	−10.6	29
2257b	713	Gray, dense, pisolitic; rhodochrosite	−10.1	29.6
2258	713	Gray, dense, pisolitic; section of light-gray cement, rhodochrosite	−13.7	29.1
2256	713	Gray, dense, pisolitic	−14.7	27.5
Well 5462				
2259	749k	Pisolite, dense, light; 62.0 m	−16.8	29.1

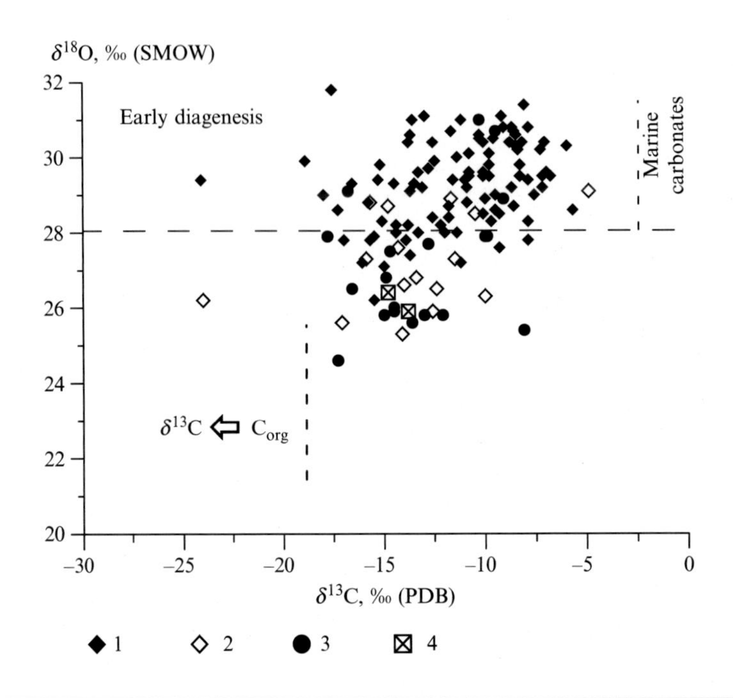

FIG. 3.11

$\delta^{13}C$ versus $\delta^{18}O$ in carbonate ores of Nikopol deposit. 1—Compact (dense?); 2—unconsolidated (loose?); 3—pisolite; and 4—secondary carbonate.

One of the noteworthy particularities of the studied rocks is that all of them without exception are characterized by a lighter isotopic composition of oxygen in comparison with that of sedimentary marine carbonates. Their $\delta^{13}C$ values fall to values close (or, in some cases, even analogous) to those of the carbon of organic matter. On the whole, the isotopic composition of the carbon of the studied samples is found in the intermediary range of $\delta^{13}C$ values coinciding, on the one hand, with carbon of normal-sedimentary marine genesis, and on the other hand, with that of biogenic (C_{org}) genesis (see Fig. 3.11).

This circumstance allows us to consider a mixed source for the carbon of carbonate ores. Evidently serving as one of the sources is carbon, oxidized up to CO_2, of organic origin. In this case the fact emerges that in the studied ores, C_{org} is practically absent—it was "burned out" during diagenesis, with the formed CO_2 going toward the formation of ore carbonate nodules. This is confirmed as well by the direct correlation between the isotopic composition of the carbon and the MnO content in the ores (Fig. 3.13).

There is no cause to doubt a past rich organic life of the ore-bearing terrigenous strata; the composition of ore nodules frequently preserves organic residues, including cores of bivalve mollusks, filled with ore matter (Fig. 3.14), as well as mineralized microbial residues (Figs. 3.15–3.17).

Another source for carbon that is characterized by a heavy isotopic composition can be only the carbon dioxide of carbonates precipitated in isotopic equilibrium with the bicarbonate of marine water. This type of CO_2 can be that of (1) mollusk shells, which by the present have been fully dissolved [in their place in the ore matter are frequently noted voids in the shape of the former valves (see Fig. 3.14B and D)] and (2) thinly dispersed carbonate matter previously scattered in the rock (chemogenic carbonate,

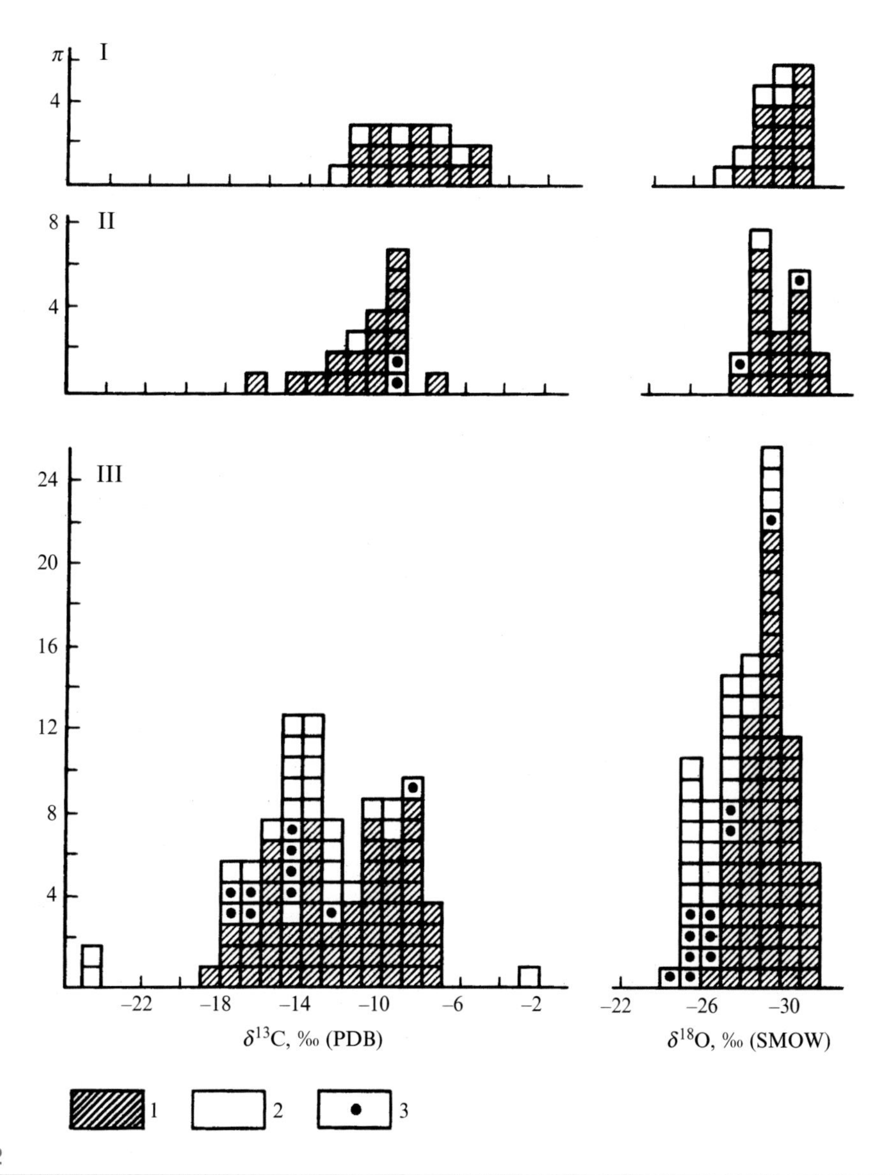

FIG. 3.12

Distribution histogram of δ^{13}C- and δ^{18}O-values in carbonate Mn ores in different areas of Nikopol basin. Mineralized areas (Ore fields?): I—Western; II—Eastern; and III—Bol'shoy Tokmak. *n*—amount of analysis; 1—stony ore; 2—friable ore; and 3—pizolite.

carbonate skeletons of protozoans, and the like). Carbonate of this source underwent dissolution and redistribution at the time of the formation of the ore at the stage of early diagenesis.

In the analysis of the distribution of the isotopic composition of the carbon inside a single sections (see Figs. 3.9 and 3.10), no stable regularity was established from the floor to the roof of the ore strata. Variations of the δ^{13}C values bear a chaotic character. We see that even when the distance between

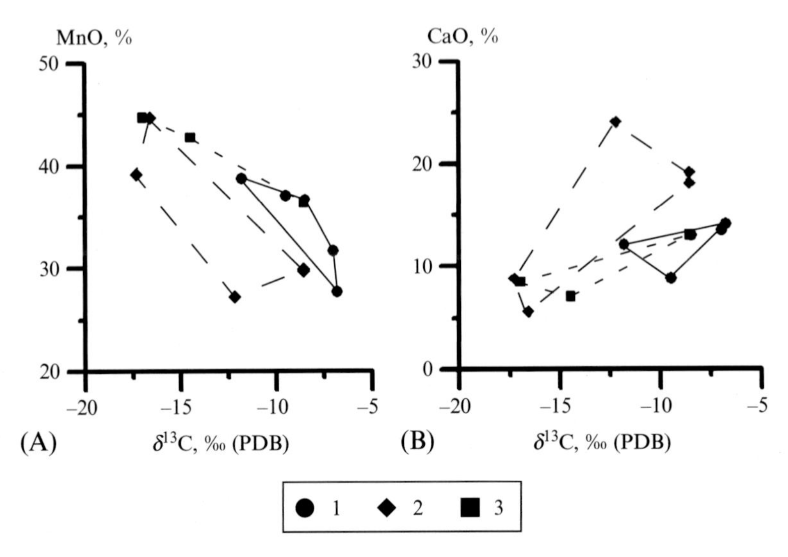

FIG. 3.13

The dependence of the carbon isotope composition and content of the MgO (A) and CaO (B) in carbonate ores of the Nikopol deposit. 1—Alexandrovsky quarry; 2—borehole 5512; and 3—borehole 1249.

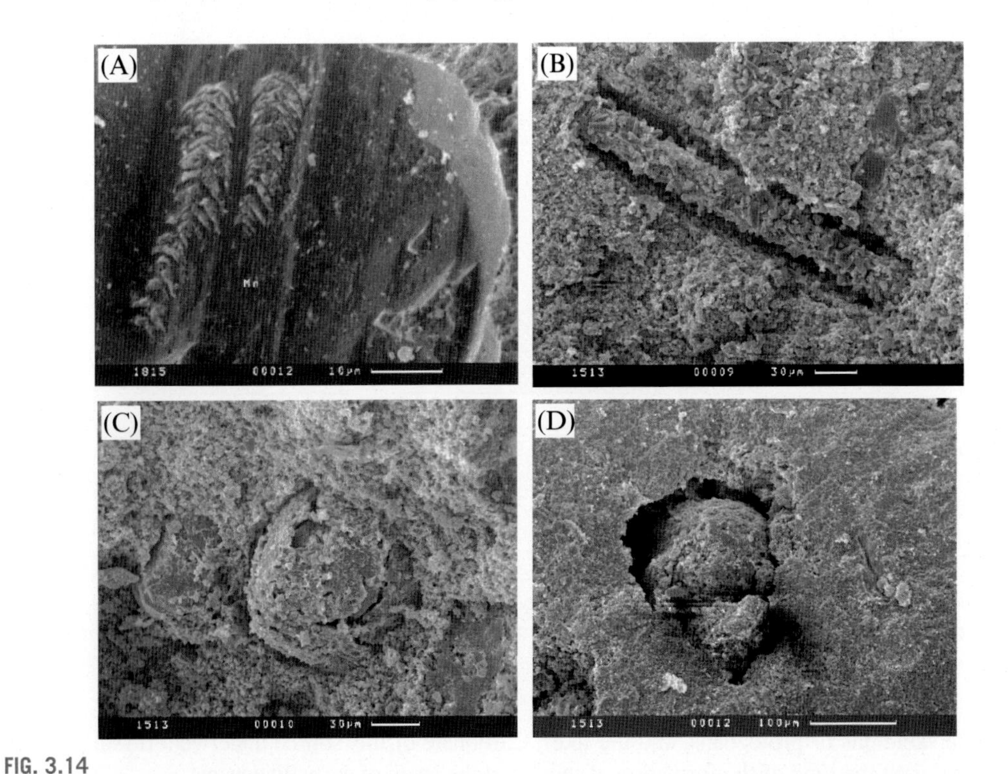

FIG. 3.14

Undeterminable small remnants of fauna composed of manganese carbonate rocks, Nikopol deposit, Alexandrovsky quarry (SEM authors: E.A. Zhegallo, E.L. Shkolnik). (A) Probe 1815 (5/83) and (B)–(D) probe 1513.

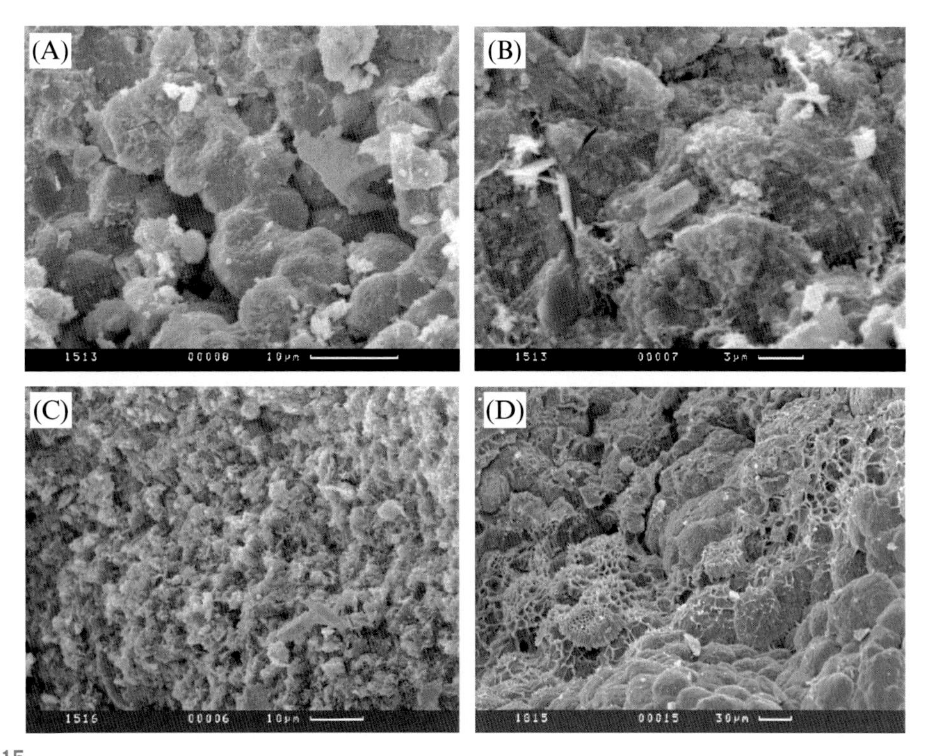

FIG. 3.15

Organogenic-microbial texture in manganese carbonate ores of the Nikopol deposit (SEM authors: E.A. Zhegallo, E.L. Shkolnik). (A) and (B)—Probe 1513; different sections at different magnifications; (C) probe (sample) 1516; and (D) probe (sample) 1815.

samples is insignificant (a few decimeters), the isotopic composition of the carbon in them is rather sharply differentiated (by 5–10‰). Substantial variations are noted as well within the boundaries of a single lump of rock (sample 713, analyses 2256–2258). This serves as evidence that the carbon dioxide at the moment of the formation of manganese carbonate within an ore horizon was not homogeneous in terms of isotopic composition. Its composition was determined entirely by local conditions, especially by the ratio between carbon of initially sedimentary origin and carbon of biogenic (C_{org}) origin.

This result is to some degree confirmed by the finding in ores from certain sequences, as noted, of a direct correlation in the distribution of contents of MnO, and an inverse correlation with the contents of CaO, with the isotopic composition of their carbon (Fig. 3.13). This correlation consists of the fact that ores with maximum contents of manganese are characterized as a rule by the lowest $\delta^{13}C$ values, while the highest concentrations of calcite are noted in ores with the heaviest isotopic composition of carbon.

In certain cases is also noted a correlation of $\delta^{13}C$ values with the contents of CO_2 in a rock—for example, in ores of the Aleksandrovskii quarry and of well 1279 (Fig. 3.17).

Evidently the observed regularity holds true only in conditions where the content of isotopically heavy carbonate, balanced with marine water, of shells of mollusks and other organisms in sediments

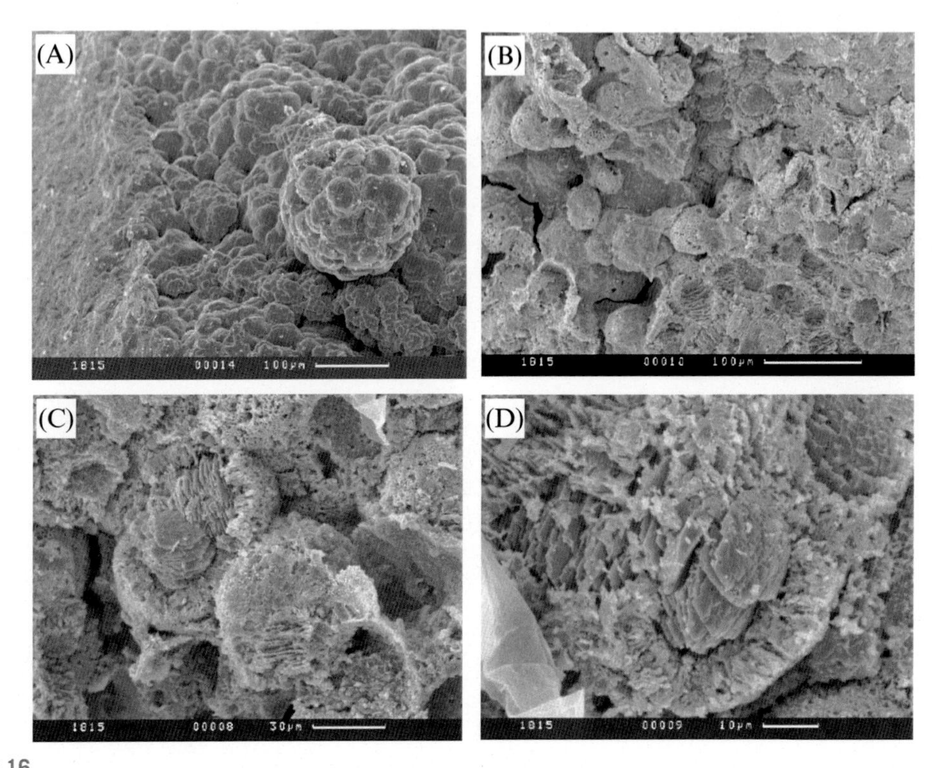

FIG. 3.16

Presumably microbial texture in manganese carbonate of the Nikopol deposit (SEM authors: E.A. Zhegallo, E.L. Shkolnik). (A)–(D) Different parts of probe 1815 at different magnifications.

has been distributed evenly. In that case, all variations in the isotopic composition of carbon will be determined by the degree of participation in the structure of the ore matter by isotopically light carbon dioxide that formed due to the oxidation of organic matter during diagenesis. To all appearances, these conditions are not always observed in natural objects.

The isotopic composition of oxygen in rocky manganese carbonate ores is relatively stable and varies mainly within the interval of 28–31‰. Such $\delta^{18}O$ values are characteristic for carbonates with oxygen located in equilibrium with that of marine water at temperatures close to normal (10–20°C) (Friedman and O'Neil, 1977).

At the same time, the bleached friable ores are characterized by a lighter isotopic composition of oxygen. This is distinctly visible in the histograms (see Fig. 3.12) and the graph (see Fig. 3.11) in the replacement of experimental points toward lower $\delta^{18}O$ values. A reason for this isotopic shift is their later origin resulting from the transformation of rocky ores evidently at the stage of late diagenesis. Formed in the sediment in these conditions are authigenic clayey minerals associated with the heavy isotope of ^{18}O, while the interstitial waters of the zone of diagenesis are accordingly enriched by the light isotope (Savin and Epstein, 1970; Yen and Savin, 1976). For this reason, newly formed carbonate in such conditions will be characterized by lower $\delta^{18}O$ values.

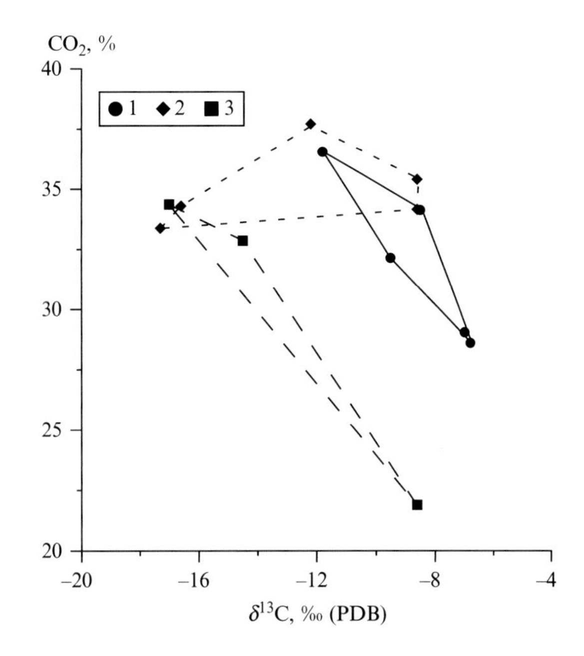

FIG. 3.17

Relationship between $\delta^{13}C$ values and CO_2 content in manganese carbonate ores of the Nikopol deposit. 1—Alexandrovsky quarry; 2—borehole 5512; and 3—borehole 1249.

Simultaneously, the isotopic lightening of late-diagenetic carbonates could be caused by an insignificant elevation of the temperature of their formation. Such a hypothesis is justified, insofar as an insignificant elevation of temperature can occur by measure of the accumulation of sediment in the marine basin and of the general submersion of the mass of sedimentary rocks.

Another potential reason for the lightening of the isotopic composition of the oxygen of friable manganese ores is their formation (more precisely, transformation) during the post-diagenetic stage of lithogenesis. A decrease in $\delta^{18}O$ values in this case could be a consequence of the presence in the samples of newly formed post-diagenetic calcite and rhodochrosite with lower isotopic ratios by comparison with those of initial diagenetic carbonate. Evidence for this is found in incidental data on the isotopic composition of secondary carbonate, which is characterized by a low isotopic composition of carbon and oxygen (see Fig. 3.11). Thus, an insignificant "elongation" of the field of isotopic ratios toward lower $\delta^{13}C$ and $\delta^{18}O$ values for various types of identified carbonates in the graph of Fig. 8.10 could be caused by the presence in the sample of isotopically light secondary carbonate.

A source for ore solutions leading to the precipitation of isotopically light carbonates could be waters of elisional origin (Pavlov and Dombrovskaya, 1993). For a clarification of the role of elisional solutions in the present case, additional isotope research is needed.

It can be supposed that it is precisely these two indicated processes that are responsible for the observed lightening of the isotopic composition of oxygen of friable bleached ores by comparison with that of "rocky" ores.

A significant proportion of carbonate pisolites are likewise characterized by $\delta^{18}O$ values lower than those of rocky ores. Taking into account the results of microscope study (in thin sections) and the isotope data, also for friable bleached ores, a post-diagenetic or elisional (catagenetic) origin for them can be considered. At the same time, for some of them the principal possibility cannot be excluded of an earlier formation in a desalinated shallow-water environment. It is possible as well that in the latter case the temperatures of formation of pisoliths could have been somewhat higher by comparison with those of the rocky ores. It is not possible to resolve this question on the basis of the isotope data presently available.

The results of mineralogical and chemical study, as well as the data on the isotopic composition of carbon and oxygen, indicate a diagenetic origin for the primary mass of manganese carbonate ores of the deposits of the Nikopol' manganese-ore basin. In certain cases, these were altered during the stages of late diagenesis and evidently during catagenesis.

This result closely agrees with the prevailing view of a sedimentary-diagenetic origin for manganese deposits in the south of the former USSR, as expostulated by Strakhov et al. (1968) and in an entire range of works of other researchers (Griaznov, 1956, 1960, 1967; Griaznov and Selin, 1959; Shniukov, 1962; Nikopol'skii et al., 1964; Varentsov, 1963, 1964; Varentsov et al., 1967). Regarding the diagenetic nature of the Nikopol' manganese carbonate ores, evidence is provided by (1) the abnormality of the development of carbonates among the terrigenous-clayey marine strata of the Oligocene; (2) the concretional character of the ores; and (3) the isotopic composition of the carbon and oxygen of the Nikopol' ores, which is characteristic for carbonates of diagenetic origin (Gautier and Pratt, 1986; Longstaff and Ayalon, 1986; Staley, 1986).

Thus, the results of the study of the isotopic composition of the Nikopol' carbonates of manganese ores indicate that against the background of a relatively narrow range of variations in values of $\delta^{18}O$ (26.4–31.8‰), values of $\delta^{13}C$ vary within a fairly wide interval—from −24.6‰ to −4.9‰. Such values of the isotopic composition of carbon are found in the intermediary range of $\delta^{13}C$ values, limited on the one hand to carbon of marine sedimentary origin, and on the other hand to carbon of biogenic (C_{org}) origin. This indicates that the carbon dioxide of manganese carbonate ores was produced due to the oxidation of organic matter in the sediment and the solution of organogenic carbonate (principally of mollusk shells). Their ratio is dependent upon local conditions and fluctuates within broad margins.

The formation of manganese carbonate ores occurred in disequilibrium between the isotopic composition of the carbon and that of dissolved bicarbonate of marine water. This provides evidence for their diagenetic origin and is a characteristic particularity of diagenetic carbonates.

The data on the isotopic composition of oxygen allow us to propose that dark-gray rocky ores by comparison with altered ores were produced at an early stage of diagenesis with the participation in their formation by oxygen in equilibrium with marine water, which is characteristic for the upper layers of the sediment (the zones of early diagenesis). The alteration of these ores, expressed in their bleaching, leaching, loosening, recrystallization, and the filling of their voids (cells, pores, fractures), by secondary crystalline carbonate occurs during the post-early-diagenetic (or catagenetic) stage(s) of life of the sediment in conditions of circulation of subterranean waters, possibly at higher temperatures and lacking a connection with natural marine waters, or later with participation by transformed solutions with a lighter isotopic composition of oxygen.

3.2.1.2 Mangyshlak deposit

The Mangyshlak manganese deposit is located in western Kazakhstan (eastern Caspian region) on the southern slope of the Karatau range. In the structural regard, it belongs to the northern flank of the Chakyrgan syncline, accompanied by the Karatau anticlinorium.

Geological setting, mode of occurrence, and types of manganese rocks and ores

Participating in the geological structure of the region are predominantly Cenozoic deposits (Fig. 3.18). More ancient rocks of Cretaceous, Jurassic, and Permian-Triassic age are exposed in the near-axial part of the Karatau anticlinorium. Manganese ores coincide with Oligocene deposits, the thickness of which in the region of the deposit constitutes approximately 80 m. In their cross section are differentiated three members (from bottom to top): (1) bluish-gray sandstone marls; (2) gray and bluish-gray sands and siltstones with the ore-bearing horizon at the top; and (3) gray, brownish-gray, and red-brown clays with inclusions of limonite and gypsum. These are respectively named the Uzunbass (Lower Oligocene), Kuyuluss, and Kendzhalin suites (Middle Oligocene) (Stoliarov, 1958). Above

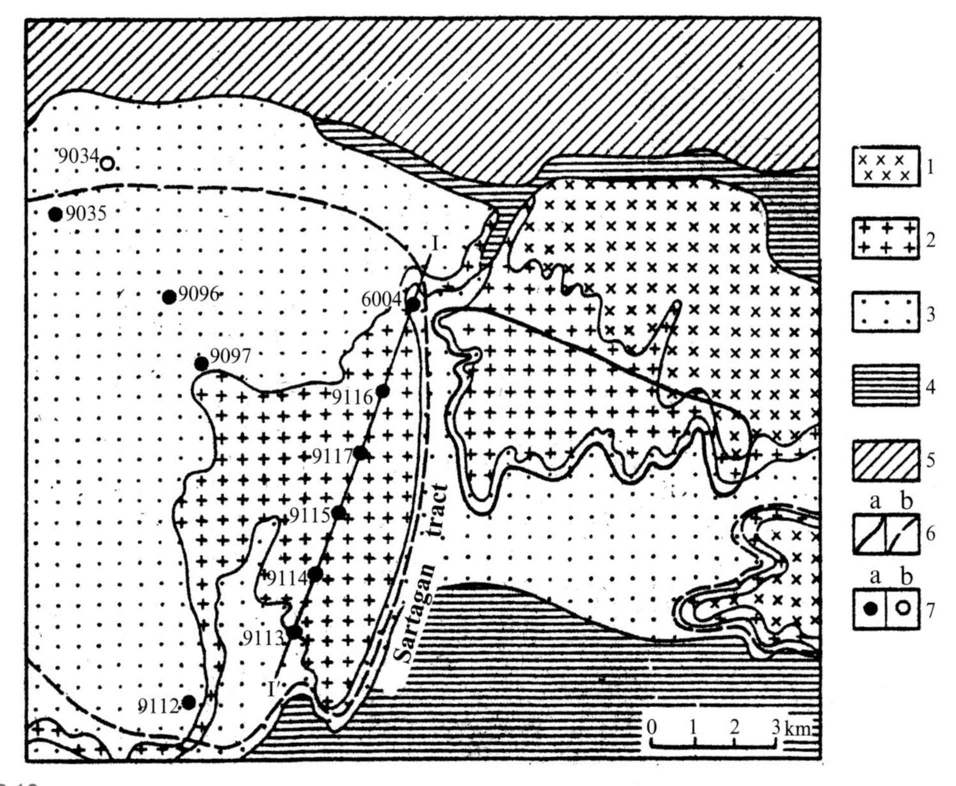

FIG. 3.18

Schematic geological map of Mangyshlak manganese deposit (fragment of a map, compiled by V.I. Lobzhanidze, 1983, as amended Zh.V. Dombrovskoy). 1—Upper Miocene, Sarmatian stage (N_1s): shell limestone, marls, clays, conglomerates at the base; 2—Middle Miocene, the Tortonian stage, Karagan and Konka horizons ($N_1kr + kn$): calcareous clays, sandstones, marls, conglomerates; 3—Oligocene, Kendzhala, Kuyulus, and Uzunbas suites, not subdivided (Pg_3): clays, silts, sands, manganese-bearing interlayers; 4—Eocene and Paleocene, not subdivided ($Pg_1 + Pg_2$): organogenic limestones, marls, sandstones; 5—Cretaceous, Danian stage (K_2d): organogenic-detritus limestones, chalk; and 6—contours of productive members (a—Mangyshlak deposit and b—areas enriched with manganese beyond the bounds of the deposit); and 7—boreholes (a—ore, b—oreless).

are transgressively superimposed Middle and Upper Miocene and Pliocene deposits with a general thickness of approximately 100 m. The north-eastern part of the deposit is destroyed by washout; in the south-west is observed a pinch-out of the ore body. In many places, the ore body is exposed by deep ravines.

Deposits of the ore-hosting Kuyuluss suite over the entire area of the deposit, with the exception of the pinch-out areas, are represented by gray friable sandstone-siltstone rocks. Clastic material here constitutes up to 70% of the volume of the rock; it is carbonaceous, barely rounded. All indications provide evidence of its arrival with the nearest landmass, which at the time of the Kuyuluss represented an island extended in a north-westerly direction (Ianshin, 1950). Clastic grains are represented principally by feldspars (40–55%), quartz (30–45%), and clasts of various rocks (10–15%). Glauconite is present in a significant quantity (10–15%). Clayey cement is composed of montmorillonite.

In pinch-out areas, rocks of the Kuyuluss suite are represented by compacted carbonate-clayey siltstones and marls with an impurity of sand. In the direction of the pinch-out of the ore body, the quantity of clastic material and its dimensions decrease while the content of clayey-carbonate cement increases.

An ore mass with a thickness reaching 20 m is associated with the Kuyuluss suite and represents a series of several (up to 12) ore layers superimposed among the host rocks (Figs. 3.19 and 3.20). Ore matter occurs in the form of isolated nodules with dimensions from 5 to 60 cm, which are frequently situated in a layer in a single row, as well as in the form of an earthy mass, occasionally forming seams with a thickness reaching 1 m and containing separate concretions. Ore material plays the role of a basal cement for the sandstone-siltstone rocks. Beyond the boundaries of the deposit is observed a very sharp and rapid (within a distance of 100–300 m) pinch-out of the productive horizon: the number of ore layers contracts to one or two and the size of the nodules decreases from the amount indicated down to a few millimeters.

Manganese ores in the Mangyshlak deposit are represented by carbonates and oxides of manganese, or by a mixture of the two. Accordingly, manganese ores are delineated into carbonate, oxide, and oxide-carbonate types.

Oxide-carbonate and oxide types of ores were produced, as shown by Betekhtin (1946), and subsequently by Strakhov et al. (1968), as a result of hypergenic oxidation of carbonate ores. They are distributed in the raised north-western part of the deposit and along the ravines exposing the ore body.

Carbonate ores are represented by dense nodules enclosed in sandstone-siltstone material or by friable carbonate matter replacing the clayey cement of the host rocks (Fig. 3.21). The nodules are close in color to that of the host rocks—gray-colored—and are distinguished from them by elevated density (Tables 8.3 and 8.4). Carbonate ores (Table 3.3) are characterized by the variable composition of their components. In part, concentrations of $MnCO_3$ and $CaCO_3$ vary within a broad range, while $FeCO_3$ and $MgCO_3$ are practically absent. The ore matter is cement. Its quantity in the rocks in the north-eastern part of the deposit constitutes 20–40%, while in the pinch-out areas, where the quantity of terrigenous impurity decreases to 20% or less, the quantity of ore carbonate grows to 83% (Table 3.3, sample 14-5).

According to the data of E.S. Tikhomirova and E.V. Cherkasova (Tikhomirova, 1964; Tikhomirova and Cherkasova, 1967), carbonate ores are composed of calciorhodochrosite with an impurity of manganocalcite and calcite, and in the pinch-out of the ore horizon by manganese calcite with an impurity of manganocalcite and relicts of calciorhodochrosite. Furthermore, the value of $MnCO_3/CaCO_3$ decreases from 1.3 to 4.4 down to 0.19. Noted in the ore member is a total absence of ferruginous carbonates.

With the total pinch-out of the ore horizon on the south flank of the Chakyrgan syncline, carbonate nodules are composed of calcite with an impurity of manganese calcite and dolomite.

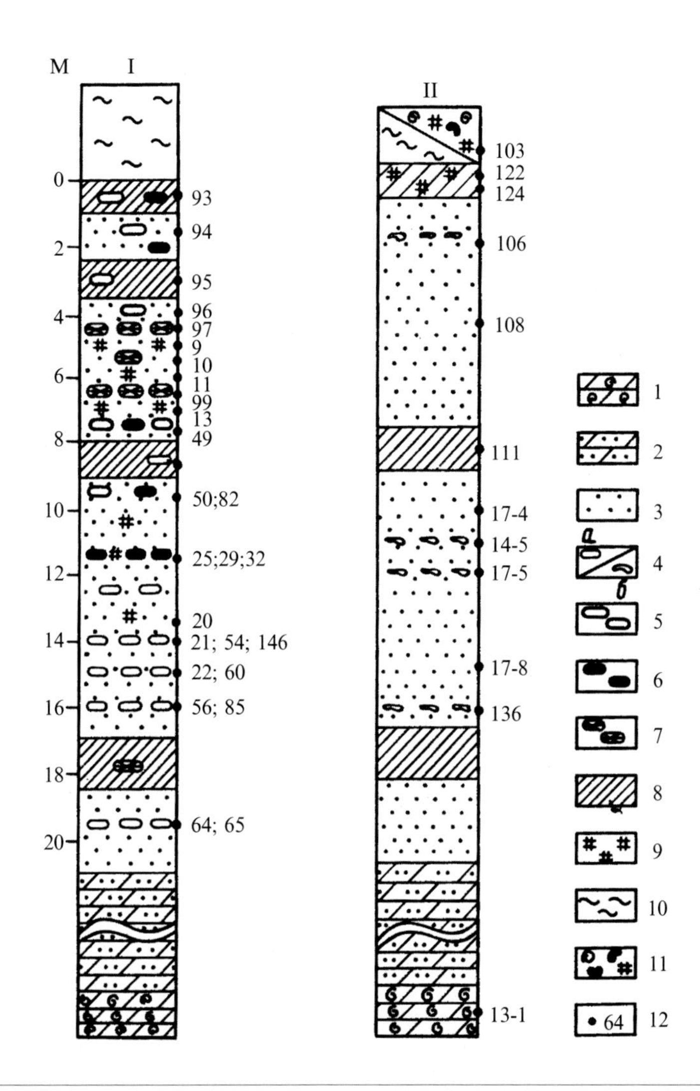

FIG. 3.19

Typical composite geological columns of the manganese-bearing beds of the Mangyshlak deposits (compiled by Zh.V. Dombrovskoy). I—Within the area deposit area and II—at the pinch-out of the ore body. 1—White and pink limestones, organogenic, chalk-like, Adaevsk suite (Pg_2^3ad); 2—blue-gray marl, sandy, Uzunbassk suite (Pg_3^1us); 3–9—ore-bearing horizon Kuyulussk suite (Pg_3^2kl): 3—sand-silty deposits, gray and bluish-gray, 4—carbonate nodules (a—predominantly calcite, b—small nodules), 5—oxide-carbonate nodules, 6—oxide nodules of the geode type, 7—ferruginous oxide nodules, 8—bedded earthy manganese ores, 9—superimposed calcitization; 10–11—overlapping rocks: 10—brownish-gray and brown limonized and gypsumited clay Kendzhalinsk suite (Pg_3^2kn), 11—conglomerate with calcareous cement with fragments of oxide-manganese ore, Tortonian stage, Konsk and Karagan horizons (N_1^2kn + kr); and 12—sample selection site on isotopic analysis and its number.

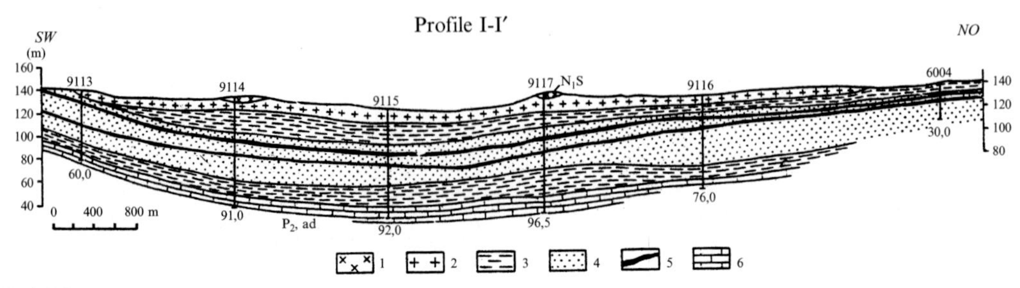

FIG. 3.20

Geological and lithological cross section along profile I-I′ (complied by Zh.V. Dombrovskaya according to data from the Mangyshlak Geological Expedition). 1—Shell limestone (N_1); 2–4—deposits: 2—clay-carbonate, 3—clay, 4—sand and silt; 5—horizons containing manganese-ore carbonate nodules; 6—chalk-like limestones (Pg_2ad).

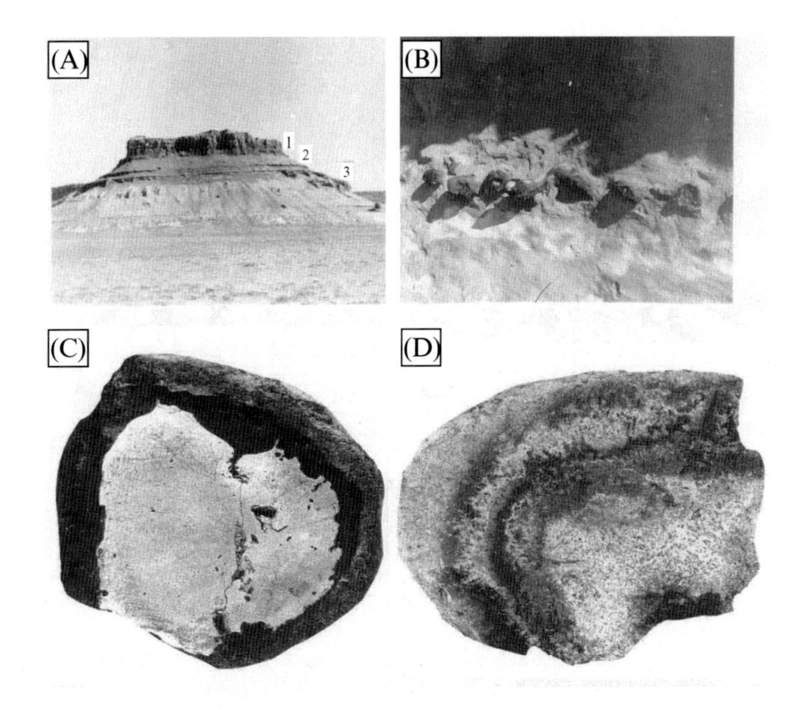

FIG. 3.21

Structure manganese-bearing horizon and types of oxidized manganese nodules on the Mangyshlak deposit (photos by Zh.V. Dombrovskaya). (A) location of the ore horizon among the Oligocene sediments: 1—brownish-gray and brown clays, Kendzhalinsk suite (Pg_3^2kn); 2—gray sand and silt deposits contain manganese concretions, Kuyulussk suite (Pg_3^2kl); 3—blue-gray marls, and Uzunbassk suite (Pg_2^3us); (B) concretionary single layer in the surrounding sand-aleurite rocks—part outcrop of the ore horizon; (C) concentrically zonal oxide-carbonate nodules with the carbonate core; and (D) oxidized “mottled” carbonate concretion with low manganese content.

Table 3.3 Chemical Composition (Mass%) of Rocks From the Mangyshlak Deposit (Compiled by Zh.V. Dombrovskaia)

Component	49	93	94	95	96	29	32	14-5	17-4
SiO_2	36.7	39.8	43.7	35.6	34.2	65.2	50.9	10.6	60.6
Al_2O_3	4.97	8.62	6.00	5.91	6.29	8.36	6.89	2.65	14.1
Fe_2O_3	0.56	3.39	2.46	2.00	2.85	2.55	2.72	Not def.	4.17
MnO	28.4	4.25	9.90	7.95	19.2	3.00	3.59	32.3	4.30
MnO_2	2.59	7.34	9.20	18.0	14.4	0.74	10.8	Not def.	None
MgO	0.89	1.86	1.29	1.18	1.58	1.00	1.03	2.02	2.04
CaO	3.78	12.3	6.86	8.53	1.99	5.42	7.77	16.9	2.36
Na_2O	1.04	4.00	2.19	2.81	1.87	2.34	2.30	Not def.	1.75
K_2O	1.56	2.20	2.01	1.74	1.81	3.17	2.75		3.33
$H_2O\uparrow^-$	1.17	1.74	3.52	2.97	1.98	1.24	1.95		2.02
H_2O^+	0.95	2.00	2.25	2.46	1.13	1.46	2.97		2.78
CO_2	15.9	11.9	11.9	9.78	11.1	1.11	1.15	33.7	2.08
P_2O_5	1.43	0.16	0.16	0.3	0.24	3.50	4.36	Not def.	0.31
Total	99.9	99.5	99.4	99.3	99.4	99.1	99.1	98.0	99.8
$MnCO_3$	41.5	6.89	16.0	10.7	28.6	0.78	0.78	52.3	5.43
$CaCO_3$	–	19.7	10.7	12.5	1.79	1.84	1.93	30.2	–
$FeCO_3$	–	–	1.00	–	–	–	–	–	–
$MgCO_3$	–	1.03	1.00	–	–	–	–	–	–0.87
Total	41.5	27.6	28.8	23.3	30.4	2.62	2.71	83.2	5.43
Manganoapatite	3.30	–	–	–	–	9.00	11.0	–	–

Lithologies: *Carbonates (49–96) and sandstone-siltstone (29, 32) cores from zonal oxide-carbonate concretions and (14-5, 17-4) from rocks of the Mangyshlak deposit.*
Analysts: *M.N. Kevbrina, E.N. Kulysova, L.S. Tsimlianskaia (IGEM AN USSR).*

The almost total absence of organic matter in the ores and host terrigenous rocks of the productive horizon, with numerous traces of past organic life, indicates intensive subjection to the processes of diagenesis.

Near the exposed surface, carbonate ores are oxidized to varying degrees (Betekhtin, 1946; Strakhov et al., 1968), resulting in concentrically zonal or spotted occurrences of Mn oxides (see Fig. 3.21E and F). Accordingly, Berdichevskaya and Rakhmanov (1981) have identified concretional-crust and lump-porous textural-structural subtypes of manganese ores. Observed here is a continuous series of transformations of carbonate ores into oxide ores through intermediary oxide-carbonate ores. As a result of the oxidation of rhodochrosite nodules are frequently formed zonal oxide-carbonate and oxide concretions of geodic type, in which the external part is composed of Mn oxides, and the core by Mn carbonates with a partial oxidation or friable sandstone-siltstone material with an almost total dissolution of carbonate (Sokolova et al., 1984). Along with sandstone-siltstone material here are commonly found an impurity of ochreous material, manganocalcite, and manganoapatite. The quantity of the latter here reaches 9–11% (see Table 8.3, samples 29 and 32).

Manganese-oxide zonal concretions in terms of structure and conditions of formation are similar to the ferruginous geodes first studied in detail by Pustovalov (1933). Sokolova and Dvorov (1987) proposed "caryolites" as their name (from the Greek for "nut").

Occurring stratigraphically below the layer of manganese-ore concretions are as a rule several concretional layers represented by calcite and containing hardly any manganese carbonate. In oxidized varieties are present in a small (4–6%) quantity Mn oxides in the form of black permeable concentrates. Calcite nodules in size and form are distinguished from manganese-ore concretions; the former have a simpler (discoid) form and as a rule are larger in size (in length up to 1 m or greater). Their surface is flat and smooth. The indicated differences serve as evidence of different conditions for the development of the upper and lower parts of the concretional cross section.

Still lower along the stratigraphic section, according to the data of Ianshin (1950), in many sections are observed a range of large concretions of sphaerosiderite and dense discs of siderite sandstone. However, these structures are developed beyond the boundaries of the ore field and are not included in our study.

Besides the concretional and primary scattered carbonate throughout the ore mass is noted secondary calcitization, manifested in the form of aggregates of crystalline calcite in the cement of the rocks hosting the ore concretions as well as in voids of oxidized manganese concretions. The quantity of calcite commonly constitutes several percentages, reaching in individual sections up to 20%; only in the north-western part of the deposit, where rocks of the ore horizon are overlapped by conglomerates of the Tortonian Stage of the Miocene, is a more intensive (40–50%) calcitization observed.

Overlapping conglomerates are also cemented by crystalline calcite, which provides evidence of a post-Tortonian manifestation of this process. Superimposed calcitization is most intensively manifest in the flanks of the Chakyrgan syncline.

In the composition of the Kuyuluss suite along the bore profile (wells 9113–9116, Fig. 3.18) are exposed two horizons of manganese carbonate concretions. The upper horizon with a thickness reaching 3 m crops out on the surface along the margins of the trough and in its central part submerges to a depth of 42 m (well 9115). The lower horizon with a thickness reaching 2 m along the margins of the trough occurs at a depth of 20–25 m and in its central part submerges to a depth of 50–52 m. The saturation of the ore horizons by concretions varies but on the whole is not high and constitutes 10–20%. In the inter-ore layers, separate concretions also occur.

Horizons of calcite concretions, characteristic for the lower parts of the sequence of the Kuyuluss suite, are absent within the boundaries of the Mangyshlak deposit itself in the described area. The content of Mn oxides in rocks of the upper horizon constitutes 0.5–3.3% and of the lower horizon 0.50–1.87%; only in nodules does it increase to 25–33% (Table 3.4).

Characteristics of the samples

Isotope research was conducted on rock material gathered in the Sartagan and Chakyrgan sections, predominantly in sections, exposed ravines, and partially from the core of wells. The distribution of the studied samples about the area of the deposit and in the vertical sequence is indicated in Figs. 3.18 and 3.19.

For the purposes of isotope research, rock material was collected that would be characteristic of all genetic and petrographic varieties of carbonates from the Mangyshlak productive manganese body. Among these, five groups were delineated.

Table 3.4 Chemical Composition (Mass%) of Manganese Carbonate Ores and Their Host Deposits (Compiled by Zh.V. Dombrovskaia)

Components	16-1p	16-4p	16-5p	17-4p	13-7	13-9	14-5r	16-4r	16-6r	17-4r	13-1
SiO_2	18.1	Not def.	15.2	10.8	51.7	49.5	52.9	41.2	55.8	53.4	14.5
TiO_2	Not def.		Not def.	Not def.	0.69	0.71	0.61	0.55	0.71	0.66	0.23
Al_2O_3	3.37		2.83	2.31	13.6	15.1	13.1	11.7	15.1	14.1	5.13
Fe_2O_3 total	2.68		2.00	2.23							
Fe_2O_3	Not def.		Not def.	Not def.	4.92	6.39	4.54	5.63	5.54	5.25	1.56
FeO					3.70	0.14	0.80	0.29	0.92	0.79	Not def.
MnO	24.9	26.9	29.5	32.9	0.45	0.58	4.08	9.18	0.50	2.93	0.39
MgO	2.23	Not def.	1.97	2.02	4.09	3.01	2.98	3.03	3.62	3.52	1.37
CaO	14.1		13.1	12.5	2.86	7.17	3.33	5.96	2.54	2.24	39.87
P_2O_5	0.59		0.29	0.47	0.17	0.13	0.19	0.22	0.18	0.25	0.30
Na_2O	1.22		0.81	0.67	1.04	1.07	1.01	0.71	1.11	1.01	Not def.
K_2O	0.96		0.78	0.61	2.84	3.04	2.71	2.12	3.08	2.85	
H_2O^+	Not def.		Not def.	Not def.	3.59	4.74	4.74	4.86	4.26	4.42	
H_2O-					1.88	2.23	2.70	3.30	3.60	3.56	
CO_2	28.6	37.5	30.6	32.5	6.60	5.35	3.75	10.40	2.60	2.95	33.30
Corg	None	Not def.	None	None	None	None	0.58	0.29	0.30	0.63	Not def.
Total	96.7	—	97.1	97.0	98.1	99.1	98.1	99.4	100.0	98.6	
$MnCO_3$	40.4	43.6	47.7	53.0	0.7	0.9	6.6	14.9	0.8	4.8	0.6
$CaCO_3$	25.74	Not def.	23.41	22.95	3.67	11.38	4.51	9.22	3.10	2.56	71.37
$MgCO_3$	Not def.		1.91	1.34	2.50	Not def.	Not def.	Not def.	1.51	Not def.	2.88
$FeCO_3$	3.64		3.33	3.86	5.96						Not def.
Total	69.8	75.0	76.4	81.1	12.9	12.3	11.1	24.1	5.4	7.3	74.9
Mn-apatite	2.79	Not def.	0.68	0.99	0.38	0.29	0.43	0.50	0.41	0.56	Not def.

Manganese carbonate ore (16-1p, 16-4p, 16-5p, 17-4p); host deposits of the Kuyuluss Suite (13-7, 13-9, 14-5r, 16-6r, 17-4r) and limestones of the Adaev suite (13-1).
Analysts: *M.I. Stepanets, S.P. Gordeeva, G.A. Granovskaia (GIN AN USSR).*

I. Manganese carbonate-ore nodules; over a large portion of the area their distribution is partially oxidized. An oxide layer is situated outside the concretions, is distinctly expressed, and has a thickness from several millimeters up to 2 cm. Inside persists a carbonate core containing a variable quantity of manganese oxides. The carbonate material is composed of cryptocrystalline aggregates cementing the sandstone-siltstone impurity. The quantity of noncarbonate impurity, that is the insoluble residue or i.r. (samples 49, 93–96), constitutes 58.53–76.75%; content of $MnCO_3$ varies from 6.89% to 41.47% and that of $CaCO_3$ from 1.79% to 19.69%. $MgCO_3$ is present in two samples in the quantity of 1.05–1.32%, while $FeCO_3$ is absent.

Carbonate matter of the manganese concretions is represented by rhodochrosite or a mixture of the latter with intermediary members of the rhodochrosite-calcite isomorphic series. In the heating curves is distinctly expressed an endothermic effect in the region of 600°C, indicative of a rhodochrosite composition of the material; occasionally additional endothermic effects in the region of 600–800°C provide evidence for an impurity of Ca-containing carbonates. Observed in X-rays are intensive reflections of rhodochrosite d_{1014} = 2.85 Å, as well as a low intensity of d_{1014} values equal to 2.90–3.01, indicating an impurity of mixed Mn-Ca-carbonates.

Friable sandstone-siltstone matter inside the oxide concretions of geodic type (Pustovalov, 1933) formed with an almost total decomposition of carbonate material (samples 29–32, Table 3.3) contains a small (~3%) quantity of carbonates, where $MnCO_3$ constitutes ~1% and $CaCO_3$ 2%. According to the data of X-ray analysis, carbonates are represented by manganocalcite with $MnCO_3$ content from 20% to 50%. Leached cores, according to X-ray analysis data and the calculations of chemical analyses, contain a significant quantity (9–11%) of manganoapatite.

II. Ore matter, developed in the western pinch-out of the ore body, is represented by small carbonate nodules and friable carbonate material scattered in the clayey-siltstone mass. Carbonate matter is cryptocrystalline, its distribution in the rock is uneven, and it frequently forms coagulations, with sections transitioning into small nodules. Carbonate matter in the small nodules (sample 14-5) constitutes 80–90%, in which $MnCO_3$ as a rule predominates over $CaCO_3$. According to X-ray data (sample 14-5), carbonate matter is represented by a mixture of rhodochrosite and manganocalcite with content of $MnCO_3$ from 20% to 50%. Friable carbonate matter is composed most frequently of calcite (samples 14-4, 17-8), but also occurring are manganocalcite with a $MnCO_3$ content of approximately 50% and rhodochrosite (Table 3.3, sample 17-4).

III. Carbonate in rocks hosting oxidized ore concretion is found in insignificant concentrations. The primary friable manganese-containing carbonates here are oxidized and are distinguished by a black sooty outlook. Secondary carbonates in the form of large-grained aggregates cement small sections of the host rock and fill the voids of leaching in the oxidized concretions. Content of carbonates in the rock (samples 9, 11, 13) constitutes 4.41–7.96% and is represented predominantly by calcite. Content of $MnCO_3$ constitutes no >1%.

IV. Carbonate concretions superimposed in the lower part of the concretional sequence. The content of carbonate matter constitutes 20.61–28.49% (samples 10, 21, 22, 65), the remainder belonging to a terrigenous impurity. In the composition of the carbonates, $CaCO_3$ sharply predominates (18.12–28.49%), while content of $MnCO_3$ commonly is less than 1%. In distinction from manganese-ore nodules, present here is a small (1.08–2.01%) quantity of $FeCO_3$. In heating curves (samples 10, 21, 22, 54, 60, 64, 65), there are broad asymmetrical endothermic effects in the 600–850°C region with deep deflections at 800°C, which provide evidence for the presence of calcite. The asymmetry of the left flank indicates an impurity of manganese-containing carbonates in the sample. In X-rays of the samples a reflection of $d_{1014}=3.019$ Å with an intensity of 10 likewise indicates the carbonate's calcite composition.

V. Carbonate cementing the oxidized eroded ores on the contact with the overlapping Middle Miocene deposits (Tortonian Stage N_1^2) is observed in the form of a solid coarse-crystalline cement, among which oxidized ores are maintained in sections irregular in form. Carbonate content (samples 122, 124) constitutes 32.28–45.33%, predominantly in the form of $CaCO_3$. In sample 122, it was determined that $MnCO_3=2.37\%$ and in sample 124 that $MnCO_3=1.85\%$. In the heating curves, there are broad endothermic effects within the 600–850°C region with a maximum at 800°C, indicating a calcite component in the sample. The asymmetry of the left flank of the effect is caused by an impurity of manganese-containing carbonates.

Analyzed for comparison were carbonate rocks from the underlying ore strata of more ancient (limestone Pg_2^3 ad, sample 13-1) and overlapping younger (carbonate cement of the supraore conglomerate, N_1^2 kr+kn, sample 103) rocks. The latter also contain clasts of nodules of oxidized carbonate ore.

Besides the rocks developed in the zone of a particular deposit, we separately studied carbonate material from the western periphery of the Mangyshlak deposit—from the pinch-out zone of the ore horizon. It has been proposed that in this zone ore-genesis of manganese is manifested in a rudimentary (embryonic) form and that processes of diagenesis occurred significantly more weakly by comparison with the areas of development of maximum thickness of the ore body. The selection of samples here was produced from the core of the wells in bore profile I-I, cut by the Mangyshlak geological-exploration crew in 1981 (see Fig. 3.18). These wells were excavated with the aim of tracing manganese ores to the west from the Mangyshlak deposit. The sweep of the profile constitutes 8.5 km, while the depth of the well shafts is 30–90 m.

For purposes of isotope research, in the western pinch-out zone were gathered rock material from the ore strata, including various petrographic and genetic varieties of carbonates. The following groups were identified: (1) solid carbonate-manganese-ore nodules; (2) friable carbonate material from their host rocks; and (3) carbonates from the barren layers of the ore strata. Analyzed for comparison and control were limestones from the lower-lying Adaev suite of the Eocene.

1. Carbonate Mn-ore nodules with dimensions from 0.1 to 7–10 cm are commonly represented by structures irregular in form with a cavernous surface. The form of the smallest of these is nearly spherical. They contain clastic materials identical with those contained in the host rock—predominantly quartz and feldspars, more rarely micas and zeolites. Clayey minerals are absent.

In carbonate nodules, the content of carbonates constitutes 70–90% (see Table 3.4). These are represented by a mixture of minerals of the calcite-rhodochrosite isomorphic series. Serving as evidence for this are the high contents of Ca and Mn as well as the broad endothermic effects in the 500–800°C temperature range in thermograms and X-ray data. These are commonly members of the right side of the calcite-rhodochrosite isomorphic series—that is, calciorhodochrosites with a variable quantity of $MnCO_3$ (43–100%)—but also occurring are manganocalcites with $MnCO_3$ content of 20–50%.

2. Friable carbonate material from rocks of the Kuyuluss suite enclosing carbonates of Mn-ore nodules is found in close mixture with clayey material, in which it is evenly scattered and forms compactions, transitioning in sections into nodules. Carbonate-clayey material encloses clastic grains of predominantly siltstone dimensions. The content of carbonate material in the host

rocks constitutes from 5% to 24% (Table 3.4). Data of chemical and spectral analyses indicate that the friable carbonate material included in the composition of the two indicated ore horizons includes manganese. Manganese-ore carbonates are composed of a mixture of minerals of the calcite-rhodochrosite isomorphic series (see Tables 3.3 and 3.4). In diffractograms in fractions less than 0.01 mm there are several systems of lines particular to members of this isomorphic series (d_{1014}=3.03, 2.994, 2.938, 2.866 Å). In heating curves of the host rocks are observed endothermic effects of clayey minerals at 100°C and 550°C and a broad endothermic effect with several conditions in the 550–750°C region, which is also characteristic for minerals of the rhodochrosite-calcite isomorphic series.

3. Friable carbonate material from the inter- and subore layers of the Kuyuluss suite. Rocks of the Kuyuluss suite within the boundaries of ore layers also contain carbonate material; however, its Mn content constitutes only tenths of a percent (see Table 3.3). While this exceeds Clarke abundance ratios of Mn for this type of rock, it does not form ore concentrations. The basic component of these carbonates is calcite.
4. Late Eocene chalk-like (pelitomorphic) limestones of white and pinkish color. They are known under the name "*beloglinka*" ("white-clay") and are widely distributed not only in western Kazakhstan, but also in the North Caucasus, and serve as a reliable horizon marker. These limestones frequently contain an impurity of clayey and siltstone material and transition into marls. In the studied sequence in these rocks sharply predominates a carbonate component (see Table 3.4, sample 13-1). Content of calcite constitutes approximately 75%.

Isotope data and their discussion

All results of isotope research are recorded in Tables 3.5 and 3.6 and are displayed in the graph of Fig. 3.22. From these data, it follows that the scattering of values of the isotopic composition of carbon and oxygen occupies a wide range. For rocks of the deposit, $\delta^{13}C$ values vary within the boundaries of –32.9‰ to 1.4‰, and those of $\delta^{18}O$ from 15.8‰ to 30.9‰; that is, they overlap with regions of isotope characteristics particular to polar groups of carbon matter of the Earth's crust—sedimentary marine carbonates and carbon of organic origin (hydrocarbons). At the same time, despite the wide boundaries of variations in isotope ratios, the separate groups of studied matter are characterized by a stability of isotopic composition.

Carbonate material of the marls hosting the small ore nodules (sample 17-8) and clayey-carbonate siltstones (sample 17-5) in the western pinch-out zone of the ore strata of the deposit, as well as the underlying Upper Eocene limestones (sample 13-1), are characterized by isotope ratios of carbon and oxygen inherent to marine carbonates of normal-sedimentary origin. Moreover, it must be noted that all remaining analyzed varieties of carbonates of the deposit are characterized by an isotopic composition distinct from that of marine carbonates.

A characteristic of the distribution of isotope ratios in the manganese carbonates of the deposit is that they are all distinguished by a significant scattering of $\delta^{13}C$ values, namely from −32.9‰ to −4.7‰. Additionally, their isotopic composition of oxygen remains fairly constant—values of $\delta^{18}O$ vary from 27.6‰ to 30.9‰.

Also characteristic for rocks of the western pinch-out zone is a broad spectrum of isotope ratios of carbon as well as oxygen; the identified groups of carbonate matter are substantially distinguished from one another.

Table 3.5 Isotopic Composition of Carbon and Oxygen in Rocks of Mangyshlak Manganese Deposit

Number of Analysis	Number of Sample	Characteristics of Sample	$\delta^{13}C$, ‰ PDB	$\delta^{18}O$, ‰ SMOW
1	2	3	4	5
Manganese-ore concretions and carbonate cores from zonal oxide-carbonate concretions				
1884	49	Carbonate core; rhodochrosite	−19	29.6
1999	50		−32.9	30.9
2002	93	Carbonate core; calcite	−1.9	15.8
		Carbonate core; rhodochrosite	−16.2	26
2611	94	Carbonate core; calcite	−9.1	22.6
		Carbonate core; rhodochrosite	−11.9	22
2003	96	Carbonate core; rhodochrosite	−20.5	29.4
1993	29	Weathered core	−8.7	19.8
1995	32		−7.8	27.8
2611	146	Carbonate (Mn-Ca) concretion	−10.1	19
2612	95	Carbonate (Mn-Ca) concretion; calcite	−5.6	18.9
		Carbonate (Mn-Ca) concretion; rhodochrosite	−6.9	27.4
Carbonate matter from rocks and concretions on tapering of ore body				
2616	106	Carbonate small nodules; calcite	−4.2	26.5
		Carbonate small nodules; rhodochrosite	−4.7	28.4
2617	108	Carbonate concretion; calcite	−8.4	27.4
		Carbonate concretion; rhodochrosite	−8.8	30.6
2618	111	Carbonate concretion; calcite	−9.9	27.9
		Carbonate concretion; rhodochrosite	−14.5	27.6
2008	14-5	Carbonate small nodules	−7.5	28.8
2009	17-4		−9.7	23.7
2624a	17-5	Siltstone with carbonate cement	−0.8	27.6
2624b	17-8	Marl, silty	0	27.1
2620	136	Carbonate small nodules; calcite	−4.4	19.3
Calcite from cement of rocks enclosing oxide concretions				
2596	9	Siltstone with calcite (4%) in cement	−5.5	18.4
2598	11	Siltstone with calcite (8%) in cement	−6.8	16.9
2599	13	Siltstone with calcite (7%) in cement	−5.8	17.9
2600	20	Calcite siltstone	−5.7	19
2603	25		−6.4	18
Calcite concretions from the lower part of the cross section				
2597	10	Calcite concretion with manganese oxides	−5.6	18.9
2601	21		−5.4	19.3

Continued

Table 3.5 Isotopic Composition of Carbon and Oxygen in Rocks of Mangyshlak Manganese Deposit—Cont'd

Number of Analysis	Number of Sample	Characteristics of Sample	$\delta^{13}C$, ‰ PDB	$\delta^{18}O$, ‰ SMOW
1	2	3	4	5
2602	22	Calcite concretion ($CaCO_3$—35%)	−5.2	19
2604	54		−3	22.6
2605	56		−3.9	19.6
2606	60	Calcite concretion ($CaCO_3$—40%)	−5.9	18.8
2607	64	Calcite concretion ($CaCO_3$—30%)	−6.9	16.6
2608	65	Calcite concretion ($CaCO_3$—28%)	−7.9	20.1
2610	85		−7	17.7
2614	99		−6.9	19.4
Carbonate from rocks and oxide ores of the north-western part of the deposit				
2619	122	Calcite cement of manganized rocks	−4.4	18.7
2005	124		−4	20.3
2615	103	Calcite from cement of supraore conglomerate	−4.3	20.8

The observed wide scattering of $\delta^{13}C$ values in the ores is evidence that the system in which the formation of manganese carbonates occurred was neither homogeneous in terms of carbon nor situated in isotopic equilibrium. The lowest isotope ratios appear to be analogous to those of carbon of organic origin. At the same time, a portion of the samples with high $\delta^{13}C$ values approach the range characteristic for marine sedimentary carbonates (see Fig. 3.22). This circumstance serves as evidence that carbon dioxide in the carbonate ores has a heterogeneous nature and represents the result of a mixture of CO_2 from two sources—isotopically light carbon of oxidized organic matter in the zone of diagenesis in the upper layers of the ooze sediment and isotopically heavy carbon imported from marine sedimentary carbonates.

The high values of the isotopic composition of carbon in the manganese carbonates and the narrow range of their variations are evidences that the environments of carbonate precipitation at low temperatures were isotopically balanced with marine water (Friedman and O'Neil, 1977).

Wide variations in isotopic composition in the manganese carbonates enable us to notice the following pattern. The area of the highest $\delta^{13}C$ values is occupied only by manganese carbonate nodules of the western pinch-out zone. At the same time, the oxide-carbonate zonal nodules of the central parts of the deposit are concentrated in the range of low $\delta^{13}C$ values. This circumstance is evidence for certain geochemical differences in their formation. It can be supposed that the increasing heaviness of the isotopic composition of carbon of the manganese carbonate nodules developed in the western pinch-out zone is associated with the higher total carbonate content of the host terrigenous-clayey sediment and an insignificant participation by organic material in the diagenesis process. On the contrary, in the near-shore lateral zone of the sedimentary paleobasin, where predominantly oxide-carbonate ores are developed, the initial carbonate content of the sediment (at the moment of sedimentation) was insignificant. Moreover, the level of activity of organic matter during the diagenesis process in this zone was much higher. This was reflected in the observed

Table 3.6 Isotopic Composition of Carbon and Oxygen of Carbonate Matter From Rocks of the Western Periphery of the Mangyshlak Manganese Deposit

		Content (wt%)		Calcite		Mn Carbonates	
Number of Analysis	Number of Sample	Mn	Ca	$\delta^{13}C$, ‰ PDB	$\delta^{18}O$, ‰ SMOW	$\delta^{13}C$, ‰ PDB	$\delta^{18}O$, ‰ SMOW
2988	14-5p	*n*–	*n*/*n*–	−6.1	30.2	−6	30.5
2888	14-6p	*n*–	*n*–	−5.9	28.7	−5.6	29.3
2987	16-1p	24.92	14.05	−5.5	30.3	−6.4	30.3
2900	16-4p	26.94	Not detected	−5.1	30.2	−5	30.2
2991	16-5p	29.47	13.1	−6.5	30.4	−6.9	30.5
2901	16-6p	*n*–	*n*–	−6.9	27.3	−7	30.1
2891	17-4p	32.9	12.49	−7.2	31.2	−7.1	30.6
Friable Mn-containing carbonate from rocks enclosing Mn-ore carbonate nodules							
2880	13-7	0.45	2.86	−5.9	29.9	−4.6	26.2
2877	14-3r	*n*	*n*–/*n*	−7.3	30.1	−4.9	28.5
2986	14-5r	4.08	3.33	–	–	−5	32.1
2887	14-6r	*n*	*n*–/*n*	−6.7	29.3	–	–
2890	16-4r	9.18	5.96	−7.1	30	−6.8	31.3
2902	16-6r	0.5	2.54	−8.4	26.1	–	–
2993	17-4r	2.93	2.24	−6.6	28	–	–
Friable carbonate from inter- and subore layers of the Kuyuluss suite							
2881	13-8	0.3	*n*–	−1.7	26.4	–	–
2884	13-9	0.58	7.17	0.5	27.3	–	–
2899	14-9	0.*n*–	*n*–/*n*	−0.1	24.8	–	–
2892	17-3	0.*n*–	*n*–/*n*	−0.3	24.2	–	–
2624a	17-5	0.0*n*+	*n*–	−0.8	27.6	–	–
2895	17-7	0.3	*n*–/*n*	1.2	30.4	–	–
2684b	17-8	0.*n*–	*n*	0	27.1	–	–
Limestone subore (Eocene)							
2623	13-1	0.39	39.87	1.4	27.9	–	–
2896	17-10	0.*n*–	*n*+	2.2	29.7	–	–
2897	1712	0.*n*+	*n*+	1.8	30.4	–	–

Contents of Mn and Ca are cited according to the data of chemical (Stepanets M.I., Granovskaia G.A.) and semiquantitative spectral (A.I. Galudzina, O.S. Zaitseva) analyses: $n^- = 1–3\%$, $n = 4–6\%$, $n^+ = 6\%$, *– = not measured.*

correlation between the isotopic composition of carbon and the contents of manganese in the ores (Figs. 3.23A, B and 3.24). Therefore, the maximum volumes of ore matter are found precisely in this near-shore section of the lateral profile of the productive member, where the diagenesis processes occurred most intensively.

Calcites from the cement of the host rock of the ore-bearing horizon and calcite concretions were found to be distinguished in terms of their isotopic composition from the above-examined groups of

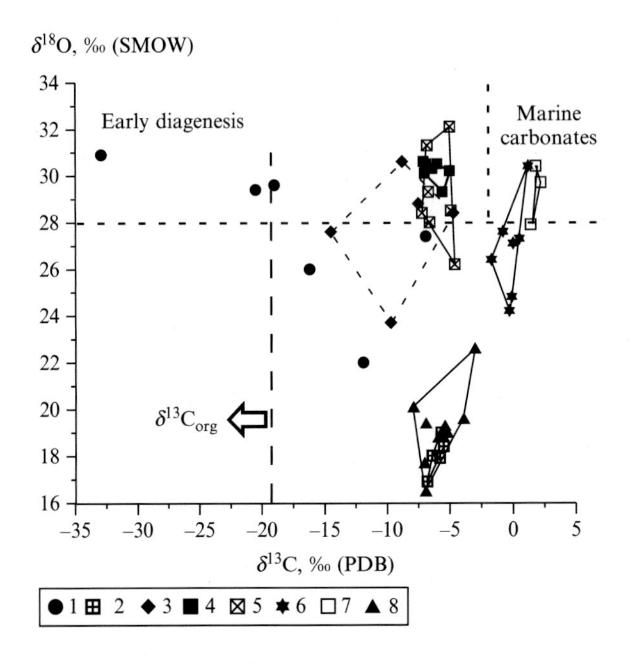

FIG. 3.22

$\delta^{13}C$ versus $\delta^{18}O$ in carbonate-manganese ores and rocks of the Mangyshlak deposit. 1—Manganese carbonate ore concretions and manganese carbonate from core zone of oxide-carbonate concretions; 2—calcite cement from rocks enclosing oxidized concretions; 3—carbonate matter and host rocks on the thinning (wedging?) out of the ore body; 4—carbonate-manganese-ore nodules of the western periphery of the deposit; 5—loose manganese carbonate from rocks enclosing manganese carbonate concretions (western periphery of the deposit); 6—loose carbonate of inter- and subjacent layers, Kuyulussk suite (Pg_3^2kl); 7—subjacent limestone's (Eocene); and 8—calcite concretions from the bottom of the section.

manganese carbonates of the deposit. These are characterized by a narrow range of variations of isotope ratios and in the graph of Fig. 3.22 form an independent area. In distinction from sedimentary carbonates of marine origin, they are characterized on the whole by a lighter isotopic composition of carbon ($\delta^{13}C = -8‰$ to $-2‰$) and oxygen ($\delta^{18}O = 16–22‰$) and are distinguished from diagenetic carbonates by lower $\delta^{18}O$ values. Such isotope ratios are on the whole characteristic for hydrothermal and catagenetic carbonates; however, the material composition, mode of occurrence, and other geological data on the deposit do not support a hydrothermal genesis for these carbonates.

The graph of Fig. 3.25 shows the correlations between the distribution of various chemical components and the isotopic composition in carbonate matter of the sandstone-siltstone rocks enclosing the oxidized manganese ores, as well as in the calcite concretions from the ore horizon and calcitized rocks from the upper part of the ore horizon. Direct correlations are noted between the isotopic composition of carbon and oxygen and the contents of CaO and CO_2 among groups as well as within the boundaries of the indicated groups. In these rocks, a direct correlation has also been established in the distribution of values of isotopic composition of carbon and oxygen. The observed regularities can most likely be taken as evidence that the formation of carbonates of the indicated groups occurred in multiple acts: at the stage of diagenesis as well as later during catagenesis.

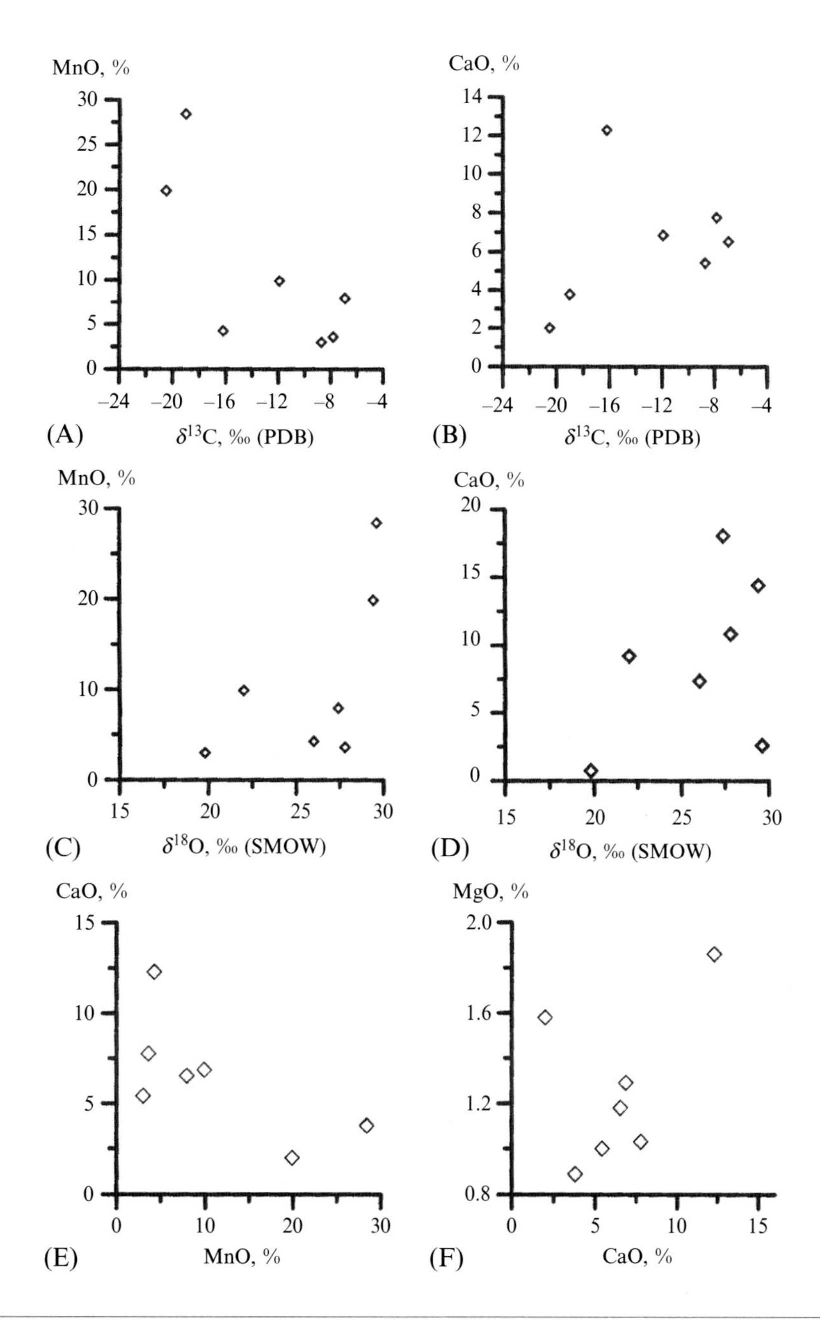

FIG. 3.23

The relationship between carbon isotopic composition and content of MnO (A), carbon isotopic composition and CaO (B), the oxygen isotopic composition and content of MnO (C) and CaO (D), the content of CaO and MnO (E) and the CaO and MgO (F) in carbonate-manganese ores of the Mangyshlak deposit.

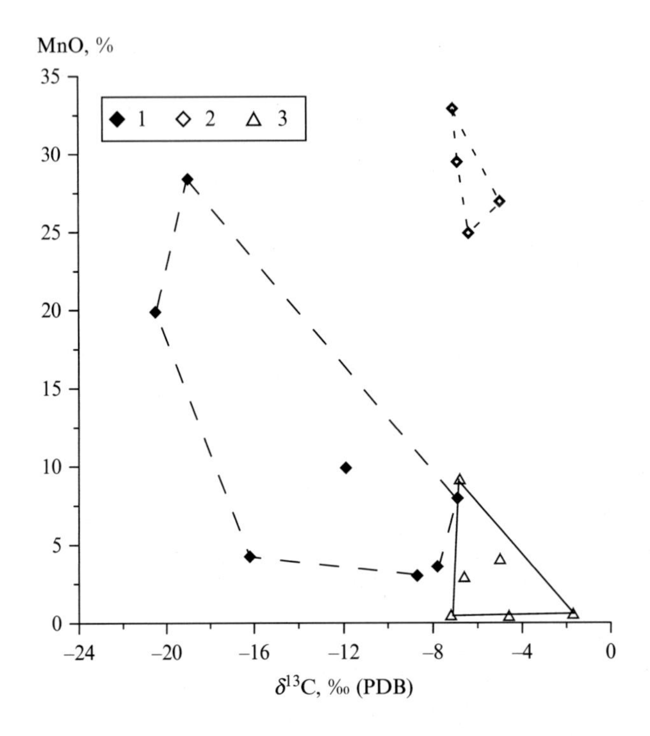

FIG. 3.24

The dependence of the carbon isotopic composition and MgO content in manganese ores and rocks, Mangyshlak deposit. 1—Ore concretions; 2—manganese-bearing nodules of the western zone of thinning; and 3—loose manganese-bearing carbonate of the western zone thinning.

On the basis of the obtained isotope data, it can be supposed that the formation of carbonate matter of cement and calcite concretions featured the participation of carbon dioxide and water of solutions of the same origin.

In carbonates of the western pinch-out zone, the lightest isotopic composition of carbon is characteristic of the Mn-containing carbonate of the ore nodules, in which the $\delta^{13}C$ values vary from –7.1‰ to –5.0‰, with values of $\delta^{18}O$=29.3–30.6‰.

A close isotopic composition is also found in Mn-containing friable carbonate from rocks enclosing Mn-ore nodules: there, $\delta^{13}C$ varies from –7.2‰ to –4.6‰, and $\delta^{18}O$ from 26.2‰ to 32.1‰. The correlation of the isotopic composition of certain components for these rocks is shown in the graph of Fig. 3.26, from which can be seen that manganese carbonate-containing matter of nodules and friable rocks forms independent fields. This provides evidence of the various conditions of its formation. In certain cases, within the boundaries of identified groups have been established correlations (eg, between the contents of CaO and MnO and the isotopic compositions of carbon and oxygen).

Calcites included in the composition of Mn-ore nodules and friable carbonates of the host rocks of the western pinch-out zone in terms of isotopic composition are mainly not distinguished from the corresponding Mn-containing carbonates: $\delta^{13}C$ for nodules varies from –7.2‰ to –5.5‰, while for friable carbonates it varies from –8.4‰ to –5.9‰. Values of $\delta^{18}O$ for calcites of these groups respectively vary from 27.3‰ to 31.2‰ and from 26.1‰ to 30.1‰.

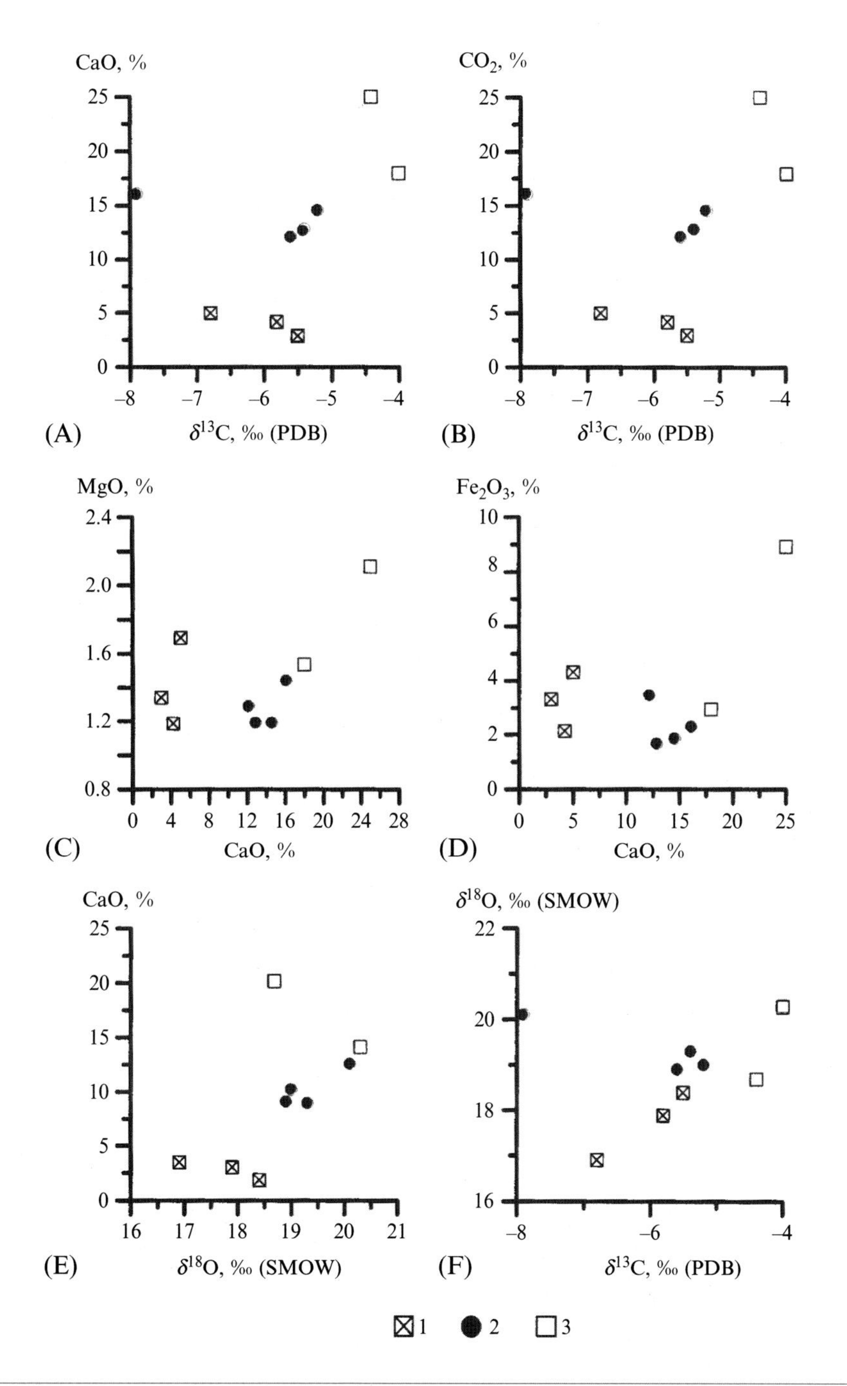

FIG. 3.25

The dependence of the content of CaO (A) and CO_2 (B) versus $\delta^{13}C$ values, CaO and MgO (C), the content of CaO and Fe_2O_3 (D), the CaO content and values of $\delta^{18}O$ (E), and the variables $\delta^{13}C$ versus $\delta^{18}O$ (F) in Mangyshlak deposit. 1—Sand-aleurolite rocks; 2—calcite concretions from the ore horizon; and 3—calcite-bearing rocks from the top of the ore horizon (north-western part of the deposit).

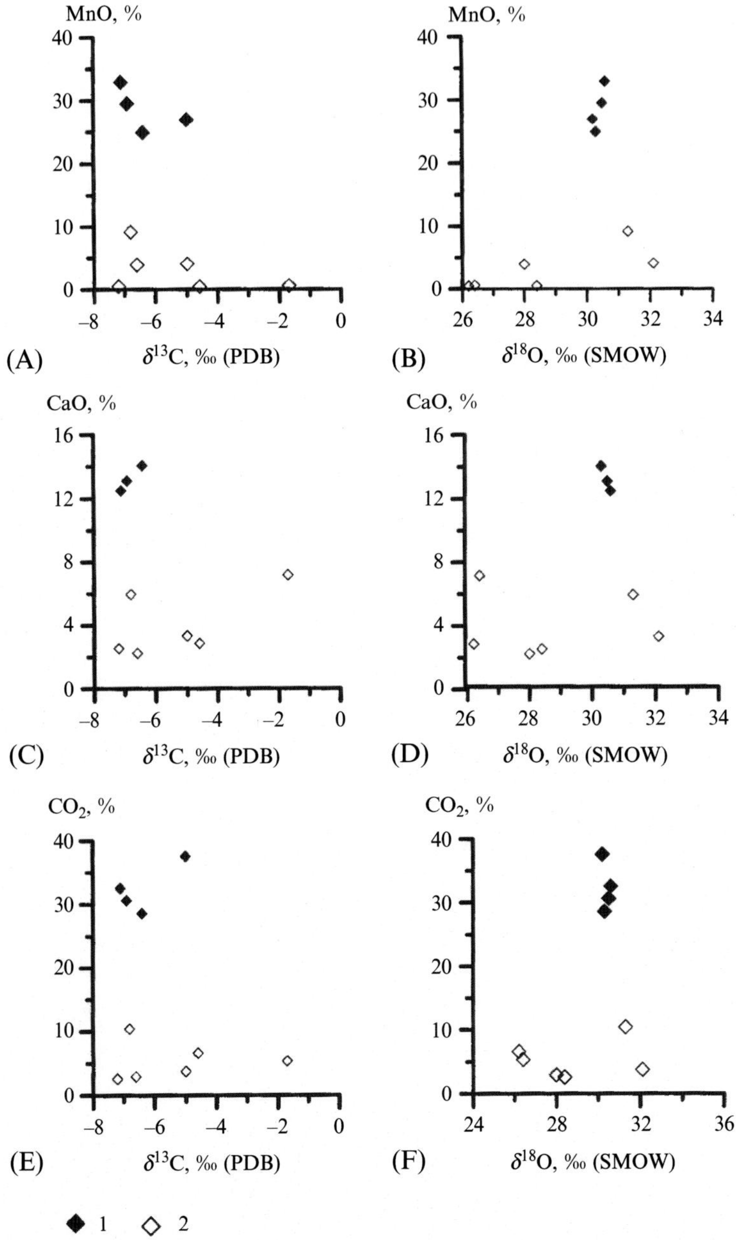

FIG. 3.26

The dependence of the MnO content and $\delta^{13}C$ values (A) and $\delta^{18}O$ values (B), the CaO content and $\delta^{13}C$ values (C) and $\delta^{18}O$ values (D), of CO_2 and $\delta^{13}C$ values (E) and $\delta^{18}O$ values (F) in carbonate-manganese nodules and host rocks of western pinching zone, Mangyshlak deposit. 1—Carbonate-manganese nodules and 2—host rocks.

The highest ratios of the isotopic composition of carbon and oxygen are characteristic for marine sedimentary limestones of Eocene age, where the range of variations in $\delta^{13}C$ values extends from 1.4‰ to 2.2‰, and that of $\delta^{18}O$ from 27.9‰ to 30.1‰.

Finally, friable carbonate matter of the inter- and subore layers of the Kuyuluss suite of the western pinch-out zone, characterized by low contents of it in the rock, occupies an intermediary position in terms of isotopic characteristics: $\delta^{13}C$ varies from −1.7‰ to 1.2‰, and $\delta^{18}O$ from 24.2‰ to 30.4‰.

Analysis of the obtained isotope data for rocks of the western pinch-out zone enables the explanation of an important particularity: the fact that all samples of carbonates containing elevated concentrations of Mn are always rich in light isotope ^{12}C by comparison with those lacking concentrations of Mn. Mn-containing carbonates in terms of the isotopic composition of carbon also are substantially distinguished from carbonates of normal-sedimentary marine origin by the former's lower values of $\delta^{13}C$. This serves as evidence that Mn-containing carbonates of the pinch-out zone of the Mangyshlak deposit, whether in nodules or scattered in the rock, have a single origin and were formed in the sediment during the stage of diagenesis with participation by isotopically light carbon dioxide of organic origin. Furthermore, specifically sedimentary carbonates characterized by elevated concentrations of Mn—that is, those that are precipitated by sedimentary means (chemogenic or organogenic) directly from the water mass and in terms of carbon and oxygen are located in isotopic equilibrium with the bicarbonate of marine bottom water (DIC—dissolved inorganic carbon)—have not been detected within the boundaries of the Mangyshlak deposit (in the western pinch-out zone as well as in within the boundaries of the ore body).

Also characteristic is the circumstance that the isotopic composition of carbon of Mn-containing carbonates does not depend on the form of localization of the carbonate in the rock (nodules or scattered matter), although the range of variations in $\delta^{18}O$ values for scattered carbonate matter is somewhat broader than that range for nodules. This serves as evidence for the fact that the process of concentration (contraction) of Mn-containing carbonate scattered in the terrigenous strata in small nodules did not produce a change in the isotopic composition of carbon. Furthermore, the initial isotopic composition of the carbon of friable carbonates characterized by an initially broader range of variations in values of $\delta^{18}O$ during the process of formation of small nodules changes insignificantly and acquires a more expressed isotopic composition, evidently in consequence of the stable physical-chemical conditions within the sediment at the moment of their formation.

Isotope data for friable carbonate matter from inter- and subore layers of the Kuyuluss suite are characteristic for carbonates of various genetic groups. Thus, samples with high values of $\delta^{13}C$ (0–2‰) and $\delta^{18}O$ (28–30‰) correspond to marine sedimentary carbonates. It can be supposed that dispersed carbonate matter with such high isotopic characteristics (sample 17-7) was formed by sedimentary means at the moment of the formation of the terrigenous strata itself directly from the water mass.

Another portion of the samples of friable carbonates is characterized by a lighter isotopic composition of oxygen (24–25‰) with fairly high $\delta^{13}C$ values (−1‰ to 1‰). These data provide evidence that carbonate matter such as calcite concretions and calcite from host rocks of the ore-bearing horizon within the boundaries of the deposit itself was formed at a later (post-early-diagenesis) stage of the geological life of the sediment with participation by waters of meteoric or catagenetic origin.

Regarding the type of catagenesis (Kholodov, 1982a,b) and naturally regarding the isotopic composition of waters participating in this process, nothing can be stated definitively on the basis of isotope data. If it is proposed that these waters have an isotopic composition close to that of the oxygen of marine water ($\delta^{18}O$ ~0), then, taking into account the isotopic equilibrium of the process, the precipitation of carbonates should have occurred at elevated (80–120°C) temperatures—a conclusion not supported by the geological observations.

At the same time, if the formation of carbonates occurred with participation by meteoric waters ($\delta^{18}O = -7‰$ to $-6‰$), then in this case temperatures of carbonate precipitation for a large portion of the studied samples should have been 40–50°C, although the general temperature interval could have been somewhat broader (30–70°C). Such temperatures are fully actual and do not contradict the geological data.

An analogous distribution of isotopic data of carbon and oxygen can be observed in this case if the formation of catagenetic carbonates occurred from the waters of elisional genesis. Elisional waters on the whole should also be characterized by elevated concentrations of the light isotope ^{16}O by comparison with marine waters (Mottl et al., 1983), while $\delta^{18}O$ values can correspond to waters of meteoric origin. Therefore, on the basis of isotope data of oxygen alone, the nature of the water of the solutions (elisional or meteoric) participating in the process of calcitization cannot be reliably established. This question could be successfully resolved with the gathering of additional (isotope and geochemical) data in conjunction with geological data.

In this way, the obtained isotope data provide evidence of the intensive development of the processes of catagenetic carbonate precipitation within the boundaries of the Mangyshlak deposit. Newly formed calcite has also evidently "saturated" the diagenetic ore carbonates—that is, it has entered the composition of manganese ores. Moreover, the ratio of initial diagenetic to later catagenetic carbonates can vary widely.

The addition of catagenetic calcite into the composition of carbonate ores produced an impact on their isotopic composition. This follows as well from the graph of Fig. 3.22, where it can be seen that all calcites in terms of isotopic characteristics are mixed with a tendency toward lighter values of $\delta^{13}C$ and $\delta^{18}O$. Their addition into oxide-carbonate zonal manganese ores is also responsible for the observed linear distribution of the isotopic ratios of carbon and oxygen within the coordinates of $\delta^{13}C$ and $\delta^{18}O$.

Regarding the genesis of the deposit

Betekhtin (1946) and subsequently Strakhov et al. (1968) and other researchers (Berdichevskaya and Rakhmanov, 1981; Sokolova et al., 1984) have established that the oxide ores of the Mangyshlak deposit are a product of the hypergenic oxidation of the initial carbonate ores. The genesis of the deposit has been examined in most detail in the above-cited work of N.M. Strakov and coauthors, who consider it a sedimentary-diagenetic genesis. According to these researchers, manganese entered the basin of sedimentation together with terrigenous material from the source area. As a result of the diagenetic redistribution of the ore matter, concretional carbonate ores were formed.

Subsequent works on the deposit have allowed researchers to express other viewpoints regarding the source of ore matter and accordingly the genesis of the deposit: Mstislavskii (1981) and Druzhinin (1980, 1981, 1986) consider it a hydrothermal-sedimentary genesis; Kholodov (1982b), Dvorov and Sokolova (Dvorov and Sokolova, 1986, 1987; Sokolova and Dvorov, 1987; Sokolova et al., 1984) have proposed an elisional-catagenetic model of manganese ore-genesis. The geological-geochemical data to varying degree of feasibility allow the coexistence of the indicated theories.

While the results of isotope research unfortunately do not indicate the source of the manganese during the formation of the ores, they do provide confirmation that manganese carbonate concretions have a diagenetic origin—that is, they were formed in the ooze sediment during the stage of early diagenesis with active participation by oxidized carbon of organic matter. Moreover, the intensity of the processes of diagenesis and consequently of the ore-formation within the boundaries of the deposit were varied. They were most actively manifested in the near-shore zones of the paleobasin—that is, in the central and particularly the north-eastern parts of the deposit, where there is an increase in the dimensions and quantity of concretional structures, their form becomes more complex, and the thickness of the ore layers grows.

Toward the more deep-water areas of the paleobasin, there was evidently an increase in the initial carbonate content of the sediment. In this same direction, there occurred a decline in the activity of C_{org} in the diagenetic processes. All of this is reflected in the increased $\delta^{13}C$ values, the decrease in the contents of manganese, and an increase of CaO in manganese carbonate ores and rocks (see Figs. 3.23, 3.25, and 3.26).

Thus, the isotope data fully support the theory of N.M. Strakhov and his coauthors regarding the formation of carbonate ores of the Mangyshlak deposit during diagenesis; manganese carbonates were formed with active participation by organic matter. Specifically sedimentary Mn-containing carbonates—that is, those that were precipitated directly from the water mass by chemogenic or organogenic means—have not been detected within the boundaries of the Mangyshlak deposit.

The intensity of the processes of diagenesis within the boundaries of the deposit was manifested to varying degrees. Most actively these processes occurred in the central and north-eastern areas of the deposit, where the greatest thicknesses of the deposits of ore matter are to be found. The activity of these processes was also reflected in a decrease of $\delta^{13}C$ values in the manganese ores.

At the same time, a significant proportion of the carbonate matter developed within the boundaries of the deposit is of a later, catagenetic origin and was superimposed on previously formed manganese ores. Associated with these are carbonates: (a) developed within the host sandstone-siltstone rocks; (b) composing concretional nodules in the lower part of the ore horizon; and (c) derived from the cement of the upper washed-out part of the ore body, as well as from the overlapping conglomerates. All of these are represented predominantly by calcite. With this stage of calcitization could be associated the formation of sandstone-carbonate nodules in the deposits of the territories adjacent to the deposit.

The process of the superimposition of calcitization at the Mangyshlak deposit occurred after the oxidation of carbonate ores in the Early Miocene. Evidence for this is provided by a wealth of geological data, particularly by the fact that oxidized manganese ores throughout incorporate secondary calcite. In the northern, washed-out part of the deposit, calcitization took place in rocks of the ore horizon as well as their overlapping conglomerates. This could serve as evidence that carbonate-forming solutions arose during the post-Early Miocene in the weakness zones, including in the zone of disconformity of the Oligocene and Middle Miocene deposits. This zone is highly permeable—that is, it is represented by siltstones and their overlapping carbonates.

Regarding the origin of the waters that participate in the catagenesis process (and consequently regarding the type of catagenesis) is difficult to judge in terms of the isotope data. They could be waters either of meteoric or with equal likelihood of elisional genesis.

3.2.1.3 Manganese deposits of Georgia (on the example of the Chiatura and Kvirila deposits)

Geological setting

Unique in terms of their reserves of high-quality manganese ores, the Chiatura, Kvirila, and a range of smaller deposits and ore-occurrences of Georgia are grouped alongside the Dzirula crystalline massif (Fig. 3.27), are situated in the shallow-water shoreline-marine deposits of the Maikop series

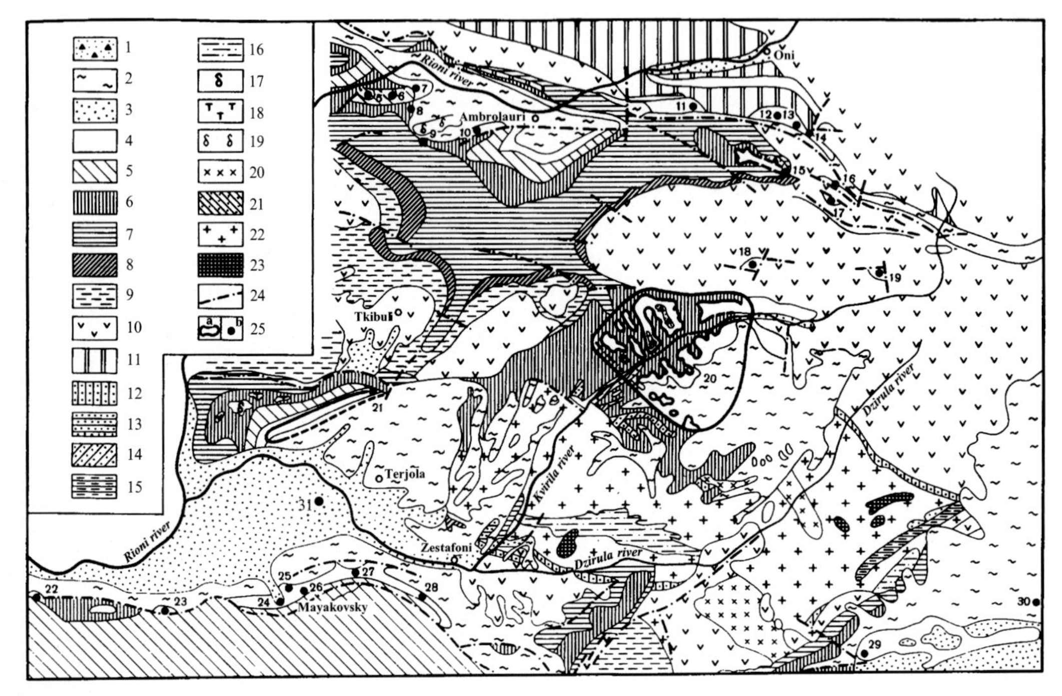

FIG. 3.27

Schematic geological map of the Dzirulskii manganese-ore area (Avaliani, 1982). 1—Landslide masses; 2—Quaternary deposits; 3—sandstones, clays, limestones, conglomerates of Middle and Upper Miocene; 4—gypsum-bearing clays, sandstones, spongolites, places manganese-bearing layers of Oligocene-Lower Miocene; 5—limestones, marls, sandstones, and tuffites of Paleocene and Eocene; 6—limestones, marls, glauconites sandstones, volcanic rocks of the Upper Cretaceous; 7—glauconite sandstones, marls, limestones, and volcanic rocks of Lower Cretaceous; 8—multicolored clays and sandstones, calcareous sandstones, limestones of the Upper Jurassic; 9—foliated schists, sandstones, coal-bearing formation, Bathonian stage; 10—porphyritic suite, Bajocian stage; 11—shales, sandstones of the Upper Lias; 12—zoogenic limestones, conglomerates, sandstones, mudstones, marls of Middle and Upper Lias; 13—lower tuffites of the Lower Lias; 14—quartz-porphyric strata of Lower and Middle Carboniferous; 15—metamorphogenic shales, phyllites of Cambrian; 16—schists and gneisses of the Precambrian and Lower Paleozoic; 17—Pliocene basalts; 18—teschenites of the Upper Miocene; 19—basalts of the Turonian; 20—granitoids of the Middle Jurassic; 21—serpentinites of Upper Paleozoic; 22—granitoids of Middle and Lower Paleozoic; 23—Paleozoic gabbro; 24—tectonic faults; 25—manganese deposits (a) and ore manifestations (b) (5–14 and 16–19—Rachinskaya group; 15—Shkmerskoe, 20—Chiatura deposit, 21—Chkhari-Adzhametskoe; 22–29—Vani-Zestafoni group; 30—Surami-Karelian group; 31—Kvirila deposit).

(Oligocene-Lower Miocene), and are represented in the region of the deposit predominantly by clays and rarely by sandstone-siltstone deposits. Toward the west and toward the east from the Dzirula massif are observed zones of greater accommodation, in which from the beginning of the Mesozoic and up to the present were accumulated multikilometer sediment strata. Here, among deposits of the Maikop series, are observed more deep-water facies.

Manganese ores are emplaced subhorizontally inside the sedimentary deposits of the Maikop series and are associated with its lower part—the Khadum horizon.

The ore horizon lies on the foraminiferal limestones or limy sandstones of the Upper Eocene (Kvirila deposit), and where those absent on limestones of the Upper Cretaceous (Chiatura deposit). Overlapping are deposits of the Maikop Tarkhanian horizon of the Middle Miocene, represented by clayey-sandstone deposits with a thickness reaching 100 m, or by marls and sandy deposits of the Chokrakian with a thickness reaching 50 m.

The *Chiatura deposit* is situated on a high plateau with absolute elevation of 430–800 m, broken by the deep canyon-like gorges of the Kvirila River and its tributaries into separate sections—the uplands. The largest of these are Darkveti, Itkhvisi, Mgvimevi, Shukruti, Perevisi, Merevi, and Pasieti.

Manganese ores occupy a vast area stretching for 12.5 km from the south-west, where the deposit is bounded by a large regional fault of north-westerly strike [the Glavnyi (Main) fault], gradually pinch out to the north-east (Fig. 3.28).

Ore bodies form a subhorizontal bed-type deposit with a thickness from 2 to 4 m in the south-western part of the deposit (near the Glavnyi fault) up to 4.5–5.5 m in the central part and up to 14 m in the eastern part. Among the host sandstone-siltstone rocks, manganese-ore mineralization forms a series of lensoidal layers of varying thickness and length, irregularly united with each other, in the vertical as well as in the horizontal direction (Fig. 3.29), creating the impression of a mesh texture for the strata (Strakhov et al., 1968).

In the deposit have been identified subore, ore, and supraore horizons (Betekhtin, 1946). The subore horizon overlies Upper Cretaceous limestones and is represented by fine-grained sands and weakly cemented sandstones with thickness from 0.1 m in the south-west up to 30 m in the north-east of the deposit. The sands have an arkose-quartz composition. The supraore horizon in the south-west is composed of silicified sands and spongolites with a thickness of 25–30 m, which in the north-east of the deposit are replaced by Maikop clays with a thickness of 70–80 m. In places in the upper part of the supraore horizon is encountered a second ore horizon lacking industrial significance (Edilashvili et al., 1973; Shterenberg, 1985).

The *Kvirila deposit* is situated to the south-west from Chiatura and coincides with and occupies a large part of the area of the depression of the same name. The Kvirila depression in the tectonic regard represents a wide graben-syncline, in the bottom of which the sediments are horizontal, whereas on the flanks they are distorted by small folds and faults. From all sides, the depression is bounded by folds: in the north, it is separated by the Sachkhere fault from the Okriba landmass; in the south, it is separated by the Surami-Gokeshur fault from the Adjara-Trialet fold region; in the east it is linked through the Chkhari fault to the Dzirula massif; and in the west, borders with the Rioni depression along the Chishura fault (Tumanishvili, 1989). The northern and southern sublateral faults, according to Edilashvili et al. (1965, 1973a,b), are long-ranging thrust faults, that superpose Jurassic and Cretaceous deposits onto Paleogene deposits.

Manganese ores are found in the lower part of the Maikop series, the thickness of which varies from 45–80 m in the eastern and central parts of the depression to 250–300 m in the western and southern parts. Similarly to the Chiatura deposit, the terrigenous component of the ore strata here is represented by sandstone-siltstone material. A specifically sandy facies is developed only in the northern part, near the Okriba land mass. Within the boundaries of the central part of the depression, the following ore sections have been identified: Terjola, Cholaburi, Rodinauli, Ajameti,

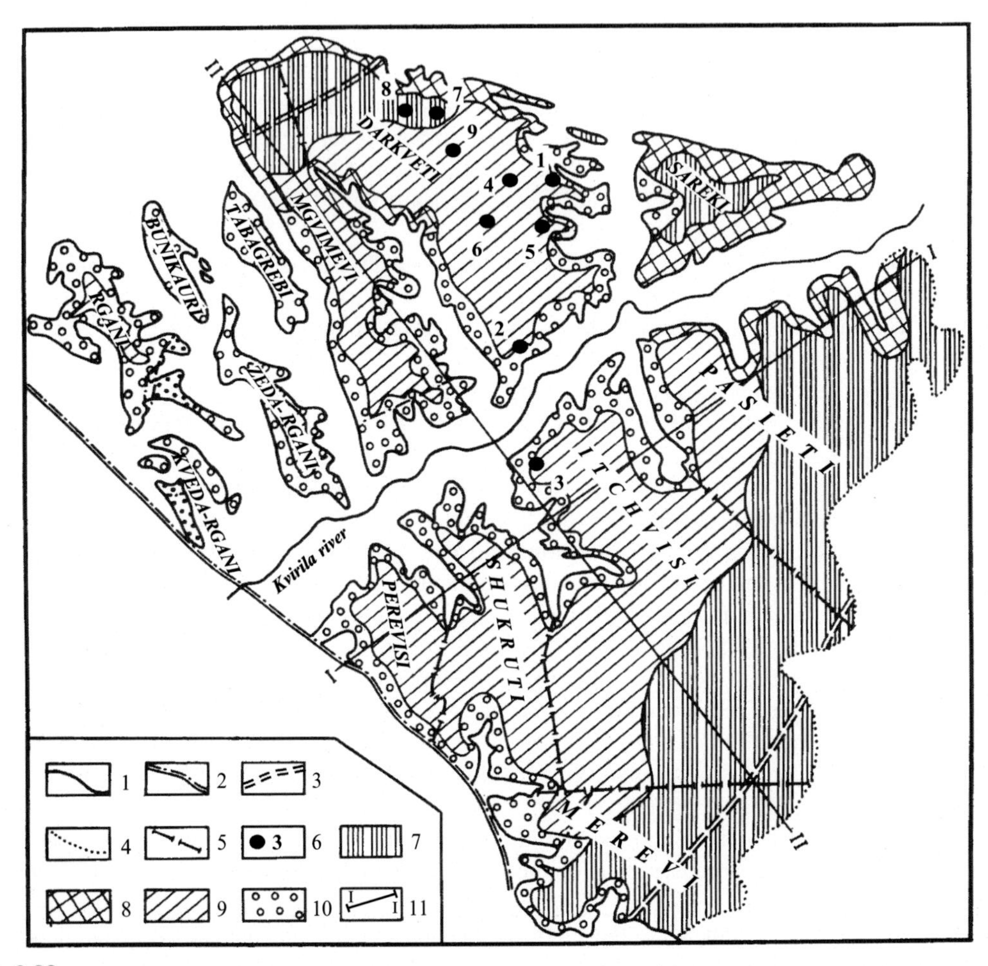

FIG. 3.28

Plan of orebody on the Chiatura manganese deposit (compiled by L.E. Shterenberg) (Chiaturskoe et al., 1964; Strakhov et al., 1968; Tabagari 1984). 1—Modern border of ore field; 2—the main deep fault; 3—estimated deep faults; 4—estimated boundary of carbonate ores; 5—estimated faults; 6—location of sampling for isotope studies (1—Darkveti Highland, sector 4, quarry; 2—Darkveti Highland, sector 8, quarry; 3—Akhali-Ithvisi Highland, adit; 4—Darkveti Highland, sector 8, adit; 5—Darkveti Highland, sector 8, adit; 6—Darkveti Highland, sector 10, adit; 7—Darkveti Highland, borehole 830; 8—Darkveti Highland, borehole 826; 9—Darkveti Highland, borehole 832); 7—the area of distribution of carbonate ores; 8—the area of distribution of oxidized carbonate ores; 9—the area of the joint dissemination of oxide and carbonate ores; 10—the area of distribution of oxide ores; 11—lines of sections.

and Rokit'i. The greatest concentration of ore matter is observed in the Rodinauli and Cholaburi sections (Fig. 3.30). The ore areas of the central parts of the depression are separated from the ore-occurrences of the Chkhari-Ajameti band by non-ore sections connected to it only in the western part in the Brolis-Kedi-Ajameti region. Judging by the low thicknesses (8–35 m) of the Oligocene deposits, the barren areas are associated with paleo uplifts.

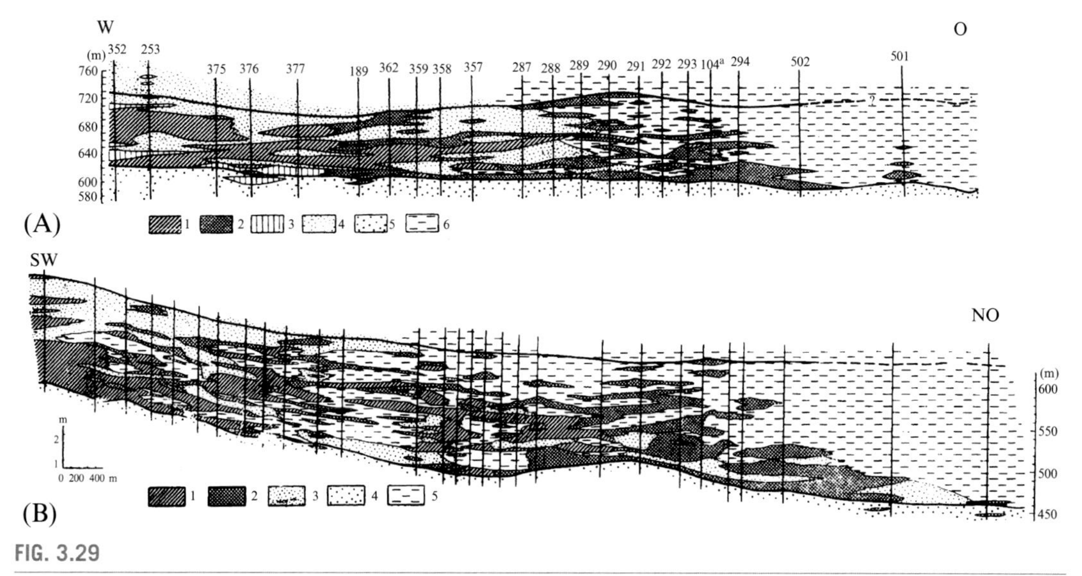

FIG. 3.29

The structure of the ore horizon on the Chiatura deposit (compiled by L.E. Shterenberg) (Strakhov et al., 1968). (A) Profile along line I-I′: 1–3 manganese ores: 1—oxide, 2—carbonate, 3—oxidized; 4—sandstones; 5—sands and sandstones (subjacent layers); and 6—clayey sandstones and clay sands. (B) Profile II–II′: 1—oxidized manganese ore, 2—carbonate-manganese ores; 3—sandstones, 4—sands and sandstones (subjacent layers); and 5—argillaceous sandstones and sandy clay.

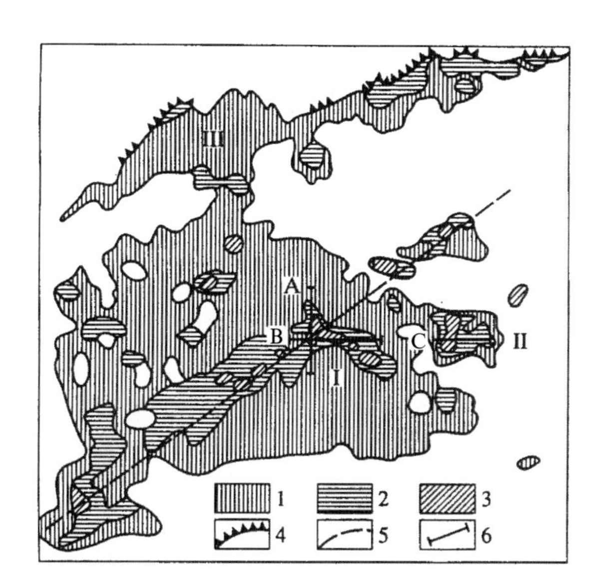

FIG. 3.30

Plan of manganese-ore horizon on Kvirila Depression (after Dolidze et al., 1980; Hamhadze and Tumanishvili 1984). 1–3—parameters of mineralization (metropercents): 1—0.1–10, 2—10–30, 3—0; 4—manganese-ore outcrops; 5—faults suggested to be ore-feeder for Cholaburi and Rodinauli sectors; 6—location of the boreholes profiles, on which was selected samples for isotope analysis [A—(from north to south) boreholes 814, 518, 824, B and C (from west to east): B—boreholes 518, 536; C—boreholes 868, 889]. I—sector Rodinauli; II—sector Cholaburi; and III—Chkhari-Adzhameti occurrence ore zone.

In the Kvirila deposit, as in the Chiatura, have been identified subore, ore, and supraore horizons. The subore horizon is composed of friable sands with thicknesses up to 3 m. The supraore deposits are represented by siliceous (spongolites) rocks, and higher by clayey rocks, the thickness of which increases from 4–5 m in the east to 21–30 m in the west.

The ore body forms a bed-type deposit with a thickness from 0.2 to 5.9 m. About the area its thickness is not constant and separate ore-occurrences (sections) are separated by non-ore or weakly mineralized sections. The greatest thickness of the ore horizon is observed in the western part of the depression; in the central part, it contracts to 2–3 m and in the eastern and southern directions, it tapers out. The ore seam is characterized by a complex structure owing to the alternation of lensoidal layers of various types of ores with layers of various quantitative content of ore and non-ore (terrigenous and authigenic) components. By comparison with the Chiatura deposit, the ore bed contains significantly fewer interbeds of host rocks.

The Kvirila deposit is distinguished from the Chiatura deposit by its post-ore history of geological development. The available data on the composition and thicknesses of their host Oligocene deposits do not constitute evidence of a multi-act process(es) of ore accumulation, and likewise of their subsequent sediment accumulation over the course of the Chokrakian, Conckian, and early Sarmatian, when the territories of the examined deposits were undergoing a slow submersion of the marine bottom and the general thickness of the Oligocene-early Sarmatian deposits constituted 200–250 m.

From the middle Sarmatian Stage, the development of the Chiatura and Kvirila basins proceeds multidirectionally. The Chiatura region as a result of the nascent uplift is aridifying. At present, the base of the ore horizon is located at a height of 630–800 m above sea level. The Kvirila depression continued its submergence and the base of the ore horizon concluded at depths of 550–600 m from the surface; that is, at 400–440 m below sea level, as a result of which the vertical range of displacement of ore beds between the Chiatura and Kvirila deposits constitutes >1000 m.

Host rocks and authigenic mineralization

As noted, manganese rocks in the Chiatura and Kvirila deposits are hosted by sandstone-siltstone rocks of the Maikop series, which are composed predominantly by quartz (sharply predominating) and feldspar (microcline, orthoclase, plagioclases). The remaining minerals (micas, amphiboles, accessory minerals), organic residues (wood, fish skeletons and scales), as well as clasts of rocks and volcanic glass, are present in insignificant quantities.

Manganese mineralization within the terrigenous deposits of the Maikop series markedly coincides with sections of intensive development of authigenic siliceous-zeolite mineralization (Makharadze, 1972, 1979; Khamkhadze, 1981; Machabeli, 1986). Manganese minerals, which are likewise authigenic, are located in close paragenetic association with it. Secondary mineralization is most intensively manifested along fault lines, where the thicknesses of such rocks reach 90 m. By measure of the distance from the faults, the intensity of mineralization decreases and a gradual transition into typical terrigenous Maikop is observed.

The authigenic mineralization of the ore-enclosing mass is extremely diverse. Most widely distributed are opal and clinoptilolite, along with frequent occurrences of montmorillonite, an alternating-layer micaceous-(glauconite-)montmorillonite assemblage, glauconite, phosphates, barite, gypsum, native sulfur, pyrite, and marcasite. In sections, occurrences of secondary mineralization are widely developed as well as hard bitumens.

Siliceous matter in rocks with authigenic mineralization is found most frequently in the form of opal, more rarely chalcedony, forming sections of opal rock, opoka (diatomite), and spongolite sandstones. Commonly the opoka and opal rock are developed in the lower part of the sequence of the ore-bearing strata and spongolite sandstones in the upper part; however, separate spicules of sponges are observed along the entire sequence. An initially chemogenic nature of the siliceous biogenic material has been demonstrated by Makharadze (1979) and Khamkhadze (1981). The opoka are very porous and light rocks. Opal-rich rocks commonly contain impurity of clayey material and have spherical textures composed of opal or clayey matter, or those and others together. Replacement by manganese is common in the more permeable zones; spongolite sandstones, in consequence of their high density and lower permeability, commonly form the roof of the ore horizon.

Another constant and widely distributed component accompanying manganese ores are zeolites. These are distributed throughout the entire ore strata; their greatest content is noted in the upper part of the ore horizon and in supraore deposits (Merabishvili et al., 1979; Gogishvili et al., 1979), where they form members with a thickness reaching 20 m with content of zeolites reaching 60–65%, represented by a highly siliceous variety—clinoptilolite. A spatial and genetic association of zeolites with volcanogenic material has not been established (Butuzova, 1964).

Authigenic mineralization is significantly more strongly manifested in the Kvirila deposit (particularly in its central and south-western parts) than in the Chiatura deposit (Makharadze, 1972; Merabishvili et al., 1979; Khamkhadze, 1981; Machabeli, 1986). It encompasses not only the ore and supraore strata, but also the subjacent Upper Eocene marls and sharply depresses the terrigenous component of the rock. The Upper Eocene marls, according to Khamkhadze (1981), are dolomitized, silicified, montmorillonitized, and zeolitized, and in the south-western margin they are almost fully replaced by opoka.

Besides terrigenous and authigenic components, in the rocks of the Lower Oligocene and also rocks of their underlying marls of the Upper Eocene of the Kvirila depression have been described the products of volcanism, represented by ash, tuffs, and tuffites (Makharadze and Chkheidze, 1971). These authors note that vitric material as a rule is argillified and replaced by montmorillonite, clinoptilolite, and opal. This has led to the loss of the initial structure of the ash material. In this connection, it is extremely complicated to establish a primary volcanogenic nature of the present in the ore strata of the designated minerals. Zh.V. Dombrovskayia (Kuleshov and Dombrovskaya, 1997a,b) have observed only separate particles of fresh unaltered volcanic glass. Reliable geological proof is lacking for a wide development of products of volcanism in the region of the examined deposits.

The structure of the ore bed and types of ores

Manganese ores are divided into two genetic categories: primary and oxidized. The present work examines only primary ores unaltered by hypergenesis. Manganese matter in these ores is represented by oxide and/or carbonate forms; accordingly, oxide, oxide-carbonate, and carbonate types of ores have been identified. In our observations, manganese in the described deposits is also contained in silicate minerals, represented by neotocite and braunite in association with hausmannite.

In the area of the Chiatura deposit, oxide ores are developed in the south-western part near the Glavnyi fault in the central part, adjoining the Kvirila riverbed. In peripheral parts of the deposit, they are replaced by carbonate ores. In the vertical sequence, until recent time have been identified two principal ore members, distinguished in terms of structure and composition (Betekhtin, 1946; Chiatura et al., 1964; Shterenberg, 1985). The lower, main ore member is composed predominantly of psilomelane-pyrolusite ores, which toward the margins of the deposit are gradually replaced by manganite ores, and subsequently by carbonate ores. The upper member is represented principally by manganese carbonates with low-yield interbeds and lenses of oxide ores. Tabagari (1980, 1984) has identified beneath these yet another ore member, composed of carbonate ores.

For the Kvirila deposit, a facies zonality is noted in the distribution of types of ores in the Chkhari-Ajameti section (Avaliani, 1982). In the southerly direction, oxide (psilomelane-pyrolusite with an impurity of manganite) ores are replaced by oxide-carbonate ores and subsequently by carbonate ores. Independent manganite layers are absent from this part of the manganese-ore basin. In the central parts of the basin in terms of bore materials the zonality is difficult to establish. In the southern Vani-Zestafoni band of ore-occurrences, only carbonate ores have been established. As in the Chiatura deposit, in the most fully developed sequences, there are observed three ore members: lower carbonate, middle oxide, and upper carbonate. The middle oxide member is frequently represented by oxide-carbonate ores. Abbreviated sequences have a unitary bed of carbonate ores.

Thus, oxide ores from the bottom, from the top, and from the side of the pinch-out areas are surrounded by carbonates. It is difficult to explain such interrelated types of ores in terms simply of their facies distributions.

In terms of structural-textural indicators, *oxide ores* are divided into (1) oolite (grained*) with varieties: with strong siliceous cement (*zhgali**) and with strong calcite cement (*mtsvari**), (2) lump-oolite (*santsirili**), in which lumps are represented by nodules of angular form, and (3) compact (beds*), in which ooliths are cemented by dense manganese-oxide material (*—local term).

Carbonate ores in terms of textural indicators in the described deposits can be divided into two varieties: compact and nodular. Compact ores have a massive texture and fine-grained structure, while nodule ores are characterized by the presence of rounded structures with diameters from 0.5 to 8 mm. The existence of the two indicated textural varieties is due, in our observations, to the primary texture of the initial rocks—massive (compact) and containing spherical inclusions of clayey-opal material (Fig. 3.31).

The composition of manganese carbonate ores varies depending upon the degree of replacement of the host rocks by carbonates. The content of the latter, according to the data of our conducted research, varies from 0% to 89%. In Tables 3.7 and 3.8 are recorded the chemical compositions of carbonate and oxide-carbonate ores with content of carbonates >20%. Predominating in the composition of carbonate matter is $MnCO_3$: its content constitutes 12.75–81.49% and exceeds the content of $CaCO_3$ by 2–15 times. In certain samples, $MgCO_3$ is present in small quantities (up to 5%). Silica forms a significant portion of the host rocks; its content constitutes 60–90%. In ores it decreases to 3.36–45.5%, owing to an increase in the content of carbonates. Content of the remaining components in carbonate ores as a rule does not exceed several percentages.

Carbonate-manganese ores of the Chiatura and Kvirila deposits are very similar in composition (see Tables 3.7 and 3.8), but in the former have been noted higher contents of CaO and BaO, while in the latter have been noted higher contents of Fe_2O_3 and P_2O_5.

Detailed research of the carbonate matter composing the manganese ores has demonstrated that they consist of a mixture of minerals of the calcite-rhodochrosite isomorphic series with a small impurity (up to 5–6%) of manganese dolomite (kutnohorite). Occasionally occurring in insignificant quantities are dolomite, oligonite, or siderite. Judging by the d_{1014} reflection intensities in X-rays (Fig. 3.32), the majority of samples represent complex mixtures, among which the most widely distributed are rhodochrosite and manganocalcite. Based on X-ray and thermal analyses, the carbonate ores should contain,

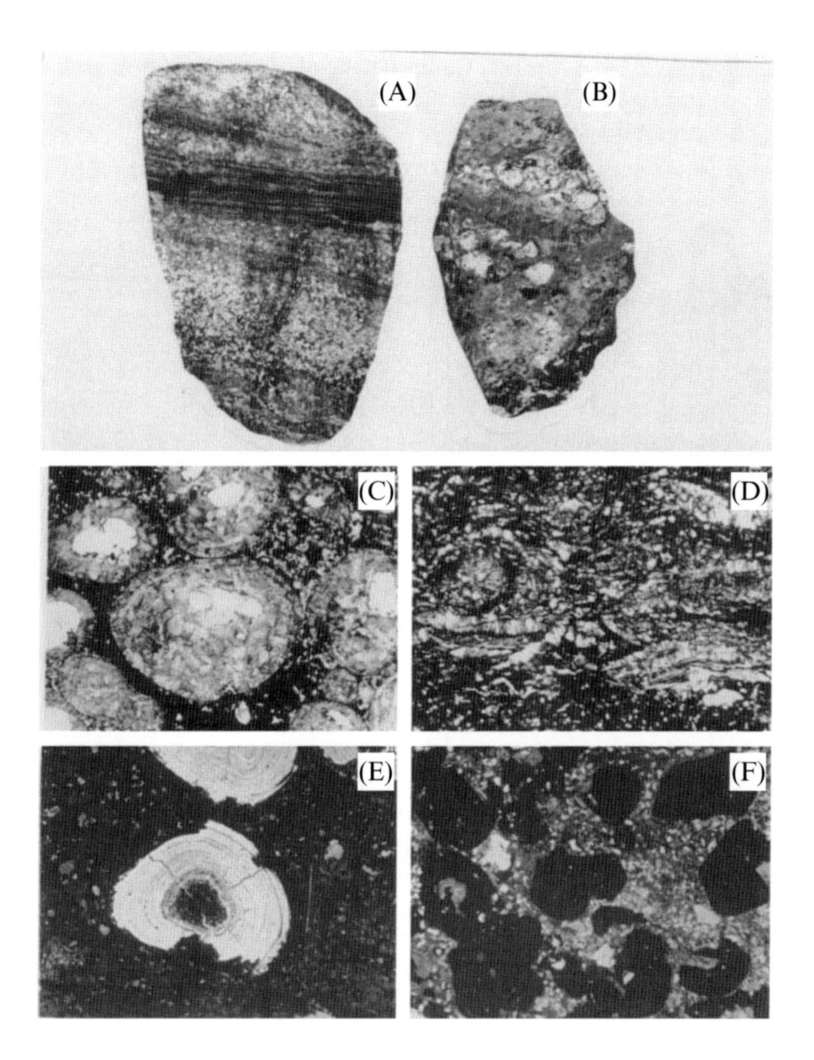

FIG. 3.31

Structural and textural features of carbonate-manganese ores of Chiatura deposit (photos by Zh.V. Dombrovskaya). (A) and (B) Disseminated ores, lump of ore; reduced 1.15 and 1.23 (white—oolitic-similar carbonate concretions, replacing the primary opal; gray—carbonate cryptocrystalline cement; black lenticules—incorporating solid bitumens); (C) spherulite aggregates of manganese carbonate, replacing opal nodules, including clastic grains of quartz; thin section, magnification 96, nicole parallel; (D) belt vein texture recrystallized carbonate ore, magnification 14.5, nicole crossed; and (E) and (F) replacement oolite manganite ore by manganese carbonate (E—polished section, magnification 14.5; F—thin section, magnification 14.5, nicole parallel).

Table 3.7 Chemical (Wt%) and Mineral Compositions of Manganese Ores and Host Rocks From the Darkveti and Itkhvisi Highlands, the Chiatura Deposit, Georgia (Compiled by Zh.V. Dombrovskay)

	Darkveti Highlands									
Lithotypes	**Carbonate**						**Oxide**	**Carbonate-Oxide**		
Sample Nos	**17**	**19**	**20**	**24**	**108**	**109**	**22**	**23**	**27a**	**28**
SiO_2	19.5	20.1	13.3	4.05	23.8	16.6	15.9	12.7	9.37	18.6
TiO_2	0.12	0.12	0.12	–	Trace	Trace	–	–	–	0.12
Al_2O_3	2.24	2.98	2.97	1.44	3.20	0.62	2.74	3.43	2.38	5.99
Fe_2O_3 total	0.35	2.75	0.74	0.15	1.13	0.67	0.91	1.62	0.74	1.6
MnO total	27.2	28.1	37.9	49.1	30.6	31.9	4.28	26.0	24.7	24.8
MgO	1.90	2.21	2.36	1.15	2.17	4.9	0.56	1.01	0.80	1.2
CaO	15.5	12.1	7.81	3.39	7.15	10.0	3.74	4.59	11.7	6.3
Na_2O	0.23	0.23	0.28	0.06	0.43	0.11	0.20	0.24	0.20	1.32
K_2O	0.43	0.72	0.49	0.20	0.46	0.15	0.47	0.24	0.37	0.72
H_2O^-	1.03	2.21	1.08	1.38	1.18	1.53	2.23	1.48	1.00	1.93
H_2O^+	0.91	0.29	0.72	5.52	1.21	1.81	2.15	7.20	7.35	5.63
CO_2	28.3	25.2	29.7	32.8	24.9	27.6	3.56	7.06	10.8	8.6
BaO	0.082	0.43	0.015	0.009	0.3	0.72	1.19	2.70	1.76	1.19
SrO	0.016	0.02	0.016	0.005	–	–	0.049	0.076	0.048	0.043
P_2O_5	–	–	–	–	0.14	0.29	–	–	–	0.08
MnO_2	–	–	–	–	1.9	2.16	61.0	29.3	37.9	22.2
$\sum$	100.0	98.9	99.3	100.1	99.4	99.4	99.7	99.5	100.0	100.1
SO_3	1.05	0.23	0.25	0.05	0.89	0.32	0.75	1.2	0.9	0.74
FePy	0.51	0.56	0.69	0.41	–	–	–	–	–	–
SPy	0.59	0.35	0.8	0.48	–	–	–	–	–	–
$MnCO_3$	42.0	40.8	61.5	78.6	49.5	51.7	1.59	8.95	4.05	13.1
$CaCO_3$	27.7	21.5	14.1	6.1	11.0	17.7	6.69	8.21	20.9	11.2
$\sum$ carb.	69.7	62.4	75.6	84.6	62.6	69.3	8.28	17.2	25.0	24.3
Manganite	–	–	–	–	–	–	8	51	51	38
Pirolusite	–	–	–	–	2	2	57	4	2	3
Neotokite	4	9	–	2	–	–	–	–	14	–
Opal + quartz	16	14	8	3	14	7	7	3	6	7
Montmorillonite + glauconite + zeolite	7	5	9	3	12	16	15	17	8	19

Analyses were performed in the chemical laboratory of the IGEM. Analysts: *L.S. Tsimlyanskaya, M.S. Obolenskaya, G.E. Kalenchuk, E.P. Frolova.* Lithotypes: *manganese ores.*

Itkhvisi Highlands									
Opal-Carb	Carbonate			Carbonate-Oxide	Opal-Carbonate		Opal	Sandstone
26	144	145	147	149	150	153	153	155
36.7	3.93	7.22	5.52	10.7	41.1	23.4	18.9	59.4
0.12	–	–	–	–	0.29	–	–	0.39
4.37	1.24	1.87	2.23	3.64	7.69	3.53	3.30	8.27
0.58	1.21	–	0.48	1.14	1.45	0.59	3.08	1.66
23.6	34.6	35.1	33.6	29.6	16.4	28.7	26.3	8.47
1.18	1.62	2.10	1.68	1.35	1.75	8.80	0.87	1.72
6.0	20.6	14.2	17.7	6.29	8.74	7.90	0.87	5.41
0.36	0.16	0.15	0.19	0.16	0.64	0.29	0.18	0.85
0.74	0.17	0.27	0.29	0.40	1.43	0.71	0.61	1.91
2.41	0.79	0.79	0.89	0.75	2.00	2.50	2.20	1.24
2.07	0.73	1.07	0.98	7.15	2.89	2.65	8.11	3.86
18.1	34.0	34.2	33.5	13.0	14.1	22.7	1.20	5.80
0.002	0.01	0.03	0.04	0.054	0.17	0.17	3.10	0.31
0.019	0.031	0.020	0.019	0.020	0.046	0.033	0.072	0.05
–	–	0.37	–	–	–	0.75	0.70	–
–	–	1.01	–	26.1	–	1.28	30.1	–
100.4	99.9	99.8	99.5	100.1	99.2	99.4	100.3	99.5
0.89	0.15	–	–	0.17	0.1	–	0.77	0.09
1.51	0.31	1.43	0.15	–	0.2	1.37	–	–
1.74	0.36	–	0.18	–	0.24	–	–	–
35.0	46.1	56.9	51.0	20.9	18.7	43.9	1.33	13.7
10.7	37.0	25.3	31.7	11.3	15.6	13.4	1.56	1.25
45.6	83.0	84.6	82.7	32.2	34.3	57.3	2.89	15.0
–	–	–	–	42	–	3.18	60.5	–
–	–	1	–	5	–	–	–	–
6	16	–	16	–	14	–	3	–
30	–	5	–	7	30	20	15	47
16	–	5	–	10	25	10	12	30

Table 3.8 Chemical (Wt%) and Mineral Composition of Manganese Ores and Host Rocks From the Kvirila Deposit (Compiled by Zh.V. Dombrovskaia

	Number of Sample	SiO_2	TiO_2	Al_2O_3	Fe_2O_3	MnO
Well 824 (interval 607.00–607.80 m)						
Op-C	27	33.9	0.23	3.56	2.90	23.17
C	26	15.4	0.27	4.08	2.88	32.61
C	25	14.8	0.23	3.44	3.70	34.50
Op-Cl	24	60.2	0.55	10.86	4.96	0.05
Well 814 (interval 1549.10–550.60 m)						
Ps	55	84.9	0.14	3.49	1.59	0.37
C	54	30.8	0.13	2.51	2.92	24.26
Op-Cl	53	60.0	0.44	9.50	6.00	0.55
G-Cl	52	45.2	0.30	8.74	21.7	3.91
O-C	51	12.1	0.06	3.14	0.98	27.68
O-C	50-1	31.0	0.25	6.46	2.66	15.73
Op-C	50-2	37.5	0.60	7.94	3.09	17.50
G-Cl	49-1	38.7	0.38	9.38	28.7	4.05
N-C	49-2	21.3	0.43	4.26	11.0	22.6
Op-Cl	48-1	58.9	0.30	12.3	9.47	0.30
Op-Cl	48-2	52.3	0.71	10.1	14.9	0.30
Well 518 (interval 576.0–577.0 m)						
Op-C	74	30.5	0.18	3.43	7.42	22.0
O-C	72	7.15	–	1.96	1.26	36.2
O-C	70	13.1	tr	3.29	1.88	31.5
G-C	69	27.5	0.40	6.92	19.3	15.6
Op-Cl	68	56.5	0.48	13.7	6.28	1.50
Well 808 (interval 560.6–561.35 m)						
C	81	18.8	0.18	2.31	3.48	26.3
C	79	9.03	0.15	2.52	4.84	32.5
Well 536 (interval 580.0–581.5 m)						
C	87	26.2	0.27	6.16	4.56	21.7
N-C	85	24.0	0.31	6.98	4.98	31.8
C	84	21.9	0.22	5.08	7.74	20.5
Well 889 (interval 584.8–587.4 m)						
Op-Cl	45	68.2	0.72	9.12	7.28	50.3
C	44	3.65	0.10	1.18	4.11	25.9
N	43	28.4	0.54	8.50	5.08	29.5
N	42	28.4	0.35	6.14	4.02	42.6
O	39	10.8	0.18	1.65	0.90	–

Sections: *Rodinauli (Wells 824, 814, 518, 808, 536) and Cholaburi (Wells 889, 868).*

	MgO	CaO	Na_2O	K_2O	H_2O^-	H_2O^+	CO_2	BaO
	1.95	8.26	0.26	0.78	0.97	1.74	20.80	0.03
	3.71	7.82	0.74	0.75	1.34	1.84	27.68	0.03
	2.58	7.88	0.59	0.50	1.43	1.26	29.12	0.02
	2.25	1.75	1.96	2.63	4.87	3.71	–	0.01
	0.88	1.07	0.57	0.55	1.73	1.57	–	–
	1.94	9.46	0.53	0.35	0.93	1.90	22.2	–
	1.18	1.70	2.19	2.34	4.68	3.48	0.50	–
	2.03	3.00	1.70	1.66	3.36	4.67	3.44	–
	2.60	7.25	0.43	0.46	0.85	5.61	16.5	–
	3.49	6.58	1.28	1.33	2.37	3.72	15.0	–
	3.23	6.20	1.63	1.47	1.45	3.62	14.8	0.05
	2.07	2.12	1.69	1.91	3.37	5.00	2.47	–
	3.42	9.38	0.90	0.83	1.47	3.41	20.7	0.02
	3.49	1.48	2.44	3.25	5.92	3.89	Sl.	–
	3.67	1.78	2.07	2.47	4.97	5.64	–	0.04
	2.60	7.07	0.56	0.46	1.35	1.17	22.6	0.01
	2.89	8.71	0.47	0.51	0.70	2.30	29.4	0.02
	2.53	6.52	0.73	0.52	1.43	3.78	21.8	0.03
	2.57	5.10	1.29	1.21	2.70	4.08	12.7	0.05
	2.40	5.44	2.20	3.35	2.29	1.62	1.84	0.08
	2.10	13.2	0.70	0.57	1.42	1.68	23.5	0.00
	4.28	11.7	0.52	0.46	1.29	1.29	31.0	0.01
	2.95	9.83	1.05	1.20	1.83	1.83	21.7	0.01
	3.42	5.26	1.09	0.90	2.08	4.03	14.2	0.05
	3.29	11.7	0.86	0.87	1.83	1.89	22.1	0.01
	1.88	0.86	1.56	2.32	3.65	3.47	–	0.02
	1.79	2.87	0.20	0.19	0.54	0.51	34.8	0.41
	8.75	0.87	0.85	0.32	4.64	8.67	4.36	0.01
	7.60	2.25	0.75	0.50	3.65	9.04	2.80	0.01
	1.39	0.32	0.40	0.19	1.33	7.98	5.36	0.01

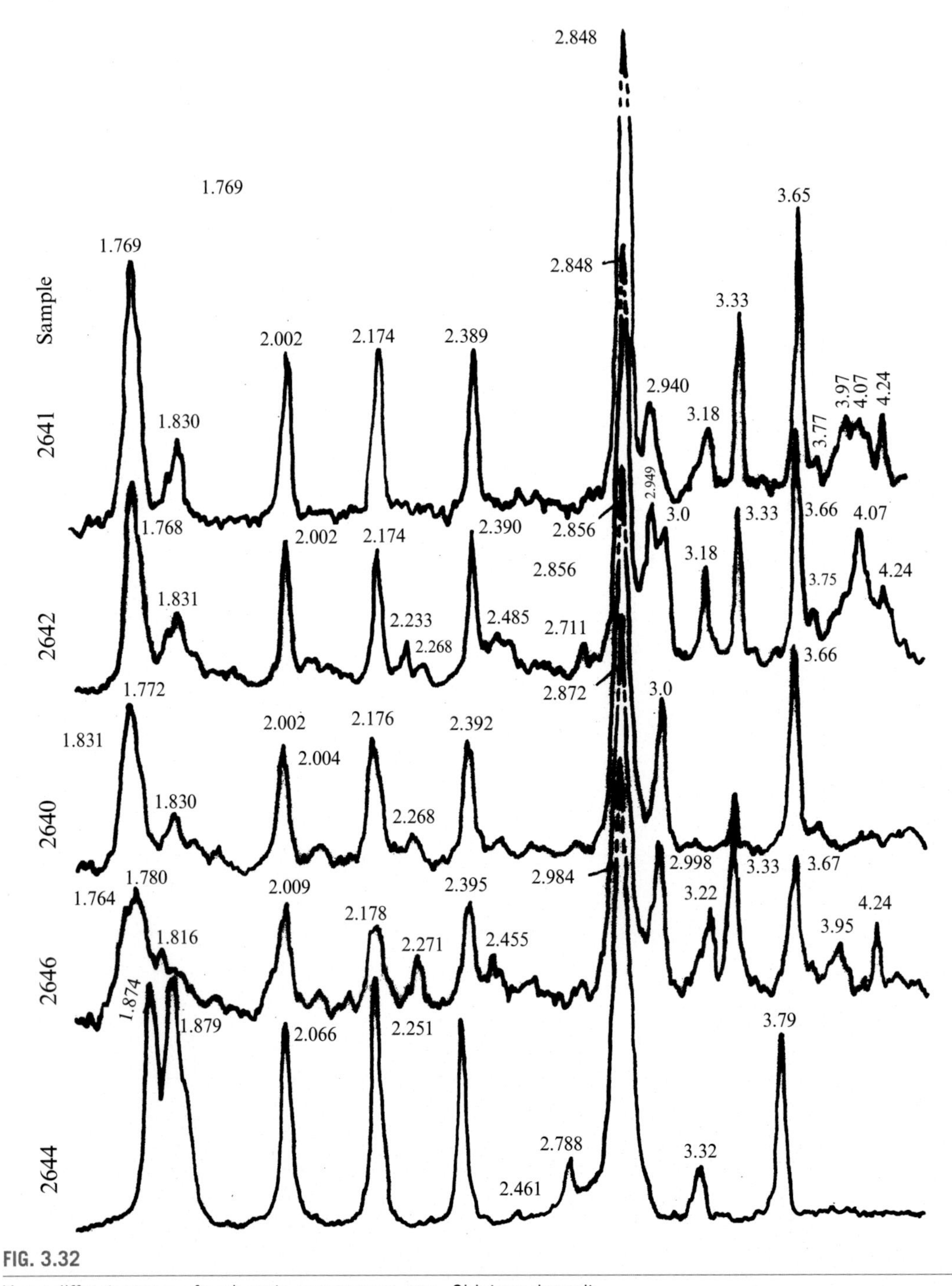

FIG. 3.32

X-ray diffractograms of carbonate-manganese ores, Chiatura deposit.

besides carbonates, terrigenous minerals—quartz, feldspars, micas, and rarely clasts of volcanic glass; and authigenic minerals—opal, zeolites (clinoptilolite), glauconite, glauconite-smectite mixed-layered assemblages, neotocite, pyrite, barite, gypsum, native sulfur, and bituminous matter.

Carbonate-manganese ores of the examined deposits in terms of mineral and chemical compositions, structural-textural particularities, and physical-chemical properties are of the same type. Characteristic for the ores are irregularities of composition, structure, and density due not only to the initial uneven distribution of carbonate material in the host rock, but also to its subsequent redistribution during the stage of diagenesis and to alteration effected by later superposition processes.

The study of carbonate ores in thin sections has enabled the identification of five varieties (groups) of carbonate matter: I—dark-gray up to black cryptocrystalline, from the basic opal-carbonate mass; II—light-gray (up to white) friable, porous, frequently recrystallized in small- to large-spherulite aggregates, from the altered basic mass; III—white, pink, greenish-gray, large-spherulite, from the spherical inclusions of pseudooolites and pseudopisolites; IV—light-gray, white, pink, crystalline from cross-veins; V—gray, pink, large-spherulite from cement of oxide-carbonate ores.

Group I. To this group is classified the earliest carbonate material found in the thin mixture with host siltstones and opal-zeolite matter. The distribution of carbonate is extremely uneven. It is cryptocrystalline and together with the host material forms dense sections of dark-gray to black color. If the host rock has nodular texture and contains spherical inclusions of opal or clayey matter, then the carbonate matter has replaced only the matrix of the rock, and the inclusions will retain their initial composition. Carbonate material of this group is commonly represented by rhodochrosite and/or calcite-rhodochrosite.

Group II includes later carbonate material, which formed by dissolution and redistribution of carbonates of Group I. In this case, the ore becomes porous, often friable. Opal and clayey inclusions are subject to leaching. Around the pores as a rule has been observed a recrystallization of the carbonate material and its purification from impurities with the formation of authigenic grains of pyrite, barite, zeolite, and others. Small-, medium-, and large-spherulite aggregates of carbonate material have formed and are distinguished by their light color against the dark-gray background. Carbonate material of this group is represented by a mixture of rhodochrosite or calciorhodochrosite with manganocalcite.

Group III encompasses carbonate material of spherical inclusions (pseudooolites and pseudopisoliths) from nodular carbonate ores. These are formed by means of replacing the carbonate material of the primary inclusions, composed of opal or clayey matter. This process is accompanied by their partial or full leaching. With that, certain cells remain empty, others maintain part of the primary material, while a third part is partially or fully filled in by carbonate. The carbonate as a rule is large-spherulite, which can be taken as evidence of its crystallization in an environment of free space.

It must be noted that in the described spherical inclusions of carbonate ores, concentrically layered textures are fully absent, which would allow their classification as oolith-pisolith structures, which in oxide form are widely distributed in the studied deposits among oxides and oxide-carbonate ores. Carbonate inclusions in this case only take the form of primary clayey-opal structures. Against the background of the enclosing mass, they are distinguished by lighter color—white, pink, and greenish-gray. Carbonate material of the inclusions is represented by rhodochrosite and/or calcite.

Group IV is represented by carbonate material from cross-veins developed in carbonate ores. Among these occur cross-veins concordant to the bedding as well as transverse veins, which are commonly represented by phanerocrystalline carbonate of light-gray, white, or pink color. In certain cases are formed feathered aggregates, rosettes, and ribbons. The thickness of the cross-veins constitutes from tenths of a millimeter to several millimeters in transverse veins, and several centimeters in cross-veins concordant to the bedding. Vein fillings concordant to the bedding have lensoidal form and pinch out. Cross-veins concordant to the bedding are commonly composed of rhodochrosite or calciorhodochrosite, while transverse veins are composed of calcite or manganocalcite.

Group V is represented by carbonate material from the cement of the oxide-carbonate ores and coincides with the middle part of the ore stratum. Carbonate matter here, alongside siltstone and opal-zeolite material, composes a matrix, among which are observed nodules with dimensions of 2–10 mm, represented most frequently by manganite and occasionally by pyrolusite, goethite, or neotocite. Oxide nodules throughout are eroded and replaced by carbonates as well as transected by carbonate cross-veins, which provides evidence of a later origin for the carbonates. Carbonate material of this group is represented by rhodochrosite with an impurity of calcite or manganocalcite.

It must be noted that the delineation of the indicated groups is to a certain degree conditional, insofar as the carbonates of the delineated groups in the thin sections are frequently located in close association with one another.

In the Kvirila deposit, owing to the insignificant quantity of samples, only two groups could be delineated. The first corresponds to the carbonates of the first group of the Chiatura deposit, and the second consists of the matter of Groups II–V.

Material for the isotope research of manganese carbonate ores in the Chiatura deposit has been gathered from quarries, tunnels, and wells of nine sections situated within the boundaries of the Darkveti and Itkhvisi uplands in the central part of the deposit (see Fig. 3.28). In the Kvirila deposit, samples were taken from the core of the bore wells drilled in the Rodinauli and Cholaburi sections in the year 1982. These are likewise situated in the central part of the deposit. Samples were gathered from various parts of the vertical section across the ore horizon—from the subore sands to the supraore spongolite sandstones.

Results of isotope research

Carbon and oxygen. All obtained isotope data for the various types of carbonate rocks of the studied deposits are recorded in Tables 3.9 and 3.10 and shown in the cross plots of Figs. 3.33 and 3.34. From these, it follows that the values of the isotopic composition of carbon and oxygen vary within a fairly wide range. The greatest scattering of $\delta^{13}C$ and $\delta^{18}O$ values is noted in the ores of the Chiatura deposit and constitutes from −34.5‰ to −8.3‰ and from 18.9‰ to 30.5‰, respectively. Moreover, the identified groups are on the whole distinguished in terms of isotopic composition ($\delta^{13}C$ and $\delta^{18}O$, ‰, respectively): for Group I (gray, cryptocrystalline, carbonate matter)—from −24.7 to −9.0 and from 24.1 to 28.5; from Group II (altered carbonate matter)—from −33.3 to −8.6 and from 22.3 to 28.8; for Group III (carbonate ooids)—from −35.4 to −9.5 and from 19.1 to 30.5; for Group IV (carbonate from cross-veins)—from −31.9 to −12.4 and from 19.3 to 28.8; and for Group V (carbonate cement from oxide-carbonate ores)—from −20.6‰ to −7.1‰ and from 18.0‰ to 28.7‰.

From the reported data, it follows that the lowest variations in isotopic composition are particular to the least altered carbonate matter (Group I), and the highest variations are characteristic of the carbonate matter of ooids (Group III) and cross-veins (Group IV). On the whole, for manganese carbonates of the Chiatura deposit no correlation is noted in the distribution of values of $\delta^{13}C$ and $\delta^{18}O$, although for the carbonate of the cementing mass (see the graph in Fig. 3.33, field 5) can be noted a fairly elongated "field" of distribution of these values in the direction from the lower left to the upper right corner of the graph. This could be evidence of a mixing of carbonate matter from different sources with different initial isotopic compositions.

Table 3.9 Isotopic Composition of Carbon and Oxygen of Carbonate Matter of Manganese Ores of the Chiatura Deposit

Number of Analysis	Number of Sample	Sample Category and Orebody	Location Detail	Year	Calcite		Rhodochrosite	
					$\delta^{13}C$, ‰ PDB	$\delta^{18}O$, ‰ SMOW	$\delta^{13}C$, ‰ PDB	$\delta^{18}O$, ‰ SMOW
I. Dark-gray cryptocrystalline carbonate from the basic opal-carbonate mass								
2652	20	Darkveti uplands	Section 4 quarry	1987	–	–	–9.8	28.1
2645	19				–	–	–16.3	26.8
2655	26				–	–	–7.7	26.0
1717	1875			1983	–	–	–25.7	26.3
1720	1879				–	–	–16.6	27.1
2641	108		Section 8 quarry	1982	–	–	–24.7	26.2
2689	1059		Section 4, quarry 5	1987	–18.8	24.1	–13.8	26.4
2688	1057				–	–	–15.5	26.6
2684	1047				–	–	–9.0	26.7
2764	1117		Section 10 quarry	1987	–11.2	26.9	–12.3	26.3
2765	1123		Section 10, well 830		–	–	–12.2	28.5
2766	1126				–	–	–10.2	27.0
2768	1130		Section 10, well 828		–19.9	28.6	–15.3	28.3
2770	1136		Section 10, well 826		–9.6	28.9	–9.6	28.1
2661	144	Akhali-Itkhvisi uplands	Tunnel	1982	–	–	–9.6	27.2
2665	145				–10.7	28.6	–11.5	27.8
2668	146				–	–	–9.1	27.8
2670	147				–10.7	27.9	–11.4	27.4
2677	152				–	–	–10.2	27.4
II. Light-gray, loose, commonly decrystallized carbonate								
2653	20	Darkveti uplands	Section 4 quarry	1982	–	–	–14.5	28.2
2643	17				–	–	–20.3	26.4
2646	18				–	–	–18.4	25.9
2648	24				–	–	–33.3	27.3
1731	1888				–	–	–13.3	21.1
1733	1889				–	–	–16.5	20.6
2649	31		Section 8 quarry		–20.5	27.2	–18.6	28.2
2640	109				–17.9	23.0	–17.2	26.2
2642	108				–	–	–31.6	26.3
2695	1076		Section 4 quarry	1987	–23.0	22.3	–18.9	23.4
2763	1056				–	–	–18.7	25.2
2685	1047				–15.4	22.3	–8.9	26.2
2700	1093		Section 3 tunnel		–17.2	24.5	–15.6	25.7
2771	1136		Well 826		–13.0	23.7	–10.9	28.0
2773	1141		Well 827		–15.2	22.1	–15.7	24.0

Continued

Table 3.9 Isotopic Composition of Carbon and Oxygen of Carbonate Matter of Manganese Ores of the Chiatura Deposit—Cont'd

Number of Analysis	Number of Sample	Sample Category and Orebody	Location Detail	Year	Calcite		Rhodochrosite	
					$\delta^{13}C$, ‰ PDB	$\delta^{18}O$, ‰ SMOW	$\delta^{13}C$, ‰ PDB	$\delta^{18}O$, ‰ SMOW
2664	144	Akhali-Itkhvisi uplands	Tunnel	1982	–	–	–16.9	27.8
2666	145				–14.8	28.1	–15.8	27.9
2669	146				–	–	–14.1	27.5
2672	147				–13.6	25.9	–16.3	27.6
2673	148				–17.7	25.4	–17.5	27.4
2674	149				–	–	–13.8	27.9
2675	150				–	–	–10.3	27.3
2676	152				–	–	–8.6	27.9
2679	153				–	–	–8.9	27.4
2680	154				–	–	–9.7	27.5
2681	155				–	–	–10.3	27.6
2682	156				–	–	–9.0	27.8
III. Carbonate ooids								
2654	20	Darkveti uplands	Section 4 quarry	1982	–	–	–17.0	25.8
2657	26				–	–	–10.6	22.0
2706	1109				–	–	–35.4	27.0
2774	1143				–10.8	19.6	–9.5	18.9
1701	1870b			1983	–20.7	20.2	–	–
1703	1871b				–	–	–20.4	28.4
1708	1872b				–	–	–15.6	26.8
1711	1873b				–	–	–17.0	24.1
1739	1893				–	–	–17.7	26.1
1744	1896				–	–	–11.8	27.0
1781	1900		Section 8		–	–	–14.0	29.7
1782	1901				–	–	–12.8	30.5
2772	1136		Well 826		–14.5	19.1	–11.6	26.4
2767	1126		Well 830		–25.7	20.1	–24.0	21.0
2662	144	Akhali-Itkhvisi uplands	Tunnel	1982	–15.1	28.8	–16.7	27.4
2667	145				–15.7	28.3	–16.3	28.0
2672	147				–14.2	27.0	–15.7	27.2

Table 3.9 Isotopic Composition of Carbon and Oxygen of Carbonate Matter of Manganese Ores of the Chiatura Deposit—Cont'd

Number of Analysis	Number of Sample	Sample Category and Orebody	Location Detail	Year	Calcite		Rhodochrosite	
					$\delta^{13}C$, ‰ PDB	$\delta^{18}O$, ‰ SMOW	$\delta^{13}C$, ‰ PDB	$\delta^{18}O$, ‰ SMOW
IV. Light-gray carbonate from cross-veins								
2644	17	Darkveti uplands	Section 4 quarry	1982	–30.0	27.5	–31.9	26.3
2647	24				–19.5	27.0	–31.1	26.5
2692	1068			1987	–21.8	25.6	–23.0	26.3
2690	1059				–21.8	20.0	–20.5	19.3
2691	1063				–16.9	25.1	–12.9	26.3
2650	31				–19.5	28.7	–21.4	28.9
2701	1093		Section 3, tunnel		–	–	–16.1	25.3
2707	1109		Section 9, tunnel		–17.3	25.4	–18.6	25.3
2769	1130		Well 826		–18.7	28.6	–12.4	27.5
2775	1146		Well 830		–19.0	26.6	–	–
V. Carbonate cement of oxide ores								
2660	23	Darkveti uplands	Section 4 quarry	1982	–	–	–20.6	23.4
1721	1880			1983	–17.4	18.8	–	–
1734	1890a				–16.5	17.5	–	–
1735	1890b				–16.3	18.5	–	–
1736	1891				–17.3	23.0	–	–
1738	1892				–	–	–16.7	21.4
1740	1894				–	–	–19.1	22.0
2658	27a		Section 8 quarry	1982	–11.2	27.3	–16.1	24.8
2659	28				–	–	–9.6	26.6
2678	153	Akhali-Itkhvisi uplands	Tunnel	1982	–	–	–9.3	27.1
2697	1083				–	–	–8.3	26.5

Dash: *absence of mineral or no data.*

Attention should be drawn to the circumstance that substantial differences have not been noted in the isotopic composition of the oxygen of the different groups of carbonate matter of Mn ores. This is a crucial point. In opposite (by different of oxygen isotope data), it might be supposed that our studied types of Mn carbonates could have different conditions of formation and a different source of CO_2, which would have been reflected in the isotopic composition of their carbon and oxygen.

Substantial isotope-oxygen differences are likewise undetected among the studied groups of carbonate material, as is clearly visible in the histograms of Figs. 3.35 and 3.36. On the whole, the $\delta^{13}C$ and $\delta^{18}O$ values and the character of their distribution are analogous for the various groups; that said,

Table 3.10 Isotopic Composition of Carbon and Oxygen of Carbonate Matter of Manganese Ores of the Kvirila Deposit

Number of Analysis	Number of Sample	Well Number	Well Depth (m)	Calcite		Rhodochrosite	
				$\delta^{13}C$, ‰	$\delta^{18}O$, ‰	$\delta^{13}C$, ‰	$\delta^{18}O$, ‰
I. Carbonate gray, crptocrystalline							
Rodinauli section							
3485	74–1	Well 518	Depth 576.10	–	–	−14.1	25.8
3491	87-1	Well 536	Depth 580.80	–	–	−11.5	28.7
3489	84-1		Depth 581.45	−15	24.6	−18.7	26.3
3488	81-1	Well 808	Depth 560. 8	−12.9	29.3	−14	28.6
3463	26-1	Well 824	Depth 607.60	–	–	−13.1	24.9
2461	25-1		Depth 607.70	–	–	−17.1	26.3
Cholaburi section							
3479	64-1	Well 868	Depth 576.80	−8	29.8	−11	27.8
3478	63-2		Depth 577.00	−7.3	27.6	−19.2	25.4
3466	34-2	Well 889	Depth 587.3	–	–	−13.1	26.3
II. Carbonate white (up to pink), crystalline							
Rodinauli section							
3486	74-2	Well 518	Depth 576.1	−0.5	24.9	−10.7	25.4
3484	72-2		Depth 576.3	–	–	−7.5	27.1
3483	70-2		Depth 576.7	–	–	−3.7	22.2
3482	69-2		Depth 577.0	–	–	−4.2	21.1
3490	85-2	Well 536	Depth 581.3	–	–	−3.6	23
3484	79-2	Well 808	Depth 561.3	–	–	−6.6	23.8
3473	54	Well 814	Depth 549.2	−6.7	22.7	−6.6	28.1
3499	53		Depth 549.3	–	–	−7.2	22.2
3472	51-2		Depth 549.4	−7.1	23.6	−7.2	26.6
3471	49-2		Depth 550.3	–	–	−5.6	28.3
3464	27-2	Well 824	Depth 607.4	–	–	−5.4	29.8
3492	27-1		Depth 607.4	–	–	−4.9	28.4
3462	25-2		Depth 607.7	−3.7	23.8	−5.6	22.2
Cholaburi section							
3481	65	Well 868	Depth 576.5	–	–	−3.3	27
3480	64-2		Depth 576.8	−0.9	24.8	−3.7	27.6
3476	61-2		Depth 577.4	–	–	−5.4	26.1
3475	60-2		Depth 577.5	–	–	−2.3	27.5
3474	59-2		Depth 577.65	–	–	−8.1	28.1
3469	44-2	Well 889	Depth 585.0	–	–	−8.4	24.9
3456	34-1		Depth 587.3	–	–	−4.2	23.9

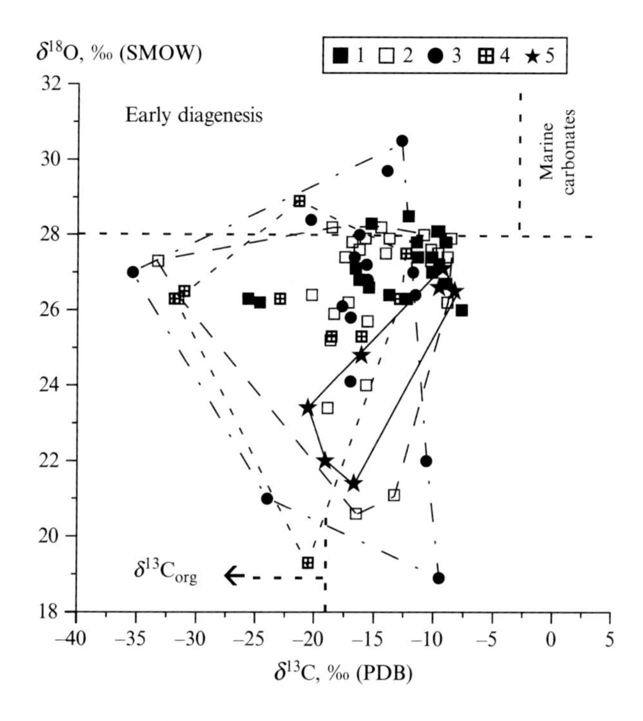

FIG. 3.33

$\delta^{13}C$ versus $\delta^{18}O$ in carbonate substance of manganese ores, Chiatura deposit. 1—Dark-gray *(to black)* cryptocrystalline from the opal-carbonate matrix; 2—light-gray *(to white)* loosening, porous, often crystallized; 3—white, pink from rounded inclusions—ooids; 4—light-gray, white from streaks; and 5—gray, pink, large-spherulitic from cement of oxide-carbonate ores. *ooid*=roundish, oval.

certain isotopic particularities can be noted that are particular to a given group. Thus, for example, for practically all of the studied rocks, the greatest number of determinations of $\delta^{18}O$ belong to the 26–28‰ interval (a "peak" in the histograms). It can be supposed that this provides evidence of consistent conditions of formation or a single source of carbon dioxide solutions from which was produced the formation of the overwhelming portion of the studied carbonates. The presence of samples with lower $\delta^{18}O$ values enables us to propose that their precipitation occurred under different physical-chemical conditions and with the presence of other sources of ore-depositing solutions characterized by lower values of $\delta^{18}O$.

Data on the isotopic composition of carbon (histograms of Figs. 3.35 and 3.36) likewise provide evidence of the heterogeneity (in the isotopic regard) of a carbon dioxide reservoir or of the presence of several sources of CO_2 taking part in the formation of the studied carbonates.

Carbonate ores of the Kvirila deposit by comparison with those of Chiatura are characterized by narrower boundaries of variations in isotopic composition. For the least altered carbonate matter (Group I) $\delta^{13}C$ values vary from −19.2‰ to −2.3‰, and those of $\delta^{18}O$ from 24.6‰ to 29.8‰ [analogues heavy carbon isotope data are characteristic for some China manganese deposits (Xie at al., 2013)]. These are close to the same values of the carbonate matter (Group I) of the Chiatura deposit. For later, altered, and superimposed carbonate matter (Group II), these variations constitute from −10.7‰ to −2.3‰ and from 21.1‰ to 29.8‰, respectively.

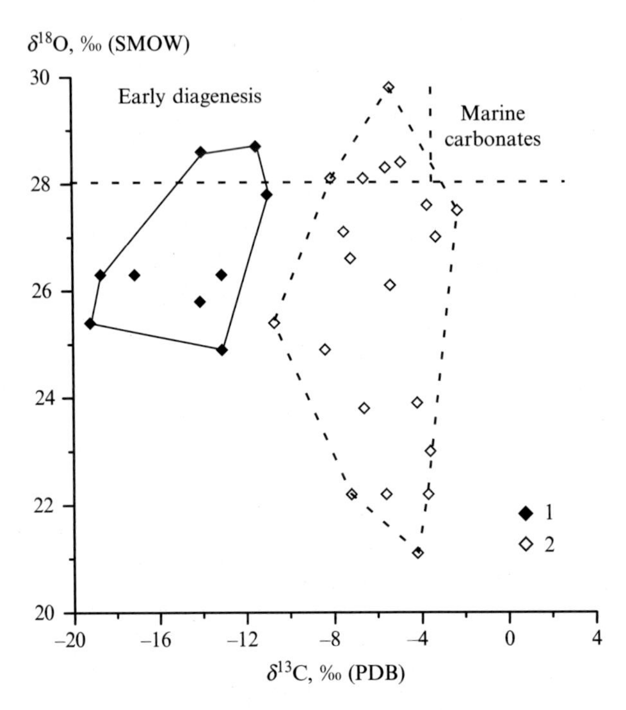

FIG. 3.34

$\delta^{13}C$ versus $\delta^{18}O$ in carbonate substance of manganese ores, Kvirila deposit. 1—Dark-gray *(to black)* cryptocrystalline from the opal-carbonate matrix and 2—light-gray *(to white, pink, and white)*, spherulitic.

In the graph of Fig. 3.34, plotted within the coordinates $\delta^{13}C$-$\delta^{18}O$, the identified groups form independent "isotope fields." Moreover, the light, crystalline varieties (Group II) are characterized by a heavier isotopic composition of carbon and wider boundaries of variations in $\delta^{18}O$ values. Within the boundaries of these groups, as for rocks of the Chiatura deposit, is noted an uneven distribution of $\delta^{13}C$ and $\delta^{18}O$ values (histograms, see Fig. 3.36), which serves as evidence for a heterogeneous source and different conditions of formation of the rocks of the identified groups.

Recrystallized light-colored varieties of manganese carbonate ores (Group II) of the Kvirila deposit are characterized on the whole by the heaviest isotope composition of carbon and do not have analogues (in the isotopic regard) in the Chiatura deposit. This serves as evidence for the substantial differentiation of the sources of matter (ore, CO_2, solutions, etc.) and conditions of formation of the studied deposits.

The isotopic composition of calcite in ores of the Chiatura deposit as a rule is close to the composition of their accompanying manganese carbonates. This is evidence of their isotope-geochemical equilibrium (or conditions close to it). However, a significant proportion of the samples of calcites differ substantially in terms of the isotopic composition of carbon (up to 11.6‰, analysis 2647) as well as oxygen (up to 7.34‰, analysis 2772). Furthermore, these distinctions are observed in lower as well as in higher values of $\delta^{13}C$ and $\delta^{18}O$ (Fig. 3.37); no definite regularity was established in their distribution on the basis of available data. However, a tendency is noted for the majority of samples toward the enrichment of calcite by light isotopes of carbon and oxygen by comparison

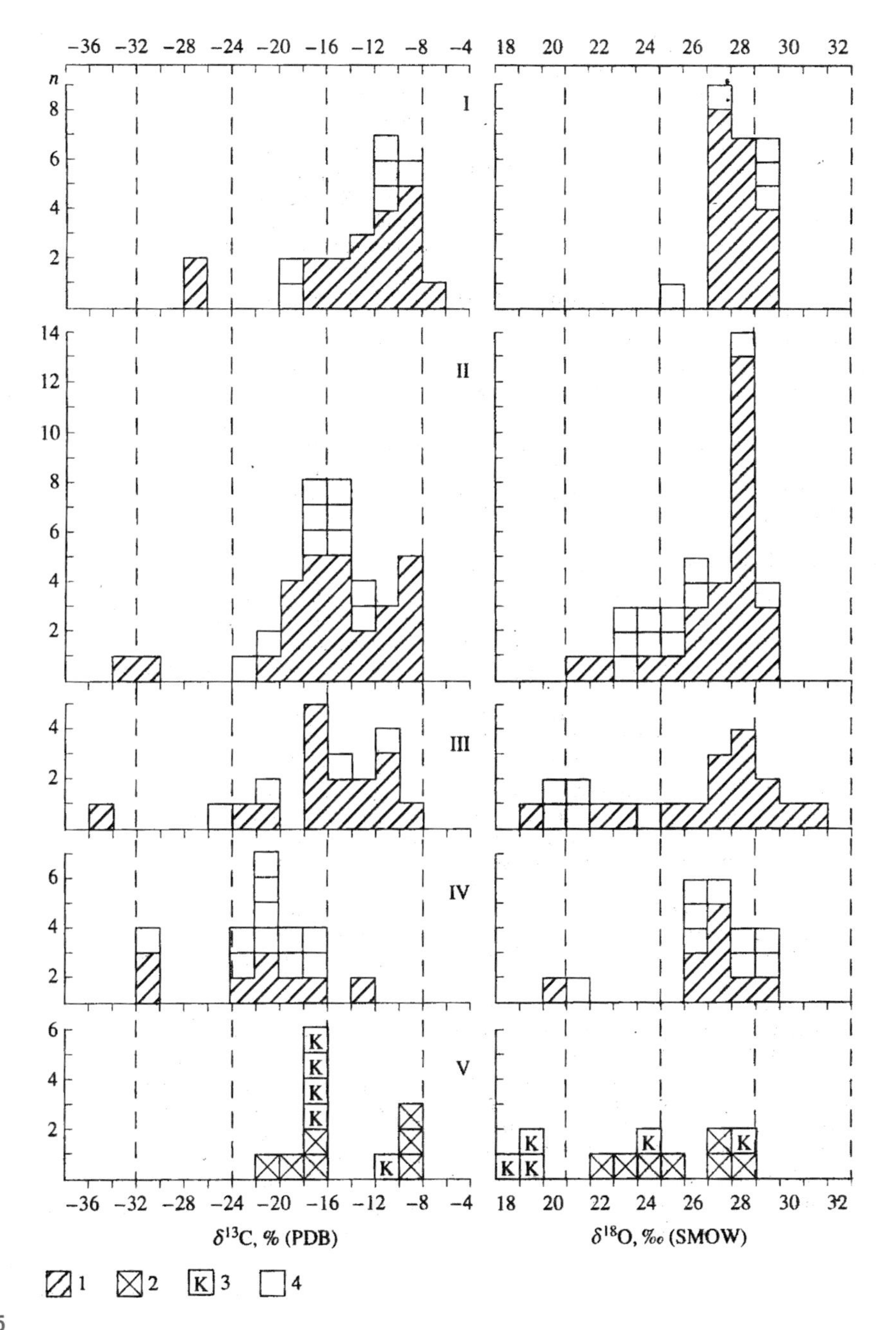

FIG. 3.35

Histograms of $\delta^{13}C$ and $\delta^{18}O$ values distribution in carbonate substance of manganese ores, Chiatura deposit. I—Dark-gray (to black) cryptocrystalline from the opal-carbonate matrix; II—light-gray (to white) loosening, porous, often crystallized; III—white, pink from rounded inclusions (pseudooooliths and pseudopisoliths); IV—light-gray, white, coarse from streaks; and V—gray, pink, from matrix of oxide-carbonate ores. 1—Manganese carbonate; 2—calcite; 3—manganese carbonate matrix of oxide-carbonate ores; and 4—calcite matrix of oxide-carbonate ores.

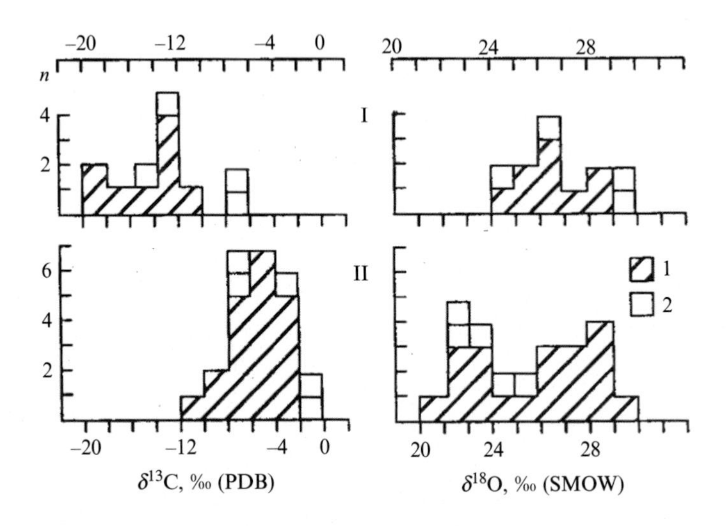

FIG. 3.36

Histograms of $\delta^{13}C$ and $\delta^{18}O$ values distribution in carbonate substance of manganese ores, Kvirila deposit. 1—Manganese carbonate and 2—calcite.

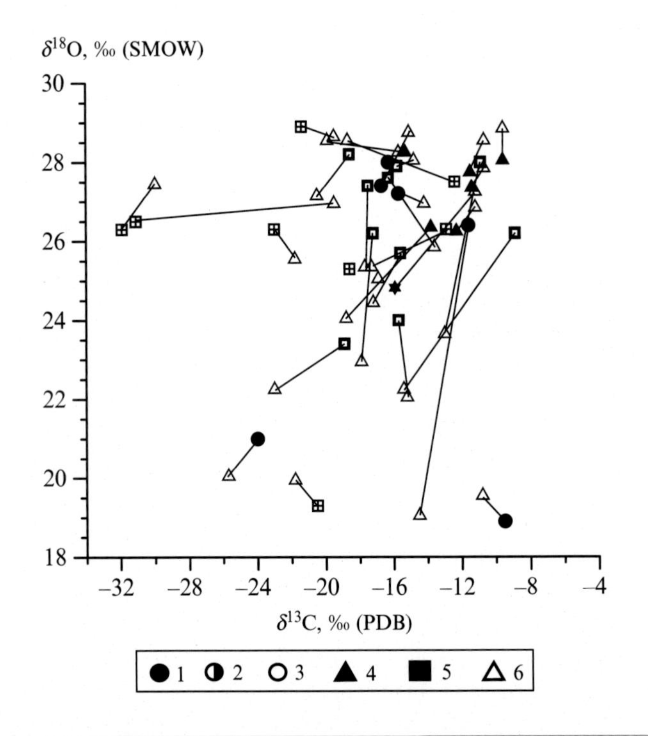

FIG. 3.37

$\delta^{13}C$ versus $\delta^{18}O$ in manganese carbonates and coexisting calcites in the ores of Chiatura deposit. 1—Dark-gray (to black) cryptocrystalline from the opal-carbonate matrix; 2—light-gray *(to white)* loosening, porous, often crystallized; 3—white, pink from rounded inclusions—ooids; 4—light-gray, white, from streaks; 5—gray, pink, large-spherulitic from cement (=matrix) of oxide-carbonate ores; and 6—coexisting calcite.

with the accompanying manganese carbonates. In the manganese rocks of the Kvirila deposit such a tendency is not noted. Here the calcites are distinguished in terms of isotopic composition either of carbon, or only of oxygen, occasionally of oxygen and of carbon simultaneously (on the part of lighter or heavier values) (Fig. 3.38). These data provide evidence that a significant proportion of the calcites have a later origin and were formed under different temperature conditions from carbon dioxide solutions substantially different in terms of isotope characteristics compared with those of manganese ores.

Sulfur. The isotopic composition of sulfur was studied in sulfides (pyrite) and sulfates (barite, gypsum) in rocks of the Chiatura and Kvirila deposits. The results of the isotope analyses are reported in Table 3.11 and are displayed in the graphs of Figs. 3.39 and 3.40. The general boundaries of variations in $\delta^{34}S$ values are fairly broad. For rocks of the Chiatura deposit, they constitute from −19.0‰ to 41.2‰. In this deposit, pyrites were the most fully studied; the scattering of their $\delta^{34}S$ values constitutes from −19.0‰ to 23.3‰.

For barite, these values vary from −4.9‰ to 41.2‰ and for gypsum (two samples) those figures constitute 2.5‰ and 7.9‰.

In the distribution of obtained $\delta^{34}S$ values for pyrite, attention is due to their fairly even distribution within the boundaries of the entire interval. However, there is observed a certain "congestion" of measured points within the intervals −20‰ to −14‰ and 4–12‰. It also stands to note that fully absent from the composition of the studied collection are samples falling in terms of isotopic characteristics into the region of "null" (mantle) values of $\delta^{34}S$.

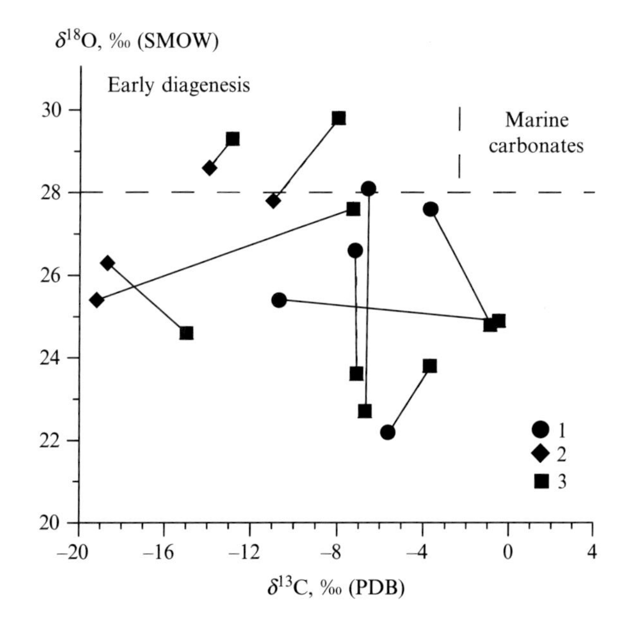

FIG. 3.38

$\delta^{13}C$ versus $\delta^{18}O$ in manganese carbonates and coexisting calcites in the ores of Kvirila deposit. 1—Manganese carbonate dark-gray (to black) cryptocrystalline from the opal-carbonate matrix; 2—light-gray (to white, pink and white) manganese carbonate, crystallized; and 3—coexisting calcite.

Table 3.11 Isotopic Composition of the Sulfur of Sulfides and Sulfates From Manganese Ores of Chiatura and Kvirila Deposits

Number of Analysis	Number of Sample	Characteristics and Location of Gathering of Samples	$\delta^{34}S$, ‰
Chiatura deposit			
I. Pyrite from carbonate ores			
2743	144	Akhali-Itkhvisi uplands, tunnel 1987 year	10.6
2745	1047	Darkveti uplands, section 4, quarry 5, 1987 year	−15
2746	1052		−17.2
2749	1076		9.8
1700	1970c	Darkveti uplands, section 4, quarry 5, 1983 year	4.5
1704	1871c		23.3
1707	1872c		4.4
1715	1874b		−6.8
1718	1877		−19
1719	1878	Darkveti uplands, carbonatized sandstone	−16
2755	1107	Darkveti uplands, section 9, tunnel 1987 year	7.2
2757	1109		2
2761	1141	Darkveti uplands, well 827, depth 22.50 m	−10.9
II. Barite secondary from vugs in carbonate-oxide ores			
2741	23	Darkveti uplands, section 4, quarry 1982 year	3
2747	1056	Darkveti uplands, section 4, quarry 5, 1987 year	41.2
2750	1076		24.2
2442	27a	Darkveti uplands, section 8, quarry 1982 year	−4.9
2756	1107	Darkveti uplands, section 9, tunnel 1987 year	24
III. Gypsum secondary			
2758	1117	Darkveti uplands, section 10, quarry 1987 year	2.5
2760	1145	Darkveti uplands, well 832, depth 120.70 m	7.9
Kvirila deposit			
1659	1857b	Well 508, depth 570.45 m. Pyrite from manganese carbonate ore	38.2
1680	1863b	Well 549, depth 583.45 m. Pyrite from manganese-oxide ore	−20.3
1692	1867c	Well 1022, depth 425.20 m. Pyrite from oolitic manganese ore	17.8

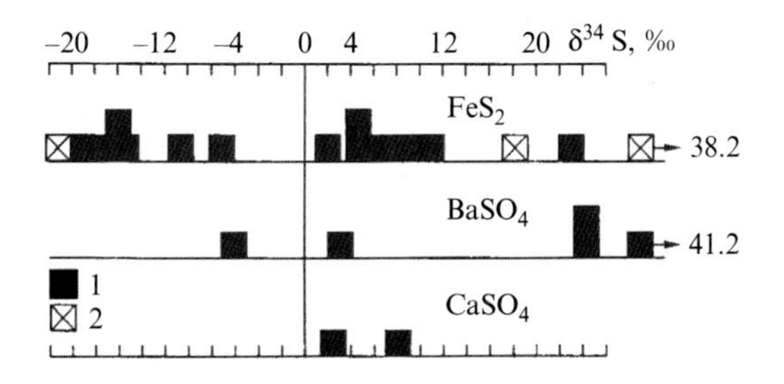

FIG. 3.39

Histograms of $\delta^{34}S$ values distribution in pyrites, gypsum, and barite from manganese ores of Chiatura and Kvirila deposits. Deposits: 1—Chiatura and 2—Kvirila.

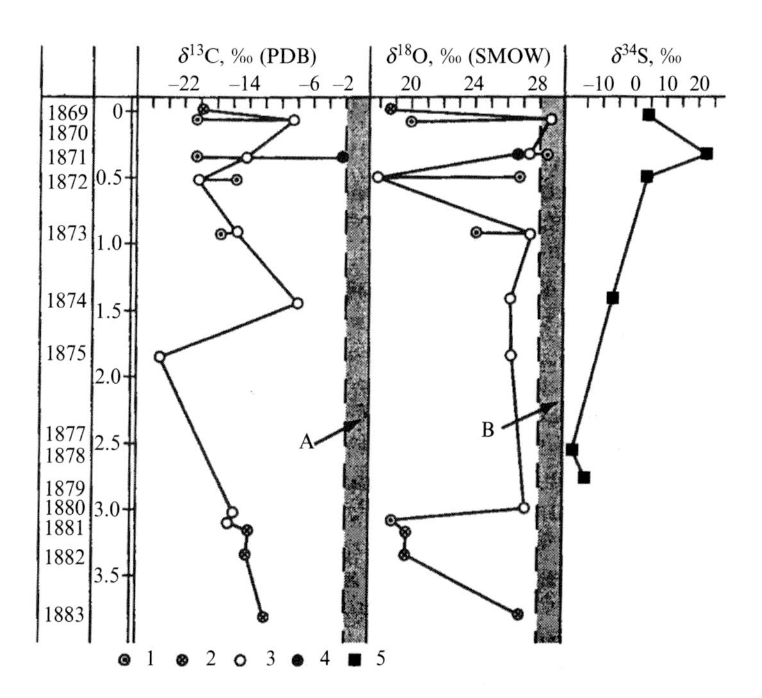

FIG. 3.40

The unique feature of the distribution of $\delta^{13}C$, $\delta^{18}O$, and $\delta^{34}S$ values in carbonate substance of Chiatura manganese-ore deposits (Highlands Darkveti, sector 4, the wall of the quarry). (A) $\delta^{13}C$ values area of normal-sedimentary marine carbonates and (B) $\delta^{18}O$ values area of normal-sedimentary marine carbonates. 1—Carbonate ooids (pseudooliths); 2—scattered calcite of host sandstones; 3—carbonate ores with rare inclusions of pseudooliths; 4—fragment of gray carbonates; and 5—pyrite.

The origin of manganese ores

The structural setting of manganese deposits and their interrelation with oil- and gas-bearing basins. The West Georgia (Dzirula) manganese-ore region, which includes the Chiatura, Kvirila, and a range of other deposits and ore-occurrences, is situated in the intermontane Riono-Kurin depression, which stretches in the sublateral direction between the fold belt of the southern slope of the Greater Caucasus and the Adjara-Trialet fold region. The intermontane depression was inherited from the Jurassic rift-related Black Sea-Caspian Sea depression (Buadze, 1991). Within it are delineated two submergence zones—the Western and Eastern zones, divided by a central uplift zone. The latter consists of the Dzirula and Okriba-Ureit subzones (Gamkrelidze, 1988). With the submergence zones are associated two large oil-and-gas-bearing structures. In the Western submergence zone, represented by the Kolkhida (Rioni) molasse depression, is situated the Black Sea oil-and-gas-bearing region with the Kolkhida, Gura, and Okriba-Racha-Ossetia oil-and-gas-bearing areas; in the Eastern zone, represented by the Kartli (upper Kura) depression, are situated the Kobystan-Kura oil-and-gas-bearing region with the Kartli, Tbilisi, and South Kakheti oil-and-gas-bearing areas (Papava, 1988). The depressions are filled with thick (up to 10–14 km) strata of Mesozoic, Paleogene, and Neogene molasse deposits, superimposed on dislocated rocks of the Paleozoic-Precambrian foundation. The surface of the foundation is submerged in the westerly and easterly directions from the central zone of the uplifts, where it (the foundation) crops out on the surface.

Taking part in the structure of the intermontane Transcaucasia depression are deposits from the Lower Jurassic to the Upper Pliocene inclusively (Buleishvili, 1972). The Jurassic complex is represented by terrigenous (Lower Jurassic), volcanogenic-sedimentary and terrigenous (Middle Jurassic), and terrigenous-carbonate (Upper Jurassic) deposits with a thickness of 2000–3500 m. The Cretaceous system in the lower part (from the Valanginian to the Aptian Stage) is represented by a carbonate complex with a thickness reaching 1000 m, and above (from the Aptian to the Albian) by clayey-marl strata containing interbeds of tuffaceous sandstones and effusive sheets with a thickness of 100–800 m. The Upper Cretaceous is represented (from bottom to top) by sandstone formations of the Cenomanian (200 m) and by a carbonate stratum of the Turonian-Senonian and Danian (200–500 m). The Paleocene and Eocene deposits are composed of marls and limestones (50–160 m). The Oligocene-Lower Miocene (Maikop) series is represented by sandstone-clayey deposits with a thickness from several hundred meters up to 3000 m. The Middle-Upper Miocene sandstone-clayey deposits (Tarkhanian, Chokrakian, Karaganian, Conckian, and Sarmatian stages) have a thickness reaching 3000 m.

The indicated maximum thicknesses of the sedimentary complexes are observed in the parts of the Georgia block that experienced the greatest subsidence; in its border parts and in the uplift zones thicknesses decrease to the full pinch-out of the deposits.

The zones of greater subsidence also contain beds that produce petroleum hydrocarbons as well as beds that serve as reservoirs, indicating their prospective oil-gas-bearing capacity. Indicators of petroleum potential have been established in various stratigraphic levels of the Mesozoic and Cenozoic. Sediments of the upper Lias of the Lower Jurassic, of the Oligocene-Lower Miocene (Maikop series), and of the Middle Miocene are distinguished by the most consistent regional oil-bearing capacity (Buleishvili, 1957). The Lias deposits, containing good quartz-sandstone reservoirs along the northern periphery of the Dzirula massif, in many places are characterized by effective oil-occurrences, which evidently are genetically connected with these deposits.

The Maikop deposits likewise contain thick beds of sands and sandstones saturated with oil over significant distances, particularly along the southern border of the Kartli depression in the Surami-Gori-Mtskheta band. Consistent oil-bearing capacity is characteristic of the Middle Miocene deposits, which contain up to seven thick (8–30 m) sandstone horizons that are heavily saturated with oil over a significant distance. Near the Dzirula manganese-ore area, the most numerous occurrences of oil are known in the northern part of the intermontane region (Upper Racha and South Ossetia). These are associated primarily with deposits of the Upper and Lower Jurassic, but a significant number of occurrences are located in deposits of the Maikop series, the clays of which are frequently bituminous and contain oil adhesions (Laliev, 1964).

To the beginning of accumulation of the Lower Oligocene sediments that contain the manganese ores of the Chiatura, Kvirila, and a range of other deposits and ore-occurrences, the main phase of oil- and gas-generation occurred in the area of the entire mass of Jurassic deposits with a thickness of 2000–3500 m (thickness of the overlying Cretaceous-Paleocene-Eocene deposits exceeded 2000 m). The process of oil-and-gas-generation lasted for the duration of the Oligocene, when a mass with a thickness reaching 3000 m was accumulated, as well as for the Neogene, the thickness of the deposits of which likewise exceeds 3000 m, gradually involving the deposits of the Cretaceous and Paleogene. Periodically occurring structural movements were responsible for the migration of oil and gases into the overlying strata.

It can be supposed that one such large restructuring on the boundary of the Eocene and Oligocene (Mstislavskii, 1985; Salukvadze, 1990) led to the movement of petroleum waters and gases from the oil-and-gas source beds into the fluid discharge zones (nonconservation zones), which were peripheral sections of the basins in those places where effective caprocks were absent (Paragenesis et al., 1990). Along the migration routes, the oil waters are enriched by various components, including manganese (Pavlov, 1989) on account of that element's leaching from volcanogenic and volcanogenic-sedimentary rocks—widely developed among Jurassic and Cretaceous deposits—with elevated contents of this element (0.12–0.52%) (Dzotsenidze, 1980). According to the data of Baskov and Pustovalova (1979), distributed in the Rioni and Kura intermontane depressions are predominantly nitric, nitric-methane, and sulfate isotherms with temperatures at the Earth's surface from 40–55°C to 80–90°C. Content of Mn in them reaches 0.1–0.3 mg/kg—that is, close to the content of this element in volcanogenic-sedimentary rocks of the region.

It is fully possible that such Mn content with the precipitation of the element on the geochemical barriers in discharge zones could have created ore concentrations. Stratal waters directly from the gas-oil deposits, in part of Dagestan (Shtanchaeva, 1984), contain in mean 10.08 mg/L manganese. A complex of microelements (Cu, Zn, Co, Ni, Cr) accompanying the manganese is typical for Oligocene manganese deposits, including Chiatura (Strakhov et al., 1968).

Serving as potential catagenetic discharge zones in the Lower Oligocene, judging by the paleographic reconstructions of this geological epoch, were the border sections of the troughs adjoining the southern slope of the Greater Caucasus and the northern slope of the Adjara-Trialet region, as well as the Central uplift zone. Coinciding precisely with these sections are numerous manganese deposits and ore-occurrences of Western Georgia. These are clustered near the Dzirula rise from the side of the Black Sea as well as from the Kobystan-Kura oil-and-gas-bearing region. In the area of the Central uplift zone at this time existed a landmass in the form of several islands (Fig. 3.41). From the southern slope of the Caucasus and the northern slope of the Adjara-Trialet fold region, these islands were separated by straits uniting the Western and Eastern submergence zones. Coinciding with the northern strait are 19 deposits and ore-occurrences of the Racha group [the northern ore-bearing band, according to Avaliani (1982)]. In the southern strait are situated Vani-Zestafon with a family of ore-occurrences, Surami-Kareli with two ore-occurrences, and Kodmani with four ore-occurrences groups, consolidated

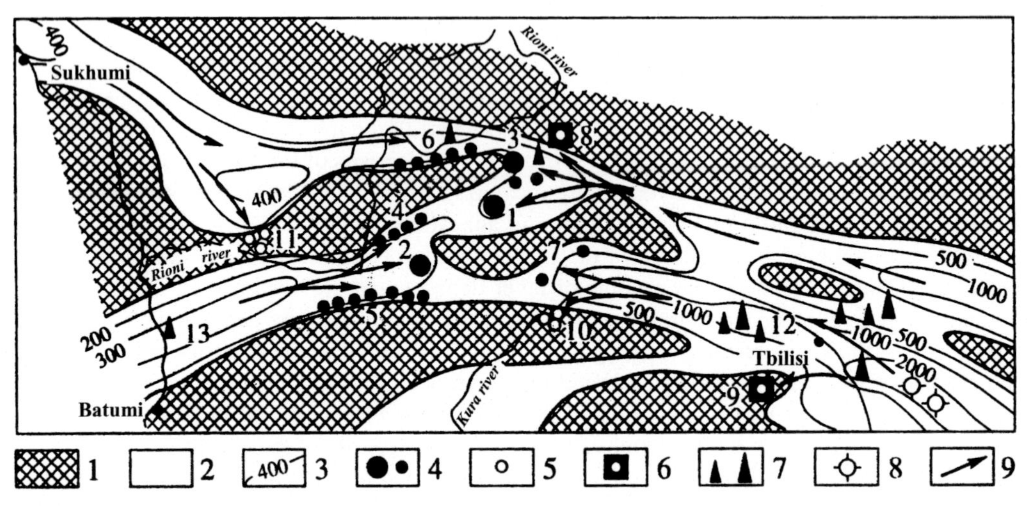

FIG. 3.41

Paleogeographic scheme of the Early Oligocene in Georgia [compiled by Zh.V. Dombrovskaja on (Атлас et al. 1967). 1—Land; 2—sea; 3—isopach of Oligocene sediments; 4 and 5—manganese deposits and Mn-ore manifestations: 4—stratiform (elisional diagenetic), 5—vein type (elisional); 6—complex ore fields; 7—oil fields and occurrences in Oligocene deposits; 8—mud volcanoes; and 9—inferred directions of oil waters migration in the Oligocene. Figures in the scheme: 1–7—elisional diagenetic manganese-ore deposits and occurrence: 1—Chiatura, 2—Kvirila, 3—Shkmeri, 4—Chckari-Adzhametian group, 5—Vani-Zestafoni group, 6—Rachinsk group, 7—Surami-Kareli; 8–9—complex ore fields: 8—Cedisi-Kvaissi (Mn, Fe, Pb, Zn, Cu, Ba, Hg, As), 9—Bolniss-Tetritskaroi (Mn, Fe, Ba, Cu, Pb, Zn); 10 and 11—elisional Mn-ore occurrence: 10—Kodmansk group, 11—Gegechkori-Tskhakaevsk group; and 12 and 13—oil fields: 12—Tbilisi oil- and gas-bearing district, 13—Supsa.

by G.A. Avaliani into a southern ore-bearing band. In the straits between the Okriba, Dzirula, and Kartli island landmasses, in the central ore-bearing band, according to G.A. Avaliani, are situated two of the richest manganese deposits—Chiatura and Kvirila (including the Chkhari-Ajamat section, usually considered an independent deposit).

The wide development of heavy bitumens represented by black amorphous material in the manganese ores and host rocks of these deposits serves as direct evidence that the Dzirula uplift area, in the central part of which are situated the Chiatura and Kvirila deposits, was in the Oligocene a nonconservation zone of the oil-and-gas basin. The bitumens form small lenses and lensoidal beds with a thickness from several millimeters to several centimeters, extending the host layers, as well as separate small (several millimeters) inclusions. Occurrences of bitumens are observed in the entire mass of the ore body in the vertical cross section as well as in the sweep of the layers; in some sections, the occurrences are more numerous, in others less. In the Chiatura deposit, they most frequently occur in the Darkveti, Shukruti, and Perevisi uplands.

Bitumens are particularly widely developed among manganite ores, where they commonly bear reactive hydroxide-ferruginous fringes. In carbonate ores and enclosing sandstones bitumens are associated with a green clayey mineral (mixed-layer hydromica—smectite). Judging by the character of the bitumens' superposition, it can be proposed that they were formed later than the manganite ores.

The connection of manganese ore-formation with faults. The displacement of catagenetic oil waters enriched by manganese, silica, and other accompanying elements to the discharge chambers could have been conducted by two paths—along the bed-conduits and along tectonic fault zones. With the formation of the Chiatura and other deposits of the Dzirula manganese area, oil waters likely initially moved along the Jurassic and Cretaceous beds up to their transecting tectonic faults, and subsequently along the faults they seeped out onto the surface.

The Georgia block is characterized by a large quantity of tectonic faults, the activation of which is connected with various phases of tectogenesis, resulting in its observed complex block-mosaic structure. The largest are the sublateral (the so-called longitudinal) tectonic faults, situated subparallel to the basic tectonic structure of the region—to the southern slope of the Greater Caucasus and the Adjara-Trialet fold region. These faults are deep, long-lived, and determined the tectonic development of the territory in the geosynclinal stage; they were also responsible for the metallogeny of the Alpine cycle (Kekeliia et al., 1979). In the Central uplift zone, large faults—the Potsirevskyi in the north and the Surami-Gokashurskyi (Gokeshurskyi) in the south—separate the intermontane rise from the geosynclinal region devoid of manganese-ore occurrences (Edilashvila et al., 1980). Yet another large longitudinal fault—Sachkhere—splits the Dzirula uplift into two blocks: the northern—Satsalik-Shk'meri, with which coincides a large quantity of ore-occurrences of manganese, and the southern—Dzirula, within the boundaries of which are situated the large Chiatura and Kvirila deposits.

Besides the above-named sublateral faults there is a large quantity of faults of north-westerly and north-easterly strike, commonly termed transverse, the activation of which began with the orogenic stage (Oligocene). These faults disrupt the monoclinal setting of the beds of the Jurassic-Cretaceous sedimentary strata, along which oil waters from source beds of oil and gas generation discharge to reservoir rocks or are lost to the surface.

The Chiatura deposit is bounded from three sides by large tectonic faults. In the north it adjoins the Sachkhi longitudinal fault, and in the south-west and east it is bounded by transverse faults (see Fig. 3.28). The amplitude of displacement along the fault in the south-west (the Glavnyi fault) constitutes >30 m; the large, south-western part of the deposit is considered to have eroded at a time of transgression by the Chokrakian Sea.

The Glavnyi fault transects Oligocene deposits. Makharadze (1972) considered that this fault is a disruption of deep type and originated in the post-Oligocene. On the other hand, it is possible that this fault is not post-Oligocene and that the observed displacement of Oligocene beds is instead connected with a pre-Chokrakian activation of an essentially earlier fault. According to Dzotsenidze (1980), the Dzirula massif was crushed in the Bajocian into blocks of deep faults, which also served as "conduits" along which post-volcanic hydrothermals emerged and reached the bottom of the Oligocene sea.

Near the fault, the ore horizon is represented by high-quality oxide ores with maximum contents of nickel, copper, and cobalt; the horizon is characterized by compactness and the absence or low quantity of inter-ore beds. The maximum concentrations of siliciclastics, phosphates, and barite are observed here. In the north-easterly direction, despite the increasing thickness of the host deposits, the ore body pinches out. Furthermore, the ore lenses and interbeds are branched as their thickness and dimensions decrease. In this connection, many studies have indicated an ingression of ore material along this fault (Ikoshvili, 1971; Makharadze, 1972, 1979; Khamkhadze, 1981, 1986; Mstislavskii, 1984; and others).

Numerous facts serve as evidence regarding the ingression of ore matter with accompanying elements along the Glavnyi fault zone. Along the plane of the Glavnyi fault a change is observed in limestones of the Upper Cretaceous, expressed in their recrystallization, dolomitization, and leaching with the formation of cavernous textures; their displacement by sediments of iron and manganese has also been established (Makharadze, 1972; Khamkhadze, 1981). Gogishvili et al. (1982) have noted that along the Glavnyi fault manganese mineralization is developed not only in the Oligocene rock mass, but also in that of the Cretaceous (Perevisi section) and Bajocian (Kvatsikhe section). Mstislavskii (1984) has described along the Glavnyi fault line (near a 14-story house in Chiatura city) in the Upper Cretaceous limestones a manganese-ore "column" of ladder type, which is considered one of the ore channels of the Chiatura deposit.

A tubular body with a diameter of several meters, composed of manganese oxides and superimposed in the Upper Cretaceous limestones 50 m below their roof (the floor of the Oligocene ore horizon), was discovered in 1987 by D.V. Tabagari and studied by Zh.V. Dombrovskaya. The body is situated between the Rgani and Kveda-Rgani uplands 1.5 km north-east of the Glavnyi fault line. Alongside the ore body the enclosing Upper Cretaceous limestones are altered—zones of clay alternation, dolomitization, and density reduction have been observed. Near the contact with the ore, the rocks are leached and cavernous up to the full disintegration and transformation into a dolomite meal. Dolomitization, as is known (Paragenesis et al., 1990), is one of the characteristic processes connected with action by magnesium-rich waters of oil- and gas-bearing basins upon the host limestones. Manganese ores here have a breccia-type outlook and represent dense lumps composed of pure pyrolusite contained in friable earthy material consisting of a mixture of montmorillonite, hydromicas, and friable manganese hydroxides. The data of spectral analyses show that in the manganese ores by comparison with the enclosing dolomites the contents of nickel, cobalt, copper, barium, molybdenum, and vanadium are higher by one or two orders and correspond to the contents of these elements in ores from the manganese-ore column (Mstislavskii, 1984), as well as from the Oligocene ores. This serves as evidence for a single source for the ore matter.

The cited examples constitute evidence that ore-bearing solutions were taken in along the ore channels—the tectonic faults (deep fault zones).

Near the Glavnyi fault in the Perevisi, Rgani, Kvela-Rgani, and Mgvimevi uplands are known numerous small normal and reverse faults with an amplitude of several meters, striking north-easterly (Chiaturskoe et al., 1964). These have been detected not only in the ore strata, but also in the underlying Upper Cretaceous limestones, which are exposed by the Kvirila River and its tributaries. Frequently observed in the outcrops exposing the underlying limestones are tectonic faults accompanied by the replacement of beds, by shatter and brecciation zones, and by seepage of sources of subterranean waters.

The existence of numerous small faults of north-easterly strike near the Kvirila River, as well as the linear character of its riverbed within the boundaries of the deposit, indicates the existence here of a large tectonic fault, situated perpendicularly to the Glavnyi fault along the Kvirila riverbed (see Fig. 3.28). The junction of these two faults determines the triangular form of the ore body. This form is repeated by the areas of the distribution of carbonate and oxide ores; peroxide ores; thickness of the manganese-bearing horizon; the absolute mass of manganese; and the contents of manganese, nickel, cobalt, and the like (Chiaturskoe et al., 1964; Strakhov et al., 1968; Avaliani et al., 1965; and others). Regarding the tectonic nature of the hydrographic network of the Kvirila River within the boundaries of the Chiatura as well as Kvirila bocks Khamkhadze (1986) has previously arrived at a similar conclusion on the basis of an analysis of the works of geomorphologists, materials of aerial photography, and geological and geophysical observations. She also considers that the ingression of the basic mass of ore matter occurred along the central axial part of the deposit.

In the structure of the Chiatura deposit has also been identified a sublateral tectonic fault situated in the northern part of the area. This is the Sachkhere thrust, the age of which, according to Gavasheli (1969), as is that of the Glavnyi fault, is pre-Cretaceous. The influence of this fault on manganese-ore deposition has not been established; near it, however, has been noted an elevated post-ore calcitization of manganese carbonate ores.

Within the Kvirila graben have also been established numerous variously oriented faults (Khamkhadze and Tumanishvili, 1984). Many of these are deep and are traced not only in the Mesozoic-Cenozoic cover but also in the Paleozoic foundation. Evidently coinciding with one of these faults of north-westerly strike, as in the Chiatura deposit, is the Kvirila riverbed. This fault divides the Cholaburi and Rodinauli sections, which differ substantially in terms of the composition of manganese ores.

In the Kvirila depression, as in the Chiatura deposit, alterations in the rocks have been observed near the faults. According to Makharadze (1979), in the north-western part of the Kvirila depression, in the area of Mt. Brolis-Kedi (Crystal Mountain), limestones and marls of the Paleocene-Eocene are dolomitized, silicified, and contain clinoptilolite and glauconite. The width of the alteration zone reaches a few dozen meters. It was observed (Dombrovskaya, 1997) that here, north-west of the outcrops of the ore bed, chalcedonies form a ridge in the modern relief that traces a tectonic fault of north-westerly trend.

Khamkhadze (1986) for the central parts of the Kvirila depression described linear zones represented by the so-called microconglomerates composed exclusively of authigenic minerals. These features consist of small nodules of silica, phosphates, and manganese carbonates and are cemented by thinly dispersed pyrite, marcasite, and montmorillonite. According to Makharadze (1972), the structure of the ore horizon indicates that the sources of ore matter were situated to the west-south-west of the Kvirila depression, in the Gura area, as well as in the northern and central parts of the axial zone of the depression itself. As evidence for this he considered: (1) the maximum thicknesses of the ore horizon and siliciclastics; (2) the maximum concentrations of iron in the form of glauconite, occasionally hematite and goethite in quantities equal to or greater than those of manganese; (3) the maximum concentrations of phosphorus; and (4) the most coarse-fragmented composition of the pyroclastic material.

To this it follows to add that south-west of the Kvirila depression is situated the Rioni depression with the Kolkhida and Gura oil- and gas-bearing area; uniting these depressions, a narrow "strait" between the Okriba and Adjara-Trialet landmass in the Oligocene evidently created favorable hydrodynamic conditions for the ingression of elisional waters from petroleum source beds into the discharge zone, which served the significantly more elevated bottom of the Kvirila basin. The modern deep position of the ore bed complicates the detection of the subaqueous channels. Analysis of the scheme of distribution of manganese ores, compiled by Khamkhadze and Tumanishvili (1984) according to drilling data, has shown that the highest values of manganese content of these ores form a northeast-trending linear zone in the axial part of the depression. Possibly corresponding to this zone is the tectonic fault that together with the Chiatura fault situated on the border with the Rioni depression are the principal ore channels.

Discussion of isotope data

The selection of material for isotope research as noted was conducted such that it would characterize most fully all types of carbonates in or associated with manganese ores. This allows the possibility to clarify definite genetic criteria with the interpretation of isotope data.

The analysis of the existing viewpoints regarding the origin of the Chiatura and Kvirila manganese deposits and sources of their matter (particularly sedimentary and, from an alternative viewpoint, volcanogenic (hydrothermal)-sedimentary) lends encouragement for finding isotopic confirmation of certain of these hypotheses. For the studied samples of manganese carbonates, a fairly broad spectrum of variations could be expected in the isotopic composition of carbon and oxygen, which would be characteristic of carbonates of initially sedimentary origin as well as isotope reference marks indicating their deep-lying nature.

Thus, for the proposed initially sedimentary ores we should have obtained high values of $\delta^{13}C$ and $\delta^{18}O$ (0–2‰ and 28–31‰, respectively) particular to carbonates of normal-sedimentary marine origin (ie, balanced with bicarbonate of marine water) (Keith, Weber, 1964; Galimov, 1968), or the low $\delta^{13}C$ (−22‰ to −30‰) and high $\delta^{18}O$ (28–31‰) values characteristic for diagenetic Mn carbonates (Kuleshov, 1986a,b; Okita et al., 1988; Kuleshov and Dombrovskaya, 1988, 1990, 1993; Kuleshov et al., 1991). At the same time, the presence among the studied rocks of secondary carbonates with all indicators of their hydrothermal origin should have given us reliable isotopic criteria for a deep (in the opinion of certain researchers, juvenile) source of carbon dioxide, with values characteristic for this source of $\delta^{13}C$ close to −8‰ to −6‰ (Galimov, 1968) or −12‰ to −10‰ (Kuleshov, 1986a), and $\delta^{18}O$ = 15–20‰, particular to carbonates of hydrothermal origin.

The results of the conducted isotope research in ores of the Chiatura and Kvirila deposits have indicated completely different isotope ratios. The obtained values and the character of their distribution enable the clarification of certain regularities determined by genetic reasons.

We shall pause to examine in greater detail the particularities of the isotopic composition of manganese carbonates in distinct studied groups.

Chiatura deposit. Dark-gray cryptocrystalline early carbonates (Group I) are characterized by the most consistent isotopic composition of carbon and oxygen. This is evidence of a single source of carbon dioxide and water from which carbonates were deposited within the boundaries of the entire deposit. Carbon dioxide of introduced solutions is characterized by an isotopic composition of carbon close to −14‰ to −8‰ (the maximum in the histogram of Fig. 3.35). The CO_2 of this source (generation) was responsible for forming a significant proportion of the manganese carbonates and other rocks of the studied groups.

In the cryptocrystalline manganese ores, the majority of accompanying calcites are located in isotopic equilibrium with Mn carbonates and share a syngenetic origin.

Dark-gray friable (frequently recrystallized) carbonates (Group II), as are the former, are characterized on the whole by fairly narrow boundaries of variations in $\delta^{18}O$ values, the predominant quantity of which belongs to the interval 26–28‰. Moreover, an insignificant proportion of the samples are characterized as well by a lighter isotopic composition (20–21‰), which is evidently due to their higher temperatures of precipitation.

The distribution of $\delta^{13}C$ values is somewhat more complex. It can be seen that inherent to a proportion of the samples are values of the isotopic composition of carbon also inherent to rocks of Group I (−12‰ to −8‰). Furthermore, particular for another proportion of the carbonates are lower $\delta^{13}C$ values, which in the histogram of Fig. 3.35 form an isotope-carbon maximum in the range of values of −18‰ to −14‰. Also noted are individual samples with the lightest isotopic composition of carbon: −34‰ to −32‰.

Such a broad spectrum of obtained $\delta^{13}C$ values indicates participation in the formation of manganese ores by carbon dioxide of three generations (or sources). The first, isotopically "heavy" generation corresponds to the source of CO_2 of Mn ores of Group I. The second generation of carbon dioxide of carbonate ores is characterized by intermediary (medium) values of $\delta^{13}C$ (−18‰ to −14‰). This generation was responsible for the precipitation of the predominant quantity of carbonate of the second group. CO_2 of the third generation, with the lightest isotopic composition (−34‰ to −32‰), is sharply suppressed and is represented in the deposit by isolated samples.

Carbonate matter of pseudooolisths and pseudopisoliths (Group III) is characterized by the least consistent isotopic composition. For carbon, the same regularities can be noted as for carbonates of Group II. In the distribution of $\delta^{18}O$ values here is also noted an isotopic maximum in the interval 26–28‰, but in distinction from the previous group the general boundary of variations in $\delta^{18}O$ values for pseudooolisths is much broader. This can be taken as evidence for a wider temperature interval of their formation and of the participation in the process of their formation by ore-bearing solutions with lower values of the isotopic composition of the water's oxygen.

Accompanying calcites in the composition of pseudooolisths and pseudopisoliths are associated with the most high temperature and isotopically light generation of CO_2 of these solutions.

Crystalline carbonates from cross-veins (Group IV) were likewise formed from various sources (generations) of carbon dioxide, under fairly similar temperature conditions. With that, the predominant generation had an isotopic composition of −24‰ to −18‰ (which is significantly lighter than carbonates of the second generation).

Carbonate minerals of the cement of oxide-carbonate ores (Group V) were formed principally from carbon dioxides of the second generation. An isotopically heavy source of CO_2 (first generation) played a sharply subordinate role in their precipitation except in a few isolated samples.

In the distribution of $\delta^{18}O$ values of the carbonate cement of the oxide-carbonate ores, three intervals can be identified (see Fig. 3.35): (1) 26–28‰, (2) 22–24‰, and (3) 18–20‰. This demonstrates the multistage nature of their formation and indicates different temperatures of their deposition and, as is apparent, a different initial isotopic composition of water of the ore-depositing solutions.

The isotopic composition of carbon and oxygen of scattered calcite in the host non-ore sandstones and the distribution of $\delta^{13}C$ and $\delta^{18}O$ values are analogous to the composition of the cement of oxide-carbonate ores of Group V.

Thus, the distribution of $\delta^{13}C$ and $\delta^{18}O$ values in the studied Mn-containing carbonates and calcites of the Chiatura deposit confirms the existence of at least three generations of carbon dioxide from which the ore was produced. Each generation was characterized by different initial isotopic compositions of carbon, due to the presence of isotope-carbon maximums in the reported histograms. The first of these led to the formation of the most isotopically heavy Mn carbonates of the Chiatura deposit, which are characterized by values of $\delta^{13}C$ in the range −12‰ to −8‰. Carbon dioxide of this generation was apparently formed as a result of the mixing of isotopically heavy CO_2 drawn from sedimentary carbonates of the host rocks ($\delta^{13}C$ ~0) and isotopically light carbon dioxide of organic origin. The latter was formed as a result of the oxidation of organic matter in the sediment strata during the process of diagenesis and catagenesis ($\delta^{13}C = -25$‰ to −22‰). The ratio of these sources in the general balance of carbon dioxide of the first generation is one-sided—isotopically heavy CO_2 of sedimentary carbonates somewhat predominates over isotopically light, biogenic CO_2.

It can be supposed that in the process of manganese ore-genesis in the Chiatura deposit, the role of CO_2 of the first generation predominated, insofar as the formation of beds of carbonate ore of this deposit (Group I) coincided with precisely this CO_2; the same goes for a substantial proportion of the carbonate component of ores of Groups II and III.

Carbon dioxide of the second generation should be characterized by values of $\delta^{13}C$ close to −22‰ to −16‰. CO_2 of this source, as of the first, represents a line of mixing of carbon dioxide of sedimentary and biogenic carbonate. However, isotopically light carbon dioxide here is substantially predominant over initially sedimentary carbon dioxide, and in certain cases the studied samples were formed fully on its account. The influence of carbon dioxide of the second generation was manifested most intensively with the formation of the presently altered carbonate matter of Group II, carbonate matter of the cementing mass, and scattered calcite of non-ore sandstones, where it constitutes an predominant proportion.

The third generation of carbon dioxide is characterized by lower values of $\delta^{13}C$ close to −36‰ to −32‰, and apparently represents the product of the oxidation of isotopically light hydrocarbons of the petroleum series. Within the boundaries of the deposit, carbonates of this generation of CO_2 are developed rather insignificantly and are represented in our collection by isolated samples in Groups II, III, and IV.

Vein carbonates (Group IV) in terms of the isotopic composition of carbon are on the whole lighter by comparison with carbonates of the second generation. It can be supposed that the development of their isotopic composition was determined by complex local conditions.

Data on the isotopic composition of the oxygen in carbonate ores of the Chiatura deposit also provide evidence of the heterogeneous conditions of their formation. This could be the result either of different temperatures of carbonate precipitation or of different sources of water of ore-depositing solutions, characterized by a different initial isotopic composition of oxygen. On the basis of the histogram of the distribution of $\delta^{18}O$ values can be identified three maximums, which are manifested most fully in the carbonate matter of the cementing mass of oxide-carbonate ores (Group V) and pseudooliths and pseudopisoliths (Group III). The maximum quantity of obtained isotope data falls in the interval 26–28‰, with which coincides the overwhelming majority of carbonates of Groups I and II. For this reason, the formation and subsequent alteration of manganese carbonate ores of the deposit can be considered a result of a single phase (or act) of input of ore-depositing solutions within a small range of fluctuating temperatures.

If we consider that the isotopic composition of oxygen of the water of ore-depositing solutions remains unchanged, then the variations in values of $\delta^{18}O$ for manganese carbonate ores of Group I (approximately 3‰) as well as carbonates of other groups (II, III, and V) falling in the interval $\delta^{18}O = 26–28‰$ should be contingent upon the fluctuations of temperatures of deposition within the boundaries of 10–15°C. To arrive at an understanding of the actual temperatures of precipitation of these carbonates, we should know the initial isotopic composition of oxygen of the water of the solutions from which their deposition was produced and, naturally, the nature of these solutions.

In our understanding, the water of such solutions can be represented by three different sources. The first, as is evident, is the buried water of the marine basin with an isotopic composition of the oxygen of the water close to "0." Mn ores from such water should have been deposited in the interval of temperatures from 20–25°C to 30–35°C (Friedman and O'Neil, 1977).

The second source of ore-depositing solutions could be the detached interstitial waters of the sediment of the zone of late diagenesis (catagenesis)—that is, the corresponding level of lithification of the rock, when the exchange processes of water between the sediment and the bottom waters already are absent. Such waters are characterized as a rule by a lighter isotopic composition of oxygen (Lawrence et al., 1975; Perry et al., 1976; Mottl et al., 1983; Gieskes et al., 1987) due to the removal from the system of heavy ^{18}O isotopes with the formation of authigenic clayey minerals. Temperatures of precipitation of Mn minerals from such waters should have been lower by 10–15°C by comparison with the first source and constitute 10–20°C. In certain cases, interstitial waters could be insignificantly enriched (up to 3‰) by heavy isotopes of oxygen (McDuff et al., 1978), in which case the temperatures of carbonate precipitation should have been higher by 10–15°C by comparison with those of the sediments deposited from waters with "null" isotopic composition.

The third source of such waters, and the most probable in our opinion, could be catagenic waters of deep deposition. Their isotopic composition varies and is determined mainly by the composition of the host rocks and their path of migration.

Altered carbonate Mn ores (Group II) are characterized, as has been noted, by a broader spectrum of variations in $\delta^{13}C$ and $\delta^{18}O$, trending toward lower values. For carbon, this is determined by the presence of the second source of carbon dioxide with lighter isotopic composition of that element. The input of isotopically light carbon dioxide evidently occurs at higher temperatures (by 10–15°C).

Evidence for this can be found in the lighter isotopic composition of the oxygen of the accompanying calcites (see Fig. 3.37), which in carbon coincides with precisely this generation of isotopically light CO_2.

Carbonate matter of pseudoooliths and pseudopisoliths (Group III) was formed out of the carbon dioxide of those same generations as were the above-examined manganese ores of the first two groups. One of the main distinctions is that here are manifest separate groups of carbonate matter, characterized by a lighter isotopic composition of oxygen ($\delta^{18}O = 17–19‰$).

Such isotopically light values are noted also in carbonate of cement (Group V), among vein carbonates (Group IV, isolated samples), and in calcite of sandstones enclosing the ore horizon (see Fig. 3.35). If they formed from the same solutions as did the carbonate matter of the first group, then they would have been deposited with higher temperatures—by 35–40°C. Thus, for marine waters these temperatures should have constituted approximately 90–100°C.

The examined mechanism (method) of formation of manganese ores owing to the heterogeneity of the ore-generating solutions led to the inability of the accompanying components of manganese carbonate formations to be located in isotopic equilibrium with each other. For example, in one of the studied sequences of rocks of the Chiatura deposit (see Fig. 3.40) gathered from the wall of the quarry (section 4, Darkveti uplands), a fairly wide variation is visible in the isotopic composition of carbon and oxygen for different sections (ore components) even within the boundaries of a single sample (lump of rock). Here carbonate matter of cement and ooids is differentiated substantially in terms of the isotopic composition of both carbon and oxygen (up to 8‰).

In this cross section is also noted a gradual increase in the weight of the isotopic composition of the sulfur of pyrites from the floor of the ore horizon toward its roof, which is caused evidently by a reduction of the initial marine sulfate of the sediment with a gradual enrichment of the remaining sulfate by heavy $\delta^{34}S$ isotopes. This could provide evidence for a local (in situ) rather than exogenous source of sulfur for the pyrites in this cross section.

Kvirila deposit. Manganese carbonate ores of this deposit, as of Chiatura, in their origin are determined by carbon dioxide of various generations (sources). They are differentiated by the initial isotopic composition of carbon as well as by the temperatures of their importing fluids.

Gray cryptocrystalline manganese ores (Group I) of this deposit apparently were also formed within a fairly narrow temperature interval (15–20°C). The source of carbon dioxide was not isotopically homogeneous, as is reflected in the wide scattering of $\delta^{13}C$ values.

Potentially indicating the multi-act carbonate-forming of these rocks is such a ratio of isotopic composition of carbon and oxygen in manganese minerals and accompanying calcites (Fig. 3.38), which are distinguished in terms of isotopic composition by higher as well as by lower values.

Recrystallized light-colored carbonate material (Group II) of the Kvirila deposit was formed, as follows from the isotope data, in two stages distinguished by temperatures. The first, low-temperature stage corresponds to the formation of analogous ores of the Chiatura deposit. In the second, higher-temperature stage the deposition of carbonates occurred at temperatures 15–20°C higher than those of the first. The accompanying calcites fully coincide with precisely the lower-temperature stage.

It can be assumed that the formation of all studied carbonate ores of the Kvirila deposit occurred at the same temperatures. In this case, the entire observed selection of $\delta^{18}O$ values will be determined by the different initial isotopic compositions of the oxygen of the water of the ore-depositing solutions. In our opinion, this is improbable, since for the phase (act) of carbonate-forming we should attribute to waters of different origin the supplying of the ore-depositing solutions (see above—the Chiatura deposit).

In this way, the isotope data support the conclusion that manganese carbonate ores of Groups I and II of the Kvirila deposit had independent sources of CO_2. The formation of the carbon dioxide of these sources occurred with varying degrees of participation by isotopically light CO_2 of biogenic origin and by isotopically heavy CO_2 of sedimentary carbonates. Moreover, the role of the latter predominates in the composition of Mn ores of Group II. These carbonates in terms of the isotopic composition of carbon are the heaviest and approach the composition of normal-sedimentary marine carbonates. Analogues of manganese rocks with such heavy isotopic characteristics have not been detected within the boundaries of the above-examined Chiatura deposit. At the same time, in distinction from Chiatura, oxidized organic matter played a simpler role in the formation of manganese carbonate ores of the Kvirila deposit. Here, we did not detect "anomalously" light (in terms of isotopic composition of carbon) Mn carbonates corresponding to the product of oxidation of hydrocarbons of the petroleum series.

Data on the isotopic composition of the sulfur in pyrites from the ores of the Chiatura and Kvirila deposits provide evidence for the initially sedimentary nature of their matter. Serving as a source of sulfur of pyrites is hydrosulfide, which forms as a result of the reduction of sulfur of marine sulfates during the process of sulfate-reduction in the sediment mass during diagenesis or later catagenetic processes (Vinogradov, 1980). None of the data on the isotopic composition of sulfur indicates a deep (juvenile) source for that element in our samples.

In this way, the isotope data provide evidence for the multistage nature of manganese ore-genesis within the boundaries of the Chiatura and Kvirila deposits. Ore-formation occurred during the stage of diagenesis of the sediments (samples with highest $\delta^{18}O$ values) and was also connected with input of carbon dioxide at multiple times by catagenic (or elisional) solutions with deep horizons of the ore-bearing sequence adjoining the oil- and gas-bearing basins.

The regions (zones) of generation of CO_2 of these solutions for the Chiatura and Kvirila deposits were various. Exogenous ore-bearing solutions were apparently characterized by a fairly wide interval of temperatures and could be distinguished by the isotopic composition of the water's oxygen. Specifically sedimentary manganese carbonates—that is, those in equilibrium with the dissolved bicarbonate (DIC) of marine water—have not been detected within the boundaries of the studied deposits.

The potential for participation by petroleum waters in the process of manganese-ore formation is supported by isotope data on the dissolved bicarbonate in such waters. Thus, Carothers and Kharaka (1980) studied the isotopic composition of dissolved HCO_3^- in 75 samples of water from 15 gas- and oil-bearing fields of the United States from deposits of Eocene and Miocene age. The values of $\delta^{13}C$ of the general sample of HCO_3^- in the surface ground waters are located within the boundaries −5.7‰ to −7.3‰; for surface formational waters (T°C < 80), these values vary within the boundaries −27.6‰ to −19.9‰ and in deep-deposited formational waters (T°C > 80) vary within the boundaries −12.3‰ to 4.8‰. These data provide evidence of an organic origin of the carbon as a result of microbial oxidation of hydrocarbons during the process of sulfate-reduction.

Particularities of genesis

The process of manganese ore-genesis in the described deposits of western Georgia was evidently of fairly long duration and was not limited to the early Oligocene. The study of the structural-textural particularities and mineral paragenetic associations of oxide- and carbonate-manganese ores and their host rocks, as well as of the isotopic composition of carbon and oxygen, has demonstrated the asynchronous and multistage nature of their formation. At the boundary of the Eocene and Oligocene in this region, the inversion of the tectonic regime began. Marked changes were produced in the composition of the biota, responding to changes in the paleographic and paleotectonic conditions. The character of sediment accumulation and of magmatism was altered, and the transition to the orogeny stage began (Salukvadze, 1990). Ore-formation began during the period of accumulation of Lower Oligocene deposits and continued after the overlap of the manganese-ore strata by overlying deposits.

In the stage of sedimentogenesis, material from two sources entered the Oligocene basins of sediment accumulation. From the close-lying landmasses drifted terrigenous, predominantly sandstone-clayey material, and along the tectonic faults through catagenic (ore-bearing) solutions were supplied manganese, silica, and accompanying elements: phosphorus, barium, iron, and microelements, which caused an authigenic mineralization of the terrigenous strata of the ore-genesis zone.

In western Georgia, according to Makharadze (1972), coinciding with the lower strata of the Lower Oligocene are numerous deposits and occurrences of the enumerated elements. These are connected exclusively with long-lived deep-lying dislocations, which are the ore channels of the ore solutions. Authigenic mineral precipitation accompanying the manganese ores always bears a strictly local development and is observed in areas featuring manganese mineral precipitation (Machabeli and Khamkhadze, 1979).

During the period of accumulation of terrigenous material in the Lower Oligocene, manganese entered the area of the described deposits together with a significant quantity of chemogenic silica, which were partially ingested by organisms (sponges) and partially deposited in the form of opal, forming opal-rich clays and opal-cemented sandstone.

Manganese in waters of gas and oil deposits, as demonstrated by Shtanchaeva (1984), was found alongside other metals in the form of metal-organic compounds and deposited initially in the form of hydroxides, under conditions of the oxidation environment prevailing on the bottom of the sea. At the stage of diagenesis under conditions of a developing oxidation-reduction of the environment occurred the redistribution of matter and the formation of the observed current structural-textural elements of the rock. Manganese hydroxides were dissolved and deposited in the form of manganite, forming oolith and pisolith nodules; terrigenous grains or spicules of sponges served as nuclei of these nodules.

A portion of the laminae of manganite ooliths is represented by opal, which constitutes evidence of their close paragenetic connection. In the non-ore interbeds, opal forms simple nodules, lacking concentric texture. In separate interbeds, similar nodules are composed of -(glauconite)-smectite mixed-layer formations, goethite, or neotocite.

Manganese carbonate ores in the described deposits are developed more widely than are manganese-oxide ores. Their formation occurred later than that of oxides, in the mass overlapped by deposits of the Miocene and Pliocene. They enclose oxide ores, surrounding them from the bottom, from the top, and replacing them downwards along the bed. Occasionally carbonates fully displace a member of the oxide ores, in which case higher and lower carbonate members are joined, forming a single horizon.

It must be mentioned as well that ores of the lower and upper carbonate members are distinguished neither by composition, nor by structural-textural indicators, and only in the central part, on the level of the manganite horizon, are observed higher contents of manganese. It can be proposed that during the period of formation of carbonate ores, there were along the Oligocene bed zones of high fluid permeability that allowed the passage of reducing, CO_2-rich fluids derived from the maturation of organic matter in underlying beds. These solutions destroyed the manganite, transferred the manganese in a dissolved form and redeposited it into a carbonate form. Added to the preexisting carbonate ores created a large volume of secondary carbonate extending into the host rocks, occasionally even into the underlying Eocene and Cretaceous or in the overlapping Maikop deposits. As a result, in many places the boundary between the Eocene and Oligocene, as well as the Oligocene and the Early Miocene, is obscured. They are shaded by superimposed ore carbonatization.

Indicating the latest formation of manganese carbonate ores with regard to oxide is the replacement by manganese carbonate of all elements of the oxide ores and their host rocks (of oolith-pisolith forms, ore cement, opal cement, opal nodules, spicules of sponges, and the like). In all cases, a full replacement is observed of these elements as well as a partial replacement. With the replacement of manganite ooliths by manganese carbonates, the former are "consumed" and with a full replacement the textural particularities of manganite ore disappear, forming the "dense" variety of carbonate ore.

With the replacement by manganese carbonates of the opal-clayey nodules, the form of the latter is frequently inherited and oolith-like carbonate nodules are formed, the presence of which creates a nodular variety of carbonate ore.

Carbonate ores, as noted above, after their formation were subjected to substantial changes, frequently manifested in leaching, with the formation of high porosity, some even cavernous, recrystallization (strengthening, spherulite structure), and redistribution of carbonate material (formation of cross-veins, purification from impurities). Spherulite structures constitute evidence of free growth in the conditions of leached space. With recrystallization occurred the purification of carbonate material with the separation of pyrite, clayey minerals, zeolites, and the like that accompanied the bleaching of the carbonate sections and the formation of additional porosity. Practically all carbonate ores are to some degree leached, porous, and cavernous. With the leaching of clayey-opal inclusions, a cellular texture was created. Effective porosity of carbonate ores at Chiatura as well as the Kvirila deposit frequently reaches 28% with bulk density of 2.12–2.30 g/cm^3.

It is interesting to note that such high (25–30%) porosity was noted previously by Gogishvili et al. (1979) for clinoptilolitization of the rocks of Transcaucasia and, in part, of the examined manganese deposits. A similar process of leaching, recrystallization, and formation of cross-veins is known as an infiltrational stage of development of elisional basins, when with the uplift of their border parts meteoric waters obtain ingress into the water-permeable sheets, displace and replace the catagenic waters previously existing there (Kholodov, 1982a,b; Makhnach, 1992).

Besides the described processes at the Chiatura deposit is widely manifested a post-ore calcitization of manganese ores and host rocks. They are expressed most clearly in the Perevisi and Rgani uplands, where in the Glavnyi fault zone near a 14-story house (Chiatura city) has been observed a dense network of cross-veins and veins with a thickness reaching 10 cm, composed of white large-crystalline calcite, transecting manganese-oxide ores contained in the Upper Cretaceous limestones. In these uplands is known a variety of manganese ores called "mtsvari," representing oxide ores cemented by calcite. Calcite not only replaces the initial friable cementing mass, but erodes and replaces ore ooliths. It is commonly large-grained, forming cross-veins and brush-type and cockade textures. It has been proposed (Betekhtin, 1946; Avaliani, 1982; Mstislavskii, 1984) that calcite has a hydrothermal origin connected with an intrusion of Pliocene vulcanites of basic composition in the Perevisi uplands. The temperature of calcite formation according to the data of decrepitation constitutes 140°C (Gogishvili et al., 1980).

According to the observations of Zh.V. Dombrovskaya, calcitization is manifested not only in oxide ores, but also in carbonate ores. Intensive calcitization of carbonate ores is observed as well in the drill cores bored in the northern part of the Darkveti uplands, where it evidently is connected with the influence of the Sachkhere-Khreit'i thrust fault, which bounds the area of distribution of manganese ores from the north. Less intensive calcitization is manifested almost everywhere in manganese carbonate ores in the form of thin cross-veins.

Besides Oligocene manganese deposits of the Chiatura type, in the Transcaucasus are known manganese ore-occurrences, considered by many researchers to be typical hydrothermal occurrences (Dzotsenidze, 1965; Gogishvili et al., 1980, 1982). Here manganese-oxide mineralization of predominantly vein type is developed in rocks of a broad age range—from Bajocian to Eocene. Gogishvili et al. (1980, 1982) considered that those and other ores are synchronous and connected with a single process of ore-hydrothermal activity manifested at the end of the Oligocene during the orogeny stage of the territory's development. Manganese mineralization is syngenetic in relation to the Oligocene deposits and epigenetic to the more ancient deposits of the Bajocian-Eocene.

It can be supposed that during the period of formation of manganese deposits in the Oligocene marine basin on the close-lying landmass along the fault zones tracing the Jurassic, Cretaceous, and Lower Paleogene deposits, the formation of manganese ores also occurred. In the first case were formed stratiform deposits; in the second, vein, lensoidal, and nest-type bodies predominantly discordant with the host rocks. These and other deposits shared a single source of ore solutions. Evidence for this is found in the identical mineral composition and mineral paragenetic associations of manganese ores and altered near-ore rocks, as well as their consistent suite of upper-Clarke microelements—Pb, Zn, Cu, Ba, Sr, As, Ni, Co, Cr, V, Mo, and P (Table 3.12).

The distinction between these two types of deposits is that manganese ores of vein deposits are preserved in their primordial form, whereas stratified deposits were subjected to the action of diagenetic processes and later elisional solutions after the overlap of ores by overlying deposits, which led to the replacement of a large part of the oxide ores by carbonate ores.

The absence of syngenetic carbonate ores indirectly confirms the total absence of carbonate facies in Maikop deposits not only of shallow-water zones, where the formation of manganese ores occurred, but also in deep-water parts of the Oligocene basin. On the bottom of the marine water body, at the contact with marine water containing oxygen, manganese can be deposited only in the form of oxides. This happens independently of the depth of the basin, as indicated by the widely known process of formation of Fe-Mn concretions at depths reaching 5–6 km on the bottom of the global ocean. With the formation of Oligocene deposits, elisional solutions bearing manganese near the surface of the landmass or marine bottom, as a result of the change of reductive environment into an oxidizing one, transplanted manganese exclusively in oxide form.

Following from the above, the origin of stratiform deposits of the examined deposits could be interpreted as elisional diagenetic, and the origin of vein deposits as elisional. Elisional solutions, as is known, are hydrothermal. However, under the latter term are commonly understood solutions connected with a volcanogenic or other magmatic source. In order to avoid confusion and to emphasize the source of ore matter of manganese deposits, as well as their connection with catagenic waters of zones of oil and gas generation, we do not use the term "hydrothermal," instead replacing it with "elisional."

Table 3.12 Content of Lesser Elements in Manganese Ores of the Chiatura and Kvirila Deposit (to 10–4%) (Compiled by Zh.V. Dombrovskaia)

Kvirila Deposit							Chiatura Deposit			
Lithotype	**Number of sample**	**NiO**	**CoO**	**ZnO**	**PbO**	**CuO**	**Lithotype**	**Number of sample**	**NiO**	**CoO**
Op-C	27	380	38	120	20	57	C	20	210	22
C	26	110	21	75	20	28	C	17	570	150
C	25	92	28	84	20	48	C	19	910	50
Op-Cl	24	290	180	330	74	160	O	22	600	30
Op-Cl	45	66	31	92	30	47	C-O	23	570	18
C	44	890	78	190	20	36	C-O	28	370	20
N	43	1100	90	250	20	63	C-O	27a	570	20
N	42	2000	160	540	41	150	C	24	240	33
O	39	1200	21	220	30	23	Op-C	2b	360	61
O	37	1200	62	150	30	12	C	144	150	16
C	34-2	81	30	41	20	40	C	145	230	32
Op-C	50-2	260	24	95	11	20	C	147	230	45
G-Cl	49-2	200	27	83	20	16	C-O	149	490	7
Op-Cl	48-2	840	100	110	20	210	Op-C	150	900	210
Op-Cl	66	1100	130	170	20	51	Op-C	153k	220	20
C	64	120	28	130	20	23	O	153m	1100	32
N-C	63	1500	120	740	20	280	S	155	250	23
C	62	1100	150	160	20	250				
Cl-C	61	760	56	72	30	71				
C-Cl	60	910	120	390	20	170				
C	59	270	27	0	2	22				
Op-C	74	280	35	54	19	73				
O-C	72	210	22	20	33	34				
O-C	70	400	34	120	20	53				
G-K	69	450	28	120	20	84				
Op-Cl	68	250	36	60	20	16				
C	81	910	120	150	20	50				
C	79	150	11	73	20	26				
C	87	260	51	150	20	87				
N-C	85	510	28	160	20	110				
C	84	270	69	130	20	130				

Lithotypes. Manganese ores: C, *carbonate;* N, *neotocite;* N-C, *neotocite-carbonate;* Op-C, *opal-carbonate;* C-O, *carbonate-oxide;* O-C, *oxide-carbonate. Basic rocks:* G-C, *goethite (hematite)-carbonate;* Cl-K, *clayey-carbonate;* C-Cl, *carbonate-clayey;* Op-Cl, *opal-clayey;* G-Cl, *goethite (hematite)-clayey;* S, *sandstone with rhodochrosite.*

The above arguments speak in favor of such a connection with regard to sedimentary deposits not only of manganese ores (Pavlov, 1989; Pavlov and Dombrovskaya, 1993), but also of many other mineral deposits; in part, stratiform deposits of lead-zinc, copper, rare-metal gold-sulfide mercury, and other ores (Paragenesis et al., 1990). For the Caucasus region, in part, the coincidence of mercury deposits and ore-occurrences, including those distributed near the described manganese deposits, near the oil-generating formations has been indicated by Kekeliia et al. (1979) on the basis of a saturation of bitumens of the ore-hosting rocks and the comparison of the isotopic composition of sulfur of sulfides and hydrosulfides of petroleum waters.

In many regions, vein silica-manganese mineralization is genetically connected with the formation of copper-lead, lead-zinc, barite, and other ores (Gogishvili et al., 1982). Frequently manganese, barite-manganese, and iron-manganese ores are replaced at depth by lead-zinc or sulfur pyrites. In the host rocks have been observed zonal near-ore alterations, in which have been established sulfides (pyrite, chalcopyrite, galena, sphalerite, cinnabar), native metals (copper, lead, zinc, iron), and others (Gogishvili et al., 1982).

Such similarity of the composition of manganese ores and their altered host rocks with the composition of sulfide polymetallic ores is with justification considered by the cited authors as evidence of a genetic connection between them.

In certain areas, the connection between manganese and other deposits is so close that they can be encompassed within a complex ore field. One such field is located on the southern slope of Greater Caucasus, in the Gagra-Djava tectonic zone in the Oni and Djava districts (Fig. 3.41). Here are known the Tsedisi, Ukivleta, Varakh-Koma, and Valkhokh manganese ore-occurrences and the Kvaisi, Varakh-Koma, and Valkhokh lead-zinc deposits. Situated nearby are the Chorda barite deposit, the Ertsoy cinnabar deposit, and others. Manganese ores of hematite-pyrolusite composition contain an impurity of copper and arsenic. Deposits are situated within a porphyritic strata of the Bajocian and limestones of the Upper Jurassic and are represented by lenses, nests, and columns. Characteristic is a metasomatic replacement with the formation of ore bodies as well as within the host rocks. All of the above enable the grouping of these deposits within a single Tsedisi-Kvaisi ore field.

A second similar field, the Bolnisi-Tetritskaro, is situated in the Somkhiti-Karabakh tectonic zone. Here, manganese ores (Tetritskaro group) likewise closely neighbor hematite ores (Poladauri group) and copper-barite-polymetallic ores (Madneuli, David-Gareji, Tsitelsopeli, Kvemo-Bolnitsi, Tamarisi, and elsewhere).

Situated in close connection with the described deposits of manganese, iron, and nonferrous metals is evidently an entire range of deposits and ore-occurrences of the so-called hydrothermal type of mercury (Arkhei, Avadkhara), arsenic (Lukhumi, Tsani), antimony (Zopkhita), as well as phosphorites, bentonites (Gumbra), zeolites, alunites, dolomites (Abana, Tkvarcheli), and others. All of these are spatially connected with the discharge zones of catagenetic (elisional) waters of oil- and gas-bearing basins of the Transcaucasus intermontane region and share common features that provide evidence of their genetic affinity.

3.2.1.4 Bezmoshitsa-ore occurrence (Cis-Timan)

The Bezmoshitsa manganese-ore occurrence is situated on the right bank of the Bezmoshitsa River, 21 km from its confluence into the Pesha River (Fig. 3.42). In the structural regard, the area of the ore-occurrence represents a shoreline-marine region of the Upper Jurassic basin, which stretches from north to south along the western slope of the Timan uplift. Its width in the middle part (on the latitude

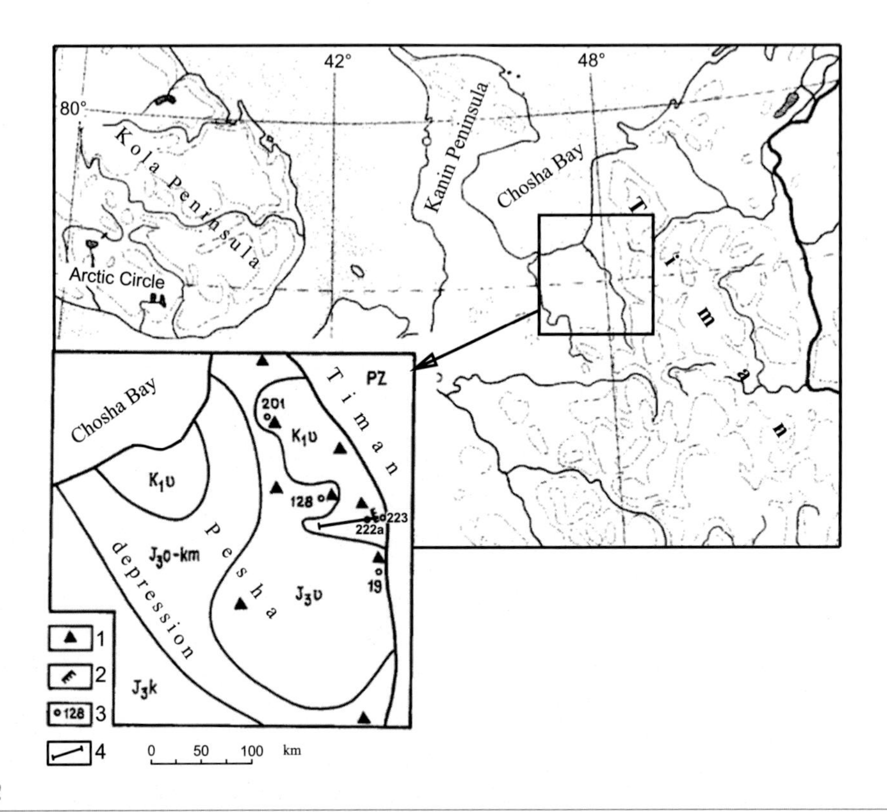

FIG. 3.42

Schematic geological location of the Bezmoshitskoe iron-manganese-ore occurrence. 1—Signs of manganese mineralization; 2—a natural outcrop of manganese ore on the right bank of the Bezmoshitsa river; 3—location and number of boreholes; and 4—profile of the geological section.

of the Volotovskaya River) constitutes 60–70 km, and its length from the shore of Cheshskaya Bay to the sources of the Pesha River is 120 km.

According to the data of petroleum exploration drilling (Danilov et al., 1983), Jurassic deposits reach the greatest thickness in the eastern part of the depression (150 m), directly adjacent to a flexure-form bluff, along which passes the border of Timan and the Mezen syneclise. In the structural regard, this border coincides with the deep fault of the Proterozoic deposit, which is traced in a meridional direction along the entire Timan.

In the southern part of the Pesha depression, in the upper reaches of the Pesha River, a band of outcrops of Jurassic rocks is sharply contracted and reaches 15 km in width. Here are developed lagoonal clayey-shale deposits with pyrite and siderite. To the north, toward Cheshskaya Bay, by measure of the widening of the depression these deposits are gradually submerged and overlapped by Cretaceous formations.

Geological setting and material composition of manganese carbonate ores

About the Bezmoshitsa-ore occurrence area are distributed shore-marine deposits of the Upper Jurassic (Oxfordian-Kimmeridgian), which in the near-surface part are strongly dislocated and fragmented as a result of glacial tectonics. In the Quaternary period, evidently under the action of ice-flow, manganese-bearing deposits were significantly disrupted and were thrust over each other in the form of separate slice-scales interstratified with morainic formations. This is clearly visible in the material of exploratory drilling and demonstrably illustrated in a schematic geological cross section (Fig. 3.43) compiled by the geologists of the Timan Geological Survey Expedition. Thus, at 10 km below the ore horizon (well 222a) were subsequently deposited tectonic slices of (from bottom to top) Quaternary, Cretaceous (K_1v), Upper Jurassic (J_3), and Cretaceous (K_1bs) deposits.

Below the dislocated strata, the manganese horizon was also exposed by the drill wells in a bedrock, undisturbed horizontal deposit.

The composition and structure of the ore horizon can be observed in a natural outcropping on the right bank of the Bezmoshitsa River (25 km upstream from the mouth of the river). Here the ore-bearing strata dips steeply (strike 270 degrees; angle 45–70 degrees) and is deposited in a detached state on the black clays and glauconite sands of the Berriasian Stage of the Lower Cretaceous. It is divided into three members: subore, ore, and supraore.

The subore member represents an alternation of sandstone layered clays of greenish-gray color; light-gray quartz sand (with residues of carbonized flora) and dark-green sandstone clays with glauconite containing a large quantity of residues of fauna (belemnites, *Aucella* sp., ammonites). Greenish-gray clays of this member contain large siderite concretions with dimensions reaching 0.3 m. The siderite is compact and cryptocrystalline. The surface of the concretions is oxidized and covered by a film of brownish-red color with a width of 2–5 mm. The thickness of this member constitutes 2.4 m.

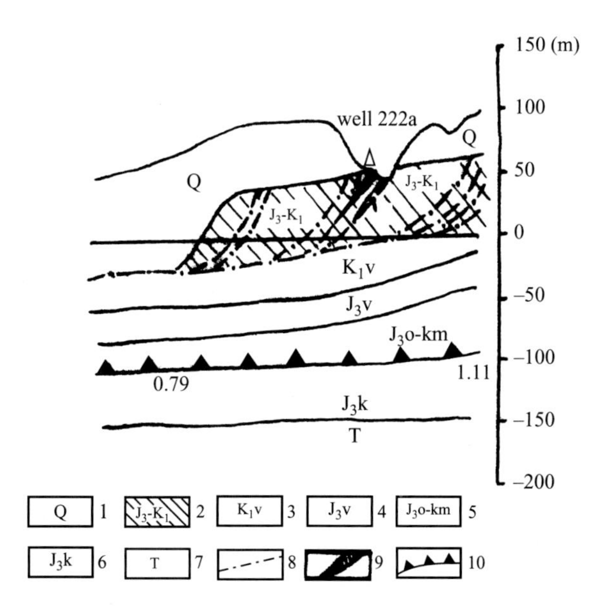

FIG. 3.43

Schematic geological section across the Bezmoshitskoe-ore occurrence (see Fig. 3.42). 1—Quaternary deposits; boulder loam, sandy loam; 2—glaciotektonical allochthon; 3—silt, sand, clay; 4—calcareous clay, marl; 5—glauconite sands and clays; 6—thin-laminated clay; 7—motley-colored sand and clay; 8—glaciotektonical faults (established); 9—manganese-ore occurrence Bezmoshitskoe (in the position of erratic mass); and 10—manganese-bearing of clay-carbonate horizon in the bedrock (0.75 and 1.11—manganese content).

The ore member is represented by an alternation of black clays with interbeds of tobacco-colored fine-oolitic Fe-Mn carbonate ores. The clays contain lenses and concretions of oligonite composition as well as concretions of massive black manganite ores. There are many residues of belemnites, Aucella, and ammonites. The thickness of this member is approximately 3 m.

In the roof of the manganese-bearing horizon, an oxidation zone is developed, reaching a thickness of 0.5 m. The thickness of the ore bed together with the oxide zone reaches 3.5 m.

In the lower part of the ore horizon predominates the carbonate type of ores, which are represented by minerals of the complex Fe-Mn carbonate isomorphic series of siderite-rhodochrosite composition; oligonite is widely distributed. A mixed oxide-carbonate type of ores coincides principally with the middle part of the horizon, and in the upper part predominate oxide Fe-Mn ores.

The supraore member is composed of greenish-gray compact limy clays with small nodules of phosphorites (10–15%). A large quantity of *Aucella* valves has been noted.

The composition and structure of the ore horizon can also be observed in detail in the core material of certain prospecting wells; it is most fully exposed by wells 22a, 201, and 128.

Well 222a

Bored in direct proximity to the bedrock outcrop (see Fig. 3.43).

Here above the ore horizon is deposited a strata of dark greenish-gray glauconite clays and siltstones with an abundant quantity of residues of fauna (aucellids and belemnites, rarely ammonites). Below runs the ore interval thus (depth, m):

8.6–8.8 Interbed of small-grained sand of brownish-green color with glauconite and small nodules of manganese-oxide ores.
8.8–9.5 Black sands and siltstones with brownish-green spots of substantially glauconite composition. Small nodules of Mn oxides with dimensions reaching 4 cm.
9.5–9.9 Alternation of black siltstones and clays with sand of greenish-brown color of substantially glauconite composition with small inclusions of Fe-Mn oxide oolites.
9.9–10.1 Dark greenish-gray siltstones and clays with lenses and nests of glauconite. Noted are inclusions of small nodules of manganese carbonate ores with oolite texture, dark-gray color, with dimensions reaching 5–6 cm; partially oxidized.
10.1–11.1 Sandy-clayey rock of black color with greenish hues, caused by the presence of glauconite. Abundant inclusions of carbonate and oxide small nodules with dimensions reaching 2–3 cm with oolitic texture. Ooliths as a rule of red-brown color with dimensions of 2–3 mm. Numerous residues of belemnite rostra.

Below are deposited black lump clays with greenish hue with separate inclusions of lenses and nests of glauconite sand.

Well 201

The ore horizon is overlapped by dark-gray siltstones and clayey sands with abundant content of residues of belemnite rostra and aucellid valves. The ore interval is represented thus (depth, m):

142.1–142.3 Glauconite sands with lump texture strongly carbonatized. Noted are separate interbeds of black clays with thickness reaching 2 cm. Rare nodules of manganese carbonate ores with oolitic texture.

142.3–153.0 Dark-gray sandstone clays and siltstones analogous to those of the supraore horizon.
153.0–153.6 Alternating sandstone clays with greenish hues, caused by the presence of glauconite, with greenish-brown weakly cemented sand. Occurring are clasts of rostra of belemnites.

Lower are deposited subore light-gray clays of the Callovian Stage.

Well 128

Ore horizon is overlapped by dark-gray and greenish-gray clays and sands with abundant content of fauna. Ore interval is composed thus (depth, m):

13.7–14.2 Black clayey sands with nodules of manganese carbonate ores with dimensions reaching 4–5 cm. Nodules with oolitic texture, from the surface as a rule are red-brown due to weathering; on a fresh shear of dark-gray color with greenish hues. Numerous belemnite rostra.
14.2–15.2 Black, dark-green clays and clayey sands, analogous to the overlying, with nodules of Mn and Fe oxides of red-brown color. Noted are lenses and "spots" of glauconite sand of dark-green color. Numerous residues of fauna.
15.2–15.8 Non-ore interbed of black clayey sandstones with glauconite. Black sandstone clays with "spotted" accumulations of glauconite. Rare carbonate small nodules with dimensions reaching 5 cm, frequently weathered, red-brown. Numerous residues of belemnites, aucellids, ammonites.

Below are exposed subore clays analogous to those described in the other wells.

In this way, ore matter in the Bezmoshitsa occurrence is represented by oxide and carbonate forms of iron and manganese. While Mn and Fe are commonly present together in the ores, the ratio of Mn/Fe (manganese module) varies within a broad interval: from 0.01 to 9.77 (Table 3.13).

The principal manganese mineral of the richest oxide concretional and lump ores, according to the data of Danilov et al. (1981, 1983) is manganite (up to 85%). Subordinate value is found for pyrolusite and psilomelane. Quartz, goethite, glauconite, and clayey minerals (montmorillonite, etc.) constitute a few percentages in these ores.

Oolitic oxide-carbonate ferromanganese ores as a rule form small nodules and nodules of various forms, but most frequently occur in near-isometric form. Ooliths in these ores are commonly composed of oxides of iron and manganese [manganite, psilomelane, goethite, and hydrogoethite (Fig. 3.44)] and evidently are developed on the initial nodules of glauconite. In certain cases, this is clearly visible in thin sections (Fig. 3.45).

The average dimensions of ooliths constitute 3–4 mm in diameter. Their form is spherical, more rarely oblate. Not infrequently in the center are observed separate grains of quartz or relicts of glauconite grains. In certain ooliths have been noted concentric interbeds of fine-grained carbonate. Ooliths as a rule are cemented by carbonate and clayey minerals.

Carbonate matter of the cement of oxide-carbonate ores has a complex mineralogical composition and represents a mixture of isomorphic minerals of the siderite-rhodochrosite-calcite series. However, judging by X-rays (Fig. 3.46) and data of thermal analysis (Fig. 3.47), the principal mineral phase of these ores is represented by manganocalcite and manganese calcite. Non-ore carbonate concretions are represented mainly by oligonite and more rarely siderite (see Table 3.13; Fig. 3.47).

Of non-ore authigenic minerals, calcite, glauconite, and clayey minerals are present in the indicated types of ores.

Table 3.13 Chemical Composition of Ferromanganese-Oxide-Carbonate Ores of the Bezmoshitsa Occurrence (Northern Cis-Timan)

Sample	2033	2034	2037	2038	2041	2042	2043	2044	2045	2051	2056	2057	2059a	2059
SiO_2	7.32	13.8	6.12	28.2	8.1	14.3	12.1	11.5	21.1	8.53	19.8	16.1	23.0	11.0
TiO_2	0.09	0.13	0.04	0.30	0.05	0.09	0.08	0.07	0.12	0.07	0.35	0.3	0.35	0.21
Al_2O_3	2.13	3.19	1.38	10.1	1.82	3.35	3.24	2.68	4.3	2.38	4.84	4.71	5.82	4.09
Fe_2O_3 total	3.29	4.97	30.8	27.4	29.9	25.3	29.0	8.65	13.6	4.52	28.7	20.0	12.0	22.9
FeO	0.13	none	23.4	0.50	0.13	0.26	0.26	1.74	0.36	None	15.4	10.8	6.74	12.2
MnO total	32.1	24.7	15.4	0.40	16.3	16.1	15.3	17.3	11.3	32.1	1.05	1.06	0.83	1.96
MgO	1.45	1.34	4.16	1.49	1.07	1.57	1.55	1.24	1.36	1.32	2.58	1.74	1.13	1.72
CaO	18.2	19.4	6.82	10.1	13.1	11.9	11.1	24.7	20.1	18.1	16.3	25.3	26.4	24.8
P_2O_5	0.35	0.88	0.04	0.46	0.3	1.81	0.54	0.99	0.4	0.82	0.27	0.58	4.43	0.33
Na_2O	0.20	0.31	0.22	1.29	0.13	0.24	0.19	0.23	0.27	0.12	0.35	0.37	0.73	0.26
K_2O	0.37	0.63	0.42	0.91	0.48	1.03	0.72	0.72	1.23	0.53	0.66	0.67	0.73	0.52
H_2O^+	0.82	3.09	0.27	8.04	6.08	5.57	6.00	3.25	–	4.22	3.74	2.62	2.32	2.85
H_2O^-	0.59	1.05	0.37	2.88	0.94	1.26	1.37	1.02	–	0.89	0.89	0.73	0.96	0.76
CO_2	34.3	26.6	37.1	7.70	20.4	17.0	18.0	29.0	20.6	24.9	22.3	26.6	21.6	29.8
Σ	101.3	100.6	101.0	99.2	98.5	99.8	99.5	101.2	100.4	100.2	100.6	99.7	99.7	99.9
NiO	0.04	0.05	0.04	0.05	0.07	0.09	0.09	0.06	–	0.08	0.04	0.03	0.02	0.02
BaO	–	0.19	0.10	–	–	0.19	0.17	0.01	0.01	0.07	0.01	0.01	0.03	0.09
ZnO	–	–	0.34	–	–	–	–	–	–	–	–	–	–	–
Cr_2O_3	–	–	–	–	–	–	–	0.03	0.03	–	0.03	0.05	0.02	0.02
MnO	32.1	23.8	15.4	0.40	16.1	16.0	15.2	17.3	11.1	25.3	1.05	1.06	0.83	1.96
MnO_2	None	1.2	None	None	None	0.13	0.13	None	0.22	8.39	None	None	None	None

Analyses conducted in the chemical laboratory of GIN AN USSR; chemist-analysts: *Stepanets M.I., Simonov I.P.*

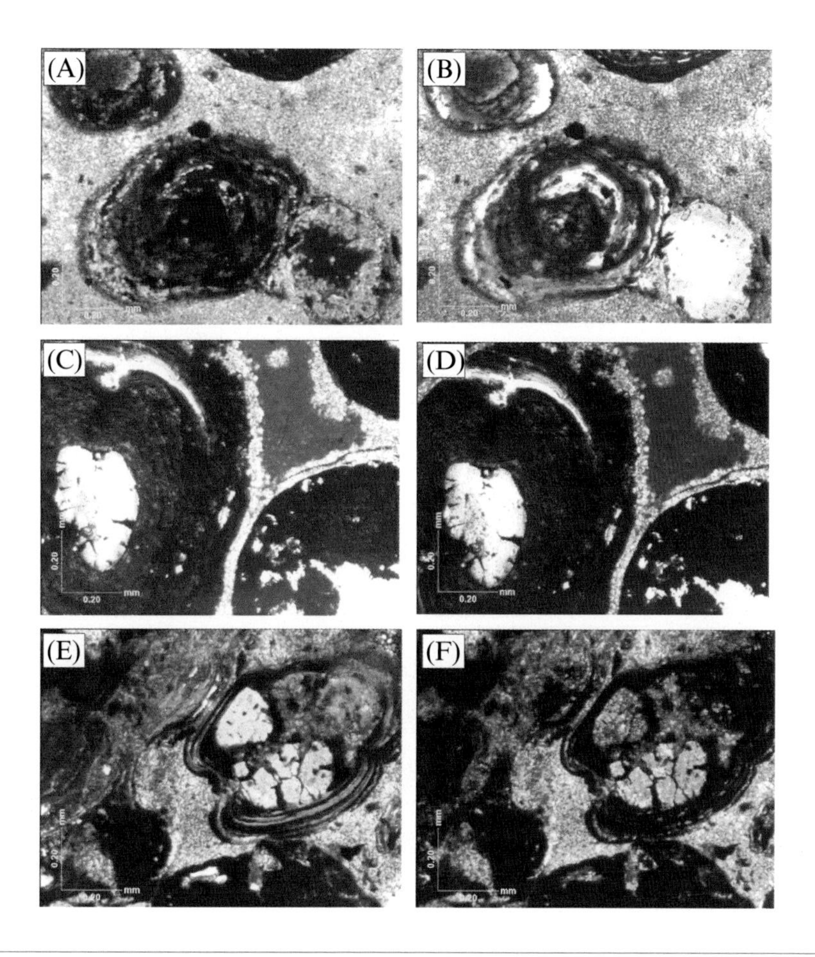

FIG. 3.44

Photos of iron-manganese oolites in thin sections from Bezmoshitskoe-ore occurrence. (A) and (B) Sample 3088 (A—nicole crossed, B—nicole parallel); (C) and (D) sample 3081 (C—nicole parallel, D—nicole crossed); (E) and (F) sample 3067 (E—nicole parallel, F—nicole crossed).

Isotopic composition and origin of manganese carbonates

All obtained isotope data are recorded in Table 3.14 and indicated in Fig. 3.48. From these data, it follows that carbonate matter of the studied manganese rocks is characterized by a fairly close isotopic composition and occupies a range of values from −9.6‰ to −2.3‰ for $\delta^{13}C$ and from 27.9‰ to 30.3‰ for $\delta^{18}O$. On the whole, the obtained $\delta^{13}C$ values are markedly enriched by light isotope ^{12}C by comparison with normal-sedimentary marine carbonates, as is characteristic for carbonates of manganese deposits. This provides evidence, as noted, that at the moment of formation of ore nodules their CO_2 carbon was not in isotopic equilibrium with the dissolved carbon dioxide (DIC) of marine bottom water. The increasing lightness of the isotopic composition of carbon is then solely a direct result of the participation in the formation of ore carbonates of carbon dioxide of organic (biogenic) origin, which is clearly visible in the graph of Fig. 3.48, where the region of $\delta^{13}C$ values for them is skewed toward organic carbon.

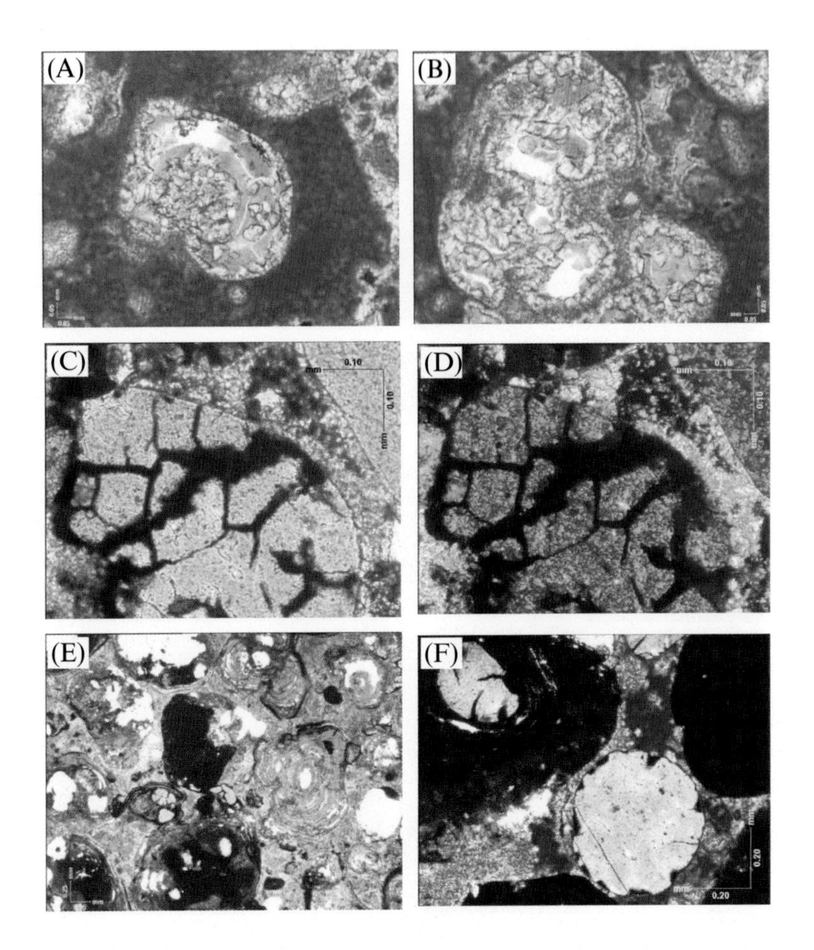

FIG. 3.45

Photos of glauconite rocks in thin sections from Bezmoshitskoe-ore occurrence. (A) sample 3985, nicole parallel; (B) sample 3077, nicole parallel; (C) and (D) sample 3081 (C—nicole parallel, D—nicole crossed); E—sample 3067, nicole parallel; and F—sample 3081, nicole parallel.

The isotopic composition of oxygen in the studied samples is on the whole analogous to the composition of marine sedimentary carbonates. The obtained $\delta^{18}O$ values provide evidence that carbonate ores in terms of oxygen are deposited in isotopic equilibrium (or close to equilibrium) with marine water under normal conditions (Friedman and O'Neil, 1977).

Thus, it can be concluded on the basis of the obtained isotope data that manganese carbonate rocks of the Bezmoshitsa occurrence were formed in the upper layers of sediment during early diagenesis. As in the deposits examined above, carbon dioxide of ooze solutions, which served as a source of CO_2 for ore carbonates at the moment of their formation, represented a mixture of two generations of CO_2—carbon dioxide balanced with dissolved bicarbonate of marine water with high $\delta^{13}C$ values (≈ 0), which correspond to carbon dioxide of sedimentary carbonates (area "A" in the graph of Fig. 3.48), mixed with CO_2 with low values of the isotopic composition of carbon. The latter represents the result of oxidation of organic matter in the ooze sediment.

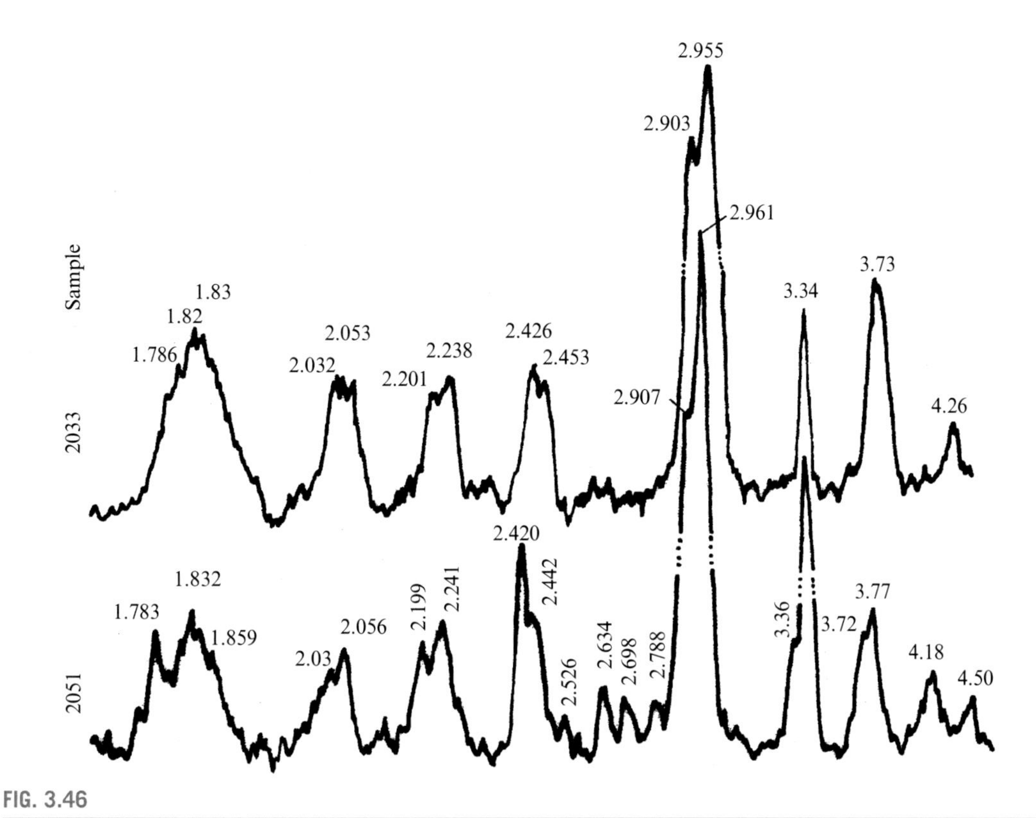

FIG. 3.46

X-ray diffractograms of manganese carbonates of Bezmoshitskoe-ore occurrence.

Consequently, the observed spectrum of isotopic ratios in the studied ores is determined principally by the degree of participation by carbon dioxide from the indicated sources.

It can be proposed that during the process of the formation of ore carbonates of the Bezmoshitsa occurrence, the main role belongs to carbon dioxide balanced with bicarbonate of marine water (carbonate of shells and other residues of organisms, chemogenic carbonate). Carbon dioxide formed due to oxidation of organic matter constitutes a subordinate value (no greater than 10–15%) in their formation.

This is demonstrably confirmed by the direct correlation between the isotopic composition of carbon and the contents of CaO in the samples: the greater the CaO, the higher the $\delta^{13}C$ values (Fig. 3.49). Consequently, the increasing heaviness of the isotopic composition of carbon and the increase in content of CaO is caused by an increase of the relative proportion of initial sedimentary (biogenic) carbonate in the manganese ores.

Among the carbonate formations of the Bezmoshitsa occurrence, the carbonate matter of non-ore nodules is distinguished in terms of the isotopic composition of carbon and oxygen. These nodules were selectively gathered from the drill core of wells 19 and 223 on the level of the development of the manganese-bearing horizon and are represented principally by minerals of the calcite-siderite isomorphic series with an insignificant content of manganese (see Table 3.13; Fig. 3.47). Values of $\delta^{13}C$ for the nodules vary from −21.1‰ to −17.7‰. Such low values likewise provide evidence of the participation

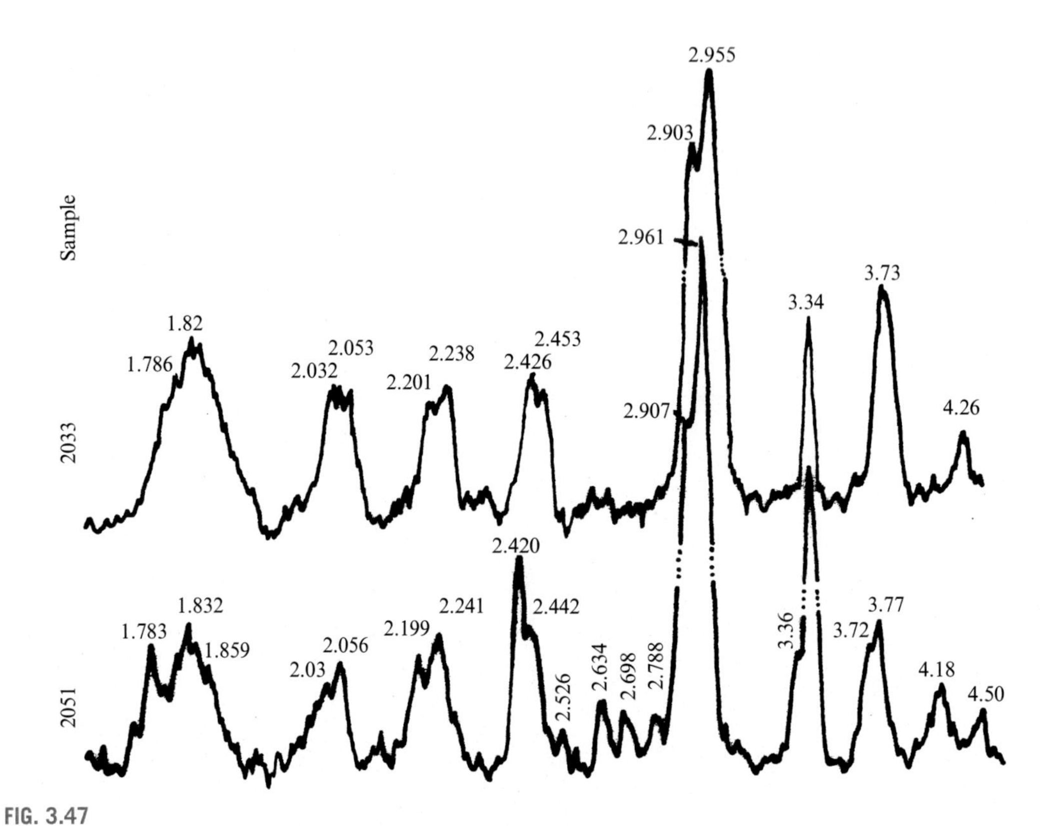

FIG. 3.47

Thermograms curves of manganese carbonates from Bezmoshitskoe-ore occurrence (see description in Table 3.13).

Table 3.14 Isotopic Composition of Carbon and Oxygen in Carbonate Rocks of the Bezmoshitsa Manganese-Ore Occurrence

Number of Analysis	Number of Sample	Well or Location	Well Depth	Sample Characteristics	$\delta^{13}C$, ‰ PDB	$\delta^{18}O$, ‰ SMOW
2031	2056	Well 222	Depth 7.2–8.1 m	Oolitic Fe-Mn—ore	−5.9	29.6
2032	3067		Depth 8.15–8.6 m		−4.4	29.3
2033	3069	Well 222a	Depth 10.0 m	Carbonate ore nodule with oolitic texture	−4.6	27.9
2034	3070		Depth 10.10 m		−5.0	28.4
2035	3072		Depth 10.45 m	Clast of clear layer of belemnite	0.4	27.9
2036	3073		Depth 10.65 m	Ore carbonate nodule	−8.0	29.5
2037	3077	Well 201	Depth 142.1–143.3 m	Oolitic carbonate ore	−9.6	29.4
2038	3078		Depth 153.1 m	Scattered carbonate in sandstone clays	1.5	30.3
2040	3079		Depth 153.1 m	Nodule of siderite	−5.1	30.3

Table 3.14 Isotopic Composition of Carbon and Oxygen in Carbonate Rocks of the Bezmoshitsa Manganese-Ore Occurrence—Cont'd

Number of Analysis	Number of Sample	Well or Location	Well Depth	Sample Characteristics	$\delta^{13}C$, ‰ PDB	$\delta^{18}O$, ‰ SMOW
2041	3081	Well 128	Depth 13.75 m	Nodule of carbonate ore	−3.6	30.3
2042	3082		Depth 14.1 m		−6.4	29.8
2043	3083		Depth 14.8 m		−4.3	30.0
2044	3084		Depth 15.9 m		−2.9	30.2
2045	3085		Depth 16.3 m		−2.3	30.1
2051	3088	Outcropping in the Bezmoshitsa River			−5.0	30.3
Carbonate oolitic ore						
2052	3086-1	Well 128	Belemnite from ore		−1.0	27.7
2047	3087-1		Oligonite nodule from the zone of oxidation		−5.3	23.5
2048	3087-2				−4.7	24.4
2049a	3087-3		Oligonite from the interior of the nodule		−5.1	23.1
2049b	3087-3		Oligonite from the exterior of the nodule		−5.4	23.3
2056		Well 19	Depth 168.4 m	Brownish-gray dense carbonate rock	−21.1	24.6
2057			Depth 168.6 m		−24.1	28.3
2059a		Well 223	Depth 161.1 m		−21.1	27.0
2059			Depth 162.0 m		−17.7	28.0

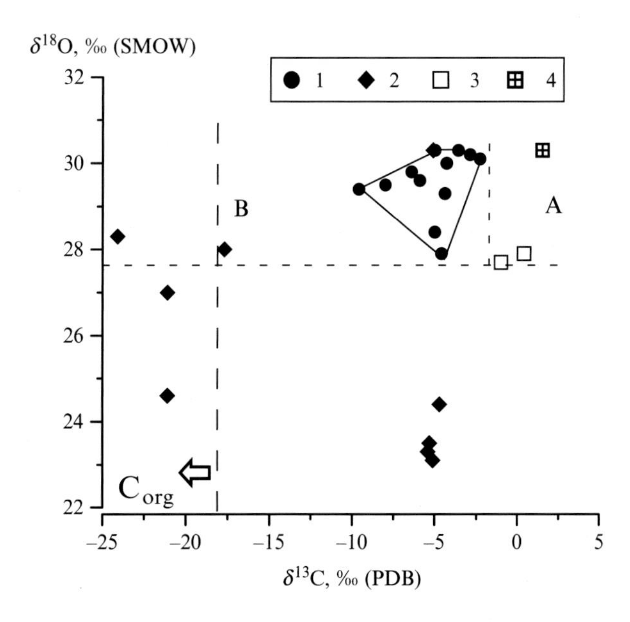

FIG. 3.48

$\delta^{13}C$ versus $\delta^{18}O$ in the carbonate substance of ore bed of Bezmoshitskoe-ore occurrence. (A) Area of marine sedimentary carbonates and (B) the field of marine diagenetic carbonates. 1—Manganese carbonates, 2—oligonites, 3—belemnite rostra, and 4—scattered carbonate in clay rocks.

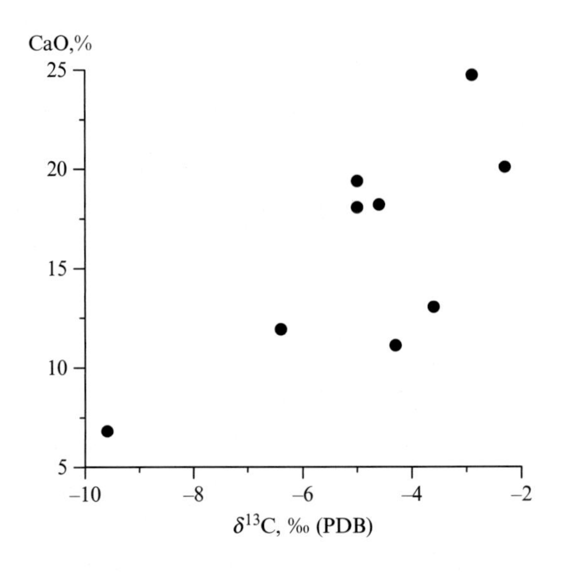

FIG. 3.49

Relation $\delta^{13}C$ values and CaO content in manganese carbonates from Bezmoshitskoe-ore occurrence.

in their formation by isotopically light carbon dioxide of biogenic origin and of the predominant role of such carbon dioxide in the process of carbonate precipitation in the ore strata by comparison with the ore nodules.

The isotopic composition of the oxygen in non-ore nodules by comparison with ferromanganese nodules likewise varies within a broader interval: $\delta^{18}O = 24.6–28.0‰$. This provides evidence of a broad spectrum of conditions of formation and sources of matter in the process of their formation. Thus, nodules with the highest values for the isotopic composition of oxygen were formed in terms of oxygen in equilibrium (or close to it) with marine water and are derived, as are ferromanganese ores, from a diagenetic origin.

Another group of samples is characterized by a lighter isotopic composition ($\delta^{18}O = 23–25‰$). Such a low isotopic composition is also found in the oligonite nodules from the ore horizon in the outcropping on the Bezmoshitsa River ($\delta^{18}O = 23.1–24.4‰$). If we consider that they were formed in the sediment at normal temperatures, then reliable values of the isotope data for oxygen could provide evidence of their later formation with participation by water of lighter isotopic composition. The latter as a rule is particular to water participating in the processes of catagenesis. On the basis of the available isotope data, nothing can be stated definitively regarding the type of catagenesis (Kholodov, 1982a,b) or naturally the isotopic composition and origin of waters that participate in this process.

It can also be supposed that the formation of these nodules could have begun immediately after the conclusion of the early-diagenetic processes. Lost at this stage of lithification of sediments is the connection with the bottom water, while its interstitial waters are enriched by the light isotope ^{16}O (Mottl et al., 1983; Savin and Epstein, 1970; Yen and Savin, 1976). Carbonate matter that forms under these conditions is also characterized by low $\delta^{18}O$ values by comparison with those formed under diagenetic conditions.

The formation of the examined non-ore nodules also could have occurred with participation by meteoric waters. In that case, we would likewise note lower values for the $\delta^{18}O$ values.

Based on the available geological and isotope-geochemistry data, it remains impossible to give preference categorically to a certain epigenetic (catagenetic) process. It can, however, be supposed that nodules with the lightest isotopic composition of oxygen were formed principally in the post-diagenetic stage of the lithification of the sediment with participation by catagenic waters of elisional or meteoric origin.

In this way, the process of the concentration of ore matter (manganese, iron) in carbonate concretions of the Jurassic deposits of the Bezmoshitsa occurrence took place only during the diagenesis stage. Subsequent processes of catagenesis to all appearances did not lead to the formation of manganese carbonate ores. Nodules of this stage are represented principally by carbonate of calcium and iron (minerals of the siderite-calcite isomorphic series) with insignificant contents of manganese not reaching ore concentrations.

3.2.1.5 Sedimentary-diagenetic deposits of manganese of various areas of the world

The *North Urals manganese basin*, in addition to the examples examined above, is without question one of the most vast basins of accumulation of sedimentary-diagenetic manganese and manganese-containing rocks. Its total predicted resources today are estimated at 120 million tons (with mean manganese content of approximately 20%) (Kontar' et al., 1999). This basin in the form of a narrow band extends almost 300 km in the meridional direction, from the latitude of Serov town in the south to the village Burmantovo in the north, prospectively continuing further to the north into the territory of Khanty-Mansi Autonomous Okrug for no fewer than 200 km. Here have been discovered 15 deposits (9 have been studied in detail: Berezovskoe, Novo-Berezovskoe, Iuzhno-Berezovskoe, Ekaterininskoe, Marsyatskoe, Yurkinskoe, Loz'vinskoe, Ivdel'skoe, Tyn'inskoe; 5 deposits have been studied up to the stage of preliminary exploration and prospecting-evaluation works: Burmantovskoe, Yuzhno-Ivdel'skoe, Visherskoe, Glukharnenskoe, Kolinskoe, and Polunochnoe, which is not counted in the balance and is considered worked-out), as well as a range of prospective areas with well profiles of within the boundaries of which are exposed beds of manganese ores reaching industrial thicknesses (Burmantovskaia, Sobianino-Shipichenskaya, Visherskaya, Losinovskaia, Mar'insko-Malinovskaia). Additionally, a range of areas (Suevat-Paul'skaia, Or'inskaya, Maslovskaya, Serovskaya, Usteiskaya, Katas'minskaya, and Kolinskaya) are also prospective for the presence of manganese ores (Rabinovich, 1971' Kontar' et al., 1999; Gal'ianov and Iakovlev, 2006) (Fig. 3.50). A significant portion of the deposits is deposited in a complex tectonic environment determined by a wide development of faults resulting from tectonic activity during the Neogene-Quaternary.

Within the boundaries of the examined basin have been discovered two manganese-bearing levels represented by two manganese-bearing horizons: the Upper Cretaceous and the Paleogene. Both are considered to have formed in the western part of the transgressive epiplatform marine basin in the Cretaceous shelf zone and are represented by glauconite-quartz deposits.

The Upper Cretaceous manganese level is traced at a distance up to 50 km from Tyn'inskoe deposit in the south to Burmantovskoe (and further to the north). It has been exposed by drill wells in the base of the Santonian-Campanian deposits in Sabianinskaya and Burtmanskaya areas and is predicted further to the north for almost 100 km.

The ore-bearing deposits are classified as a terrigenous glauconite-quartz formation and are represented by sandstone clays, clayey sands, siltstones, sandstones with clayey-rhodochrosite cement, and

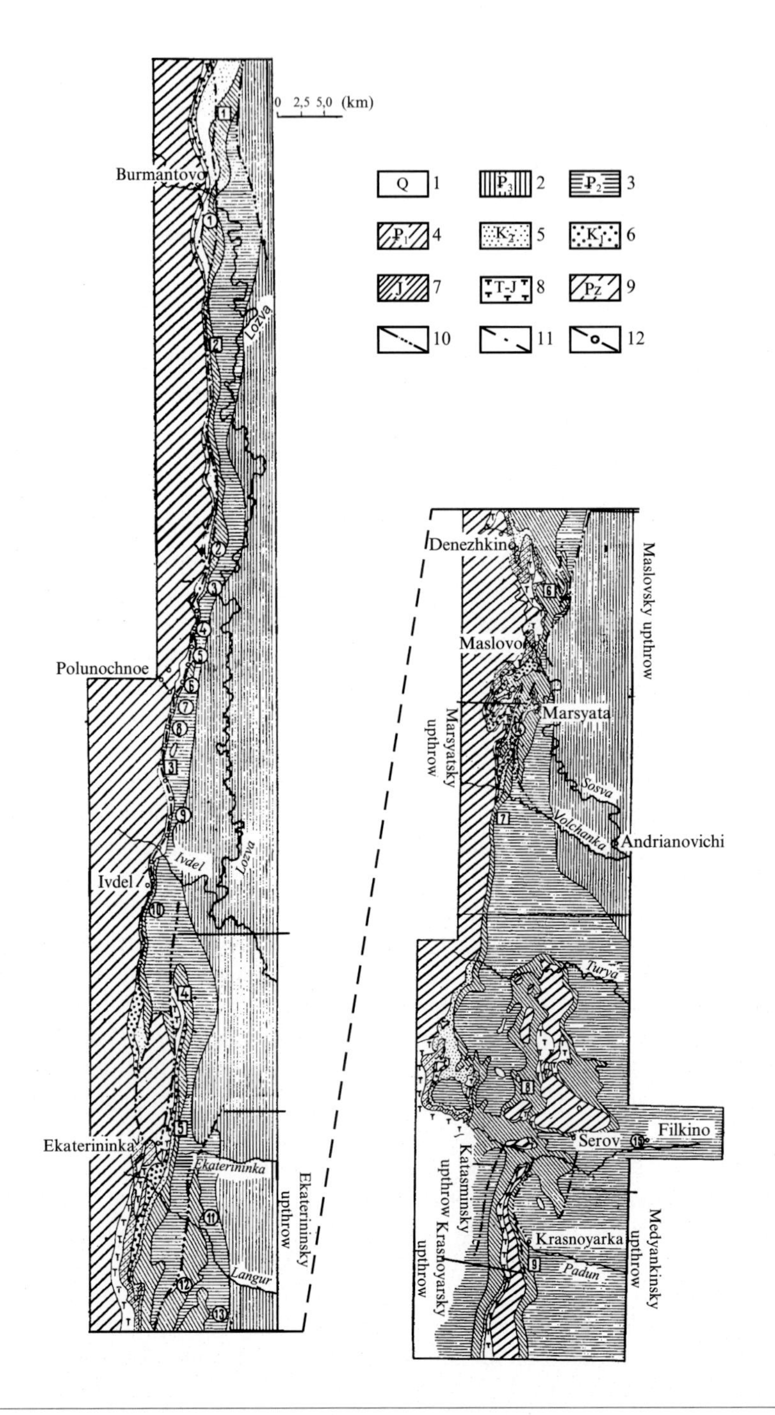

FIG. 3.50

See the legend on opposite page.

interbeds of manganese carbonate ores in the basal horizon of the Zaikovskaya suite of Santonian-Campanian age. The thickness of the ore horizon varies from 0.5 to 2.7 m. Carbonate ores are clayey-sandstone and concretional-clayey; in terms of composition, they are rhodochrosite with manganocalcite, with content of manganese from 10% to 26% (in mean approximately 16%). Below the Quaternary deposits, they are commonly oxidized and transformed into oxide psilomelane-pyrolusite ores.

In the base of the ore horizon are commonly deposited clayey-sandstone deposits; the horizon is overlapped by marine, predominantly clayey rocks of the upper part of the Zaikovskaya suite, and in the western area by Quaternary deposits.

The Paleogene ore horizon is also connected with terrigenous deposits of the quartz-glauconite type, separated in the composition of the Polunochnoe member by the Marsiatskaya suite (Lower Paleogene). Deposits of the member were formed in the shelf zone in the epoch of a transgressive Paleogene sea following the regressive stage of the end of the Cretaceous Period. The ore-bearing formation is traced from the latitude of Serov town in the south to Burmantovo village and further to the north for a distance >300 km.

Predominating in the composition of the Polunochnoe member are sands, siltstones, glauconite-quartz sandstones, quartz-glauconite siltstones, gritstones, and conglomerates with interbeds of sandstone clays and manganese carbonate ores.

A characteristic particularity of the ore-bearing horizon is its rhythmic structure, featuring from one (Kolinskoe deposit) to three (Tyn'inskoe deposit) rhythms of sediment accumulation with manganese ores; occasionally they are present in full capacity within the boundaries of a single deposit. The number of rhythms increases in the western part of the deposit (closer to the shoreline of the paleobasin). In the eastern part of the Cretaceous shelf zone the ore-bearing horizon is commonly represented by a single rhythm.

FIG. 3.50

Schematic geological map of the North Ural manganese-ore basin (composed by S.D. Rabinovich, E.I. Danilovoy et al., 1961–64) (Rabinovitch, 1971). 1—Lozvinskaya suit (boulders, loams of ancient Quaternary Valley; 2—Chegan suite (mudstone and beydellitic clay); 3—Irbit and Serov suites (diatomite light color, flask, rare—the sand, sandstone); 4—Ivdelskaya and Marsyatskaya suites, Polunochnaya pattern (clays, mudstones, diatomite, sandstone, manganese ore; 5—Fadyushinskaya (Gankinskaya), Kamyshlovskaya (Zaikovskaya), Nadsurpiyskaya, Surpiyskaya, Mysovskaya, Samskaya suites (sea—argillite, montmorillonite, glauconite, sandstones, rarely—manganese ore; continental—kaolin clays, silts and sands, siderite, brown iron ores, bauxite); 6—Khanty-Mansiyskaya, Shapshinskaya and Tyninskaya suites (sea—clay kaolin-hydromica; continental—kaolin clay, silt, sand); 7—Chapinskaya, Visherskaya and Maninskaya beds (kaolin clay of different colors with seams of coal, sand and silt, brown iron ores and siderite); 8—weathering crust of Paleozoic rocks in the south—lateritic iron ore, nickel ore; 9—effusive-sedimentary strata Middle Devonian (serpentinite, gabbro, diorite, and others igneous rocks); 10—upthrow open (facing surface); 11—upthrow closed (projected on the surface); and 12—the eastern boundary of thrust. The numbers in the circles marked deposits: 1—Burmantovskoe, 2—Tyn'inskoe, 3—Lozvinskoe 4—Yurkinskoe 5—Polunochnoe, 6—Novo-Berezovskoe, 7—Berezovskoe, 8—Yuzhno-Berezovskoe, 9—Ivdelskoe, 10—Yuzhno Ivdelskoe, 11—Eketerinenskoe, 12—Gluharnenskoe, 13—Visherskoe, 14—Marsyatskoe, and 15—Kolinskoe. The numbers in the squares designated exploration areas: 1—Burmantovskaya, 2—Sobyanino-Shipichenskaya, 3—Maninsko-Malinovskaya, 4—Orinskaya, 5—Losinovskaya, 6—Maslovskaya, 7—Yuzhno Marsyatskaya, 8—Serovskaya, and 9—Krasnoyarsk sector.

The ore horizon in the east is replaced facially by more deep-water clayey deposits of the Marsiatskaya suite.

In the structure of the deposits are substantially developed tectonic faults of Neogene-Quaternary age (Rabinovich, 1971). These have submeridional strike and are represented by dip-separation thrusts and reverse faults with an amplitude of replacement from a few dozen meters up to 100 m, which break the ore-bearing bed into flexure-type folds.

The primary ores are represented by sandstone, sandstone-clayey, siliceous, concretional-clayey, and clayey types (Kozhevnikov and Rabinovich, 1961; Rabinovich, 1971). The former two have the widest distribution and in total constitute 48–53%, the siliceous constitute from 6 (Visherskoe deposit) to 13.6% (Ekaterininskoe deposit), concretional-clayey constitute up to 11.6% (Losino-Visherskaya area), and clayey constitute up to 27% (Ekaterininskoe deposit).

The ores are classified as carbonate ores of industrial type and consist of calciorhodochrosite (with various ratios of Ca and Mn, including an isomorphic impurity of Ca and Fe), manganocalcite (content of $CaCO_3$ exceeding 25%), and rarely oligonite (ferrorhodochrosite, with content of $FeCO_3$ exceeding 15%).

Non-ore minerals are represented by quartz, opal, glauconite, and clay. The principal terrigenous impurity is quartz with dimensions from pelitic up to large-grained, more rarely gravel. Mean content of quartz in the ores constitutes 10% (varying from rare grains up to 35%). Glauconite is present in quantity from 1% to 30% and is represented by grains with dimensions from 0.005 to 2 mm. Its greatest quantity is noted in sandstone, clayey-sandstone, and clayey-concretional ores. In carbonate-siliceous ores opal forms exceptionally thin (1–2 mm) interbeds; in concretions, it is present in the composition of rhodochrosite-siliceous cement in a quantity from 10% to 50%.

Clayey minerals are characteristic for all varieties of ores and are represented by beidellite and montmorillonite, frequently interstratified with opal. Maximum contents of clayey minerals are noted in concretional-clayey-carbonate ores.

Of phosphorous minerals are present calcium phosphate, dahllite, and collophane. Occurring as well are phosphates in the form of films and incrustations in the grains of quartz, glauconite, and rhodochrosite. Maximum contents of phosphates have been detected in sandstone and clayey-sandstone ores of deposits formed in the marginal part of the Cretaceous shelf zone (Marsiatskoe, Kolinskoe, Tyn'inskoe).

Content of manganese in ores strongly varies and reaches a maximum value of 28.5%. The richest are "sandstone ores" with content of manganese from 21.71% to 28.5%, and the poorest are concretional-clayey ores (content of manganese not exceeding 23%).

In a range of deposits (Marsiatskoe, Tyn'inskoe, Yuzhno-Berezovskoe, and others) are present manganese-oxide ores, formed due to oxidation of the initial carbonate ores in the hypergenesis zone. Manganese content in these varies from 10% to 36% with a mean of 27.7%. The depth of distribution of oxide ores does not exceed 25 m from the surface.

The principal minerals of oxide ores in the deposits are pyrolusite, more rarely manganite and vernadite. Characteristic for manganite ores is a thin layering due to the alternation of thin (tenths of a millimeter), dense, and loose interbeds of manganite (possibly also psilomelane). Noted rarely are iron hydroxides developed along fissures and taking the form of thin rims around the pyrolusite and psilomelane.

In carbonate ores, pyrite occurs in isometric form, while in oxide ores marcasite occurs rarely in the form of irregular accumulations, cross-veins, and isolated grains. The quantity of sulfides constitutes on average 1–2%.

The principal features of the geological structure of the deposits of the North Urals manganese-ore basin can be seen in the deposits that have received the most study—Tyn'inskoe, Marsiatskoe, and Polunochnoe. The first of these at present has been explored in detail (Rabinovich, 1971; Kontar' et al., 1999; Gal'ianov and Iakovlev, 2006).

The *Tyn'inskoe deposit* is situated in Ivdel' district of Sverdlovsk Oblast. Its length from north to south constitutes approximately 1.5 km, with a width reaching 0.5 km. In the east, the border of the deposit passes along a large subvertical normal fault, and in the west the ore-bearing horizon is bounded by the distribution zone of the Paleozoic and Lower Cretaceous deposits and is overlapped by Quaternary formations. From the north and south, the distribution zone of ore-bearing deposits is bounded by blocks of rocks raised during the Neotectonic stage, where these deposits have been eroded (Fig. 3.51).

Ore-bearing deposits of a terrigenous glauconite-quartz formation are represented by sandstones with clayey, opal-clayey, rhodochrosite cement, glauconite-quartz sands, siltstones, and sandstone clays.

Sandstone clays have the following composition: a pelitic fraction (up to 80% of the volume of the rock) is represented by minerals of the group of montmorillonite (87%) and opal (10%); and a sandstone-siltstone fraction (composed of quartz, carbonate of siderite-rhodochrosite composition, clasts of rocks of clayey-siliceous and siliceous composition, glauconite, and grains of zeolites). In the heavy fraction, carbonates of the siderite-rhodochrosite series predominate (up to 90%).

The primary ores in terms of content of impurity of non-ore minerals are subdivided into concretional-clayey (72%) and sandstone-clayey (28%). Ore matter is represented by rhodochrosite and manganocalcite. The content of manganese in carbonate ores varies from 10.04% to 25.84% (mean 15.62%), phosphorus impurity is on average 0.292%.

In the western part of the deposit are distributed oxide psilomelane-pyrolusite ores with mean content of manganese 14.09%, of phosphorus 0.26%, and of sulfur 0.04%.

The thickness of the manganese-ore horizon within the boundaries of the deposit varies from 0.5 to 9 m, bears a rhythmic structure, and is moderately pitched to the east at an angle of 5–10 degrees.

The deposit features from one to three rhythms, which are represented by a regular replacement (from bottom to top) of sands that are variously grained, glauconite-quartz, and quartz-glauconite with numerous grains of rhodochrosite by diatom clays with diagenetic concretions of ore-bearing quartz-glauconite sandstones with rhodochrosite and siliceous-rhodochrosite cement.

In the roof of the ore-bearing horizon are deposited clayey deposits of the Marsiatskaya suite, and in the floor, argillitic clays of the Zaikovskii suite composed at 99% from material of the pelitic fraction—minerals of the group of montmorillonite (67–76%), clasts of opal, and more rarely residues of diatom algae (20–25%). The siltstone fraction (approximately 1%) is represented by quartz, glauconite, clasts of opal, feldspar, carbonates of spherical form, and minerals of the heavy fraction (ilmenite up to 32%, epidote up to 39%, carbonates up to 12%, magnetite up to 3%, and limonite up to 5%, as well as grains of zircon, garnet, tourmaline, apatite, sphene, and hornblende in total up to 1%).

In all varieties of carbonate ores, clastic material is represented by quartz, plagioclases, muscovite, glauconite, tourmaline, and epidote. The quantity of clastic material in the siliceous massive ores is insignificant; in the concretional-siliceous ores it is approximately 2% and in sandstone ores it is 10%. The dimensions of clastic material are 0.3–0.5 mm; grains are commonly not rounded; glauconite takes spherical forms. Carbonates have a reddish-brownish color. Small "kidney-shaped" formations of rhodochrosite frequently have a fringe of ferruginous carbonate.

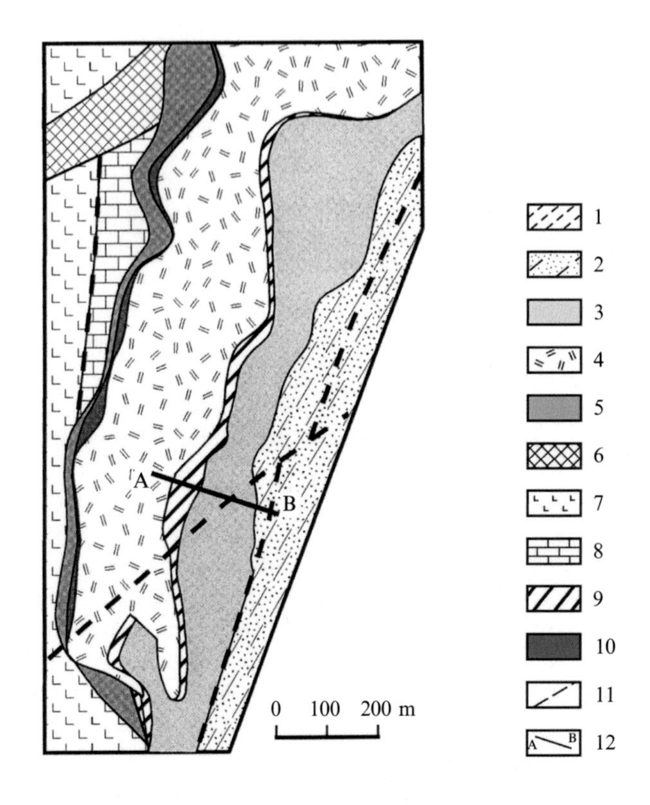

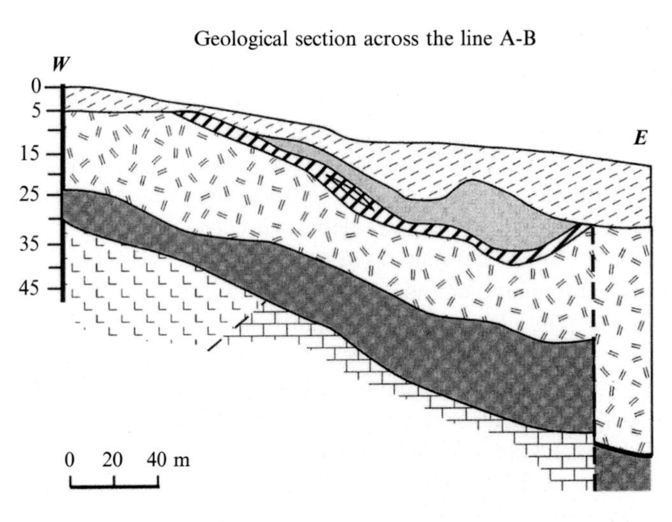

FIG. 3.51

See the legend on opposite page.

With the recrystallization of carbonates during the post-early-diagenetic stage, an intensive microstreaking occurred with the formation of filiform and micronest-like structure. Recrystallized carbonate (rhodochrosite) commonly forms spherulites (0.5 mm and larger). Manganous dolomite is present. A transection by micro-cross-veins of carbonate grains of glauconite is observed. A thin encrustation of marcasite occurs in the form of spherical-"echinoid" formations (Kontar' et al., 1999).

Predominating among primary carbonate ores of the Polunochnoe member are sandstone-clayey (60%) and concretional (25.6%) varieties; sandstone varieties constitute 19.8%, while siliceous varieties constitute 6%. Content of manganese in sandstone-clayey varieties varies within the boundaries 15.34–28.4%, in concretional-clayey varieties from 10.13% to 25.23%, and in sandstone varieties from 14.36% to 25.32%. The mean content of manganese in ores about the deposit constitutes 20.22%; mean thickness of the ore horizon constitutes 2.3 m. Mean content of SiO_2 in carbonate ores constitutes 28.1%, CaO—4.02%, P_2O_5—0.55%, and Fe—2.48%. In oxide ores content of manganese constitutes 33.5%, CaO—2.75%, P_2O_5—0.37%, and Fe—2.05%.

The deposit is divided into categories B_1, C_1, and C_2. The total reserves of manganese ores in the deposit constitute 579.4 thousand tons [the Paleogene ore horizon has been evaluated in terms of categories B_1+C_1: their combined total of primary carbonate ores constitutes 449.5 thousand tons, while for oxide ores the figure is 52 thousand tons; reserves of primary carbonate ores and oxide ores in the Upper Cretaceous horizon for categories C_1+C_2 constitute 77.9 thousand tons (Kontar' et al., 1999)].

Carbonate-manganese ores of the deposits of the examined basin have been classified as sedimentational-diagenetic (Shterenberg, 1963; Rabinovich, 1971). Continental runoff has been proposed as a source of manganese due to the destruction of Silurian and Devonian vulcanites of the Urals in humid conditions.

The studied manganese ores of the Tyn'inskoe deposit are represented by glauconite-sandstone rocks with cement of manganese carbonates (Fig. 3.52). Carbonate matter is represented by colloform formations evidently as well as altered organogenic and microbial residues.

A diagenetic origin of manganese carbonates is supported by the isotope data. Values of $\delta^{13}C$ and $\delta^{18}O$ in these carbonates vary within narrow boundaries: −25.9‰ and 26.9‰ for sample 5993; −25.1‰ and 25.6‰ for sample 5994; and −24.3‰ and 26.8‰ for sample 5995. Such low $\delta^{13}C$ values are evidence that the anion group CO_3^{2-} was formed practically entirely from oxidized carbon of organic matter, as is characteristic for authigenic carbonates of the diagenesis zone.

Values of $\delta^{18}O$ are characterized by lower values with regard to carbonates formed in common marine conditions in the diagenesis zone. To all appearances, the formation of the studied manganese carbonates occurred in a marine basin with insignificant desalinization at somewhat elevated

FIG. 3.51

Skech map of the Tyn'insky deposit (by K.P. Saveleva) and cross section along the line A–B (of D.A. Kostromin and K.P. Saveleva) (Kontarev et al., 1999). 1—Quaternary deposits; 2—opoka, opoka-like clays, Serovskaya suite, Eocene; 3—montmorillonite clays with glauconite, Marsyatskaya suite, Lower Paleogene; 4—montmorillonit-beydellitic clays, Zaikovskaya Formation of Upper Cretaceous; 5—kaolin clays, sands with plant detritus, Lower-Upper Cretaceous; 6–8—deposits of Middle Devonian: 6—apotuffite shales, 7—volcanic rocks of median composition, 8—limestones; 9, 10—manganese ore horizons: 9—in deposits of Marsyatskaya suite, Lower Paleogene, 10—in deposits of Zaikovskovskaya suite, Upper Cretaceous; 11—faults; 12—line of geological cross section.

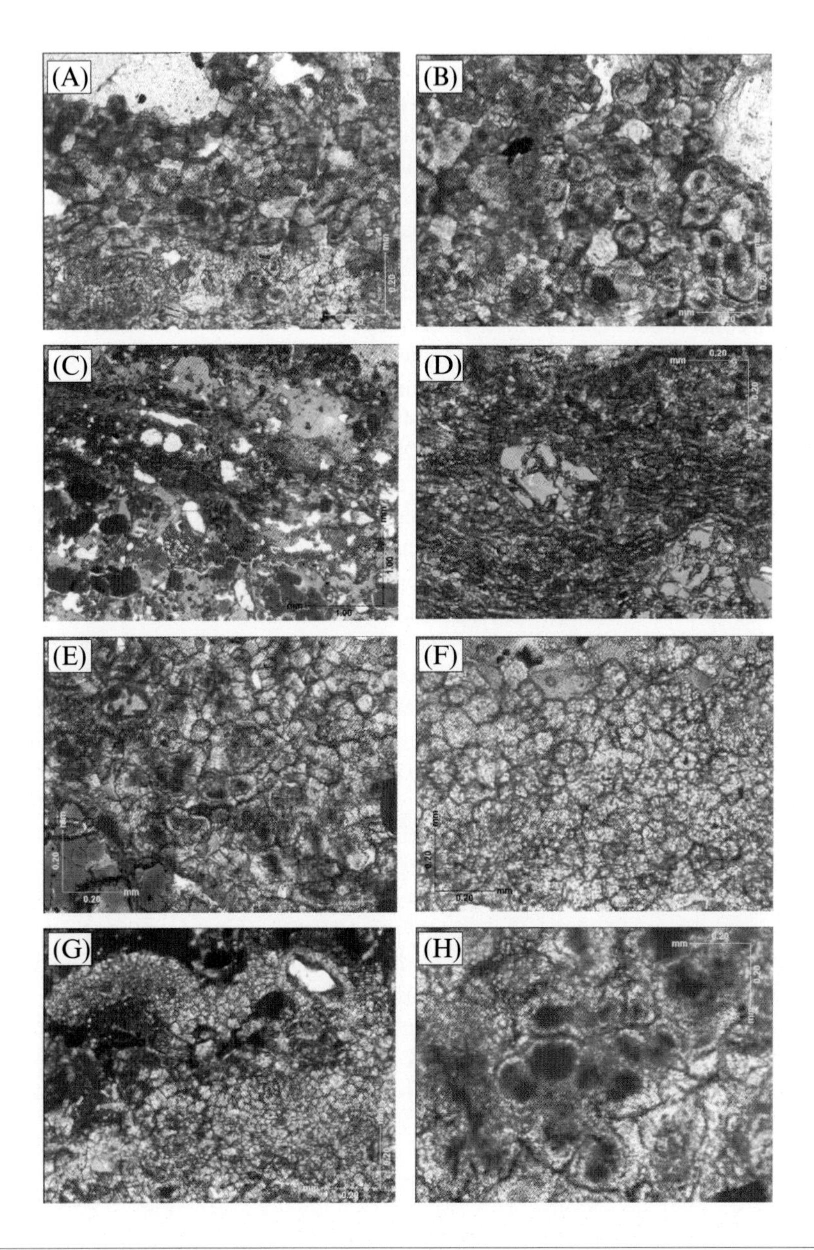

FIG. 3.52

Photos of the thin sections of manganese carbonate ores, Tyn'inskoe deposit. (A)–(D) Different parts of the sample 5994 (A–C—nicole parallel, D—nicole crossed; A, B, D—an increase of 10, C—an increase of 20); (E)–(H) different parts of the sample 5995 (E, F, H—nicole parallel, G—nicole crossed; an increase of 10).

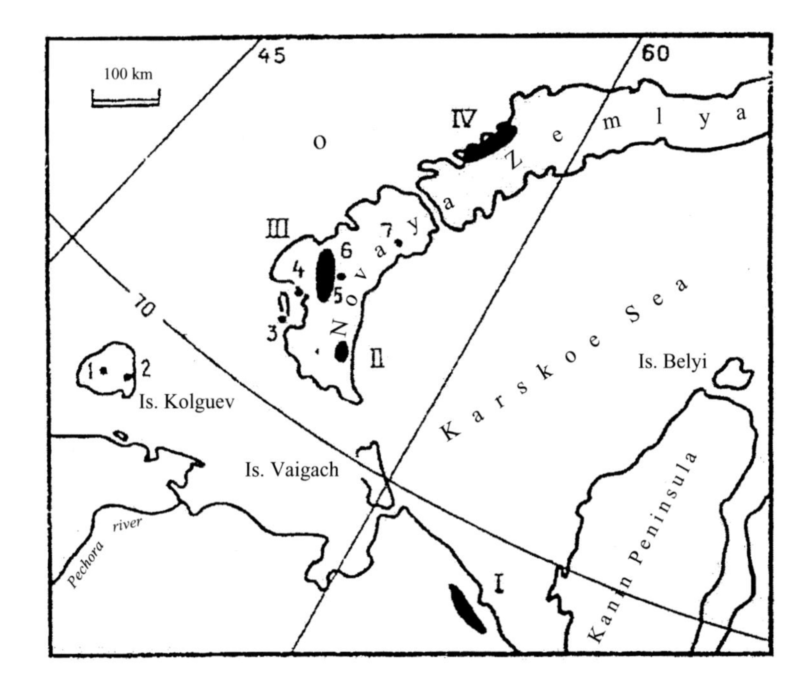

FIG. 3.53

Location of manganese areas and key sections of Paykhoi-Novaya Zemlya manganese basin (Stolyarov et al., 2009). I–IV—areas: I—Paykhoisky, II—Kolodkinsky, III—Rogachev-Severotayninsky, and IV—Sulmenevsky. 1—Well Izhimka-Tarskaya; 2—well Peschanoozerskaya; 3—foreland Costin nose; 4—peninsula Vypucliy; 5—brook Margantseviy; 6—brook Posudniy; and 7—river Krasnaya.

temperatures. This does not preclude the possibility that, judging by the isotope data, a portion of these could have been formed during the post-early-diagenetic stage of the lithification of the sediment (possibly with participation by catagenic solutions).

One of the largest manganese-ore basins and the richest in terms of resources is without question the *Pai-Khoi-Novaya Zemlya* basin (Fig. 3.53). It developed through the end of the Paleozoic, when (in the Permian epoch) it ran from the Northern Cis-Urals in the East, Pai-Khoi in the south, and encompassed the island of Novaya Zemlya (Voiakovskii et al., 1984; Rogov and Galitskaya, 1984). By some estimates, the potential resources of manganese here constitute a few billion tons (Platonov et al., 1992; Mikhailov, 1993; Sharkov, 2000; Stoliarov et al., 2009).

At present, three distribution areas of manganese-containing carbonate rocks have been established in the Novaya Zemlya archipelago: the Rogachevskii and Kolodkinskii areas on the southern island and the Pribrezhnii area on the northern island.

The most studied is the Rogachevskii area in the central part of the southern island. Manganese-bearing horizons within the boundaries of this island are traced in an area with a length of 70 km and width of 30 km. The ore-hosting mid-Sokolovskaya subsuite of Lower Artinskian age of the Lower Permian is represented by black argillites and siltstone argillites interlaid with manganese carbonate ores; which are in part pyritized. In the stratigraphic succession have been distinguished from two to five manganese-bearing horizons, which are subdivided into two types. Characteristic for the first type is a thin interstratification of argillites and manganese carbonates, with a predominance of the latter. Characteristic for the second type is an interstratification of those same varieties of rocks, but non-ore (siltstone-argillite) interbeds predominate. Mean content of manganese in the richest horizons (River Vadegi basin) constitutes 12.4% (at a thickness of 20 m) with variations in content in different parts of the horizons according to trench-sampling data from 11% to 14.7%.

Productive horizons of other manganese-bearing areas of Novaya Zemlya are not substantially distinguished from those of the Rogachevskii area. A productive member of interstratification of rhodochrosite rocks and argillites of a visible thickness of approximately 10 m in the Kolodkinskii area is traced in an area of approximately 40 km^2.

In the Pribrezhnii area on the western shore of Novaya Zemlya in the strata of argillites have been established interbeds of manganese-containing carbonate rocks with a thickness reaching 0.5 m. The visible thickness of the productive part constitutes approximately 150 m. Manganese-containing rocks here are the dense fine-grained rocks consisting predominantly of manganese carbonates (50–80%; rhodochrosite, calciorhodochrosite); dolomite is noted occasionally (up to 5%). Hydromicas, montmorillonite, chlorite, and silica are present in variable quantity. A clump of sulfides (pyrite, sphalerite) is noted with dimensions reaching 2–3 cm. Manganese content reaches 27%.

In the Rogachevskii area, insignificant occurrences have been uncovered of manganese-oxide ores of pyrolusite-psilomelane composition, controlled by tectonic fault zones. Manganese content in trench samples reaches 45%.

In the genetic regard, according to available data, the accumulation of manganese here occurred in a stagnant basin with free hydrogen sulfide in the bottom water (Rogov and Galitskaya, 1984).

For the purpose of elucidating the genesis of manganese carbonates, we have conducted preliminary isotope research on rocks from a well (well 7) and from an outcrop ("B," samples 670, 689a, 697, 719, 718c, 747; samples from the collection of V. Kh. Nasedkina). The results are recorded in Table 3.15 and displayed in Fig. 3.54. From these it follows that in the area of common marine carbonates (area "A" in the graph) not a single one of the studied samples occurs. At the same time, as for many of the deposits examined above, a portion of the samples is characterized by an isotopic composition particular to authigenic carbonates of the diagenesis zone. That said, in the distribution of values of $\delta^{13}C$ and $\delta^{18}O$ is noted an inverse linear correlation evidently due to processes of their subsequent (post-early-diagenetic, catagenetic) transformation (most distinctly manifested in rocks of the outcrop).

It can be supposed that carbon dioxide of the transformed catagenic solutions was characterized by a heavier isotopic composition of carbon and a lighter isotopic composition of oxygen.

It should be noted that the light values of the isotopic composition of pyrite (sample B-492, $\delta^{18}S = -34.2‰$) are also evidence for a potential early-diagenetic origin of primary manganese carbonates.

Lower Permian manganese carbonate ores have also been uncovered in Pai-Khoi, on the right bank of the Kara River. The ores are represented by manganocalcite and manganese calcite; the content of manganese in samples falls in the range of 1.03–12.74% (Rogov and Galitskaya, 1984).

Also potentially classified to the examined sedimentary-diagenetic type of manganese deposits are evidently the deposits genetically and spatially connected with formations of black shales. Serving as a potential example of deposits associated with black-shale deposits are the *Molango basin deposits (Hidalgo state, Mexico)*, with a length of over 50 km. Its resources are estimated at greater than 1.5 billion tons (Laznica, 1992).

Table 3.15 Isotopic Composition of Carbon, Oxygen, and Sulfur (Pyrite) in Carbonate Rocks of the Novaya Zemlya Manganese Occurrence (Samples of V.Kh. Nasedkina)

Number of Analysis	Number of Sample	Location	Sample Characteristics	$\delta^{13}C$, ‰ PDB	$\delta^{18}O$, ‰ SMOW
2358	7-23	Well 7	Rhodochrosite-siliceous rock	−9.1	25.6
2359	7-36			−10.4	26.0
2360	7-45			−11.2	29.3
2361	7-54			−8.2	29.2
2362	7-61			−7.5	23.7
2363	7-70			−20.2	28.7
2365	7-120			−8.5	24.2
2366	7-139			−10.5	29.8
2367	7-157			−9.8	23.7
2368	7-160			−8.2	27.0
2369	B-670	Outcrop	Rhodochrosite-siliceous rock	−3.5	25.9
2370	B-689a			−9.3	29.3
2371	B-697			−7.9	29.3
2372	B-710			−5.1	25.9
2373	B-718c			−1.6	25.5
2377	B-747			−9.0	27.3
2378	492		Pyrite	$\delta^{34}S = -34.2$‰	

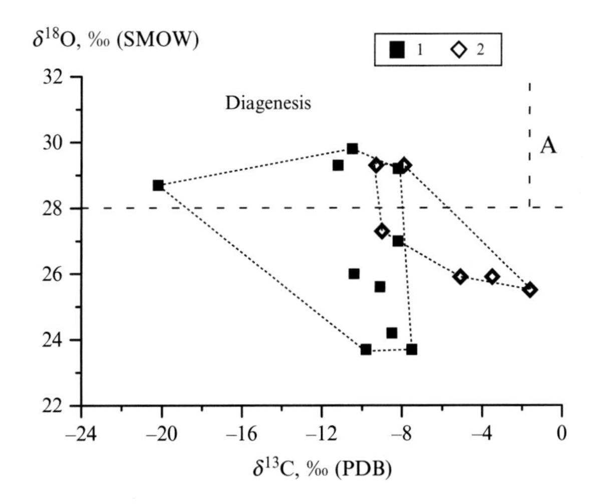

FIG. 3.54

$\delta^{13}C$ versus $\delta^{18}O$ in manganese carbonates of Novaya Zemlya deposit. A—Area of normal-sedimentary (marine) carbonates; 1—rhodochrosite-siliceous rock, well 7; and 2—the same, from outcrop.

Manganese ores here are enclosed in the sequence of manganese-containing carbonate (predominantly limestones) rocks of the Upper Jurassic (Kimmeridgian, Chipoco facies of the Taman group), forming an ore-bearing strata with a thickness reaching 50–60 m. This strata is deposited on carbonatized black shales of the Santiago formation and is overlapped by a strata of dark-gray and black micrite non-manganese limestones, interstratified with their clayey varieties and spiculites.

The primary ores are represented by lamino-(sheet)-like fine-grained manganese carbonates (rhodochrosite, kutnohorite, manganocalcite); in the weathering crust are developed high-quality oxide-manganese ores, useful for the production of dry-cell batteries.

Detailed isotope research has been conducted by Okita et al. (1988) with the aim of clarifying the genesis of manganese and manganese-containing carbonates. The obtained data have allowed these authors to conclude an early-diagenetic formation of Mn carbonates. For the studied rocks has also been established a direct correlation between the isotopic composition of carbon and the content of manganese (Fig. 3.55), which provides evidence of active participation by isotopically light carbon of oxidized (during the diagenesis process) organic matter. This closely matches the examples for other deposits reported above (Nikopol', Mangyshlak, and others).

It should be noted that rocks of the examined deposit were subject to substantial post-early-diagenetic (catagenetic) transformation resulting in changes to their initial isotopic composition. Thus, using the data (Okita et al., 1988), it can be seen (Fig. 3.56) that manganese carbonates [using only the data of analyses obtained after the reduction of the samples over the course of 70 hours (Okita et al., 1988, Table 2)] in terms of the isotopic composition of oxygen are represented by two groups: (1) initial early-diagenetic manganese carbonates, with high δ^{18}O (28‰ and higher) and (2) secondary manganese carbonates (area "B") with lighter isotopic composition of oxygen (23–26‰). Into this area "B"

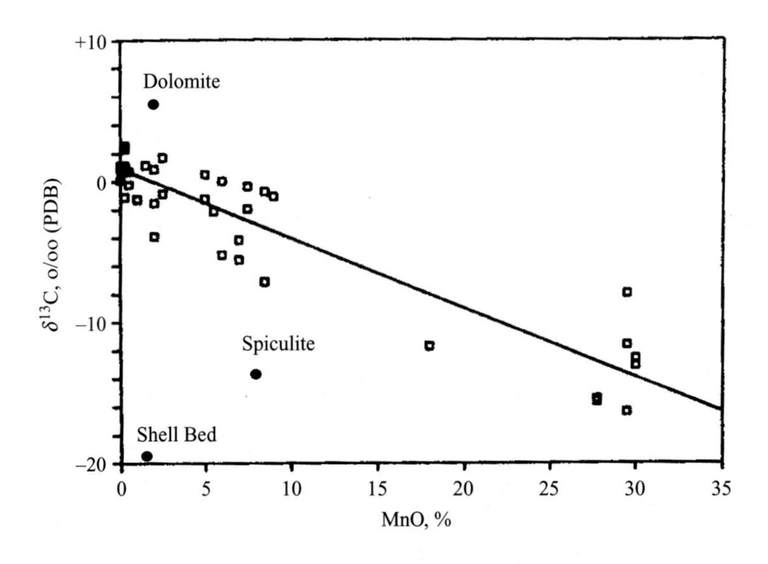

FIG. 3.55

Dependence of the content of manganese and δ^{13}C values total rock sample *(hollow squares)* field Molango Mexico (Okita et al., 1988). For comparison, the data spikulitam and dolomite *(black circles)*.

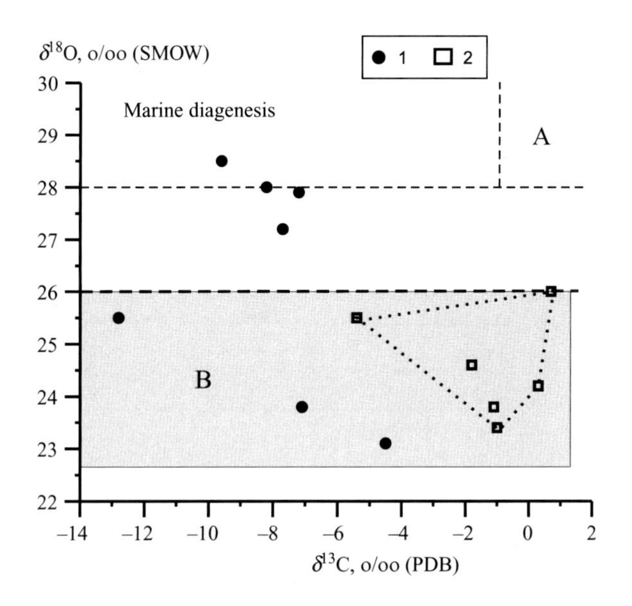

FIG. 3.56

$\delta^{13}C$ versus $\delta^{18}O$ in manganese carbonates of Molango deposit, Mexico [composed on data (Okita et al., 1988; Table 2)]. A—Area of normal marine carbonates; B—Area of secondary carbonates; 1—manganese carbonate (rhodochrosite, kutnahorite); and 2—calcite.

fall also calcites—data for the product of the reduction of samples in the first 4 hours (Okita et al., 1988, Table 2), which in the graph occupy the area of highest $\delta^{13}C$ values.

These findings have been corroborated by subsequent, more detailed isotope research conducted at this deposit (Okita and Shanks III, 1992). Calcite on the whole is characterized by higher values of $\delta^{13}C$ and lower values of $\delta^{18}O$ by comparison with Mn carbonates (Fig. 3.57A). Enrichment of carbonates by light isotopes of oxygen, and in certain cases by heavy isotopes of oxygen (up to 36.7‰, SMOW), in the opinion of the cited authors could be caused by the interaction of the rock with fluids during the stage already of lithification of the sediment. It follows to note as well that the distribution of isotope ratios in the rocks of various deposits of the Molango basin bears its own particularities.

The data on the isotopic composition of sulfur obtained by P. Okita and W. Shanks (Okita and Shanks III, 1992) also provide evidence for potential widely occurring post-early-diagenetic (catagenetic) processes of transformation of initial rocks and authigenic mineral precipitation within the boundaries of the examined basin (as is clearly seen in Fig. 3.57C and D). All studied rocks of the manganese-ore strata have been categorized by mineral composition into four groups: calcite, manganocalcite, rhodochrosite-manganocalcite, and rhodochrosite. For each of the groups, the isotopic composition of sulfur has been studied (see Fig. 3.57A). For comparison the authors have referred to the isotopic composition of sulfur of marine sulfate for Late Jurassic time [according to Claypool et al. (1980)].

The lightest values of $\delta^{34}S$ and correspondingly the highest of $\delta^{13}C$ characterize non-ore carbonates (see Fig. 3.57C). This serves as evidence for the formation of pyrites during diagenesis (the microbial process of sulfate-reduction; values of $\delta^{34}S$ for pyrites—S^{2-}—approach equilibrium with marine sulfate—SO_4^{2-}).

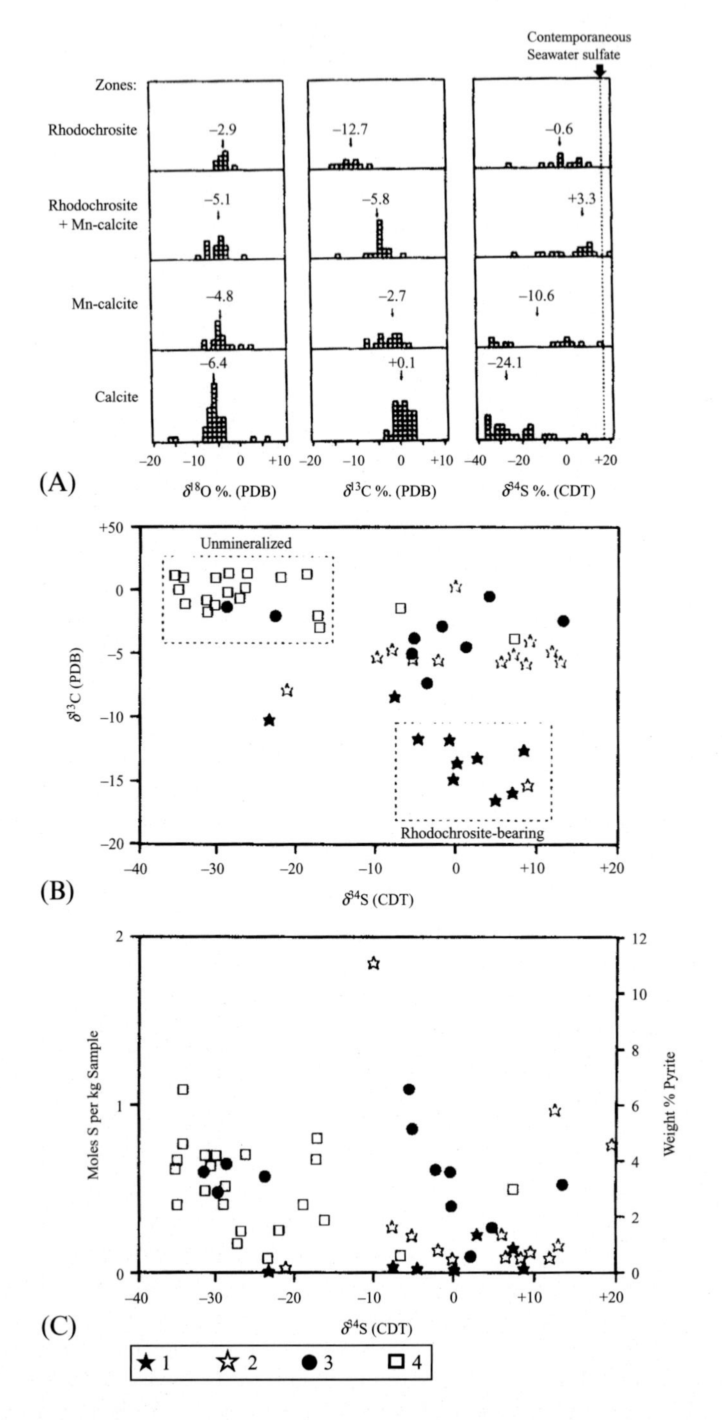

FIG. 3.57

Isotopic characteristics of rocks and ores of Molango deposit (Okita and Shanks, 1992). (A) Values $\delta^{13}C$, $\delta^{18}O$, and $\delta^{34}S$ in different zones of the ore bed, (B) dependent values of $\delta^{13}C$ and $\delta^{34}S$ in barren and rodohrozit-containing rocks, (C) the dependence of $\delta^{34}S$ and sulfur content of the rock. 1—Rhodochrosite, 2—a mixture of manganocalcite and rhodochrosite, 3—manganocalcite, and 4—calcite.

At the same time, pyrites from horizons containing manganese carbonates, especially rhodochrosite, are characterized by the heaviest isotopic composition of sulfur—positive values up to 19.7‰. This signifies that the processes of sulfate-reduction occurred with a deficit of marine sulfate—that is, the entry of dissolved SO_4^{2-} from bottom waters into the sediment was stopped. Forming under these conditions hydrosulfides and correspondingly sulfides become enriched in the heavy isotopes through Rayleigh "distillation," the preferential removal of the light isotope, leaving a heavy residual of S^{34} (Grinenko and Grinenko, 1974; Vinogradov, 1980). This follows as well from the correlation between concentrations of sulfur and values of $\delta^{34}S$ (Fig. 3.57D).

In this way, the formation of manganese carbonates in deposits of the Molango basin occurred at a completed stage of early diagenesis and in the beginning of late diagenesis (the beginning catagenesis stage of submergence).

Additionally, it has been proposed (Okita and Shanks III, 1992) that the rocks were subjected to some transformation during the later stages of the geological life of the sediment, leading to the formation of secondary calcite and in certain cases to a change in the isotopic composition of the oxygen.

Also classified to the examined genetic type of manganese deposits are the *deposits of the Úrkút basin* (*Hungary*). The Úrkút deposits to date have been received detailed study particularly in the scientific and practical regards (Polgari et al., 1991, 2009, 2012; Biro, 2009). Manganese rocks and ores here are enclosed in marine sedimentary deposits of the Late Triassic-Early Jurassic, which are represented predominantly by limestones, radiolarian marls, and black-shale deposits of dark-gray and black color (Polgari, 2009). The development zone of manganese rocks in the area stretches for 12 km with a width of 4–6 km.

Manganese mineralization is limited to two horizons of carbonates of the Upper Lias: the lower horizon (level 1) with a thickness of 8–12 m and the upper horizon (level 2) with a thickness of 2–4 m, which are divided by strata of radiolarites and high-clayey marls with a thickness from 10 to 25 m. The content of manganese in rocks of the ore zone constitutes from 17% to 37%, which with conversion to $MnCO_3$ constitutes from 36% to 77%. Contents of total carbon and sulfides (C_{org} and S^-) in this zone are insignificant (less than 0.4% and 0.5%) by comparison with the host rocks (up to 4.2% and 4.0%, respectively).

Detailed geochemical, mineralogical, isotope, and microscope (scanning electron microscope (SEM) and other) research conducted by Polgari et al. (1991, 2000, 2009) have enabled basic regularities to be established for the polychronous formation of manganese ores. Primary metal-bearing sediments were formed in a marine shallow-water basin under photic aerobic conditions with intensive accumulations of bacterial-algal organic material and with insignificant entry into the sediment by pyroclasts (tuffs). The source of manganese is considered to be hydrothermal. Carbonate-manganese mineralization is of a biogenic-bacterial (locally hydrothermal) origin; rocks were formed in marine conditions. During the formation of manganese ores and the enrichment of the sediment by ore elements, a principal role was played by bacterial activity.

Manganese carbonate in the ores is represented by rhodochrosite; kutnohorite has not been detected. Present as well are calcite and dolomite. The gypsum present in the rock, as proposed (Polgari et al., 1991), is a product of the hypergenic reduction of pyrite. Present throughout in various quantities are quartz, illite, and goethite.

The conducted isotope research demonstrates that the general scatter of values of $\delta^{13}C$ for carbonates of the lower horizon constitutes from −30.8‰ to −1.2‰, and for the upper horizon from −18.0‰ to −7.3‰. Such light values of the isotopic composition are evidence, as for the deposits

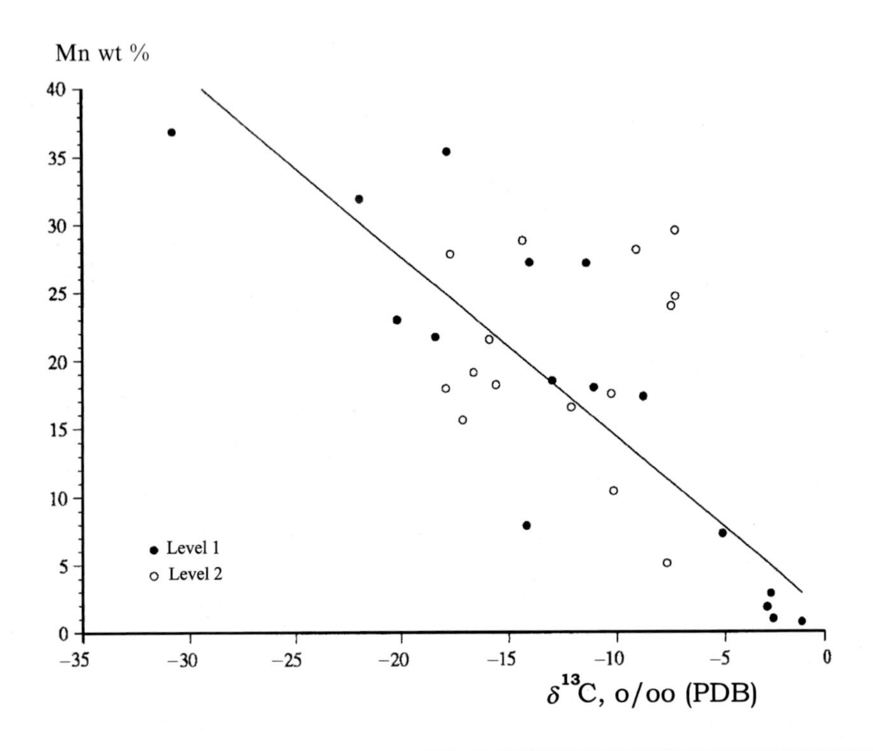

FIG. 3.58

The dependence of manganese content and $\delta^{13}C$ values in the manganese carbonates of Urkut deposit, Hungary (Polgari et al., 1991).

examined above, of participation in the formation of manganese carbonates by carbon of microbially oxidized organic matter. Here also is noted an inverse correlation between $\delta^{13}C$ values and manganese content (Fig. 3.58).

A characteristic particularity of manganese carbonate ores of the examined deposit is a direct correlation between the total content of C_{org} in the rock with the isotopic composition of carbon, and an inverse correlation between the contents of C_{org} and manganese (Polgari et al., 1991, Fig. 5).

The isotopic composition of oxygen is characterized by fairly high $\delta^{18}O$ values, which vary within the interval from 24.8‰ to 32.5‰ and are on the whole characteristic for authigenic carbonates of the diagenesis zone (Fig. 3.59). Lighter values are evidently the result of the presence of secondary carbonate in the sample.

It should be noted that rhodochrosites of the Úrkút deposit are characterized on the whole by higher values of $\delta^{18}O$ by comparison with manganese carbonates of the deposits of sedimentary-diagenetic genesis examined above. It can be supposed that one of the reasons for this is the heavier isotopic composition of water of the basin of sedimentation (possibly a shallow-water marine basin with intensively developed processes of evaporation in semiarid or similar conditions). This does not preclude the likelihood that an insignificant inflow of isotopically heavy catagenic waters that underwent isotope-exchange processes with the host carbonates was conducted into the system of the zone of diagenesis.

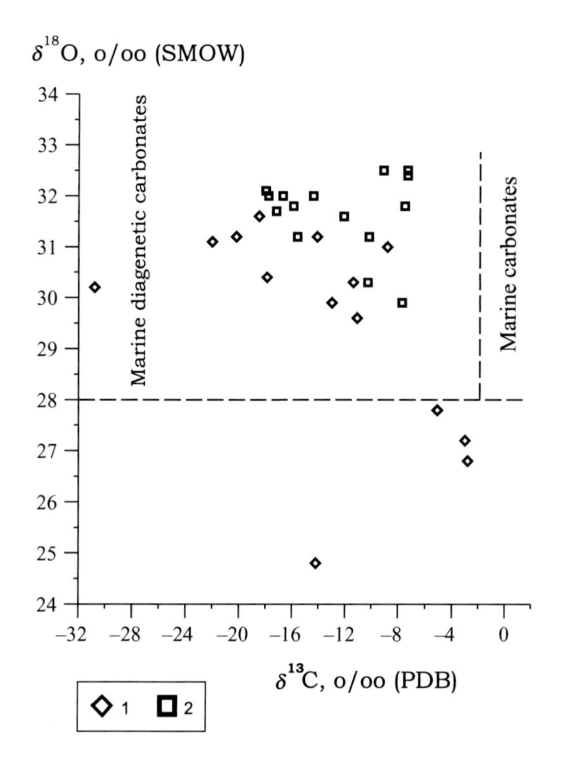

FIG. 3.59

$\delta^{13}C$ versus $\delta^{18}O$ in manganese carbonates of Urkut deposit, Hungary [composed on data (Polgari et al., 1991; Table 3). 1—level 1, 2—horizon 2.

Close values of $\delta^{13}C$ and $\delta^{18}O$ have been obtained by H. Öztürk and J. Hein (Öztürk and Hein, 1997) for rhodochrosites and kutnohorites of the *Ulukent deposit* (south-western Turkey) deposited in black shales of Turonian-Cenomanian age. Values of $\delta^{13}C$ for 10 samples have been found within the boundaries −15.2‰ to −10.6‰, and of $\delta^{18}O$ within 25.1–27.4‰, as is characteristic for authigenic carbonates of the diagenesis zone. However, by comparison with Mn carbonates of the Úrkút deposit, the examined kutnohorites and rhodochrosites are characterized by a lighter (by 4–5%) isotopic composition of oxygen. This is due evidently to higher temperatures (by 1.5–2°C) of their formation or by a lighter isotopic composition of oxygen of the interstitial waters.

A significant group of deposits of manganese of sedimentary-diagenetic genesis is distributed within the boundaries of *China*. All manganese deposits of the PRC (Fig. 3.60) (Editorial Committee of Mineral Deposits of China, 1995; Fan and Yang, 1999) have been categorized by Chinese geologists into six genetic types, among which approximately 71.4% is occupied by specifically sedimentary deposits (or sedimentary-diagenetic in our classification) (Ye et al., 1988; Editorial Committee of Mineral Deposits of China, 1995; Fan and Yang, 1999; Mineral'nye et al., 1999).

Manganese rocks and ores of the deposits of China of sedimentary (sedimentary-diagenetic) genetic type are further divided into three main subtypes: (1) siliceous deposits (siliceous-argillitic and siliceous-shale-limy associations) of Dounan (Triassic) and Wafangzi (Middle Proterozoic); (2) black-shale

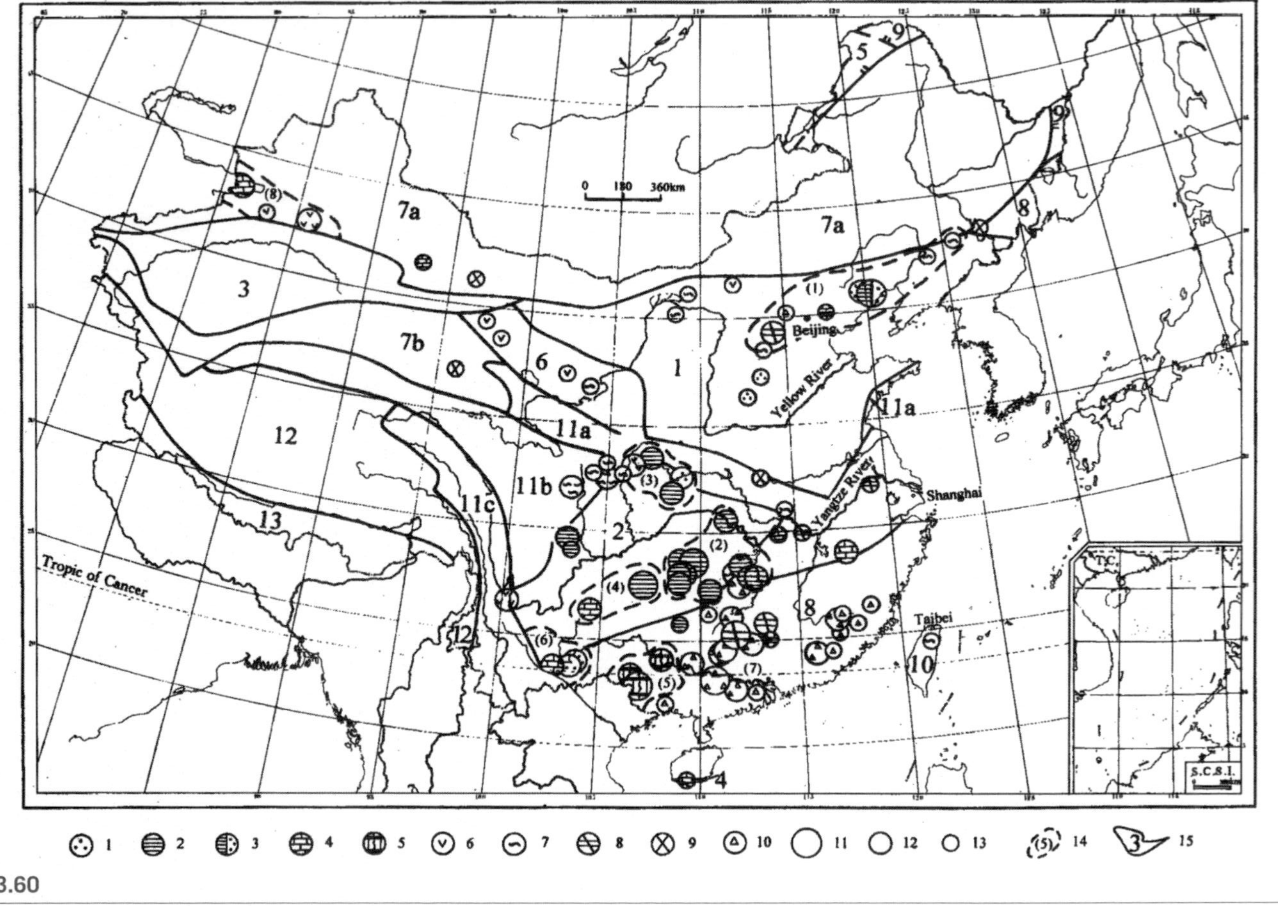

FIG. 3.60

Distribution of genetic types of manganese deposits in China [Mineral Deposits, 1995; Fan and Yang, 1999). 1–5—Sedimentary deposits: 1—prisoners in shales, 2—enclosed in a black-shale series, 3—enclosed in black-shale clay-series, 4—enclosed in carbonate rocks, 5—enclosed in a silica-clay-carbonate rocks; 6—volcanic-sedimentary; 7—metamorphosed; 8—hydrothermally altered; 9—hydrothermal; 10—weathering crusts; 11—large deposits (reserves of more than 20 million. tons); 12—medium-scale deposits (reserves of 2—20 million. tons); 13—small deposits (reserves of less than 2 million. tons); 14—metallogenic regions (1—Yanliao 2—confluence of Hunan-Sichuan-Guizhou-Hubei Provinces, 3—Southern Shaanxi—Northeaster Sichuan, 4—central Guizhou—eastern Yunan, 5—south-west Guangxi, 6—south-east Yunnan; 7—Hunan-Guangxi-Guangdong-Fujian, 8—Xinjang); and 15—tectonic units: A—platforms: (1)—North China (2)—Yangtze, (3)—Tarim, (4)—the South China Sea; B—Fold-Belts: (5)—Ergun, (6)—Qilian, (7a)—Altai, Junggar, Tianshan, Inner Mongolian-Great Hinggan, Jilin-Heilongjian, (7b)—eastern and western Kunlun, (8)—South China, (9)—Upper Heilongjiang, Nadanhada, (10)—Taiwan, (11a)—Qinling, (11b)—Songpan-Garze, (11c)—Sandjian (12)—Karakorum-Tanggula, Gangdise-Nyanqengtalha (13)—Himalayan; S.C.S.I.—The South China Sea Islands.

deposits (associations: black-shale, black-shale-clayey, black-siliceous-shale, black-shale-dolomite, and black-siliceous-shale limestone): Xiangtan, Minle, Yanglizhang, Datangpo, Gucheng, etc. (Early Sinian), Zunyi (Late Permian), Taojiang (Middle Ordovician), Mugui (Late Devonian), Gaoyan (Late Sinian), Jiaodingshan (Late Ordovician), Dongxiangqiao (Early Permian); and (3) specifically carbonate deposits (dolomite-limy, siltstone-clayey-dolomite, and siliceous-siltstone-limy associations): Xialei (Late Devonian), Longtou (Early Carboniferous), Baixian (Triassic), Zhaosu (Early Carboniferous), and Dongshuichan (Middle Proterozoic).

These deposits coincide with rocks that were formed in the platform and marginal seas, rarely in geosynclines, over the duration of a lengthy geological period beginning from the Middle Proterozoic and continuing practically until the Quaternary time.

Carbonate-manganese ores of many deposits of China have been rather fully characterized in the isotope regard. The *Wafangzi deposit*, classified as ferromanganese, is one of the largest and most studied in northern China (Liaoning province) (Fan et al., 1999). It is represented by a group of ore bodies that occupy an area of approximately 80 km^2, coinciding with deposits of the Middle Proterozoic—the Tieling formation in the Jixian system.

An ore strata with thickness of approximately 42 m is composed of red, dark-brown, and black frequently carbonate-containing shales, siliciclastics, and siliceous limestones, frequently with oolitic texture. The lithological composition of rocks of the ore district varies about the area; in its southern part are distributed manganite ores enclosed in red- and dark-brown clays, simultaneously as in the northern part are developed rhodochrosite ores coinciding with black and calcite-containing shales. In the eastern part of the district, a formation of manganese ores occurred in shoreline shallow-water, well-aerated conditions.

In the composition of strata have been identified three ore bodies coinciding with shales and siliciclastics. The thickness of the ore horizons constitutes from 0.5 to 2 m for the lower and 1–3 for the middle. The upper ore body is not present everywhere; the thickness on the whole is substantially less than that of the lower-lying body.

Primary ores, as proposed by the authors (Fan et al., 1999), have a sedimentary-diagenetic origin. Oxide ores of the facies are represented predominantly by manganite; hematite and braunite are present in subordinate quantity. Primary carbonate ores are composed of rhodochrosite as well as ferro- and calciorhodochrosite. Present as well are siderite and dolomite.

In near-contact zones with transecting sills of andesite porphyrites and dikes of diabases of Jurassic-Cretaceous age, rocks of the ore strata have undergone intensive alteration with the formation of contact-metamorphic oxide and carbonate ores. Oxide bixbyite-braunite ores are composed predominantly by braunite, bixbyite, and jacobsite (I-III); present are magnetite, hematite, tephroite, bustamite, and the like. Metamorphic (contact-metamorphic) carbonate ores are represented predominantly by rhodochrosite, Ca-rhodochrosite, and manganopyrosmalite. Frequently present in their composition are spessartine, andradite, magnetite, manganous calcite, and other minerals. In the hypergenesis zone, on the initial manganese rocks are developed secondary manganese oxides—pyrolusite, psilomelane, vernadite, goethite, and hydrogoethite (Fan et al., 1992).

Manganite ores commonly have massive oolitic textures and are represented by interstratified oolitic manganite horizons and siliceous shales and siliciclastics. Occasionally noted in them are stromatolite-like structures. In rhodochrosite ores have also been noted spheroid textures, in which are present residues of microbial structures (Yin, 1988; Fan et al., 1999).

For carbonate ores was studied the isotopic composition of carbon and oxygen (10 samples). Values of $\delta^{13}C$, as in the majority of manganese deposits, are low, varying from −10.5‰ to −3.9‰, and constitute evidence of participation in their composition by oxidized carbon of organic origin (Fan et al., 1999). Values of $\delta^{18}O$ (SMOW) occupy a fairly broad range—from 21.3‰ to 26.0‰—significantly lower than what is characteristic for manganese carbonates of diagenetic origin. To all appearances, such low values of $\delta\ ^{18}O$ are secondary and provide evidence of post-sedimentational hydrothermal transformations. It can be supposed that secondary hydrothermal transformations disrupted the initial geochemical characteristics. In part, in manganese carbonates having initial diagenetic origin, a correlation has been noted between the isotopic composition of carbon and the manganese concentration and contents of SiO_2, CaO, and the like.

Serving as an example of deposits of sedimentary-diagenetic type is the *Xialiei deposit*, which is situated in Guangxi Zhuang Autonomous Region, 53 km north-west from Daxin city.

This deposit coincides with the deposits of the Wuzhishan formation of the Late Devonian (Famennian) and is classified, as noted, to the third subgroup of sedimentary deposits. Here were studied different types of manganese carbonates and host rocks: argillite, siliceous, and manganese-containing limestones and pisolitic, massive, layered, and other manganese carbonate ores and concretions (Zheng and Liu, 1999). The deposit is classified to the stratiform type, ore bodies are enclosed in the sequence of siliceous-carbonate rocks. Ores are represented by three types: carbonate, silicate-carbonate, and oxide. Manganese carbonates in silicate-carbonate ores coincide with rhodonite, stilpnomelane, actinolite, chlorite, manganese epidote, biotite, and manganous-ferruginous antigorite. These minerals in varying ratios form thin interbeds, cross-beds, pellet-type structures, and ooliths. Noted are intervals rich in ooliths and pisoliths.

The ore deposit is represented by three ore bodies with a thickness of 0.36–1.2 m, interstratified with limestones of similar thickness that include a terrigenous impurity of siltstone dimensions; it is also underlain by limestones with a terrigenous impurity of analogous dimensions and overlapped by siliceous rocks.

In the structure of the deposit have been noted three zones: (1) internal, composed of manganese carbonates, silicates, and oxides; (2) transitional, represented by manganese carbonates and stilpnomelane; and (3) external, characterized by manganese carbonates with quartz, illite, chlorite, and pyrite. The deposit is considered by Zheng and Liu (1999) to have formed in shallow-water marine conditions (carbonate platform) with subsequent hydrothermal restructuring.

The isotope data provide evidence for a complex history of formation of carbonate-manganese ores. Thus, $\delta^{13}C$ values for the host limestones in Fig. 3.61 occupy the area of highest values and are close to normal-sedimentary marine carbonates.

At the same time, practically all studied manganese carbonates are characterized by a lighter isotopic composition of carbon. This indicates the participation in their formation by oxidized isotopically light carbon of organic matter. In the opinion of the cited authors (Zheng and Liu, 1999), the quantity of carbon C_{org} in the composition of carbonates constitutes from 30% to 60%.

Drawing attention are the very low $\delta^{18}O$ values in the manganese carbonates (from 19.7‰ to 27.0‰) as well as in the host limestones (from 22.4‰ to 25.8‰). To all appearances, these values are secondary and provide evidence of subsequent processes of transformation of the initial rocks.

Processes of secondary transformation are reflected as well in the isotopic composition of sulfur in pyrites of the host rocks and of the manganese carbonate ores. The general scatter of $\delta^{34}S$ values for them is wide: from −24.2‰ to 32.2‰. The lowest value (−24.2‰) corresponds to sulfide sulfur of the

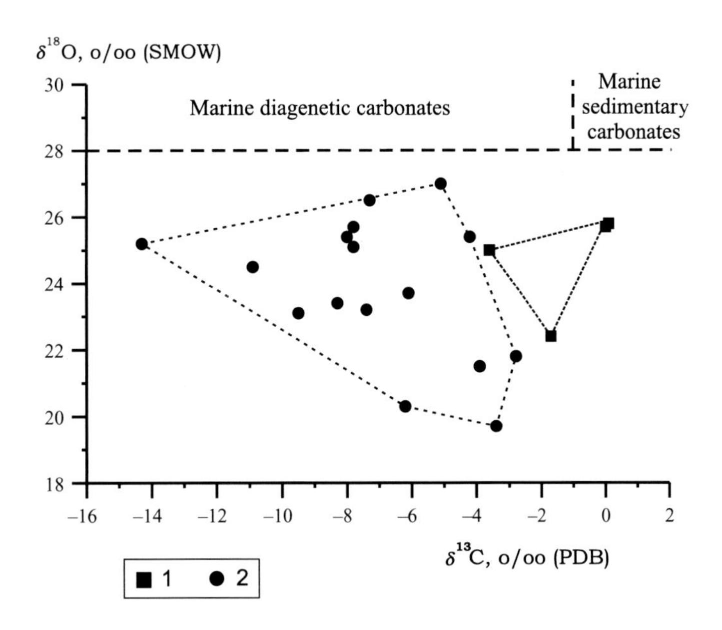

FIG. 3.61

$\delta^{13}C$ versus $\delta^{18}O$ in manganese carbonates of Xialei deposit, Chaina [composed on data (Zheng and Liu, 1999, Table 2)]. 1—enclosing limestone and 2—manganese carbonates.

diagenesis zone and is characteristic of pyrite collected from the carbonatized clayey shale enclosed in the ore member. Practically the entire remainder of studied samples have a heavier sulfur isotopic composition, not typical of the diagenesis zone, which provides evidence for the processes of sulfate-reduction in deficit conditions of marine sulfate ("closed" conditions relative to the bottom waters) or for intensively occurring post-early-diagenetic (catagenetic, epigenetic) processes.

A significant group of the manganese deposits of China are classified to the black-shale subtype of the black-shale-clayey association. Their typical representatives are the deposits of Taojiang, Zunyi, Gaoyan, Minle, Wafangzi, and others.

The *Taojiang deposit*, situated 75 km north-west from Chang-Shah in the north of Hunan province (southern China), has been under development for over 20 years. Its reserves constitute 6 million tons with a mean content of manganese of 20%.

Manganese-ore mineralization is confined to the rocks of the Modaoxi black-shale formation of the Middle Ordovician and is represented by a strata of Mn-containing limestones and rhodochrosite ores with a thickness of approximately 5 m, divided by interstratified non-ore black and gray shales, and deposited in the highly carbonaceous and pyrite-containing shales of the Hule formation (Okita and Shanks III, 1992). Isotope research conducted by these authors has allowed the establishment of a broad variation in the isotope ratios of carbon, oxygen, and sulfur. Thus, values of $\delta^{13}C$ are contained within the boundaries −17.8‰ to −5.8‰ for manganese carbonates, simultaneously as carbonate of the non-ore underlying dark-gray and black shales is characterized by $\delta^{13}C = -15.1‰$. Values of $\delta^{18}O$ vary within the boundaries 9.7–23.2‰. Substantial variations in $\delta^{34}S$ have been established in pyrites: the lightest in non-ore gray and black shales, and the heaviest in horizons with manganese carbonate (Fig. 3.62).

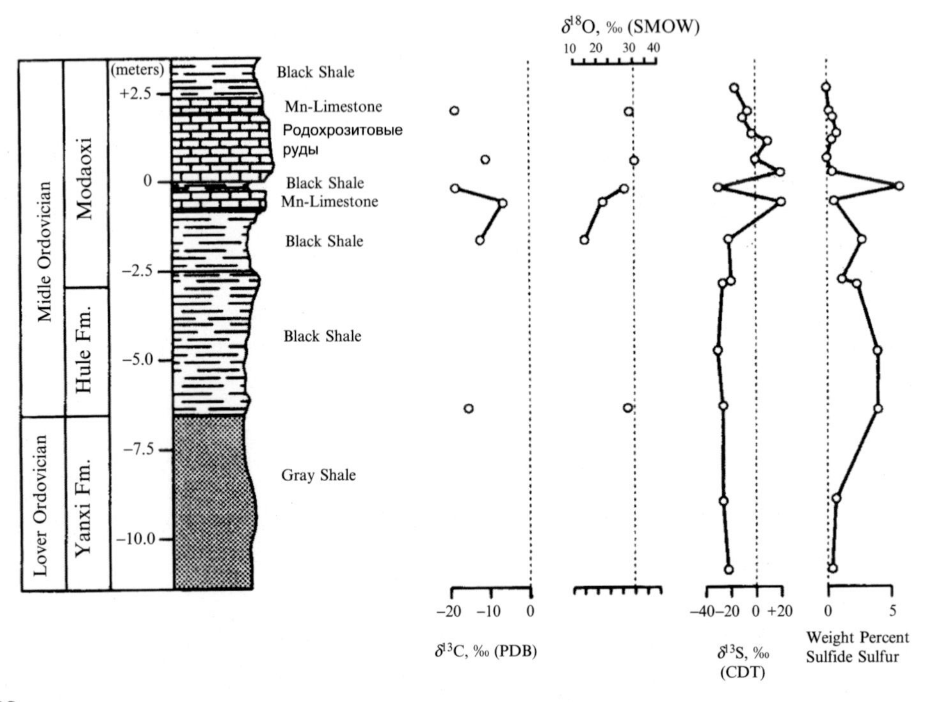

FIG. 3.62

Isotope data and sulfide content plotted against generalized stratigraphic column for the ore section at the Taojiang deposit, Chaina (Okita and Shanks, 1992).

The results of isotope and mineralogical research have allowed the authors (Okita and Shanks III, 1992) to conclude the host sequence was deposited under normal marine conditions. The formation of carbonates of the ore strata occurred with participation by isotopically light carbon that was remobilized from organic matter during the process of its microbial oxidation by the oxygen of manganese oxides and sulfates. High values of $\delta^{34}S$ and low concentrations of sulfide sulfur in the horizons with manganese carbonates are explained by a bacterial process of oxidation of sulfides by oxygen of manganese in anoxic closed and semi-closed conditions with the formation of sulfate, magnetite, and Mn^{2+}.

Constituting evidence in favor of the occurrence of an intensive process of sulfate-reduction in closed conditions relative to sulfate of bottom waters is the direct correlation of $\delta^{34}S$ values with those of $\delta^{13}C$ (Fig. 3.63A) and the inverse correlation of $\delta^{34}S$ values with the content of sulfur in the rock (Fig. 3.63B). Moreover, manganese carbonates by comparison with calcite are characterized by a heavier isotopic composition of both carbon and sulfur. This can be taken as evidence for their later, apparently post-early-diagenetic (catagenetic) formation.

The *Gaoyan deposit* is classified as a phosphorus-containing manganese deposit, with resources of approximately 15 million tons (Fan et al., 1999). It is situated 11 km west from Chengkou city (Editorial Committee of Mineral Deposits of China, 1995). In the geological regard, the deposit occurs in rocks of the Late Sinian Doushantuo formation in Sichuan province as part of the south-western limb of the Chengkou synclinorium of the Dabashan fold belt. The area of the deposit is composed of terrigenous-carbonate deposits of the Upper Proterozoic (Sinian) and Lower Cambrian; magmatic and volcanic rocks are unknown.

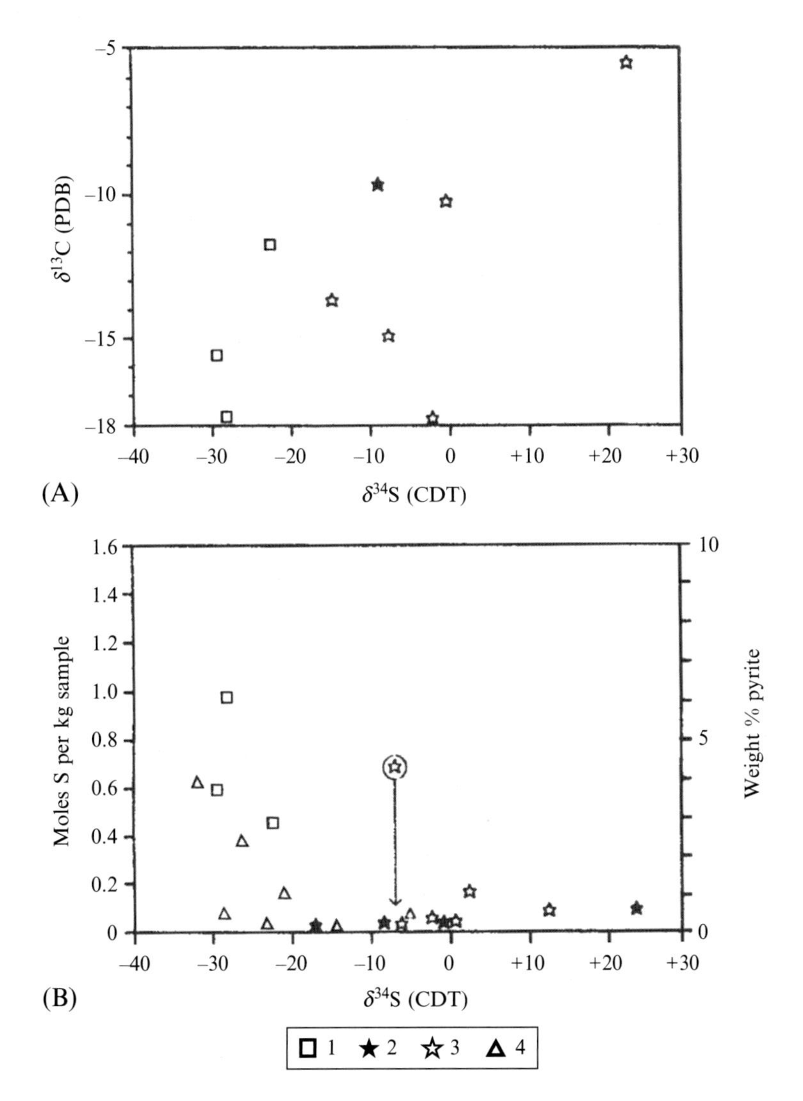

FIG. 3.63

Relationship between $\delta^{13}C$ and $\delta^{34}S$ (A) and sulfur content versus $\delta^{34}S$ (B) from samples of Taojang deposit (Okita and Shanks, 1992). 1—Calcite; 2—rhodochrosite; 3—Mn-calcite; 4—silicite (without carbonate).

The ore strata coincides with terrigenous-sedimentary rocks of the upper part of the Doushantuo formation, represented by purple arkosic sandstones, purple-red kaolinites, clayey (illite) shales, dolomites, siliceous-argillite, limy, and other dolomites, and limestones. In the structural regard, the deposit is classified to the stratiform type.

Manganese ores are highly phosphatic. They are represented by initial manganese-containing argillitic dolomites and manganese carbonates (Mn content 13–25%), as well as by secondary manganese oxides (Mn content 25–33%) of the hypergenesis zone on the initial ores. The hypergenesis zone in the deposit constitutes approximately 13–26 m, occasionally reaching depths of 50–60 m.

Characteristic for manganese carbonate ores of the Gaoyan deposit are spheroid textures, high contents of C_{org}, and the presence of mineralized microbial residues (Fan et al., 1999).

The authors of the cited work have studied the chemical and isotopic composition of carbon and oxygen in manganese carbonates along the cross section of the ore strata. A direct correlation (Fig. 3.64) has been noted between the isotopic composition of carbon and oxygen and the manganese content, as well as a very distinct inverse correlation between the former and the calcite content. Thus, samples with the highest manganese content are characterized by the heaviest isotopic composition of carbon and oxygen. Such distribution is not characteristic for sedimentary-diagenetic manganese carbonate ores and is evidently due to secondary processes.

Attention is called to the fact that the overlying manganized pyrite dolomites are characterized by the lightest isotopic composition of oxygen ($\delta^{18}O=21.9$–$25.7‰$). This same tendency is noted for manganese carbonates—that is, samples with the lowest concentrations of MnO (Gy33 and Gy34) are characterized by the lowest values of $\delta^{18}O$. This could be taken as evidence that as a result of secondary processes (hydrothermal-metasomatic?) of transformation there occurred a "dilution" of manganese ores with an increasing lightness of the isotopic composition of both oxygen and carbon of manganese carbonates of the ore strata as well as of the host (overlying) dolomites. In this case, the transformed solutions should have been enriched by carbon dioxide with a lighter isotopic composition of carbon by comparison with the initial manganese carbonates. For verification of this proposal, additional isotope research is needed.

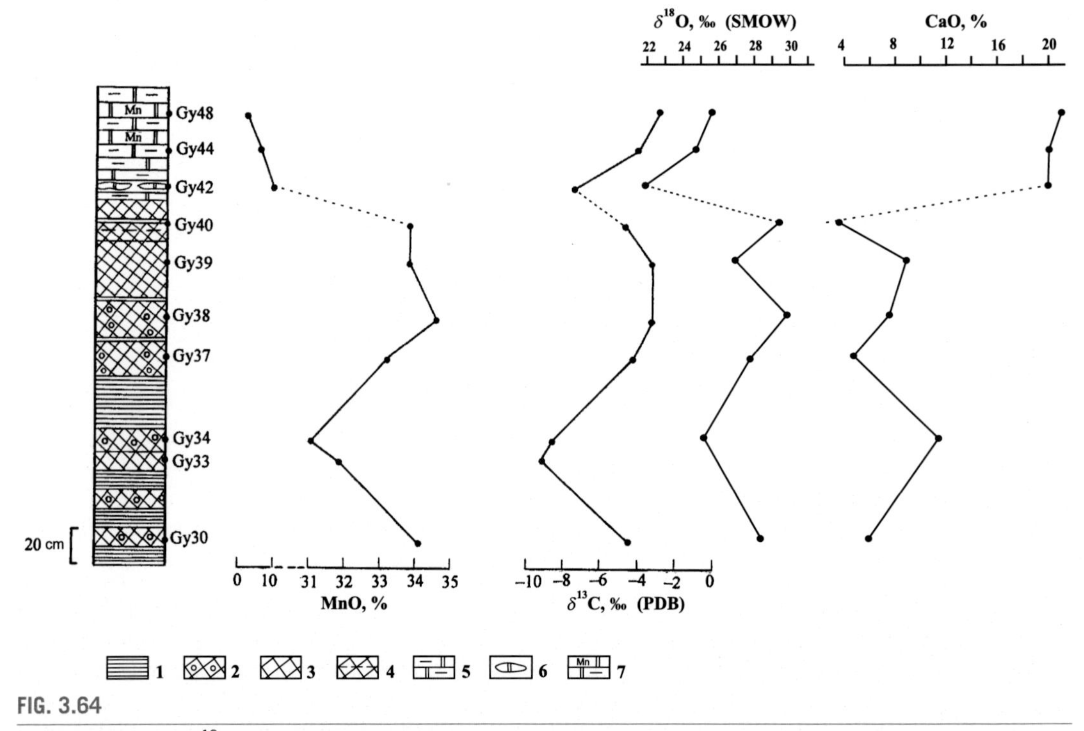

FIG. 3.64

The distribution of $\delta^{13}C$ and chemical composition of the manganese carbonate of Gaoyan deposit (China) (compiled on Fan et al., 1999). 1—Black-shale; 2—spheroidal ore; 3—massive ore; 4—laminated ore; 5—interlayered black-shale and dolostone; 6—dolostone lens; and 7—dark-gray micritic dolostone, manganized.

Intensive post-early-diagenetic transformations of manganese deposits are characteristic as well for the *Minle deposit* (China). It is situated approximately 400 km west-north-west from Changsha city and 28 km south-east from Huayuan city. Manganese ores here are characterized by a high content of phosphorus and low content of iron; the ratio of Mn/Fe is 7.7.

Manganese ores are contained in rocks of the Minle black-shale formation which was deposited in the Neoproterozoic as an interglacial sequence between the Sturtian and Marinoan glaciations. Zircon dates from tuffs indicate an age of about 660 Ma for the base and 654 for the top of the interval (see Li et al., 2012). In China, this part of the Precambrian is assigned to the Lower Sinian. The ores comprise two horizons with a thickness of 1.07 and 2.17 m, with a length of 1500 and 4250 m, with a width of 1290 and 1829 m, respectively. Manganese content constitutes 16.0–25.8% with a mean of 19.9%. The ores are composed predominantly by rhodochrosite, Ca-rhodochrosite, Mg-Mn-dolomite, and manganiferous dolomite.

A characteristic of this deposit, as of the Gaoyan deposit, is the presence of a large quantity of mineralized residues of microorganisms (blue-green algae), which were mineralized by manganese carbonate characterized by a light isotopic composition of carbon. This serves as evidence of participation in its formation by carbon of oxidized organic matter (Fig. 3.65); in rhodochrosites the variations in values of $\delta^{13}C$ constitute from −13.0‰ to −8.6‰, PDB, and of $\delta^{18}O$ from 21.7‰ to 26.4‰, SMOW (Tang and Liu, 1999).

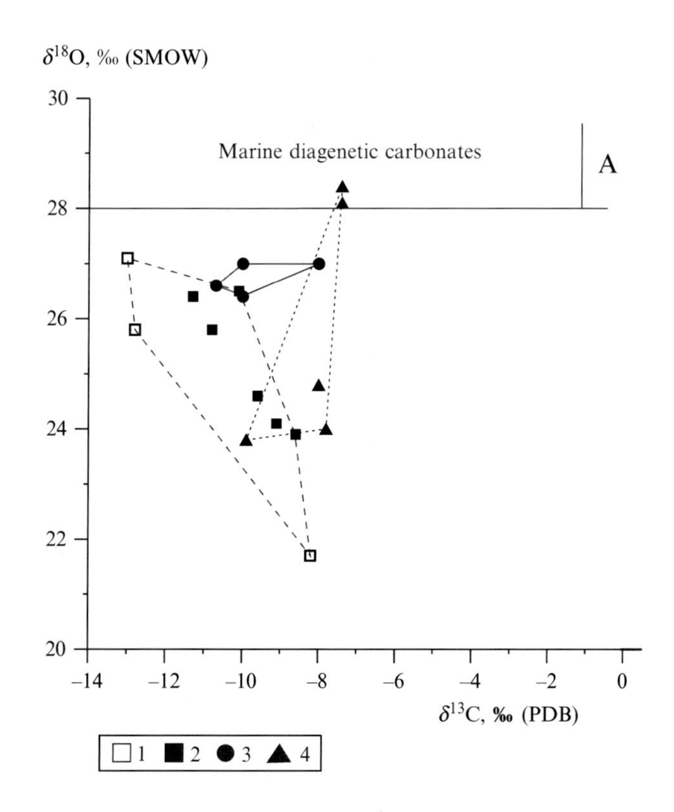

FIG. 3.65

$\delta^{13}C$ versus $\delta^{18}O$ in carbonate-manganese ores of Menle, Datangpo, and Yanglizhang deposit (China) (compiled from the Tang and Liu, 1999). 1 and 2—Minle deposit, 1—host dolostone, 2—rhodochrosite; 3—rhodochrosite, Datangpo deposit; and 4—rhodochrosite, Yanglizhang deposit.

An insignificant increase in the lightness of the isotopic composition of oxygen (26–28‰) in rhodochrosites of the Minle deposit (as well as Datangpo deposit) is possibly due to the isotope-geochemical properties of the basin of deposition (elevated temperatures, possibly a lighter isotopic composition of oxygen of the water of the paleo-water bodies by comparison with the modern ocean). Authors of the article (Tang and Liu, 1999) have proposed that the basin of sedimentation represented a gulf or lagoon on the paleocontinental coast that was separated from the open sea by barriers. If the water of the basin was initially low in dissolved sulfate, pyrite formation, which, if sedimentation rates are slow, removes very light sulfur (Maynard, 1980) and hence drives residual sulfate to more positive values.

At the same time, the presence of very low $\delta^{18}O$ values (up to 21.7‰) allows us to consider the possibility of intensively occurring subsequent processes of transformation. The anomalously heavy isotopic composition of sulfur—the values $\delta^{34}S$ in sulfides of black-shale and rhodochrosite rocks vary from 46.6‰ to 58.6‰—could also be explained this way. Such $\delta^{34}S$ values are characteristic for residual sulfate subjected to processes of sulfate-reduction in a closed system once sulfate-reduction is nearly complete [90% and greater (Vinogradov, 1980)]. Subsequent investigation has shown that this pattern of very heavy sulfide sulfur is displayed by most Neoproterozoic manganese deposits in China (eg, Liu et al., 2006; Feng et al., 2010; Li et al., 2012, and may be a world-wide phenomenon). Li et al. (2012) argued that the sulfur isotope pattern was imprinted during early diagenesis and thus indicates a stratified ocean with $\delta^{34}S$ of sulfate varying with water depth.

Classified to this same genetic type are the *deposits of Zunyi type* (Tongluojing Fengjiawan, Gongqinghu, Heshangchang etc. ore fields), associated with black shales and siliceous limestones of the Longtan formation of Permian age. A broad spectrum of variations in values of $\delta^{13}C$ (−8.09‰ to −2.97‰, PDB), $\delta^{18}O$ (24.7–27.8‰, SMOW), and $\delta^{34}S$ (−24.5‰ to 24.6‰ for pyrites, −22.5‰ to 19.5‰ in alabandite) (Editorial Committee of Mineral Deposits of China, 1995), characteristic for manganese ores of these deposits, also provides evidence for complex processes of formation of authigenic manganese carbonates and sulfides in early diagenesis and subsequently.

Also classified to the examined sedimentary-diagenetic type of manganese carbonate are the little-studied in the scientific (including isotope) regard carbonaceous shales of the *Franceville Basin* (*Gabon*), particularly the Moanda deposit (Gabon). Rhodochrosite-containing black shales here do not of themselves present practical significance. Industrial ores of this deposit are represented by manganese oxides of the weathering crust (laterites), which will be examined below (see Section 3.2.7.1).

The isotopic composition of rhodochrosite from carbonaceous shales (ampelites) of the Moanda deposit, surveyed by J. Hein and coauthors, is characterized by values of $\delta^{13}C$ (approximately −15.4‰) and $\delta^{18}O$ (approximately 27.1‰) close to those of the Mn carbonates examined above [reference cited in the work Polgari et al. (1991)]; see also Hein et al. (1999).

Classified by Chinese geologists to the *phosphorus-manganese subtype* of the sedimentary type (Editorial Committee of Mineral Deposits of China, 1995; Hein et al., 1999) is the *Tiantaishan deposit* (Shaanxi province). It coincides with the terrigenous-carbonate deposits of the Tananpo formation of the Lower Cambrian. The host rocks are represented by black phyllites and dolomites. The ore strata is deposited in terrigenous deposits of Precambrian age, composed of sericite and sericite-quartz shales with gritstones, frequently of black color, with lenses of dolomites, and is overlapped by metamorphosed deposits of the Lower Cambrian: crystallized limestones with sericite-quartz horizons, massive quartzites and garnet-containing sericite and sericite-quartz shales. Phosphorites form stratiform bodies (31 bodies) with a length of 140–2970 m and thickness from 0.6 to 8.1 m. Content of P_2O_5 varies from 11.9% to 28.3% (with a mean of 16.6%).

Manganese ores compose five bodies with manganese content (in the ore bodies) from 13.2% to 20.4% (with a mean of 18.1%). Iron content varies from 0.43% to 3.33% (with a mean of 1.59%), and P_2O_5 content from 0.16% to 11.5% (a mean of 2.25%).

The ore bodies are composed predominantly by rhodochrosite and Ca-rhodochrosite, dolomite, and include Mg-kutnohorite, manganese dolomites, quartz, sericite, apatite, and an insignificant quantity of pyrite, alabandite, manganite, neotocite, spessartine, and the like. Rhodochrosite ores predominate and are characterized by oolitic, spheroidal, and grained structures. In certain manganese carbonates and sulfides are preserved stromatolitic textures. Spheroidal textures, filled by apatite and manganese carbonate, evidently represent cyanobacterial formations. In rocks of this deposit have been established seven types of residues of bacteria (Hein et al., 1999). This indicates the formation of phosphorus-manganese ores in conditions of rich organic life.

The isotope composition of carbon and oxygen in manganese carbonate rocks and ores has been studied; the general boundary of variations in values of $\delta^{13}C$ (PDB) varies from −11.2‰ to −4.1‰, and that of $\delta^{18}O$ (SMOW) varies from 17.3‰ to 22.7‰ in manganese carbonates; in non-ore interbeds, those values vary from −12.5‰ to −0.4‰ and from 16.4‰ to 19.9‰, respectively (Fig. 3.66). These data, as for the manganese ores examined above, provide evidence of participation by oxidized carbon of organic matter in the formation of the carbonates of the deposit and evidence of intensively occurring processes of subsequent transformation. Proof of this is found in (1) lower values of $\delta^{18}O$ in carbonate

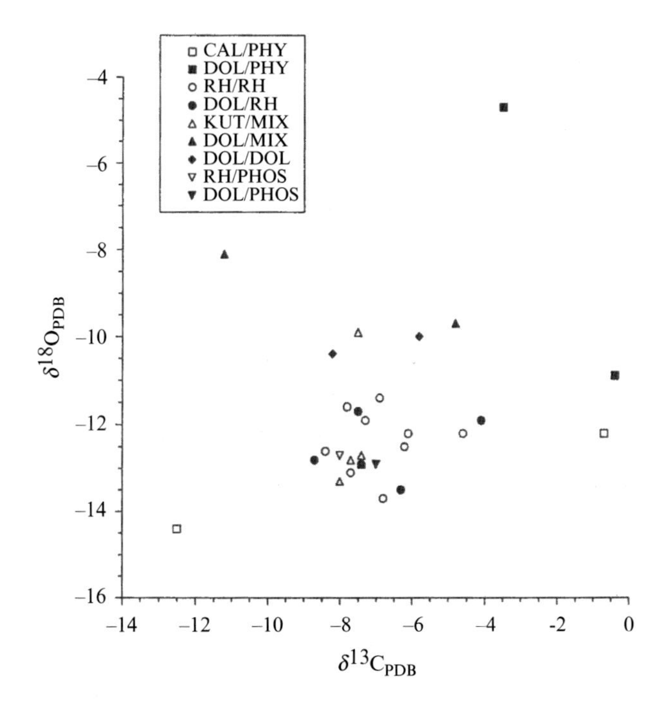

FIG. 3.66

$\delta^{13}C$ versus $\delta^{18}O$ in carbonate rocks of Tiantaishan manganese deposit (Shaanxi Province, China). *Cal*, calcite; *dol*, dolomite left or dolostone rifht; *kut*, kutnagorite; *rh*, rhodochrosite left or rhodochrostone right; *phy*, phyllite; *mix*, mixed carbonates; *phos*, phosphorite.

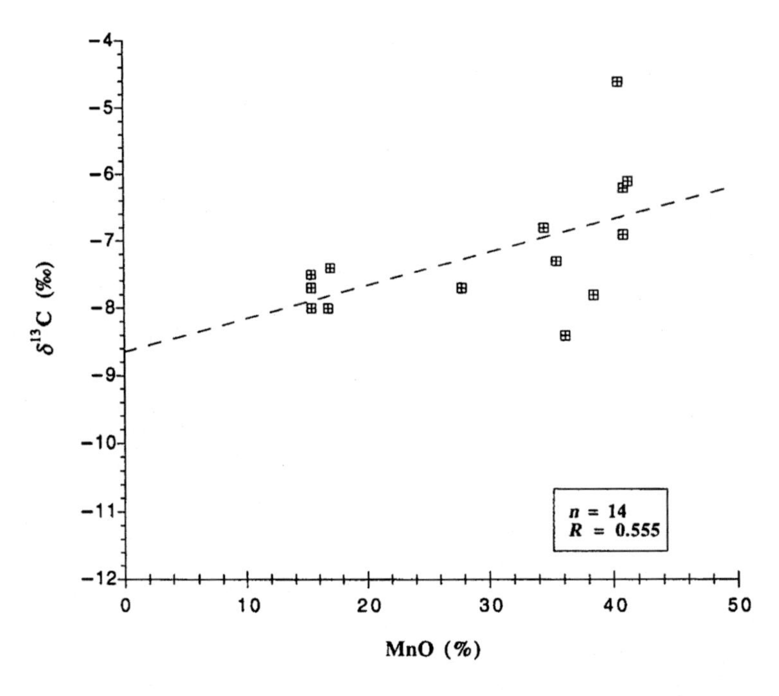

FIG. 3.67

Scatter plot of Mn contents versus $\delta^{13}C$ for rhodochrosite ores of Tiantaishan manganese deposit (Shaanxi Province, China).

rocks of the deposit, (2) analogous variations in the isotopic composition of ore and non-ore carbonates, and (3) the inverse correlation between values of $\delta^{13}C$ and the content of manganese in the rock (Fig. 3.67).

In this way, the cited examples of deposits of manganese ores of sedimentary-diagenetic origin demonstrate convincingly the necessary participation in the formation of manganese and manganese-containing carbonates by oxidized carbon of organic matter (oxidation of C_{org} during the process of sulfate-reduction or microbial oxidation caused by the oxygen of manganese oxides). Their formation occurred predominantly at the stage of early diagenesis, and in many cases also later in late diagenesis (the beginning stages of submersion catagenesis).

At the same time, despite the general regularities of manganese-ore carbonate precipitation, in each deposit that process is characterized by specific conditions contingent upon paleoclimatic, paleogeographic, tectonic, and other factors. Therefore, the general scheme of formation of manganese ores classically occurring in deposits of the South Ukrainian basin (Nikopol', Bol'she-Tokmak), is not always developed in full measure within the boundaries of other basins. For example, certain deposits feature well-developed coastal shallow-water facies with ores of manganese oxides (Chiatura, Georgia). In other cases (deposits of the North Urals manganese-bearing basin, Molango, and elsewhere), coastal oxide ore facies are completely absent; formed instead were predominantly diagenetic manganese carbonates.

A carbon dioxide-water medium of authigenic manganese carbonate genesis was dependent upon the paleogeographic environment and the type of basin of sedimentation. This was manifested in the diverse initial isotope ratios of the oxygen of unaltered carbonates: lower values of $\delta^{18}O$ for those desalinated and at higher temperatures of mineral precipitation, and higher values of $\delta^{18}O$ for basins of sedimentation of semiarid climate (with intensive processes of evaporation) and lower temperatures of the environment of formation of authigenic carbonates. A definite influence on the isotopic composition of carbon in certain cases could have been rendered by a saturation of the sediment by organic matter and sulfides (sulfates), which could have led to an increase in the lightness of the forming carbonates by light isotopes of O^{16}.

3.2.2 DEPOSITS OF HYDROTHERMAL-SEDIMENTARY GENESIS

Deposits of hydrothermal-sedimentary and hydrothermal genesis in rocks of the Earth's crust are distributed very widely. Small deposits, ore-occurrences, and points of manganese-ore mineralization are noted practically in all fold regions of the development of volcanogenic and volcanogenic-sedimentary rocks of various ages in countries of the former USSR (Chukhrov, 1940; Margantsevye et al., 1966, 1978, 1982; Gavrilov, 1967, 1972; Margantsevo-zhelezisto-kremnistaya et al., 1981; Avaliani, 1982; Sokolova, 1982a,b; Vulcanogenno-osadochnye et al., 1985; Zheleznye et al., 1985), as well as in other states (Simposium et al., 1956; de Villiers, 1960; Geology, 1980; Roy, 1981). However, by comparison with sedimentary-diagenetic deposits, world-class accumulations of manganese do not form by hydrothermal and hydrothermal-sedimentary means; the deposits as a rule are small and medium, rarely large. Initial ores are commonly ferromanganese.

It follows to note that the degree of mineralization (manganese concentration) in a unit of an area in deposits of this type is by 5–20 times higher than in sedimentary-diagenetic deposits, and that of phosphorus, accordingly, is lower by one to two orders of value (0.001–0.01%) than in ores, for example, of the Oligocene deposits of southern Ukraine (0.1–0.8%).

Deposits of the hydrothermal-sedimentary type are distinguished by the variety of geological structure and mineral composition of the ore and host rocks and have received only inadequate study, despite over a century of research history. The dual character of the processes of ore-genesis with their formation—of host and ore rocks—determines the complex combination in their indicators that is characteristic for sedimentary and volcanogenic-sedimentary as well as for specifically hydrothermal-sedimentary rocks. Additionally, in many cases hydrothermal-sedimentary ore deposition (the accumulation of ore-bearing oozes) is frequently replaced by specifically hydrothermal ore formation, when the discharge of ore-bearing fluids passes into already compacted sediment. In the latter case forming stratiform deposits as well as vein ore bodies.

Factors including the complex correlations of exogenic and endogenic processes of mineral formation, the changing distance of fixation (in various deposits) of ore matter (ore deposition) from the source of entry of ore components in the aquatic basin, the geological interactions of the ore strata, of volcanogenic and volcanogenic-sedimentary rocks of various ages, the source of manganese (volcanogenic, hydrothermal, subaqueous weathering of volcanic rocks etc.), and subsequent processes of superimposed hydrothermal ore and non-ore mineralization have led to the formation of intermediary (mixed) or transitional genetic types of rocks. This is the reason for various categories of deposits connected with hydrothermal activity: exhalation-sedimentary, volcanogenic-sedimentary, hydrothermal-sedimentary, specifically hydrothermal, and all of their possible varieties.

The fullest classification of deposits coinciding with basins where processes of volcanism, hydrothermal activity, and volcanogenic-sedimentary sedimentation are manifest has been recorded by Shatskii (1954) and Roy (1981). Roy in the cited work delineates three groups of deposits connected with corresponding manganese-bearing formations.

1. Deposits of the greenstone group; *formations*: *greenstone* (Middle Eocene deposits of the Olympic Peninsula, Washington State, United States; Lower and Middle Devonian deposits of the Magnitogorsk synclinorium, eastern slope of the South Urals: Kusimovskoe, Klevakinskoe, and others; the Upper Cretaceous-Middle Eocene deposit of Tarantana, Priente province, Cuba; the group of Middle Eocene-Middle Miocene deposits of Viti-Levu island in the Fiji archipelago, South Pacific Ocean; the group of Paleozoic and Mesozoic deposits of Japan on the islands Hokkaido, Kyushu, and Shikoku; the Permian-Jurassic deposits of the Northland and Auckland areas in New Zealand); *jasperoid* (Carboniferous-Permian deposits of the mountains Ashio and Kitakami and the Tambo uplift, Japan; the Parsettens and Faletta deposits of the Upper Jurassic, Swiss Alps; the group of Upper Jurassic-Upper Cretaceous deposits of the Fransiscan formation, California, United States; and elsewhere), *siliceous-shale* (the Birimian system of Lower Proterozoic deposits of Ghana, Cote d'Ivoire, Burkina Faso, eastern Liberia, and Guinea; the Precambrian Ansongo series in Mali; the Lower Carboniferous Kellerwald deposit, Germany; and elsewhere).
2. Porphyry-siliceous group; *formations*: *porphyry* (group of Precambrian deposits of the Värmland province in central Sweden: Långban, Harstigen, Pajsberg, Sjö, and others; Late Precambrian deposits of Tiouine, Migouden, Oufront, and Idikel, Morocco; the San Francisco Middle Tertiary deposit, Mexico; the group of Early Cretaceous deposits of the Arqueros and Quebrada-Marquesa formations of Coquimbo province, Chile; the Pozharevo, Klisura, and other Early Cretaceous deposits of Bulgaria, Sofia province; the Late Cretaceous manganese deposits of the Kafan and Alaverdi tectonic zones of Armenia; the Lower Cambrian Durnovskoe deposit, Salair ridge, Western Siberia; the group of Early Carboniferous deposits of the Iberian Peninsula, Spain and Portugal; and elsewhere), *siliceous-carbonate* (Late Devonian-Early Carboniferous deposits of the Atasui area of central Kazakhstan: Karazhal, Bol'shoi Ktal, Zhairem, Zhomart, Kamys, Ushkatyn-I, -II, and -III, and elsewhere).
3. Associations of manganese and base metals in metamorphosed volcanogenic-sedimentary deposits. Here are classified deposits of the hypergenesis zone developed on polymetallic ores (predominantly lead and lead-zinc), in the composition of which manganese is contained in subordinate, albeit fairly substantial quantities (the Early Proterozoic deposit of Broken Hill, Australia; the Precambrian deposits of Franklin and Sterling Hill, New Jersey, United States; the ore deposits of Långban in Sweden and Noda-Tamagawa in Japan; the lead-zinc deposits of the Atasui region of central Kazakhstan, and elsewhere).

The classification reported above is based on associations between host rocks of various types and ore matter and on the whole does not rest on a genetic basis. The principal distinctions in the genesis of manganese rocks and ores of the identified formations, the source of manganese, the conditions and time of formation of ore minerals, and their stadiality remain unclear.

In many cases, carbonates and sulfides of manganese are developed alongside oxides and hydroxides of manganese (and iron). These minerals form under essentially different conditions and can be of a different origin and a different source of matter (for iron and manganese as well as for carbon dioxide of carbonates).

The formation of manganese (ferromanganese) ores of the examined type in many cases is determined by the submarine discharge of metal-bearing hydrothermal solutions. Ore matter within defined geochemical barriers precipitates into the sediment; the forming metal-bearing oozes are subject as a rule to corresponding diagenetic transformation with the formation of some type of hydrothermal-(volcanogenic)-sedimentary rocks.

The composition of the host sediments (rocks) of the basin of sedimentation (carbonate-terrigenous-volcanogenic-sedimentary rocks) is not of principal significance in terms of the process of ore deposition. The ore and non-ore matter supplied to the sediment with hydrothermal solutions under conditions of submarine discharge can be superimposed on any facies of volcanogenic, volcanogenic-sedimentary, terrigenous-sedimentary, and carbonate and other rocks as well as on their paragenetic associations.

Moreover, sources of ore components of metal-bearing hydrothermal solutions discharged directly into the hydrosphere of the basin of sedimentation likewise can differ in origin; they can be derived from rocks of various composition and geneses of the host sequence—vulcanites, sedimentary rocks—as well as supplied from petroleum waters, with brine waters, or the like. In any case, the formation of initial ore sediments will be essentially identical—that is, hydrothermal-sedimentary with subsequent transformation in the diagenesis zone. In many cases, later during the post-early-diagenesis stage of lithogenesis is observed a metasomatic and hydrothermal transformation of the ore-bearing sedimentary rock, frequently with a supply of new portions of ore matter.

In the isotope regard, manganese rocks of the deposits of the examined type, with certain exceptions, have received only insufficient study. Our works have been concerned with the study of the isotopic composition only of carbonates. It is evident that the isotopic composition of oxides and hydroxides of iron and manganese ($\delta^{18}O$, δD) as well as that of sulfides ($\delta^{34}S$) and of the ore-hosting rocks likewise contains important genetic information.

As model examples of deposits of manganese ores of hydrothermal-(volcanogenic)-sedimentary genesis, we turn to certain deposits of the Magnitogorsk synclinorium in the South Urals and the Parnok iron-manganese deposit in the Cis-Ural region, among the most studied in the isotope regard, as well as the Devonian deposits of central Kazakhstan.

3.2.2.1 Hydrothermal-sedimentary deposits of the South Urals

From the second half of the 19th century, within the boundaries of the western border of the Magnitogorsk paleovolcanic belt in the South Urals were discovered several dozen manganese deposits unified in the so-called Primagnitogorsk or eastern Bashkiria group (Fig. 3.68). The overwhelming majority of researchers classify these objects to the hydrothermal-sedimentary genetic type (Kheraskov, 1951; Varentsov and Rakhmanova, 1974; Kalinin, 1978; Kontar' et al., 1999; Mikhailov, 2001). This conclusion has been confirmed by numerous geological and petrographical materials and for practically all researchers is now beyond doubt.

In the composition of manganese ores of certain deposits (Kozhaevskoe, Urazovskoe, Bikkulovo, Kazgan-Tash, Kyzyl-Tash, South Faizuly, and others), carbonates are present in the capacity of primary minerals—rhodochrosite, manganese calcite, and kutnohorite. These minerals are not characteristic for hydrothermal manganese deposits of the modern ocean, which consist principally of the oxide phases. Hence, the carbonates were formed most probably during the post-sedimentational stages of development of the deposits. For confirmation of this proposal, as well as for the refinement of the conditions of genesis of deposits on the whole, it was necessary to establish the nature of the carbon dioxides

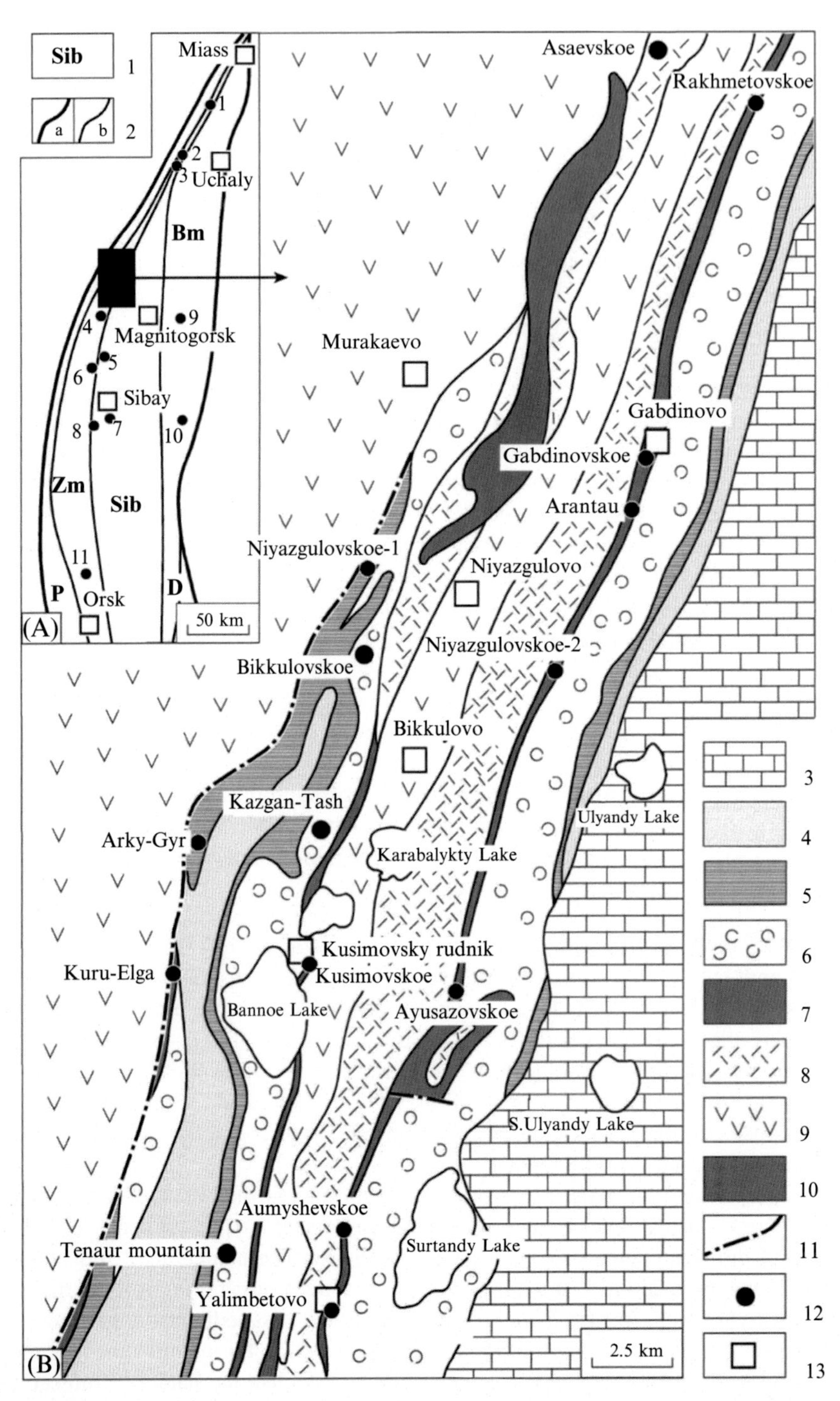

FIG. 3.68

See the legend on opposite page.

concentrated in manganese-containing rocks. With this aim, the isotopic composition of carbon and oxygen was studied in carbonate-containing manganese ores of the most studied deposits—*South Faizuly, Kyzyl-Tash, and Bikkulovo* (Kuleshov and Brusnitsyn, 2004, 2005; Brusnitsyn et al., 2009).

The results of isotope research have enabled the elucidation of a new mechanism of the formation of carbonate-manganese ores in the South Faizuly deposit. Its main particularity consists of the participation in the formation of rhodochrosite by carbon of oxidized methane.

South Faizuly and Kyzyl-Tash deposits

Survey of geological characteristics. The geological structure and petrography of these deposits, as well as the models of sedimentation of the ore-bearing deposits, have been characterized in detail in the literature (Brusnitsyn et al., 2000; Starikova et al., 2004; Brusnitsyn and Zhukov, 2005). Therefore, we will mention here only certain important points.

The deposits coincide with the strata of vulcanites and volcanogenic-sedimentary rocks accumulated in the environment of the island arc (western Magnitogorsk) and adjoining it from the eastern interarc basin (Sibaiskii). The volcanogenic strata has a two-stage structure (Fig. 3.69) (Seravkin et al., 1992; Brusnitsyn et al., 2009). The lower, Emsian-Eifelian stage of the Lower and Middle Devonian was formed during a stage of active volcanic eruptions and is composed predominantly by effusives of rhyolite-basalt and andesite-basalt formations. The upper, Eifelian-Lower Carboniferous stage was formed during a period of attenuation of magmatic activity in the course of the erosion of volcanic buildups and the associated processes of sedimentation of edaphogenic, hydrothermal, and organogenic material. In its composition have been identified siliceous, tephro-terrigenous, terrigenous, and carbonate-terrigenous complexes. Coinciding with rocks of the lower structural stage are copper-zinc-sulfide deposits; those of the upper stage constitute a large number of small-scale hydrothermal-sedimentary deposits of manganese.

The South Faizuly deposit is localized in the siliceous deposits of the Bugulygyr horizon (D_2ef), accumulated in the deep-water (below the level of carbonate compensation) depressions of the interarc

FIG. 3.68

Schematic location of the major manganese deposits in the Magnitogorsk paleovolcanic belt (A) and geological setting of the Bikkulovskoe deposit (B). (1) Paleovolcanic zones: (S) Sakmara region (accretionary prism fragment); (WM) West Magnitogorsk island arc, (SIB) Sibai interarc basin, (EM) East Magnitogorsk island arc, (D) Dombarovka back-arc basin; (2) boundaries: (a) Magnitogorsk paleovolcanic belt, (b) paleogeodynamic zones; (3–9) stratified sediments: (Z) Zilair (D_3fm-C_1t), Berezovka (C_1t), and Kizil (C_1v) Formations (limestones, siltstones, sandstones, and shales); (4) Koltuban Formation (D_3fm) (volcanogenic conglomerates, sandstones, and siliceous sheles); (5) Mukasovo Horizon (D_3fr) (siliceous shales); (6) Ulutau Formation (D_2zv-D_3fr) (volcanomictic sandstones and gritstones of the acid and intermediate compositions); (7) Bugulygyr Horizon (D_2ef_2) (jaspirites, jaspers, and siltstones); (8) Karamalytash Formation (D_2ef) (basalts, Rhyolites, and their tuffs, and jasper interbeds); (9) Irendyk Formation (D_1em-D_2ef) (basalt and basaltic andesite porphyrites); (10) serpentinites; (11) large faults; manganese-ore deposits; and (13) towns and settlements. Numerical designations of deposits: 1—Kozhaevskoe, 2—Tetrauk, 3—Urazovskoe, 4—Kyzyl-Tash, 5—Mamilinskoe, 6—Gubaidulinskoe, 7—Yanzigitovskoe, 8—North, Middle, and South Faizulinskoe, 9—Bakhtinskoe, 10—LisHi Gory, 11—Repino-Krutorozhinskoe. Plans drawn by A.I. Brusnitsyn on materials: A—E.S. Kontarev and L.E. Libarovoy (1997), E.S. Kontarev et al. (1999); used—O.A. Nestoyanova and D.G. Zhiganova (Map, 1958) as amended.

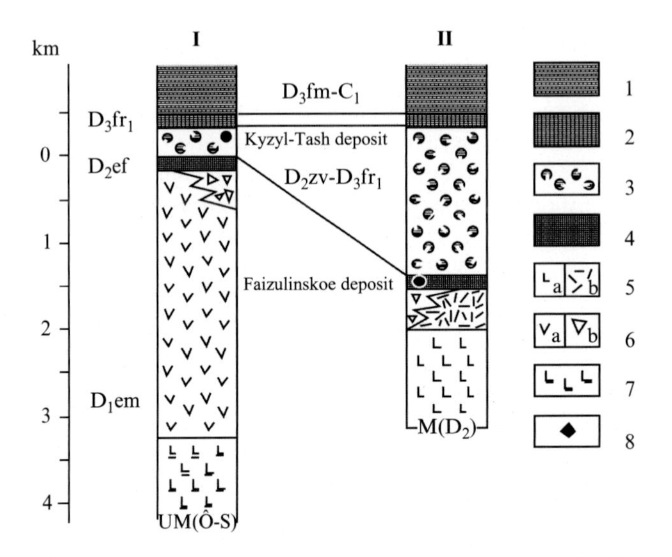

FIG. 3.69

Position of the South Faizuly and Kyzyl-Tash deposits in lithological sections of volcanic zones of the Magnitogorsk Belt [compiled by A.I. Brusnitsyn (Kuleshov and Brusnitsyn, 2005)]. 1—7 Stratified units; 1—Zilair and Koltuban suites: terrigenous and terrigenous-carbonate complex, 2—Mukasovso Horizon (cherty rocks), 3—Ulutau suite (tephra-terrigenous complex), 4—Bugulygyr Horizon (chertyrocks), 5—Karamalytash suite: a—basaltic complex, b—rhyolite-basalt complex, 6—Irendyk suite: a—andesite-basalt complex, b—olistostrome volcano-sedimentary complex, 7—Polyakovka suite (basalt complex), and 8—manganese deposits. Volcanic zones: I—Irendyk zone of the West Magnitogorsk paleoisland arc, II—Karamalytash zone of the Sibai interarc basin. Type of basement: *UM*, ultramafic; *M*, mafic.

basin (Zaikov et al., 2002). The Kyzyl-Tash deposit coincides with deposits of a higher stratigraphical level—the tephro-terrigenous sediments of the Ulutau suite (D_2zv–D_3fr_1).

A characteristic structural feature of these deposits is the close spatial and genetic connection of lensoidal manganese-ore deposits with bodies of breccia-type hematite-quartz rocks—jaspilite (Fig. 3.70). In the modern understanding, jaspilite are considered lithified analogues of siliceous-ferruginous deposits forming on the surface of the marine floor in places of percolation of low-temperature solutions (Zaikova and Zaikov, 2003; Crerar et al., 1982; Ashley, 1989; and others). The association of manganese ores with jaspilites is considered one of the indicators of an accumulation of ore-bearing sediment near a hydrothermal source.

At the South Faizuly deposit, jaspilites form a bond of small (2–3 m) in thickness lensoidal bodies composing the foundation of the Bugulygyr horizon. Above are deposited thin-banded wax-red jaspers, which then upwards through the section and lateraly are gradually replaced by gray-colored siliceous siltstones. The general thickness of the siliciclastic member in the area of the deposit reaches 15–20 m.

Manganese mineralization has been established in two sections of the deposit: southern and northern. The greatest interest is presented by the southern section. Manganese ores here compose a compacted-lensoidal, bed-like body with dimensions of 220 m along strike, >150 m along the fall, and up to 3 m in thickness. The northern flank of the manganese deposit directly overlaps jaspilite and is

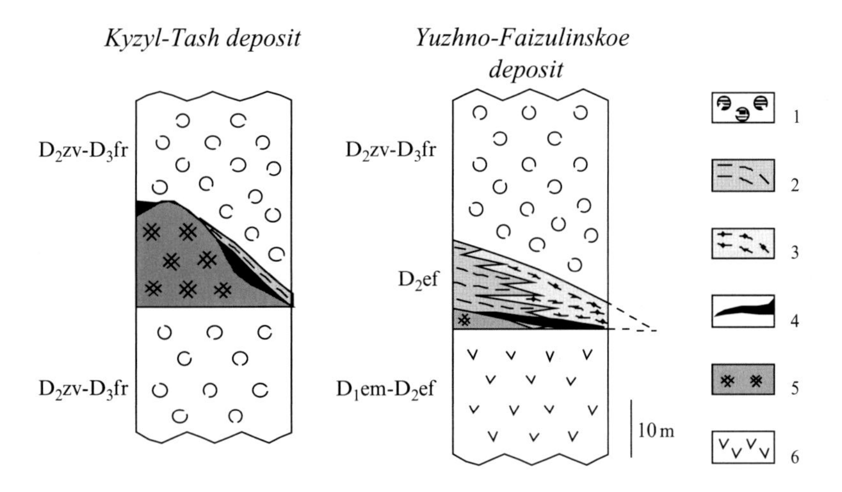

FIG. 3.70

Stratigraphic columns of manganese deposits [compiled by A.I. Brusnitsyn (Kuleshov and Brusnitsyn, 2005)]. 1—Ulutau suite (the rhythmic alternation of volcaniclastic sandstones, siliceous siltstones and mudstones; 2–5—Bugulygyr horizon (D_2ef_2): 2—red jasper, 3—gray cherty siltstone, 4—manganese ore, 5—jasperite; and 6—Irendyk suite (porphyry basalt and andesite-basalt, their lavobreccia, hyaloclastites).

itself overlapped by waxy jaspers; and in the southerly direction, where red-colored siliciclastics rapidly pinch out, the ore body is continued fully in gray siliceous siltstones. The second (northern) section is essentially smaller in scale. It represents a small (up to 0.5 m in thickness) fragment of a monotone jasper member containing thin (up to 1.5 cm) small lenses and interbeds of braunite (braunite ores, rare for the given deposit, are not considered in the present work).

An interesting particularity of the South Faizuly deposit is a find of well-preserved relicts of near-hydrothermal fauna (Zhukov et al., 1998; Zhukov and Leonova, 1999). Organogenic residues have been uncovered in a small section in the roof of a jaspilite lens. They are represented principally by hollow casts of shells of crinoids, brachiopods, gastropods, tetracorals, tabulates, orthoceratites, and the like, around which have been observed hematite-quartz crusts of colloform and spherulite structure—a characteristic indicator of the benthic bacterial encrustation of organisms.

At the Kyzyl-Tash deposit, jaspilite composes a large lensoidal body with apparent dimensions of 350 m along strike and up to 60 m in thickness, concordantly deposited in the strata of volcaniclastic sandstones. In the central part, the roof of the jaspilite deposit is complicated by two swells demarcated by narrow troughs; on account of the latter, the morphology of the ore-bearing body corresponds to that of a hollow hummocky buildup with two apexes. Manganese mineralization is concentrated in the roof of the jaspilite buildup. The ore deposits have a deformed lensoidal form and are localized in depressions of a paleohydrothermal hill—in the bends of relief of the southern and northern slopes, as well as in its central part in the trough between the two apexes. At present, five ore lenses are preserved in the deposit. The length of the largest of these reaches 60 m with a thickness of up to 2 m.

Thus, both deposits are examples of zonal paleohydrothermal buildups with a periphery of iron-rich chert (jaspilite) core and manganese deposits typical for the South Urals and many other fold regions.

Petrography of manganese ores

In all deposits, manganese ores are represented by weakly oxidized in the near-surface zones oxide-carbonate-silicate rocks (Brusnitsyn et al., 2009). This is a fine-grained formation, the structure and composition of which bear the indicators of several geological processes. On the one hand, predominating in the ores are typically sedimentary-diagenetic textures and structures—banded, lensoidal-layered, pelitic, globular, colloform, spherulite, relict-organogenic, and their transitional varieties. On the other hand, also observed in the ores are clear indicators of transformation under conditions of elevated temperatures and pressure (metamorphism). Occurring in association are the presence of mosaic, granoblastic, and sheaf-like structures, vein-latticed and taxitic textures, as well as a wide development of manganese silicates (rhodonite, tephroite, garnets, and the like).

Predominating in the composition of manganese ores are hausmannite, carbonates, quartz, and numerous silicates. With that, the contents of carbonates and silicates in the examined deposits are substantially differentiated owing above all to the different concentrations of the principal components in the initial ore-bearing sediment.

The mineralogical diversity of ores of the South Faizuly deposit is controlled primarily by the distribution in the ores of two sharply dominating elements—manganese and silicon. The presence of other elements is manifested principally in the occurrence of secondary and accessory phases. The leading manganese silicates in these ores are pyroxmangite, caryopilite, ribbeite, and tephroite with rarer occurrences of rhodonite, alleghanyite, spessartine, parsettensite, pyrosmalite, and others. Among carbonates, rhodochrosite sharply predominates while calcite and kutnohorite are classed with the rare minerals.

In relation to the principal minerals, several varieties of ores have been identified. The richest ores are composed of hausmannite, tephroite, ribbeite, and rhodochrosite. With a gradual decrease of manganese concentrations and an elevation of silicon concentrations, the ore initially acquires a caryopilite-ribbeite-rhodochrosite-tephroite composition, then rhodochrosite-pyroxmangite, and finally rhodochrosite-pyroxmangite-quartz or more rarely pyroxmangite-quartz-caryopilite. Additionally, there are occurrences of practically monomineralic rhodochrosite and pyroxmangite rocks. The most widely distributed in the deposit are rhodochrosite-pyroxmangite and rhodochrosite-pyroxmangite-quartz ores (Brusnitsyn et al., 2009).

Also characteristic for ores of the Kyzyl-Tash deposit, apart from the high contents of manganese and silicon, are significant concentrations of iron and calcite as well as aluminum in separate sections. In a number of silicates here, rhodonite, tephroite, caryopilite, and andradite predominate, while grossular, parsettensite, manganaxinite, and others are less widely distributed. From carbonates rhodochrosite, calcite, and kutnohorite are present in almost equal quantity. Also classed with the primary minerals are hematite and more rarely magnetite. In terms of mineral composition, several varieties of ores have been identified in the deposit. The most widely distributed ores are caryopilite-carbonate-tephroite, rhodonite, calcite-rhodonite-quartz, andradite-rhodonite, and rhodonite-hematite-andradite. Also occurring are caryopilite, rhodonite-grossular-caryopilite, caryopilite-rhodonite-tephroite, and epidote-rhodonite ores.

The chemical composition of carbonates of each of the deposits is represented in a three-part diagram (Fig. 3.71) (Brusnitsyn et al., 2009). In the South Faizuly deposit, rhodochrosites sharply predominate (content of manganese is most frequently higher than 90% $MnCO_3$) with minor impurities of calcite, magnesium, and iron. At the same time, broad variations in the quantities of manganese and calcite have been established in carbonates of the Kyzyl-Tash deposit. The concentration of manganese in rhodochrosites does not exceed 83% $MnCO_3$.

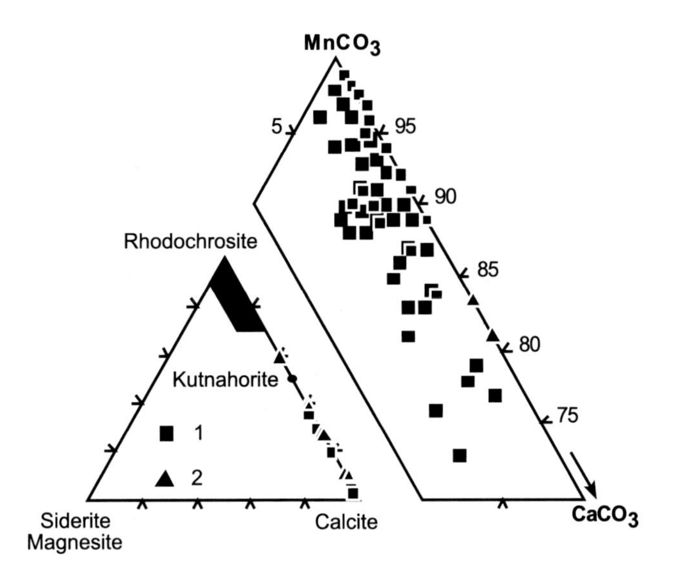

FIG. 3.71

Composition of carbonates from manganese deposits of the Southern Urals [compiled by A.I. Brusnitsyn (Kuleshov and Brusnitsyn, 2005)]. Deposits: 1—South Faizuly. 2—Kyzyl-Tash.

The study of spatial-temporal interrelationships among minerals enables a reconstruction of the crucial stages of the formation of manganese ores. The results of the observations indicate that manganese was initially accumulated predominantly in oxide form, as has been observed in modern hydrothermal deposits. At the diagenesis stage of the ore-bearing sediment the basic mass of oxides was replaced by carbonates. Relicts of initially sedimentary manganese oxides are observed in the rock in the form of isometric or irregular in configuration (with dimensions of 1–3 cm in diameter) isolates of hausmannite intensively corroded from the surface by rhodochrosite. Concentrations of hausmannite occur only in ores with the maximum content of manganese (up to 72% MnO); in all other sections, oxides are fully replaced by carbonates.

The particularities of their morphology also serve as potential evidence for a post-sedimentational (evidently early-diagenetic) genesis of carbonates. In this regard, the most indicative is the South Faizuly deposit. Characteristic for it are rhodochrosite aggregates of pelitic, micro-grained mosaic or significantly more frequently spherulitic structure (Fig. 3.72). In the latter variant the volume of the rocks is almost completely filled by relatively large (0.3–2 mm in diameter) spherulites, which are formed by filliform formations of rhodochrosite diverging from the center. Spherulites frequently have a concentrically zoned structure expressed in rhythmic waves of the content of calcite and manganese in the carbonate, as well as in the manifestation of occasionally very thin (approximately 0.01 mm) quartz fragments.

A formation of similar spherulites of rhodochrosite occurred evidently in the same manganese-containing sediment after its deposition. The environment of metamorphism is not favorable for the growth of filliform individuals, insofar as needles of carbonates should with extended thermal exposure have recrystallized into larger grains. The growth of spherulites commonly occurs in a viscous (colloidal) medium and with high supersaturation of mineral-forming solution (Krasnova and Petrov, 1997). The diagenesis processes correspond to similar conditions.

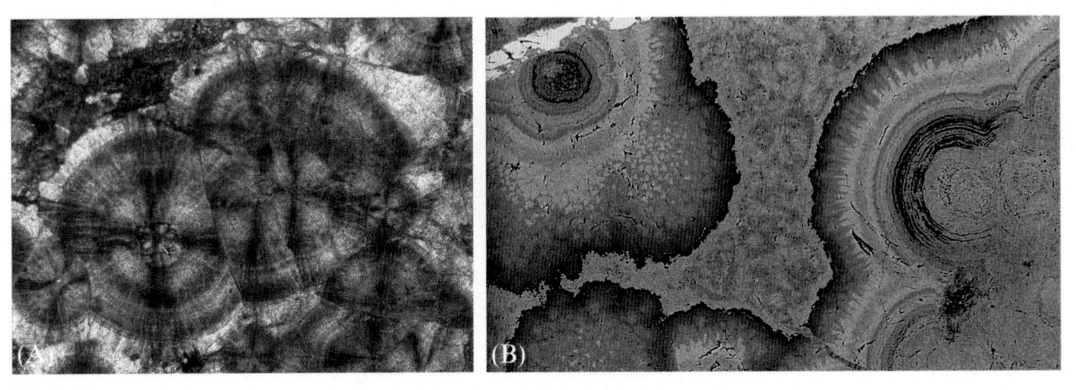

FIG. 3.72

Rhodochrosite spherulites [photos A.I. Brusnitsyn [Kuleshov and Brusnitsyn, 2005)]. (A) Thin section, crossed polars and (B) SEM image (black areas—quartz, dark-gray—enriched with calcium rhodochrosite, light-gray—enriched rhodochrosite, homogeneous gray mass—pelitic rhodochrosite).

This result additionally is supported by the occurrence of precisely such rhodochrosite spherulites in the manganese ores of the Obrochishte deposit (Bulgaria), the formation of which was completed during the diagenesis stage (Aleksiev, 1959). To all appearances, carbonate spherulites at the South Faizuly deposit also have an early-diagenetic origin, and their high state of preservation indicates the relatively low degree of metamorphism of the manganese rocks.

At the Kyzyl-Tash deposit rhodochrosite spherulites have not been detected. Carbonates here most frequently form aggregates of pelitic or mosaic structure. The interrelationships of rhodochrosite with hausmannite are the same as those of the South Faizuly deposit. Therefore, Kyzyl-Tash rhodochrosite likewise has an evidently early-diagenetic nature.

With deeper metamorphogenic transformations, numerous manganese silicates were crystallized in the ores. By our estimates (Kuleshov and Brusnitsyn, 2005), the maximum values of temperature and pressure of the metamorphism of manganese deposits constitute 250°C and 2.5 kbar, respectively. Such figures closely match the general PT-level of transformation of volcanogenic-sedimentary strata of the region, which does not surpass the borders of the prehnite-pumpellyite facies (Mednokolchedannye et al., 1985).

Isotope data and their interpretation

The obtained isotope data for carbonates of the examined deposits are recorded in Tables 3.16 and 3.17 and displayed in Fig. 3.73. From these data it follows that the studied calcites and rhodochrosites are characterized by broad variations in values of isotopic composition, which for $\delta^{13}C$ (‰, PDB) constitute from −51.4 to −10.8 and for $\delta^{18}O$ (‰, SMOW) from 14.4 to 21.4. The isotope characteristics of carbonate minerals of various deposits differ substantially from each other.

The lowest values of $\delta^{13}C$ (from −51.4‰ to −28.9‰) and $\delta^{18}O$ (from 15.5‰ to 18.7‰)—that is, the lightest isotopic composition—are characteristic of rhodochrosites of the South Faizuly deposit. Such isotope characteristics of carbon today are not particular to the distributed groups of carbonate matter of the Earth's crust. By that indication, rhodochrosites of this deposit can be considered anomalous.

Isotope characteristics of carbon. On the whole, low values of $\delta^{13}C$ (up to −51.4‰) are inherent to authigenic carbonates formed on account of carbon dioxide of microbial origin. Such CO_2 is a product of the oxidation of organic matter and methane in the strata of ooze sediment at the stage of early diagenesis or later, during the process of late diagenesis (catagenesis) of the sedimentary rock.

Judging by only one set of isotope data, it is difficult to identify the precise stage of formation or transformation of ore-bearing deposits at which occurred the formation of manganese carbonates rich in light isotope ^{12}C. However, relying on petrographic observations, we can suppose that their crystallization most probably occurred in the strata of the ooze sediment at the early diagenesis stage.

Values of $\delta^{13}C$ in rhodochrosites of the South Faizuly deposit are substantially distinct from those of manganese carbonates of all the examples analyzed above. In the majority of the studied samples, values of $\delta^{13}C$ are contained within a range below −30‰. Such low values of $\delta^{13}C$ are characteristic only for a rare group of authigenic carbonates that form on account of the carbon of oxidized methane. In the modern ocean, carbonates of similar genesis, rich in isotopically light carbon, have been established in sections of benthic discharge of CH_4-containing fluids and in regions of gas-hydrate development.

For example, such carbonates are known today among sediments of the North ($\delta^{13}C$ = −57.2‰ to −36.6‰), Norwegian ($\delta^{13}C$ = −28.9‰), Black ($\delta^{13}C$ = −48‰ to −34‰), Okhotsk ($\delta^{13}C$ = −47.5‰ to −49.2‰), and Arabian ($\delta^{13}C$ = −44.8‰ to −38.7‰) seas and other sections of the global ocean (Hovland et al., 1987; Lein at al., 2000; Belen'kaya, 2003; Von Rad et al., 1996; and others). An analogous mechanism of crystallization has been proposed for carbonates from Pleistocene sandstones of the north-western continental coast of the United States, in which values of $\delta^{13}C$ vary within boundaries from −60.0‰ to −52.1‰ (Hartway and Degens, 1969).

The cited examples and many others are classified as non-ore substantially calcite, aragonite, or protodolomite sediments. In manganese deposits to date, carbonates sharply enriched by isotopically light carbon ^{12}C have been noted only once; low values of $\delta^{13}C$ (from −54.0‰ to −16.6‰) have been reported in rhodochrosites of several small hydrothermal-sedimentary occurrences of California, coinciding with the deep-water argillite-siliceous deposits of the Franciscan series at the Ladd, Buckeye, and Double-A Mine deposits (Hein and Koski, 1987). In terms of geological environment and mineralogy, these objects are highly similar to the deposits of the Magnitogorsk belt; hypothetically, they would also share similar conditions of genesis.

Rhodochrosite in the Californian deposits is considered a product of the diagenesis of manganese-oxide sediments. The lower values of $\delta^{13}C$ in the carbonate are explained by the participation of methane in the formation of rhodochrosite. Moreover, American researchers propose that methane in the given case did not enter the ore-bearing deposit from the outside but was formed in place as a result of the anaerobic oxidation of organic matter buried in the sediment.

In this way, the obtained isotope data allow us to suppose that methane (in part, ethane, propane) for rhodochrosites of South Faizuly deposit serves as the principal or even the only source of carbon. In the opposite case, as has been noted in the majority of known deposits of manganese carbonates contained in sequences of sedimentary and volcanogenic-sedimentary strata, we should have observed a broader spectrum of variations in $\delta^{13}C$ in favor of positive values relative to those actually established.

Carbonates of the Kyzyl-Tash deposit by comparison with those of South Faizuly are characterized by a heavier isotopic composition of carbon: values of $\delta^{13}C$ vary from −19.7‰ to −10.8‰ in calcites and from −28.1‰ to −12.8‰ in rhodochrosites. Such isotope characteristics on the whole are close to

Table 3.16 Chemical (Mass%) and Isotopic (‰) Composition of Various Types of Manganese Ores of the South Faizuly Deposit, South Urals (Compiled by A.I. Brusnitsyn)

	Rhodochrosite-Tephroite-Ribbeite-Hausmannite				Caryopilite-Ribbeite-Rhodochrosite-Tephroite	
Samples	**Fz-13-97**	**Fz-14-97**	**Fz-22-98**	**Fz-65-00**	**Fz-61-00**	**Fz-63-00**
SiO_2	17.8	13.4	10.5	22.9	25.3	24.4
TiO_2	0.02	0.05	0.03	0.03	0.05	0.03
Al_2O_3	0.45	0.55	0.29	1.50	0.90	1.10
Fe_2O_3 total	0.58	0.64	0.84	0.65	0.62	0.88
MnO total	62.4	72.2	66.7	65.5	58.7	58.3
MgO	1.50	2.90	1.80	0.87	1.10	0.74
CaO	1.30	1.00	1.60	1.50	2.30	2.00
Na_2O	<0.20	<0.20	<0.20	<0.20	<0.20	<0.20
K_2O	<0.01	<0.01	<0.01	<0.01	0.1	<0.01
P_2O_5	0.17	0.14	0.17	0.07	0.07	0.12
LOI	16.0	9.00	17.9	7.10	10.9	12.5
Total	100.2	99.9	99.8	100.1	100.0	100.1
$\delta^{13}C$ PDB	−44.8	−36.4	−1.4	−29.2	−28.9	−30.2
$\delta^{18}O$ SMOW	16.7	16.6	18.3	15.5	15.9	15.5

Analyses conducted in the VSEGEI laboratory of chemical and spectral analysis by the b X-ray-spectral fluorescence method with a SRM-25 spectrometer; analysts V.V. Petrov and B.A. Tsimoshchenko.
Samples: *Fz-13-97, Fz-14-97, Fz-22-98, Fz-65-00—rhodochrosite-tephroite-ribbeite-hausmannite rocks; Fz-61-00, Fz-63-00—caryopilite-ribbeite-rhodochrosite-tephroite rocks; Fz-9-97, Fz-64-00, Fz-67-00, Fz-20-98, Fz-21a-98, Fz-21b-98—rhodochrosite-pyroxmangite rocks (Fz-9-97—aggregate enriched by rhodochrosite, Fz-64-00—aggregate enriched by hematite and spessartine, Fz-67-00—aggregate, enriched by spessartine and chalcopyrite, Fz-21b-98—aggregate enriched by parsettensite); Fz-50-99, Fz-68-00—rhodochrosite-pyroxmangite-quartz rocks; Fz-66a-00b Fz-66b-00—rhodochrosite-quartz rocks (Fz-66a-00—aggregate enriched by quartz, Fz-66b-00—aggregate enriched by rhodochrosite); Fz-62-00—pyroxmangite-caryopilite- quartz rock; Fz-36-99—hematite-braunite-quartz rock.*
With calculation of lithological-petrochemical indicators, the contents of elements in mass% have been cited in atomic quantities; a minimal content of TiO_2 was taken as 0.01 mass%.

the data of carbonates of the above-examined Phanerozoic manganese deposits of Georgia, Ukraine, Russia, China, Mexico, and elsewhere and correspond to ores of diagenetic and, as will be indicated below (Section 3.2.4), metasomatic origin.

Evidently serving as a source of carbon for carbonate minerals of the Kyzyl-Tash deposit is carbon dioxide of mixed origin. Its basic part is associated with the microbial decomposition of organic matter buried in the sediment that supplied predominantly the light isotope ^{12}C. The possibility cannot be excluded that the process of diagenesis was accompanied by generations of small quantities of methane. Another portion of carbon rich in the heavy isotope ^{13}C entered together with chemogenic and/or biogenic calcite (carbonate shells of benthic organisms) still during the sedimentation stage of the ore deposits. This is reflected in the correlations between chemical and isotopic composition: samples with lower content of MnO (Fig. 3.74A) and higher content of CaO (Fig. 3.74B) are also characterized by a heavier isotopic composition of carbon.

Rhodochrosite-Pyroxmangite			Rhodochrosite-Pyroxmangite			Rhodochrosite-Pyroxmangite-Quartz		Rhodochrosite-Quartz	
Fz-9-97	**Fz-64-00**	**Fz-67-00**	**Fz-20-98**	**Fz-21a-98**	**Fz-21b-98**	**Fz-50-99**	**Fz-68-00**	**Fz-66a-00**	**Fz-66b-00**
19.9	32.9	43.5	37.5	41.1	49.7	63.0	60.7	60.1	11.4
0.18	0.22	0.20	0.05	0.02	0.03	0.05	<0.01	<0.01	0.09
2.80	3.20	3.40	0.96	0.58	1.00	5.10	0.92	0.38	1.80
1.80	3.40	3.60	1.30	1.10	1.40	1.30	1.00	0.68	2.30
49.2	39.5	38.1	49.5	49	37.7	16.9	23.5	21.2	45.6
3.60	4.70	4.00	1.40	1.10	0.78	1.20	1.90	0.69	2.50
2.50	2.50	2.10	3.40	2.70	1.80	2.90	2.70	3.30	7.00
<0.20	<0.20	<0.20	<0.20	<0.20	<0.20	<0.20	<0.20	<0.20	<0.20
<0.01	0.33	0.22	<0.01	<0.01	<0.01	0.07	<0.01	<0.01	0.04
0.10	0.10	0.11	0.13	0.06	0.09	0.06	<0.05	<0.05	<0.05
19.6	13.1	4.90	5.85	4.69	7.84	9.70	9.20	13.9	29.5
99.7	100.0	100.1	100.2	100.3	100.3	100.3	99.9	100.3	100.2
−37.3	−34.4	−42.4	−33.3	−33.8	−51	−40.2	−46.1	−30	−20.4
16.1	16.2	18.1	18.5	18.1	17.7	17	17.4	18.7	17.4

Isotope characteristics of oxygen. The isotopic composition of oxygen of authigenic carbonate minerals that formed in the early diagenesis zone should be close (or analogous) to the composition of sedimentational and organogenic carbonates (Kuleshov, 2001a). Our established values of $\delta^{18}O$ in manganese carbonates of this deposit are significantly lower than those commonly particular to diagenetic carbonates but are fairly typical for minerals of metamorphic (or metasomatic) rocks (Kuleshov, 1986a, 2001c). In part, an isotopic composition of oxygen similar to that of carbonates of the South Urals is found in rhodochrosites and manganocalcites of the Usa manganese deposit of Kuznetsk Alatau (Kuleshov and Bych, 2002). Therefore, it can be supposed that the isotopic composition of oxygen of the carbonates of the South Faizuly and Kyzyl-Tash deposits was formed in the course of the metamorphism of the ore-bearing strata.

This proposal closely matches the results of mineralogical research: indicators of metamorphogenic transformations are distinctly manifested in the mineral composition as well as in the structural-textural

Table 3.17 Chemical (Mass%) and Isotopic Composition (‰) of Carbonate-Silicate Manganese Ores of the Kyzyl-Tash Deposit (Compiled by A.I. Brusnitsyn)

Sample	KT-1-1	KT-1-7	KT-5-6	KT-5-7	KT-41	KT-43-1	KT-43-2	KT-46-1	KT-46-2	KT-107/12	KT-109/1	KT-1090/3	KT-113	KT-115/4	KT-116	KT-140
SiO_2	61.3	47.4	19.6	24	46.1	41.5	41.9	39.3	40.4	39.6	33.3	33.4	30.2	25.3	18.6	32.9
TiO_2	<0.01	0.02	0.25	0.06	<0.01	0.05	0.06	<0.01	0.03	0.11	0.14	0.17	0.1	<0.01	0.12	0.06
Al_2O_3	0.48	2.50	1.40	1.20	0.90	0.98	0.55	1.00	1.20	6.90	3.40	4.10	2.60	0.55	0.91	1.80
Fe_2O_3 total	0.88	1.20	3.80	1.50	1.80	5.20	6.80	12.4	5.80	3.10	4.50	6.30	7.60	5.40	3.70	4.00
MnO total	14.6	31.6	53.1	56.8	37.5	32.9	35.3	26.4	31.3	37.2	39.2	30.2	36.7	24.1	54.4	41.8
MgO	0.37	0.68	0.54	0.64	0.33	0.59	0.62	0.53	0.53	0.96	1.60	0.98	1.10	0.72	0.55	0.94
CaO	14.1	11.2	10.1	9.00	11.8	14.7	11.5	18.4	13.7	7.8	11.4	19.3	10.5	28.3	13.3	12.00
Na_2O	<0.20	<0.20	0.53	0.70	<0.20	<0.20	<0.20	<0.20	<0.20	<0.20	<0.20	<0.20	<0.20	<0.20	0.26	<0.20
K_2O	0.05	0.92	<0.01	0.02	<0.01	<0.01	<0.01	<0.01	<0.01	0.03	0.02	0.02	0.07	0.05	<0.01	<0.01
P_2O_5	<0.05	<0.05	<0.05	<0.05	0.05	0.07	0.07	<0.05	<0.05	0.06	0.07	0.07	0.07	<0.05	0.07	0.07
LOI	7.60	2.90	7.70	4.80	0.97	3.70	2.80	2.10	6.40	3.80	6.10	5.40	10.7	14.9	7.10	6.40
Total	99.4	98.4	97.0	99.96[a]	99.4	99.7	99.6	100.1	99.9	99.6	99.7	99.9	99.6	99.3	99.0	100.0
$\delta^{13}C$ (PDB) Cct	−19.1	−19.7	–	–	−15.4	−15.8	−15.9	−15.3	−14.4	–	−15.2	−18.1	–	−10.8	–	−15.9
$\delta^{13}C$ (PDB) Rhd	–	–	−28.0	−28.1	–	–	–	−16.1	–	−16.6	−15.8	−17.1	−12.8	–	−21.6	–
$\delta^{18}O$ (SMOW) Cct	15.6	16.7	–	–	16.4	20.5	16.7	18.8	20.7	–	19.5	17.5	–	16.2	–	15.8
$\delta^{18}O$ (SMOW) Rhd	–	–	14.7	14.4	–	–	–	21.4	–	17.2	18.7	17.8	16.00	–	15.1	–

Analyses conducted in the VSEGEI laboratory of chemical and spectral analysis by a X-ray-spectral fluorescence method on a SRM-25 spectrometer; analyst V.V. Petrov.

Mineral types of ores: *KT-116, KT-5-7, and KT-20—caryopilite-carbonate-tephroite; KT-113 and KT-44—caryopilite; KT-107-12—rhodonite-grossular-caryopilite; KT-140—caryopilite-rhodonite-tephroite; KT-41 and KT-1-7—rhodonite; KT-43-1, KT-43-12, and KT-115—andradite-rhodonite; KT-46-1 and KT-46-3—rhodonite-hematite-andradite; KT-1-1—calcite-rhodonite-quartz.*

With calculation of petrochemical indicators the contents of elements by mass% have been cited in atomic quantities; a minimal content of Al_2O_3 and TiO_2 was taken as 0.10 and 0.01 mass% respectively.

Density of the rocks (g/cm^3): sample KT-5-7—3.86; KT-43-1—3.47; KT-41—3.29; KT-1-1—3.20.

[a] *In sample KT-5-7 was established 1.20 mass% BaO.*

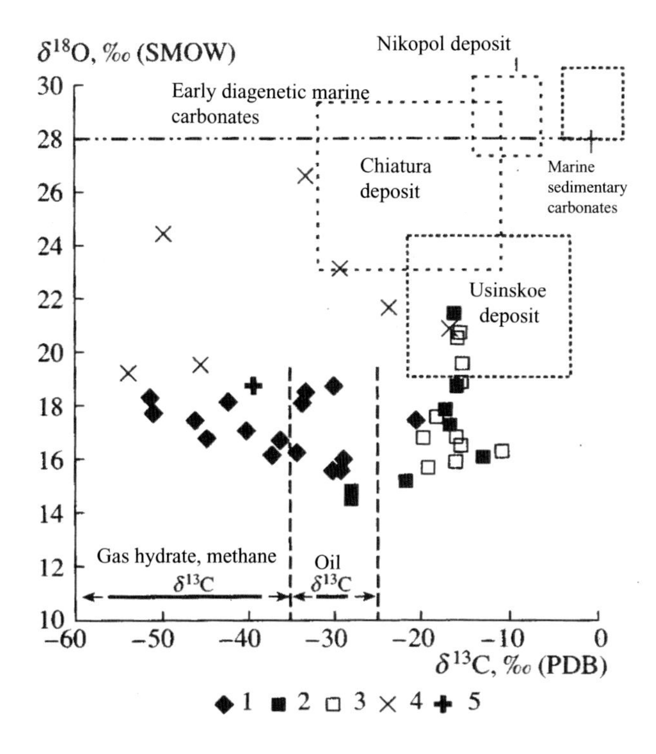

FIG. 3.73

$\delta^{13}C$ versus $\delta^{18}O$ in carbonates from manganese deposits [compiled by A.I. Brusnitsyn (Kuleshov and Brusnitsyn, 2005)]. 1—South Faizuly deposit; 2–3—Kyzyl-Tash deposit: 2—rhodochrosite, 3—calcite; 4—Ladd and Backeye (California); 5—Dubl A Mine (California). The data for the fields of California are taken from Hein and Koski (1987).

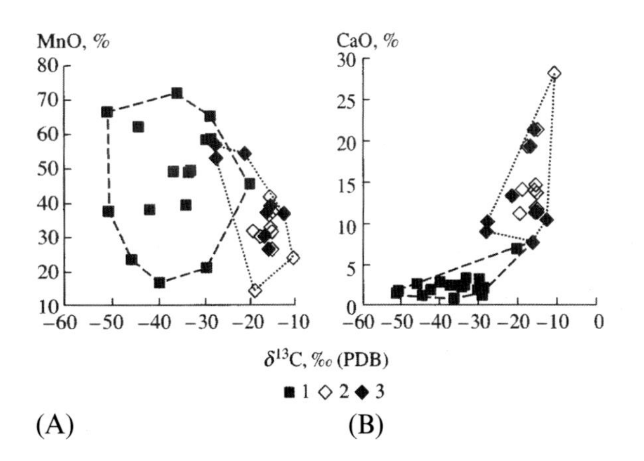

FIG. 3.74

Relationship of $\delta^{13}C$ with MnO (A) and CaO (B). 1—South Faizuly deposit and 2–3—Kyzyl-Tash deposit: 2—rhodochrosite, 3—calcite.

particularities of manganese ores. At the same time, traces of an intensive metasomatic restructuring of volcanogenic-sedimentary strata accompanied by a wide scale displacement of matter in large blocks of rocks have not been established in any of the examined deposits. Only a narrowly localized area of redistribution by local interstitial solutions—manganese, silicon, and carbon dioxide—has been observed. These are expressed in the form of small zones of development of vein pyroxmangite, rhodonite, rhodochrosite, and quartz mineralization, which commonly coincide with sections of tectonic deformations of the ore body and do not occur in the host siliciclastics. The basic volume of the manganese deposits underlying and overlapping their rocks has maintained without substantial changes the main features of the structure and chemical composition of the initial sedimentary strata.

The character of the distribution of $\delta^{18}O$ values (see Fig. 3.73) provides the basis to propose that rhodochrosites of the South Faizuly deposit were formed within a narrower temperature interval with participation by fluids of a more consistent isotopic composition of oxygen by comparison with carbonates of the Kyzyl-Tash deposit. Accompanying calcite and rhodochrosite in the rocks of the latter deposit are evidently syngenetic and were formed (or transformed) in conditions close to isotopic equilibrium.

If it is supposed that the isotopic composition of metamorphic fluids was close to the composition of marine water ($\delta^{18}O \approx 0$), then the temperature of transformation of carbonates could constitute 90–100°C (Friedman and O'Neil, 1977). If the same fluids were characterized by higher values of $\delta^{18}O$, then these temperatures should have been substantially higher. The transformation of the studied carbonates could have been effected under temperatures of approximately 250°C (the peak of metamorphism of deposits of the South Urals). In this case, for the acquisition of carbons with the reported values of $\delta^{18}O$, the isotopic composition of oxygen of the fluids should have been close to 8–10‰, which does not contradict the isotope data for waters of metamorphogenic genesis.

Another variant might be proposed that the isotopic composition of oxygen was maintained from the moment of formation of the rhodochrosites in the early diagenesis zone and did not change in conditions of subsequent metamorphogenic processes. In this case, we should accept that manganese carbonates at the outset were formed at elevated temperatures with participation by ooze solutions of a lighter isotopic composition of oxygen than that of marine water. Such a possibility is in our view unlikely, though not entirely precluded.

Bikkulovo manganese deposit (South Urals)

The Bikkulovo deposit is located in the South Urals 40 km to the west of Magnitogorsk city, 3.5 km north-west of the village Bikkulovo. The Bikkulovo deposit along with South Faizuly and Kyzyl-Tash is almost universally classified to the hydrothermal-sedimentary genetic type (Betekhtin, 1946; Khersakov, 1951; Khodak, 1973; Mikhailov and Rogov, 1985; Magadeev et al., 1997; Ovchinnikov, 1998; Kontar' et al., 1999; Salikhov et al., 2002; Brusnitsyn et al., 2009; and others). Localization of ore beds among siliceous or ferrosiliceous deposits is a particularity characteristic for deposits of this type. This is typical on the whole for deposits of the South Urals. The structure and models of such deposits have been treated in a large number of works (Kheraskov, 1951; Roy, 1986; Starikova et al., 2004; Brusnitsyn and Zhukov, 2005; Bonatti et al., 1976; Crerar et al., 1982; Huebner et al., 1992; and others). The lithology of the Bikkulovo deposit is characterized by its own particularities: siliciclastics here have limited development, and the basic volume of manganese ores is limited to volcaniclastic rocks. Correspondingly, the conditions of accumulation of metal-bearing sediments in the present case possess their own specifics deserving of special consideration.

Brief outline of the structure of the deposit. The deposit is situated, as are South Faizuly and Kyzyl-Tash, on the western border of the Magnitogorsk paleovolcanic belt (see Fig. 3.68).

Identified in the structure of the sedimentary strata in the area of the deposit from bottom to top along the sequence are the Bugulygyr siliceous horizon (D_2ef), the Ulutau tephro-terrigenous suite (D_2zv-D_3fr), the Mukasov siliceous horizon (D_3fr), the Koltuban carbonate-terrigenous suite (D_3fm), and the Eilair terrigenous suite (D_3fm-C_1t).

Manganese mineralization is localized in deposits of the upper part of the Ulutau suite near their contact with overlying shales of the Mukasov horizon. The host deposits are represented by limy tuffites, volcaniclastic sandstones, and gritstones with small interbeds of siliceous shales and limestones. The ore-bearing member is composed of specifically manganese ores and their closely associated jaspilites and ferruginous, manganous, and ferrosiliceous tuffites, which in the complex form a unified stratiform deposit with sharply expressed vertical and lateral zonality (Fig. 3.75) (Brusnitsyn et al., 2009).

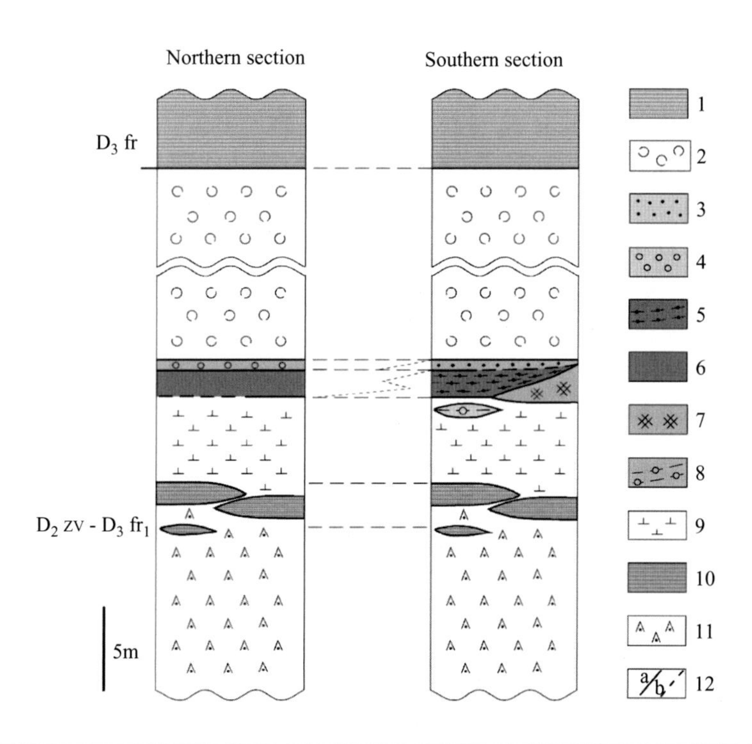

FIG. 3.75

Stratigraphic columns of the Bikkulovskoe deposit (Brusnitsyn et al., 2009). 1—Mukasovo Horizon (siliceous shales and siltstones); 2–11—Ulutau suite: 2—gray-white coarse-clastic volcanomictic sandstones and gritstones; 3 and 4—brown-red ferruginous-siliceous tuffites (jasper tuffites): 3—with a small amount volcanomictic material, 4—enriched in volcanomictic material; 5 and 6—manganese ores: 5—heterogeneous ore composed of andradite, caryopilite, rhodonite, hematite, quartz, and calcite; 6—homogeneous parsettensite-hematite-quartz-andradite ore; 7—hematite-quartz and hematite-andradite-quartz jasperites; 8—manganiferous tuffites; 9—cherry red ferruginous fine- and medium-clastic tuffites; 10—limestones; 11—violet-green medium- and coarse-clastic calcareous tuffites; 12—boundaries: (a) distinct and (b) gradual.

The association of manganese ores with jaspilites is considered one of the indicators of an accumulation of ore-bearing sediment near a hydrothermal source (Kheraskov, 1951; Brusnitsyn et al., 2000; Zaikova and Zaikov, 2003, 2005; Crerar et al., 1982; Ashley, 1989; Gutzmer et al., 2001; Grenne and Slack, 2003; and others). At the Bikkulovo deposit, jaspilites compose a relatively small lensoidal, hummocky body with apparent dimensions of 10 m along the strike and a thickness reaching 2.5 m.

A bed of manganese ores on the southern flank is deposited directly on the jaspilites, and in the northerly direction after a rapid pinch out of hematite-quartz rocks it overlaps ferruginous tuffites. The length of the ore bed constitutes 340 m, the thickness from 1 to 3 m, and in slope the bed is traced for 50 m.

On the manganese ores is deposited a small (approximately 0.5 m in thickness) layer of bright-red ferrosiliceous tuffites (jasper-tuffites)—a rock intermediary in composition and structure between jaspers and volcaniclastic sandstones. In the southern part of the deposit jasper-tuffites are distinguished by insignificant contents of clastic material. In mineral composition, these are predominantly quartz-hematite rocks, in separate sections rich in calcite. Not infrequently these rocks are connected with the underlying ores by gradual transitions.

By volume, ferruginous rocks in the deposit sharply predominate over manganese rocks. The basic mass of the ferruginous deposits is concentrated stratigraphically below the manganese ores, while above them is situated only a thin bed of jasper-tuffites.

Lateral zonality is manifested in the changes in the composition of the metal-bearing deposits in the direction from south to north. The southern section of the deposit is characterized by a maximally full sequence of the productive member (from bottom to top): ferruginous (in sections manganous) tuffites → jaspilites → manganese ores → jasper-tuffites. The northern section possesses a transected sequence, from which jaspilites fall.

In the structural regard, the deposit represents an anticline overturned to the west, composed in the hinge part by folds of smaller orders (Figs. 3.76 and 3.77) (Brusnitsyn et al., 2009). The axis of the anticline is gently inclined to the north-west, consequent to which in the south-easterly direction more ancient rocks gradually emerge on the surface. Somewhat to the west along its axis the anticline is broken by the plane of a thrust of submeridional strike with a slope toward the east at a 40–50 degrees angle. As a result of the thrust deformations, the deposits of the anticline's western flank in a large area of the deposit have been buried under rocks of the core and eastern limb of the fold.

Mineral composition of manganese rocks. The mineral composition of the basic volume of the ore-bearing deposits was formed during the process of regional metamorphism. Manganese mineralization forms two types of oxide-carbonate-silicate rocks (Brusnitsyn et al., 2009): manganous tuffites, deposited in the form of small lenses in the base of the ore bed among iron-rich volcaniclastic deposits (ferruginous tuffites); and manganese ores, composing a particularly productive bed. From the surface, both types of oxide-carbonate-silicate rocks are replaced by hypergenic oxides and hydroxides of manganese. However, the scale of this process in the Bikkulovo deposit is quite insignificant, and therefore the mineralogy of the oxidation zone is not considered in the present work.

Specifically manganese ores represent fine-grained rocks heterogeneous in mineral composition, textural outlook, and color. The principal minerals are andradite, rhodonite, caryopilite, parsettensite, hematite, calcite, and quartz (Tables 3.18 and 3.19). Moreover, present in insignificant quantities in the sections rich in volcanogenic material are piemontite, manganese epidote, and pumpellyite-Mn. In the capacity of secondary and accessory phases are diagnosed tephroite, ilvaite, johannsenite, shirozulite, neotocite, rhodochrosite, hausmannite, barite, apatite, and native copper.

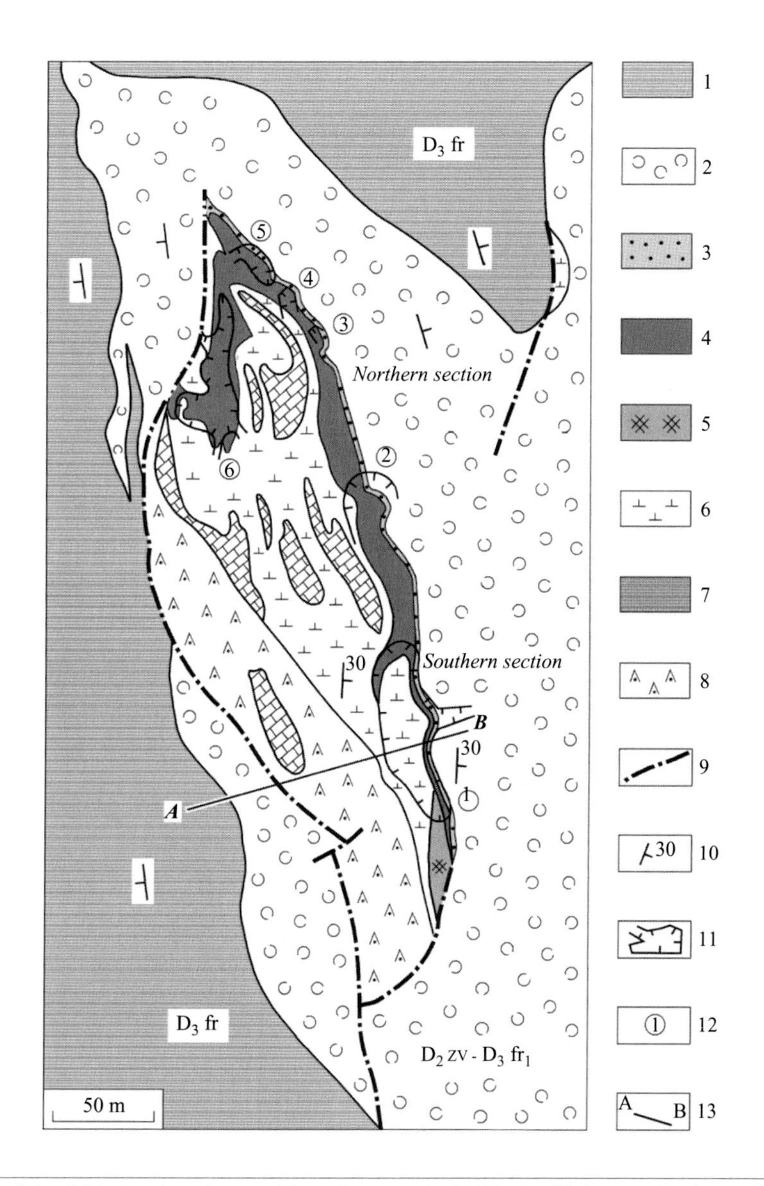

FIG. 3.76

Geological map of the Bikkulovskoe deposit (Brusnitsyn et al., 2009). 1—Mukasovo Horizon (siliceous shales and siltstones); 2–8—Ulutau suite: 2—gray-green coarse-banded volcanomictic sandstones and gritstones; 3—brown-red ferruginous-siliceous tufites (jasper tuffites); 4—manganese ores; 5—jaspirites; 6—cherry red ferruginous fine- and medium-clastic tuffites; 7—limestones; 8—violet-green medium- and coarse-clastic calcareous tuffites; 9—faults; 10—bedding elements; 11—quarry contours; 12—quarry numbers; and 13—profile presented in Fig. 3.77.

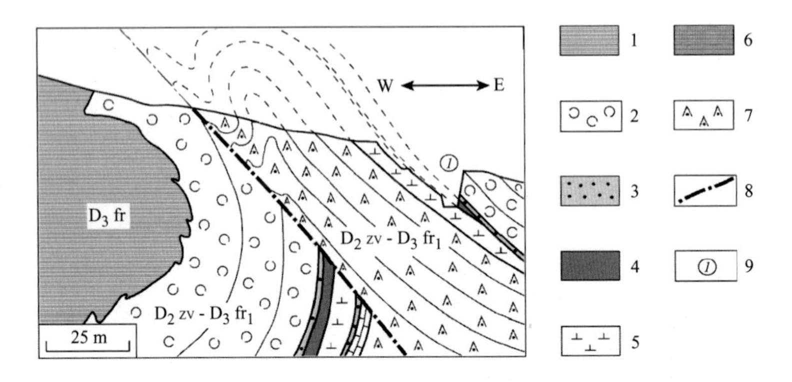

FIG. 3.77

Geological cross section of the Bikkulovskoe deposit (Brusnitsyn et al., 2009). 1—Mukasovo Horizon (siliceous shales and siltstones); 2–7—Ulutau suite: 2—gray-green coarse-banded volcanomictic sandstones and gritstones; 3—brown-red ferruginous-siliceous tuffites (jasper tuffites); 4—manganese ores; 5—cherry red ferruginous fine- and medium-clastic tuffites; 6—limestones; 7—violet-green medium- and coarse-clastic calcareous tuffites; 8—fault; 9—quarry number.

By relation to leading minerals among manganese ores, Brusnitsyn et al. (2009) have identified several varieties connected by gradual transitions (in the names of the rocks the minerals are listed in sequence by increasing measure in content in total %): rhodochrosite-caryopilite-tephroite, pumpellyite-piemontite-andradite-caryopilite, hematite-calcite-andradite-caryopilite, quartz-rhodonite-andradite, and parsettensite-hematite-quartz-andradite.

A lateral zonality, typical for the deposit on the whole has been established in the distribution of the enumerated varieties of ores, manifested in the change in composition and simultaneous decrease in the mineralogical diversity of the manganese ores in the direction from south to north within the boundaries of a unitary bed. Thus, in the southern, jaspilite-adjoining section of the deposit, the ore bed has a very heterogeneous internal structure, a result of the uneven content of the primary petrogenetic elements (Si, Al, Fe, Mn, Mg, and Ca) in the initial deposits. Hematite-calcite-andradite-caryopilite and quartz-rhodonite-andradite ores predominate: their total quantity constitutes approximately a total of 70%.

At the deposit, fragments with high manganese content are composed of rhodochrosite-caryopilite-tephroite ores, while aluminum-rich occurrences are composed of pumpellyite-piemontite-andradite-caryopilite ores. At the same time, by measure of the distance from the jaspilites, the composition of the manganese-bearing deposits rapidly becomes homogenous. Already 100 m to the north of the jaspilites in the structure of the ore bed, parsettensite-hematite-quartz-andradite rocks sharply predominate (a total of 90%); among these, there are rare occurrences of small lenses (approximately 1 m in length and up to 20 cm in thickness) of rhodochrosite-caryopilite-tephroite aggregates. Still further north the ore body is composed entirely of parsettensite-hematite-quartz-andradite rocks.

In this way, in the southern and northern sections of the deposit, manganese ores have a different composition not only in terms of the selection of minerals but also in terms of the character of their distribution. The southern section by comparison with the northern is distinguished by smaller dimensions and a significantly larger variability in the mineralogy of the ores.

Isotope data and their interpretation. The results of isotope analyses are recorded in Table 3.20 and indicated in Fig. 3.78. The isotopic composition of carbon of the studied carbonates is characterized by

Table 3.18 Chemical Composition (Mass%) of Host Rocks of the Bikkulovo Deposit (Brusnitsyn et al., 2009)

	Jasperites			Underriding			Overlapping			Tuffites		Piemontite tuffites (a)				
Component	**Bk-21**	**Bk-22**	**Bk-105**	**Bk-20**	**Bk-24**	**Bk-25**	**Bk-23**	**Bk-26**	**Bk-27**	**Bk-28**	**Bk-29**	**Bk-38**	**Bk-39**	**Bk-73**	**Bk-74**	**Bk-100**
SiO_2	75.9	88.6	77.9	46.5	39.9	58.8	73.4	63.0	51.5	52.6	36.6	35.3	38.5	55.5	43.9	43.6
TiO_2	<0.01	<0.01	<0.01	0.37	0.37	0.29	0.03	0.09	0.43	0.40	0.61	0.47	0.57	0.46	0.49	0.82
Al_2O_3	0.74	0.38	0.52	7.12	12.7	6.16	0.9	1.58	9.09	20.0	13.7	12.5	16.1	15.6	14.7	12.1
Fe_2O_3 total	15.6	9.18	14.6	30.7	24.1	22.6	22.8	23.6	26.9	5.18	5.69	14	8.36	5.24	5.63	8.70
MnO total	5.22	0.19	0.65	3.37	4.61	2.62	0.16	0.35	0.87	0.37	0.55	18.5	6.56	2.21	14.5	17.00
MgO	1.29	1.20	1.24	2.02	2.88	1.72	1.33	1.42	3.33	7.74	3.56	2.29	2.16	3.06	1.73	2.50
CaO	1.03	0.16	4.85	7.08	11.7	5.79	0.49	5.56	1.87	3.26	17.4	12.3	20.2	14.0	14.8	10.7
Na_2O	<0.05	<0.05	<0.05	<0.05	<0.05	<0.05	<0.05	<0.05	<0.05	3.28	1.45	<0.05	<0.05	<0.05	<0.05	0.05
K_2O	0.10	0.08	0.08	0.10	0.10	0.12	0.10	0.12	2.72	1.30	3.61	0.27	0.19	0.19	0.35	0.50
P_2O_5	0.05	<0.05	<0.05	0.07	0.06	0.61	0.10	<0.05	0.08	0.11	<0.05	0.09	0.12	0.13	0.53	0.08
LOI	0.84	0.15	0.17	2.27	3.7	1.55	0.24	4.21	3.13	5.69	16.4	4.41	7.15	3.52	3.42	4.14
Total	100.77	99.94	100.01	99.6	100.12	100.26	99.55	99.93	99.92	99.93	99.57	100.13	99.91	99.91	99.95	100.19
Lithological-petrochemical modules																
(Fe+Mn)/Ti	2685	1175	1917	94	80	89	718	362	65	14	10	74	28	17	45	34
(Al+Ti) ×102	1.46	0.76	1.03	13.9	25.4	13.1	1.8	3.2	18.4	39.7	27.6	25.1	32.3	31.2	29.4	24.8
(Fe+Mn) ×102	26.9	11.8	19.2	46.1	36.6	31.9	28.7	39.8	34.9	7	7.9	44.1	19.7	9.7	27.5	34.8
Al/(Al+Fe +Mn)	0.05	0.06	0.05	0.24	0.4	0.29	0.06	0.07	0.34	0.85	0.77	0.51	0.62	0.76	0.51	0.41
Mn/Fe	0.38	0.02	0.05	0.12	0.22	0.13	0.01	0.01	0.04	0.08	0.11	1.49	0.88	0.47	2.9	2.2

Analyses were conducted in the VSEGEI laboratory of spectral analysis by X-ray-spectral fluorescence method with an ARL-9800 (Switzerland) spectrometer; analyst: *B.A. Tsimoshchenko.*

Samples: *sample Bk-105 andradite jasperite. In sample Bk-73 was analyzed a section enriched by pumpellyite. In the remaining samples were analyzed the basic mass of the rock. With calculation of the lithological-petrochemical indicators the contents of elements in mass% have been cited in atomic quantities; a minimal content of Al_2O_3 and TiO_2 was taken as 0.10 and 0.01 mass% respectively.*

Table 3.19 Chemical Composition (Mass%) of Manganese Rocks of the Bikkulovo Deposit (Brusnitsyn et al., 2009)

Mineralogical Group	c	d	e		f				g					h		
Sample	**Bk-32**	**Bk-72**	**Bk-1**	**Bk-7**	**Bk-6**	**Bk-51**	**Bk-60**	**Bk-77**	**Bk-50**	**Bk-70**	**Bk-71**	**Bk-76**		**Bk-31**	**Bk-34**	**Bk-35**
SiO_2	24.4	34.4	32.4	26.9	38.6	38.7	41.5	35.9	60.8	47	56	35.8	28.7	39.1	45.1	39.4
TiO_2	0.04	0.3	0.15	0.06	0.02	0.03	0.03	0.07	0.02	0.04	0.04	0.02	0.41	0.29	0.16	0.16
Al_2O_3	0.89	8.44	1.37	2.11	1.49	0.6	0.63	2.36	0.89	1.99	2.2	2.09	4.01	5.34	5.51	5.21
Fe_2O_3 total	2.93	13.1	12.5	13.6	15.1	4.67	5.56	10.2	7.83	11.1	9.06	5.65	28.3	17.7	20.2	20.4
MnO total	64.4	23.1	32.5	24.2	21.3	34.2	36.5	27.2	15.5	27	17.2	23.3	26.4	19.1	14.6	14.5
MgO	1.25	1.98	1.47	1.25	1.45	1.33	1.4	1.57	1.25	1.71	1.58	1.7	1.64	1.58	1.68	1.5
CaO	1.48	13.8	12	19.4	19.9	15.3	12.1	18.8	10.8	12.3	10.9	22.8	6.76	13.5	10.2	15.8
Na_2O	<0.05	<0.05	<0.05	<0.05	<0.05	<0.05	<0.05	<0.05	<0.05	<0.05	<0.05	<0.05	<0.05	<0.05	<0.05	<0.05
K_2O	0.09	0.21	0.11	0.11	0.08	0.08	0.09	0.09	0.1	0.18	0.13	0.18	0.19	0.2	0.29	0.23
P_2O_5	0.1	0.09	0.1	0.1	<0.05	0.08	<0.05	<0.05	<0.05	<0.05	<0.05	<0.05	0.1	0.21	0.16	0.23
LOI	4.73	4.41	7.27	11.8	1.26	5.09	1.42	3.99	2.41	<0.10	2.75	8.3	3.49	2.82	2.08	2.4
Total	100.3	99.8	99.9	99.5	99.2	100.1	99.2	100.2	99.6	101.3	99.9	99.8	100.0	99.8	100.0	99.6
Lithological-petrochemical modules																
(Fe+Mn)/Ti	1887	129	323	648	1629	1350	1459	567	1054	1038	7112	1329	142	136	229	230
(Al+Ti)102	1.8	16.93	2.88	4.22	2.95	1.22	1.28	4.72	1.78	3.95	4.36	4.13	8.37	10.83	11	10.42
(Fe+Mn)102	94.4	48.9	61.4	51.1	48.9	54.0	58.4	51.1	31.6	51.9	35.6	39.9	72.6	49.0	45.8	45.9
Al/(Al+Fe+Mn)	0.02	0.25	0.04	0.07	0.06	0.02	0.02	0.08	0.05	0.07	0.11	0.09	0.1	0.18	0.19	0.18
Mn/Fe	24.78	1.99	2.93	2	1.59	8.25	7.4	3	2.22	2.74	2.14	4.65	1.05	1.22	0.81	0.8

Analyses conducted in the VSEGEI laboratory of spectral analysis by X-ray-spectral fluorescence method on an ARL-9800 (Switzerland) spectrometer; analyst: B.A. *Tsimoshchenko.*

c–h—mineral varieties of ores: c—rhodochrosite-caryopilite-tephroite, d—pumpellyite-piemontite-andradite-caryopilite, e—hematite-calcite-andradite-caryopilite, f—andradite-rhodonite, g—rhodonite-quartz-andratite, h—parsettensite-hematite-quartz-andradite. In sample 76 analysis 12—the basic mass of the rock, analysis 13—section enriched by hematite. With the calculation of lithological-petrochemical indicators the contents of elements in mass% were cited in atomic quantities.

Table 3.20 Isotopic Composition of Carbon and Oxygen in Rhodochrosite and Calcites From Manganese Rocks of the Bikkulovo Deposit (Southern Urals)

Number of Analysis	Number of Sample	Minerals	$\delta^{13}C$, ‰ PDB	$\delta^{18}O$, ‰ SMOW
Bikkulovo deposit, s. Urals				
5357	Bk-32-01	Rhodochrosite	−15.1	16.6
5358	Bk-72-00	Calcite	−16.9	25.4
5359	Bk-7-00		−19.5	16.8
5360	Bk-1-00		−22.2	17.2
5361	Bk-70-00		−15.8	19.8
5362	Bk-71-00		−23.0	23.0
5363	Bk-51-00		−13.7	19.5
5364a	Bk-39-01		−13.9	20.2
5365	Bk-77-02		−19.1	19.0
5366	Bk-76-00		−29.3	18.1

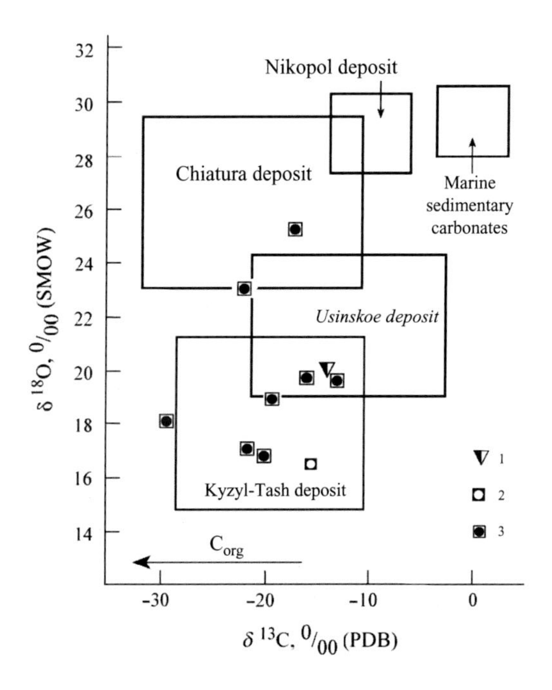

FIG. 3.78

$\delta^{13}C$ versus $\delta^{18}O$ values in carbonates of the Bikkulovskoe deposit [compiled by A.I. Brusnitsyn (Brusnitsyn et al., 2009)]. 1—Calcite from manganese-bearing tuffites, 2—rhodochrosite, and 3—calcite of manganese ores.

low values of $\delta^{13}C$ (‰, PDB): from −29.3 to −13.7 in calcite and −15.1 in rhodochrosite. Such values of $\delta^{13}C$ are particular to authigenic carbonates formed, as in the above-examined deposits, on account of carbon dioxide representing the result of microbial oxidation of organic matter in the strata of ooze sediment at the early diagenesis stage or later, in the process of catagenesis of the sedimentary rock.

Serving as potential evidence for this is the wide development in the studied carbonates of typically sedimentary-diagenetic structure (pelitoform, colloform, spherulite, and the like), which, considering the wide scattering of $\delta^{13}C$ values, enables us to propose that their crystallization most probably occurred in the strata of ooze sediment at the early diagenesis stage.

Thus, the most likely source of carbon for the carbonate minerals of the studied deposit as well as those examined above is the carbon dioxide generated with the microbial decomposition of organic matter buried in the sediment, which supplies the light isotope ^{12}C. Manganese minerals (primarily oxides) acting in the capacity of an oxidant transform the organic matter of the surrounding sediment into CO_2. Manganese is thus reduced to the divalent state and, reacting with carbon dioxide, forms carbonates. Relicts of the initial oxides of manganese are occasionally preserved in the form of isometric isolations of hausmannite. Characteristically, this mineral is observed only in maximally manganese-rich ores. In all remaining varieties, the oxides have been fully replaced by carbonates and silicates.

The isotopic composition of the oxygen of authigenic carbonates formed in the early diagenesis zone should be close (or analogous) to the composition of sedimentational and organogenic phases. In the carbonates of the Bikkulovo deposit, however, values of $\delta^{18}O$ (‰, SMOW) from 16.8 to 25.4 are significantly lower than those commonly determined in diagenetic carbonates but fairly typical for minerals of metamorphic (or metasomatic) rocks (Kuleshov, 1986, 2001b; Zagnitko and Lugovaya, 1989). Most likely, the isotopic composition of the oxygen of carbonates of the Bikkulovo deposit, as for South Faizuly and Kyzyl-Tash, was formed in the course of the metamorphism of the ore-bearing strata. This proposal matches well with the results of mineralogical research: the indicators of metamorphogenic transformations are distinctly manifested in the mineral composition as well as in the structural-textural particularities of the manganese ores and their host rocks.

Particularities of the genesis of manganese deposits of the South Urals

The results of isotope research in conjunction with geological and mineralogical data are suitable for interpretation within the framework of modern understandings of the hydrothermal-sedimentary ore-genesis of manganese, to which are classified South Faizuly, Kyzyl-Tash, and Bikkulovo.

Accumulation of ore-bearing sediment occurred in the discharge zone of hydrothermal solutions circulating in the strata of rocks of the oceanic crust, which probably had a thermoconvectional nature. By analogy with the hydrothermal sources of the modern oceans, it has been proposed (Brusnitsyn et al., 2009) that the solutions were low temperature ($T \leq 100°C$) and featured a manganese-iron-silicon specialization. In the benthic zone as a result of the difference of physical-chemical parameters of the medium, a geochemical barrier was created in which the ore components were deposited. By measure of the mixing of heated and acidic (pH 4–6) solutions with cold and weakly alkaline (pH 7.5), marine water was distinctly manifested a differentiation of elements—the first to transition into the sediment are iron and silicon, while manganese as a more mobile metal is deposited in the uppermost, most diluted part of the stream. As a result, zonal deposits form with a ferrosiliceous base and manganese rocks on the periphery. Insofar as the content of iron and silicon in the solution as a rule is substantially higher than that of manganese, ferrosiliceous rocks in the ore-bearing deposits commonly predominate sharply over manganous rocks.

The precipitation of manganese occurred predominantly in oxide form, which subsequently during the stage of diagenesis was transformed into the carbonate form in the course of a reaction with the carbon of organic matter and methane. Evidence in favor of a diagenetic origin of carbonates is found in the character of their interrelationships with hausmannite as well as the particularities of the morphology of aggregates of rhodochrosite.

The most likely source of carbon for the formation of carbonates of the Kyzyl-Tash deposit is the organic matter buried in the ore-bearing deposits. The presence of organogenic material in the form of faunistic residues and their casts in the composition of manganese deposits coinciding with the deep-water siliceous and edaphogenic sediments could be explained by their localization near a hydrothermal source.

At present it has been firmly established that in deep-water conditions the existence of benthic organisms is largely controlled by the distance from hydrothermal sources. Consequently, the presence of C_{org} in the composition of manganese deposits can be explained by their localization near a hydrothermal source, while at a distance from the deposits the biological productivity of the medium is sharply decreased.

Additionally, at the South Faizuly deposit, judging by the isotope data, the principal source for the carbon of rhodochrosites was methane (possibly also ethane and propane). In the course of the diagenesis of the sediments, manganese oxides reacted with methane acting in the capacity of an oxidant; the metal was reduced to Mn^{2+} and bonded to rhodochrosite. Schematically the interaction of manganese oxides with methane can be presented in the form of the reaction: $MnO_2 + 3CH_4 + 5H_2O \rightarrow MnCO_3 + 2CO_2 + 11H_2$.

The principal point of discussion in the given case is the question of the origin and the mechanism of the entry of methane (ethane, propane) into the ore deposit.

Methane could have been present in the sediment (eg, in the form of gas hydrates) prior to the entry of manganese oxides from a hydrothermal source. It could also have infiltrated into the surface of the marine bed in the form of low temperature or cold seeps, thus feeding the already existing but still unconsolidated manganese-containing ooze sediments. Finally, it could have been generated in the sediment itself in the course of the microbial decomposition of organic matter, as has been proposed for the deposits of the Franciscan complex in California.

The available geological data at present do not allow for unequivocal preference for any of the enumerated sources of methane.

Proving any of the variants for the entry of methane into the ore-genesis zone from the underlying rocks encounters difficulties. As is known, in modern marine basins the overwhelming majority of fields of CH_4-saturated sources are distributed in areas of development of thick sedimentary cover with concentrations of hydrocarbons (petroleum, gas, gas hydrates, and the like). In the structure of the Magnitogorsk paleovolcanic belt, there are no thick sedimentary strata capable of producing deposits of hydrocarbons. Moreover, as noted, drawing from analysis of the geological environment in the given region, no researchers have suggested any probability for the presence of hydrocarbon deposits here. In this situation, the isotopic composition of rhodochrosite is unlikely to serve as all-encompassing (or even indirect) evidence for the potential existence among the rocks underlying the South Faizuly deposit of even a small concentration of hydrocarbons. Consequently, the underlying strata might not be a source of the methane of hydrothermal solutions.

Another origin of methane that is not connected with concentrations of hydrocarbons—that is, an abiogenic origin—also seems to us unlikely: such methane as a rule is characterized by a heavier isotopic composition of carbon by comparison with the isotopic composition we have established here.

In part, abiogenic methane from the hydrothermals of the black smokers of the Rainbow field in the Atlantic Ocean has values of $\delta^{13}C$ from −13.4‰ to −13.0‰ (Lein et al., 2000b).

Nevertheless, the entry of isotopically light methane from subaqueous volcanic emanations cannot be fully precluded. Such fluids have been established, for example, in the Piip volcano in the Bering Sea (Torokhov et al., 1991). Here in the apical part of the subaqueous volcanic cone have been discovered thermal sources with gas shows of carbon dioxide-nitrogen-methane composition, in the discharge zone of which occurs the formation of secondary calcite. Values of $\delta^{13}C$ for methane here constitute −48.7‰ and values for calcites from −36‰ to −29‰. An insignificant increase in heaviness of the isotopic composition of calcites is evidently caused by the presence of carbon dioxide characterized by lower values of $\delta^{13}C$ (−21.9‰) in the composition of the discharging source of gas generation and its role in the formation of ore-bearing carbonate-manganese sediments.

It can also be supposed that methane was not added to the manganese deposits from the external sources, but was generated directly in the strata of sediment by microbial breakdown of organic matter developing near the hydrothermal source. This scenario is fully plausible: in the South Faizuly deposit, relicts of near-hydrothermal macrofauna with traces of bacterial encrustation have been found in the roof of the jasper lens (Zhukov et al., 1998; Zhukov and Leonova, 1999). This provides a basis for proposing that low values of $\delta^{13}C$ in the given case are also indicators for the existence near a hydrothermal source of a specific bacterial community, which in the process of its burial becomes capable of generating methane with isotopically light carbon.

The latter model eliminates the need to search for deep sources of methane and therefore seems at first glance preferable to the other models. But this viewpoint cannot be accepted unconditionally; in contradiction stand the following facts. Manganese ore-genesis in the upper layers of the ooze sediment (zone of diagenesis) at present is fairly well studied; without exception, however, in all of the isotope data for manganese carbonates from modern lacustrine, marine, and oceanic sediments, as well as from ancient deposits of various ages (Paleozoic, Mesozoic, and Cenozoic), deposited in strata of sedimentary rocks, there are no indications of any carbonates that formed on account of the carbon of oxidized methane.

On the other hand, it has been firmly established that isotopically light carbonates can be actively manifested in the diagenesis zone with a microbial destruction of organic matter on account of released methane. For example, as noted, protodolomite and calcite of concretions of the Gulf of California from the uppermost layers of sediment (5–45 cm) are characterized by values of $\delta^{13}C$ equal to −40.0‰ and −38.0‰ (Lein et al., 1975). Moreover, the general boundary of variations in $\delta^{13}C$ values in this zone is very broad and constitutes from −40.0‰ to 1.7‰. Carbonate matter of the underlying layers of sediment (45–280 cm) is characterized by higher values of $\delta^{13}C$, spaced far from the range of values for carbon of oxidized methane (−20.0‰ to 2.4‰). The isotopic composition of the rhodochrosite of the South Faizuly deposit is fundamentally different: regardless of the wide scattering, $\delta^{13}C$ values there never exceed −20.4‰.

In this way, according to modern data, the quantity of methane in the diagenesis zone released with the microbial destruction of organic matter is very low. Such methane cannot ensure crystallization of large masses of carbonates with isotopically light carbon, and therefore $\delta^{13}C$ values of the manganese carbonate of the sediments are commonly determined by the contribution of carbon from another origin.

Finally, the proposal regarding an input of hydrocarbons in the ore beds at the stage of cata- or metagenesis is likewise improbable in our view. Judging by petrographic observations, manganese rocks and silicates syngenetic with them inherited the initially sedimentary structure. The metasomatic phenomena accompanying the metamorphism have a local character, with which the migration of ore components was carried out at a very small distance, not exceeding the thickness of the deposits. Absent from the ores are indicators of intensive restructuring of the deep hydrothermals delivered from the outside. Additionally, in the history of the geological development of the western part of the Magnitogorsk paleovolcanic belt, no periods of generation in the strata of volcanogenic-sedimentary rocks of hydrocarbon-rich fluids have been detected, as is proposed for certain petroleum-gas-bearing regions. The petroleum-bearing area nearest the deposit is located in western Bashkiria, within the boundaries of the Permian Cis-Urals trough. There is no basis to connect its development with those rocks found among the volcanogenic deposits of the western flank of the Magnitogorsk paleovolcanic belt.

The formation of the Bikkulovo deposits apparently occurred somewhat differently (Fig. 3.79) (Brusnitsyn et al., 2009). In the given case, the discharge of solutions with the precipitation from them of iron and silicon began not on the surface of the oceanic floor, but beneath this level, already in the near-bottom strata of volcanoclastic sediments. In such a way was formed the member of ferruginized volcanogenic-sedimentary rocks—protoliths of subore tuffites. Moreover, in the place of the most intensive hydrothermal stream, a portion of the solution nonetheless reached the surface of the bottom without substantial changes in its composition. At the mouth of the source occurred a final discharge of hydrothermals by the standard mechanism with the manifestation of a ferrosiliceous (protojaspilite) hill with peripheral manganese deposits of heterogeneous composition.

In the modern erosional section, corresponding to the epicenter of the hydrothermal field is the southern section of the deposit, characterized by a maximally full selection of rocks (ferruginous tuffites → jaspilites → manganese rocks). Simultaneously, in the remaining (greater) territory of the hydrothermal field the hydrothermal stream was of lesser thickness. The surface of the bottom here was reached only by "depleted" solutions, already almost devoid of iron and even of a certain amount of manganese, although the basic mass of the latter was still preserved in the solution. Mixing with fresh marine water, the solutions ultimately lost the metals: iron and manganese passed into the sediment, giving rise to an ore bed. By similar means was formed a "transverse" sequence of the northern section of deposit (ferruginous tuffites → manganese rocks) with a more homogeneous chemical composition of rocks and a less contrasting differentiation of iron and manganese.

After the discharge of the hydrothermals was completed, the protojaspilite hill was partially washed out by near-bottom flows. Its composing ferrosiliceous material was transferred to lower relief, giving rise to protolith for jasper-tuffites.

It follows to note that the contents of volcanogenic matter in supraore cherts, as well as in manganese rocks, grow rapidly by measure of their distance from jaspilites.

Examples of manganese occurrences analogous to the Bikkulovo deposit in other regions of the world are known in the area of Watson's Beach in New Zealand (Kawachi et al., 1983); serving as their modern analogues are hydrothermal buildups of the Galapagos rift in the eastern Pacific Ocean (Lisitsyn, 1993; Corliss et al., 1978; Honnorez et al., 1983; Marchig et al., 1987; Herzig et al., 1988; and others). Here have been established chains of hummocky hydrothermal buildups with a height from 5 to 20 m above the level of the bottom and with a diameter from 25 to 100 m—the so-called nontronite hills. The basic volume of sediments of the hills is composed of ferrosiliceous material represented by mixed-layer clayey phases of nontronite-celadonite composition. Manganese oxides and hydroxides (todorokite, birnessite, δMnO_2, amorphous phases) form crusts with a thickness reaching 1.5 m and layers of concretions on the surface of the nontronite hills—in the zone of mixing of hydrothermal solutions with open marine waters.

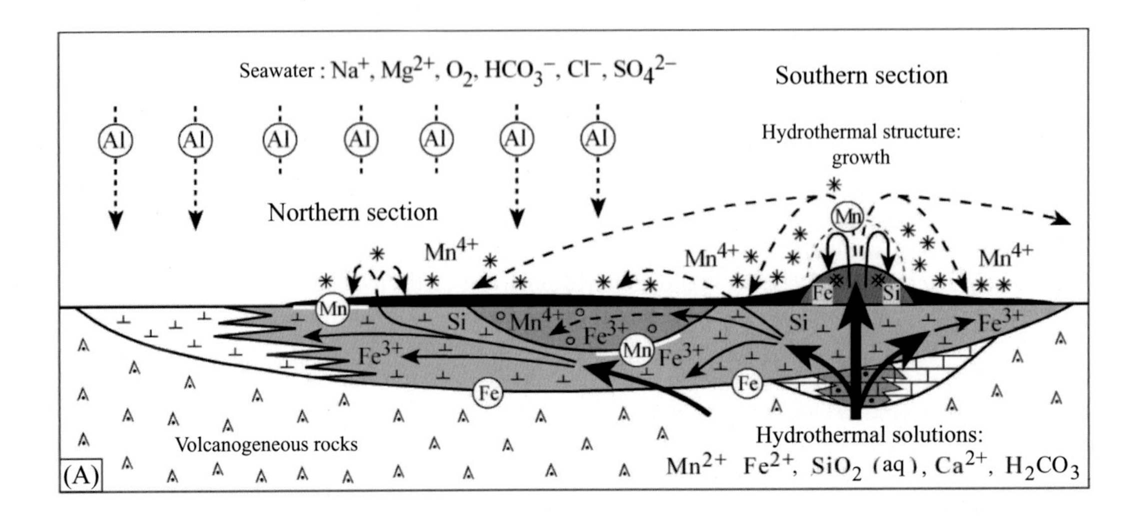

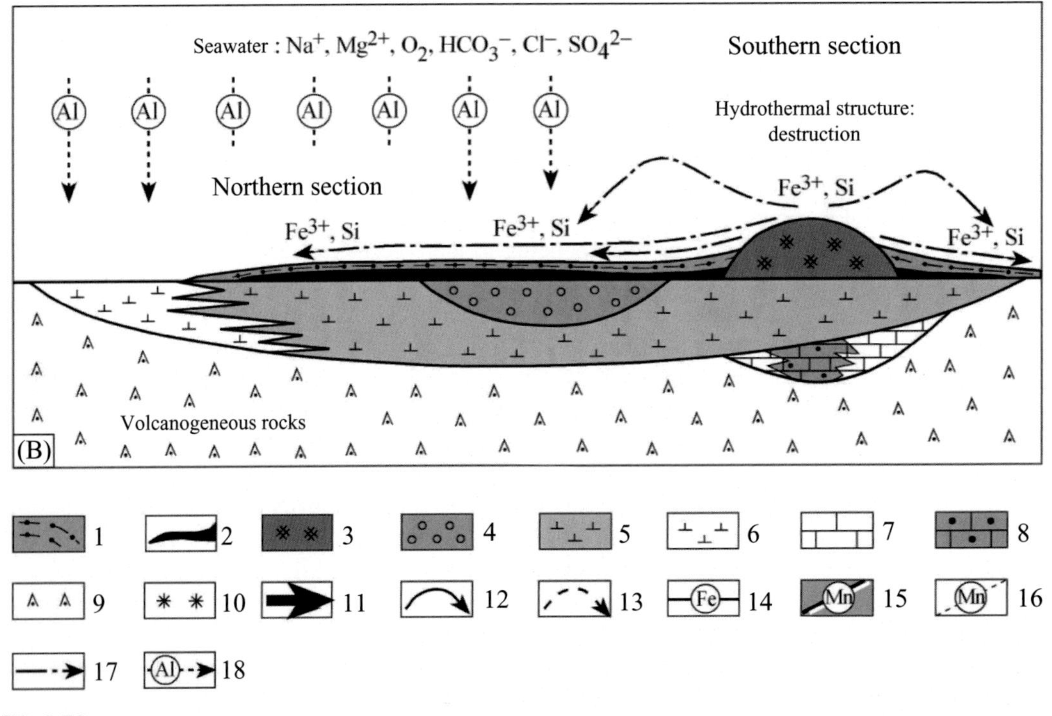

FIG. 3.79

Formation model of ore-bearing rocks in the Bikkulovskoe deposit in the phase of (A) active and (B) terminated hydrothermal activity [compiled by A.I. Brusnitsyn (Brusnitsyn et al., 2009)]. (1–6) Rocks: (1) ferruginous-siliceous rocks with volcanomictic admixture (protolith for jasper tuffites); (2) manganese rocks; (3) ferruginous-siliceous rocks (protojaspirites); (4) manganized volcanomictic rocks (protomanganiferous tuffites); (5) hematitized volcanomictic (protoferruginous tuffites); (6) loose volcanomictic rocks unaltered by hydrothermal processes; (7) limestones; (8) limestones altered by hydrothermal processes (skarnized); (9) massive volcanomictic rocks; (10) finely dispersed suspension of manganese minerals; (11–13) hydrothermal solutions: (11) before discharge, (12) Fe-loosing, (13) Mn-loosing; (14–16) geochemical barriers: (14) for Fe, (15, 16) for Mn: (15) below the bottom surface, (16) above the bottom surface; and (17) direction of the washout of ferruginous-silicate sediments; (18) directions of the input of terrigenous material.

In this way, in the Galapagos hills alongside a tight association of ferruginous and manganese hydrothermal deposits are strongly manifested processes of differentiation of ore elements. Moreover, the mixing of hydrothermal fluids with marine water and its partial discharge occurs not on the surface of the marine bed, but earlier—already in the strata of the pelagic sediments supplied by the fluid. Ferruginous deposits predominate in the composition of the hills; manganese deposits have substantially lesser volume and are localized in the periphery of the hydrothermal edifice. According to these indicators, the Bikkulovo deposit ought to be considered an analogue of the ore-bearing deposits of the Galapagos buildups preserved in a mineral state. Not incidentally, the basic mass of carbonates in the Bikkulovo deposit is concentrated immediately in the rocks of the southern section, which is interpreted as the near-mouth zone of the paleohydrothermal field (Brusnitsyn et al., 2009). In the northern section, the carbonates are merely accessory phases.

It is also interesting that not only rhodochrosite, but also manganese calcite is rich in isotopically light (biogenic) carbon; manganese calcite is significantly more widely distributed in the deposit than is rhodochrosite. To all appearances, the source of calcite was those same hydrothermal solutions that carried manganese, iron, and other elements into the near-bottom part of the marine basin. Carbon dioxide was generated in the strata of the still-unconsolidated ore-bearing sediment owing to the oxidation of buried organic matter. Seepage of solutions through such sediment enabled the precipitation of calcium in carbonate form. The given schema supports the distribution of calcite within the boundaries of the ore bed: concentrations of this mineral are typical precisely for the southern "near-mouth" section of the deposit, where high hydrothermal activity created elevated concentrations of calcium and indirectly carbon dioxide.

3.2.2.2 Parnok deposit of ferromanganese ores

Another example of deposits of hydrothermal-sedimentary genesis, in our opinion, is the Parnok deposit of ferromanganese ores, which is well studied and important not only in the scientific, but also in the practical regard.

This deposit is situated 70 km south-east of Inta city (Komi Republic, Russia), in the middle course of the Parnoka-Yu River (tributary of the Lemva River), in the western foothills of the southern part of the Polar Urals. In the geological regard, it is situated within the boundaries of the Lemva structural zone (Fig. 3.80), having imbricate-thrust structure. Here have been identified three independent isolations: a parautochthon and two nappes (Khaiminski and Grubeinski) that consistently override each other (Shishkin, 1989), respectively composing three structural stages (lithological-stratigraphical complexes): lower, middle, and upper (Shishkin and Gerasimov, 1995; Gerasimov, 2000).

The lower stage corresponds to the parautochton and is represented by deposits of the Nadotamyl'ski suite of Lower-Middle Devonian age: black quartz sandstones, dark-gray siltstone-clayey shales, gray-green siltstones, and in subordinate quantity limestones. The thickness of the suite is more than 1000 m.

The middle stage is formed by the Khaiminsky nappe, which is composed of deposits of the Kachamyl'ski (O_{2-3}km) and Kharotski (S-D_1hr) suites. The Kachamyl'ski suite on the whole has a terrigenous-carbonate composition and in terms of lithological particularities is divided into three strata: Pokoinikshorski, Pacheshorski, and Parnok.

The Pokoinikshorski strata is represented by gray-green limy sandstones and siltstones with rare interbeds of sandstone limestones. The thickness is estimated at 100–150 m. The Pacheshorskaya strata is composed of a monotone strata of green and gray-green limy siltstones with lenses of limy sandstones and thin-detrital sandstone limestones. Its thickness constitutes 400–600 m. The Parnok strata

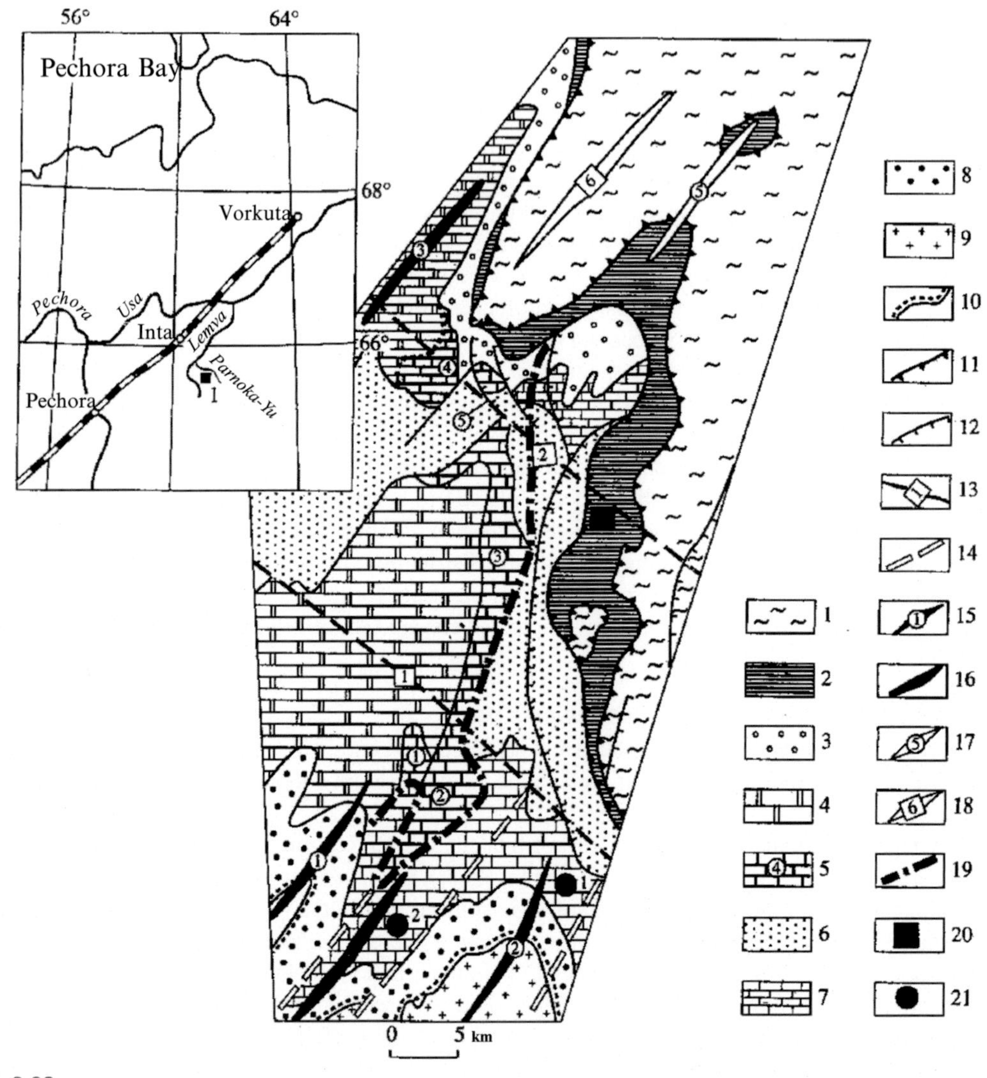

FIG. 3.80

Position of Parnok iron-manganese deposit in the structures of the Lemvinsk South District (Shishkin and Gerasimov, 1995). 1–2—Lemvinsk allochthon: 1—siliceous clastic continental slope formations of Early Ordovician-Carboniferous age composed Grubeinsk nappe, 2—siliceous-terrigenous-carbonate formations deep shelf of Ordovician-Devonian age as part Hayminsk nappe; 3–9—autochthonous and parautohtonous: 3—Early Jurassic molasse formation, 4—shallow shelf carbonate formations of Middle Ordovician—Early Devonian and Late Devonian-Early Carboniferous age on the outer edge zone of the shallow shelf (1—Late Ordovician age reef "Badia," 2—Silurian: Balbanyusky reef, 3—Early Devonian: Lemvinsk Reef, 4–5—Late Devonian-Carboniferous: Bolshenadotinskaya oolitic Megabank and the reef "Olys'a"), 6—Middle Devonian transit clastic formation, 7—siliceous-carbonate formations of deep shelf of Ordovician and Late Devonian-Carboniferous age, 8—Early Ordovician rift formation of basal horizons of Uralides; 9—pre-Paleozoic basement; 10—boundary

See the legend on opposite page

is ore-hosting, has a thickness of 150–300 m, and is represented predominantly by limestones. In the composition of the strata have been identified three members (from bottom to top): C-3, C-2, and C-1, which are unevenly distributed in the sequence of rocks of the deposit as well as along the lateral, and on the strength of this can be considered as lithofacies.

Member C-3 is composed of dark-gray anastomosing layers and lenses of siltstone limestones, detrital limestones (greenstones) with indicators of gradational layering, and lensoidal-layered limestones. Coinciding with a bed of more large-detrital limestones are rare interbeds with ferruginous oolites, more rarely greenalite and magnetite laminae. The thickness of the member is 130–200 m.

Member C-2 is formed by carbonaceous-clayey-limy shales and black carbonaceous limestones rhythmically interstratifying one another. The rocks are pyritized. Occasionally present are interbeds of magnetite ores and manganous rocks. The thickness of the member is 1–40 m.

Member C-1 is ore hosting; it is represented by gray pelitoform manganous limestones with beds of manganous siltstones, manganese and iron carbonate ores. A particularity of the composition of the rocks of the member is the practically full absence of terrigenous impurity and crinoidal or other detritus. The thickness of the member varies from a few meters to 60 m.

The Kharotski suite is composed of black carbonaceous-siliceous, frequently phosphate-containing (content of P_2O_5 in separate horizons can reach 5–9%) shales and phthanites with members and interbeds of pelitoform limestones. The thickness of the suite constitutes 150–200 m.

Rocks of the Grubeinski nappe form the upper stage. In its composition are included the deposits of the Grubeinski (O_1gr) and Kharbeishorski ($O_{1\text{-}2}chs$) suites, composed predominantly by terrigenous formations: multicolored strata of hematite-containing shales, green exfoliated siltstones, feldspathic-quartz siltstone sandstones. The thickness is more than 1500 m.

It should be noted that earlier proposals as to the presence in the area of a large quantity of faults has not found support and in some cases has been called into doubt by subsequent research and geological-exploratory works (Lemeshev et al., 2009). The majority of contacts among various material complexes have begun to be considered as normal (concordant) geological boundaries. The structure of this area of the deposit was represented in the form of a synclinal fold with westward-recumbent flanks. It has been suggested that manganese mineralization is localized in two stratigraphic levels: the Middle Ordovician in the eastern flank of the deposit and the Middle Devonian in the western flank.

Within the boundaries of the area of the deposit are developed intrusive formations represented by diorites of the Ust'-Khaiminski (O_3?) complex and gabbro-diabases of the Orang-Yugansk-Lemva (O_1) complex. Diorites of the Ust'-Khaiminski complex are developed to a limited extent within the

FIG. 3.80, CONT'D

of angular and structural unconformity at the base of Uralides; 11–14—tectonic faults: 11—gentle thrusts, limiting the same name of thrust nappes, 12—other thrusts, 13—large basement faults (1—Kozhimsky, 2—Parnoksky), 14—updated faults of Uralides rift stage; 15–18—the axis of the fold structures: 15—anticlines (1—Maldy-nyrd, 2—Rossomakha, 3—Rugoza Musyur), 16—Balbanyu syncline, 17—Khayminsko-Grubeinsky antiform, 18—Malodotinsky synform; 19—boundary position of the Eletsk and Lemva structure-formation zones in the autochthon; 20—Parnok deposit; and 21—large polymetallic ore occurrences: 1—Kozhimskoe, 2—Pelengicheyskoe.

boundaries of the Khaiminski nappe in the form of small subtabular bodies at 3–5 km south-east from the Parnok deposit proper; their tearing contacts have been established only with terrigenous-carbonate deposits of the Kachamyl'ski series.

Gabbro-diabases of the Orang-Iugansk-Lemva complex are present only in the deposits of the Grubeinski nappe in the form of thin sills among rocks of the Grubeinski and Kharbeishorski suites.

The structure of the Parnok deposit is defined by the predominance of the imbricate-thrust dislocations above the fold dislocations (Fig. 3.80). Within the boundaries of the central part of the deposit have been identified three basic slices included in the composition of the Khaiminski nappe: the lower, composed of siliceous and carbonaceous-siliceous shales of the Kharotski suite with separate lenses of manganese and iron ores; the middle, formed by limestones, manganous siltstones, and ores of the Parnok strata; and the upper, represented by limy siltstones and sandstones of the strata. All the deposits of the central part of the deposit are located in overturned bedding.

On the whole, rocks in the deposit have a steep slope (30–70 degrees) of south-easterly direction (AЗ 110–140 degrees).

In the modern erosional section, the productive strata is exposed in several relatively isolated sections (see Fig. 3.81): Magnitnyi-1, Magnitnyi-2, Vostochnyi-1, Vostochnyi-2, Vostochnyi-3, and Vostochnyi-4. Additionally, with drill boring manganese and iron ores have been struck in the Ust'-Pachvozh, Dal'ni, and Upper Pachvozh sections.

Beds of manganese and iron ores within the boundaries of the Parnok ore-bearing strata are grouped in several ore deposits of stratiform character, which are subconcordant with the inner structure of the Khaiminski nappe and almost monoclinally deepened at an angle of 30–50 degrees to depths exceeding 800 m (Fig. 3.82).

Each deposit consists of several interstratified lensoidal beds of manganese and iron ores (with a thickness from 0.8 to 5.8 m), which are separated by alternations of limestones. Along the dip and strike the separate beds are pinched out. In plan, the orc deposits have an oval or more frequently elongated (ribbon-like) form. The deposits on the whole are consistent, but there are noted as well pinches and taperings, which within the boundaries of the deposit result in the presence of a series of ore sections (Magnityi-1, Magnityi-2, Ust'-Pachvozh, Dal'ni, Vostochnyi).

Within the boundaries of the deposit, the upper ore deposit, localized in the limestone member C-1, is the most continuous and is developed almost everywhere. The remaining deposits are developed only within the boundaries of sections and are correlated to one another fairly conditionally.

The basic part of the area of the Parnok deposit is situated within the boundaries of the Pachvozh depression of the Late Mesozoic deposit. Fragments of the bottom of this depression have been preserved along both borders of the Pachvozh stream valley in the form of level areas (absolute elevations of 350–400 m). To these areas are associated preserved portions of the weathering crusts (thickness reaches 70 m) hypothetically of Late Paleogene age, developed predominantly on rocks of the Kachamyl'skii series (including on the primary manganese and iron ores), with a formation of iron and manganese "caps" (sections Magnitnyi-1 and Magnitnyi-2; in the remaining sections, the products of the weathering crusts have mostly eroded and oxidized ores are developed to a limited extent).

Characteristics of manganese ores

Among manganese ores of the Parnok deposit have been identified four principal varieties (Shishkin and Gerasimov, 1995; Gerasimov, 2000): carbonate, rich-oxidized, poor-oxidized, and semi-oxidized. Additionally present in a subordinate quantity are oxidized ferromanganese and primary iron ores.

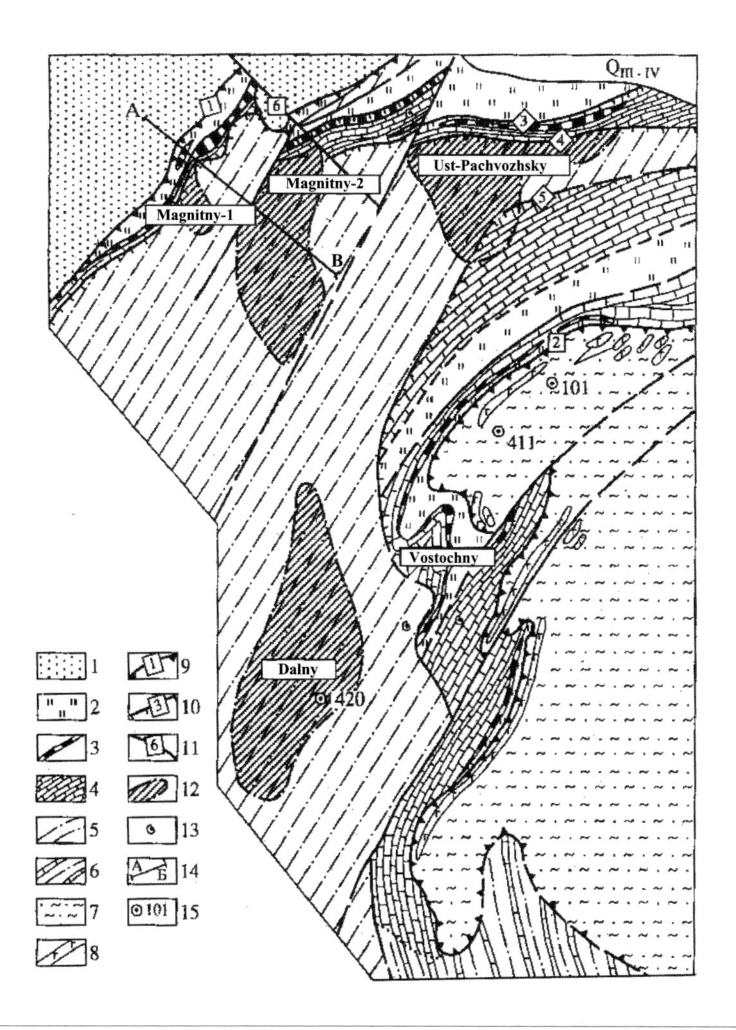

FIG. 3.81

Schematic geological and structural map of the Parnok iron-manganese deposit area (Shishkin and Gerasimov, 1995). 1—paravtohton: Nadotamylsk suite ($D_{1\text{-}2}$nd); 2–8—Lemvinsk allochthon: 2–6—Khaiminsk nappe: 2—Kharotsk suite (S-D_1hr), 3—deposits of iron and manganese ores, 4—Parnok strata ($O_{2\text{-}3}$pr), 5—Pacheshorsk strata (O_2psh), 6—Kachamylsk series (undivided) ($O_{2\text{-}3}$km); 7–8—Grubeinsk nappe: 7—Grubeinsk and Kharbeyshorsk suites (combined) (O_1gr + $O_{1\text{-}2}$chs), 8—diabase sills of Orang-Yugansk-Lemva complex (O_1ojl); 9—main thrusts (1—Khayminsky, 2—Grubeinsky); 10—other thrusts restricting individual scales inside the covers (3—Podrudny, 4—Nadrudny, 5—Vostochny); 11—faults (6—Northern); 12—Magnetic anomalies associated with iron- and manganese-ore bodies at depths from 50 to 800 m; 13—sampling locations of organic residues; 14—line of geological cross section (see Fig. 3.82); and 15—drilling wells.

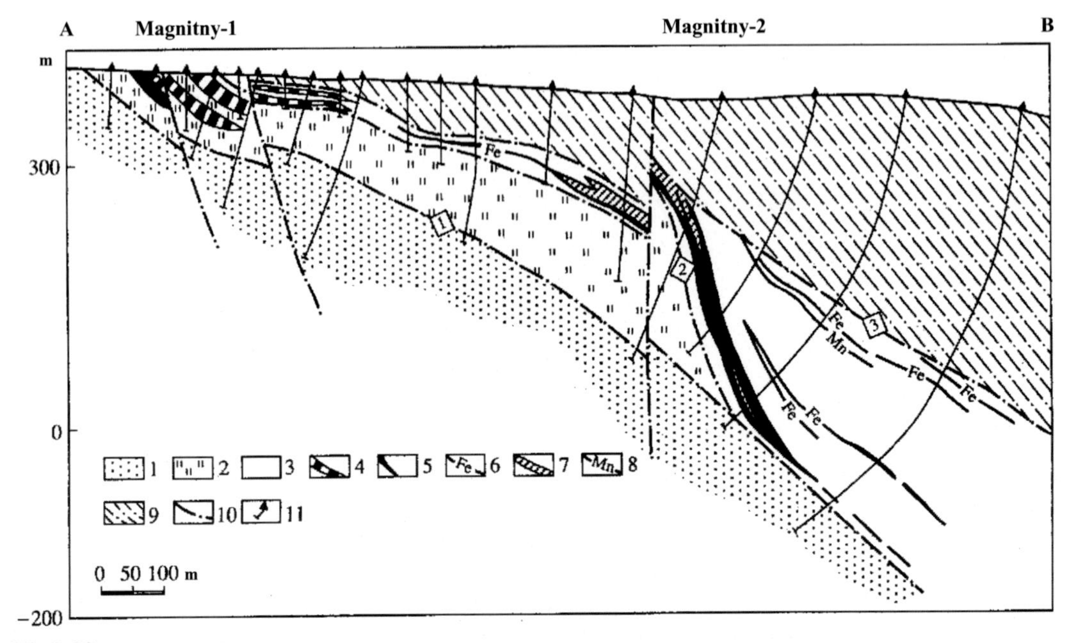

FIG. 3.82

Schematic geological cross section through the "Magnetic-1, A 'and' Magnetic-2, B" sectors of the Parnok deposit (location see Fig. 3.81) (Shishkin and Gerasimov, 1995). 1—Parautochthonous: Nadotamylsk suite (D_{1-2}nd); 2–9—Lemvinsk allochthon, Hayminsk cover: 2—Kharotsk scale: Kharotsk suite (S-D_1hr); 3–8—Parnok scale: 3—Parnok strata (O_{2-3}pr), 4—body of oxidized manganese ores, 5—body of magnetite ores, 6—they are out of scale, 7—body of carbonate-manganese ores, 8—they are out of scale; 9—Pashechorsk scale, pacheshorsk strata (O_2psh); and 10—thrusts (1—Hayminsky, 2—Podrudny, 3—Nadrudny); 11—boreholes.

Carbonate ores are the main industrial and technological type for the Parnok deposit. They represent dense fine-grained formations of dark-gray, red-brown, and cream color; the texture is massive, thin-layered, lensoidal-layered, or breccia-type. The structure of the ores is commonly cryptocrystalline, pelitomorphic, fine-grained, and more rarely coagulation-spherulite and fine-grained. The main rock-forming mineral is rhodochrosite, while kutnohorite and manganocalcite are present in subordinate quantity.

As a result of previous processes of metasomatism expressed in the replacement of rhodochrosite by manganese silicates, the most widely distributed of the secondary minerals are rhodonite, tephroite, bementite, spessartine, manganese stilpnomelane.

The content (%) of manganese in carbonate ores (Table 3.21) varies from 21% to 38% (on average in the deposit—24); MnO_2 is absent; Fe—0.8–3.5; P—less than 0.05; SiO_2—5–24; Al_2O_3—1.5; MgO—0.5–3.5; CaO—10–18; and CO_2—16–25.

Rich-oxidized ores represent an industrial type of ores of the oxidation zone and are developed as noted on the primary carbonate ores. Externally they are distinguished by their black color. The textures of the ores are massive, rubbly, foliated or lensoidal-foliated, nodular, and more rarely sinter-colloform. In terms of mineral composition, the ores are divided into hausmannite, hausmannite-psilomelane, nsutite, and nsutite-cryptomelane. In nature, these ores have fairly sharp geological contacts with the host rocks, forming isolated beds with a thickness from 0.5 to 11 m.

The content (%) of manganese in this type of ores varies from 15 to 57 (on average throughout the deposit—30.5); MnO_2—from 35 to 78; Fe—from 1.5 to 10; P—from 0.02 to 0.3; SiO_2—from 2.2 to 30; CaO, MgO—from 0.5 to 2.5; CO_2—0.5; Al_2O_3—from 3 to 8; S_{total}—0.003–0.004.

Poor-oxidized ores are present together with rich-oxidized ores. They are characterized by dark-gray and chocolate-brown color; textures are earthy and earthy-rubbly (on the strength of which they are called "chocolate coagulations" at the deposit). Minerals of manganese are predominantly X-ray amorphous, with psilomelane evidently as the principal of these minerals; todorokite, vernadite, pyrolusite, and birnessite are noted in subordinate quantity. Non-ore minerals present include quartz, hydromica, muscovite, goethite, chlorite, and montmorillonite.

Poor-oxidized ores were formed primarily on account of the oxidation of manganese-containing host rocks—manganous siltstones and limestones. On the strength of this they feature gradual transitions into manganese-containing coagulations. The content (%) of manganese varies from 10 to 15, MnO_2—8, Fe from 1.5 to 11, P 0.15–0.35 (occasionally up to 1–3), SiO_2 from 20 to 50, CaO and MgO do not exceed 1, and Al_2O_3—8–10. S_{total} is practically absent. The thickness of the ore bodies varies from 1–2 to 5–10 m.

Semi-oxidized ores in terms of outlook are massive or rubbly, black, or spotted, which is caused by uneven distribution in the rock of relict residues of carbonates and newly formed manganese silicates. The mineral composition of semi-oxidized ores is the same as that of rich-oxidized ores, with the addition of rhodochrosite and manganese silicates. The ratio of quantities of primary and secondary minerals varies along the strike of the beds.

The chemical composition of the ores (%): Mn—from 20 to 42, MnO_2—from 10 to 50, Fe—1.5–8, P—0.05–0.2, SiO_2—10–25, CaO can reach 16 (owing to impurity of carbonate minerals), MgO—1–3, S_{total}—0.02–0.1, and Al_2O_3—0.2–8.

Oxidized ferromanganese ores are developed in the infiltration zone and represent a result of the supply of manganese oxides along the fractures into the beds of iron ores. Externally, the ores represent breccia-type spotted-ochreous formations with nests and cross-veins of manganese oxides (principally pyrolusite and psilomelane). The content of manganese in these ores varies from 10% to 30%, of iron from 15% to 35%, P—0.03–0.2%, and SiO_2—15–20%.

Iron ores are closely associated with manganese ores and are substantially manganese-rich. The main rock-forming mineral is magnetite, while greenalite and ferruginous stilpnomelane are present in subordinate quantity. In the upper part of the oxidation zone are developed martite, hematite, goethite, limonite, and other iron hydroxides. The ore bodies are characterized as a rule by distinct borders with the host rocks.

Chemical composition of the ores (%): Fe—from 20 to 75, Mn—2–13, P—0.09–9.4, SiO_2—6–12, CaO and MgO—0.2–0.4, Al_2O_3—0.5–2, and S_{total}—0.04–0.1.

Isotope data and their discussion

The isotopic composition of carbonate rocks of the Parnok deposit varies within broad boundaries: for manganese ores $\delta^{13}C$ (‰, PDB) varies from −17.1 to −1.3, while values of $\delta^{18}O$ (‰, SMOW) range from 19.2 to 23.9; in host carbonates, the values are from −4.4 to 1.2 and from 17.6 to 20.3, respectively (see Table 8.21; Table 3.22, Fig. 3.83). In terms of isotope characteristics, they are close

Table 3.21 Chemical (Mass%) and Isotopic (‰) Composition of Rocks of the Parnok Deposit (Compiled by A.I. Brusnitsyn)

	Manganese Ores											
Components	M1-1/1	M1-1/6	M1-1/7	M1-1/13	M1-1/16	M1-2/5	M2-1	M2-1a	M2-2	M2-3	M2-4	K-25b/5 rose
SiO_2	12.4	3.99	29.9	23.4	12.8	11.9	20.6	13.4	5.47	16.5	9.32	3.62
TiO_2	0.06	0.06	0.03	0.24	0.17	0.14	0.04	0.03	0.04	0.05	0.05	0.08
Al_2O_3	0.38	1.25	1.29	2.56	2.87	4.94	0.21	0.08	<0.05	2.37	<0.05	1.03
Fe_2O_3 total	0.55	1.87	0.34	1.38	2.16	1.17	0.42	0.49	0.49	1.19	0.88	1.93
MnO	65.3	58.7	63.5	56.8	15.8	53.15	64.3	64.0	58.8	55.9	56.2	55.6
MgO	0.20	1.19	0.16	0.58	1.63	4.60	0.38	0.46	0.25	2.01	0.06	2.24
CaO	1.67	2.52	1.19	2.17	37.6	12.7	3.21	1.63	4.57	4.09	4.47	4.02
Na_2O	0.27	0.26	0.42	0.28	0.21	<0.05	0.29	0.28	0.25	0.26	0.56	0.30
K_2O	0.07	0.06	0.09	0.07	0.06	0.06	0.08	0.07	0.06	0.08	0.06	0.07
P_2O_5	<0.05	<0.05	<0.05	<0.05	0.05	0.07	<0.05	<0.05	<0.05	<0.05	<0.05	<0.05
LOI	18.7	29.9	2.74	12.1	26.7	11.3	9.84	19.2	29.7	17.1	28.4	30.7
Total	99.7	99.9	99.7	99.6	100.1	100.0	99.4	99.7	99.7	99.6	100.1	99.6
Corg	1.14	0.58		0.65					1.42		1.39	0.20
$\delta^{13}C$ (PDB)	−17.1	−12.8	−11	−15.9	−1.3	−10.3	−16.5	−15	−10.9	−16.8	−13	−11.4
$\delta^{18}O$ (SMOW)	20.5	23.3	23.9	21.2	20	20.6	20.3	20.5	20.4	20.3	20	21.4

Mineralogical varieties of rocks: *Samples M1-1/1, M1-1/6, M2-2, M2-4, K-25b/5rose, K-25b/56beigeb, K-18d/10-1—rhodochrosite; samples M1-1/7 and M2-1—tephroite (±rhodochrosite), samples M1-1/13 and M2-3—tephroite-rhodochrosite; sample M1-1/16—caryopilite-tephroite-calcite (±spessartine); M1-2/5—calcite-hausmannite; M2-1a—sonolite-tephroite-rhodochrosite; K-25b/7—quartz-rhodonite-rhodochrosite; K-29/1—quartz-rhodochrosite (±siderite); M1-1/17, M1-1/18, and M1-1/20—limestones; B2-2 and B2-3—carbonaceous-clayey-limy shales; B2-4 and B2-7—carbonaceous-siliceous-limy shales.*

to manganese carbonates of the Primagnitogorsk deposits of the Urals group (Kyzyl-Tash, Bikkulovo, and Kusimovskoe) and the Usa deposit (Kuznetsk Alatau).

In terms of the isotopic composition of carbon, carbonates of the host rocks are close to carbonates of "normal" marine sediments. Values of $\delta^{13}C$ in pelitoform limestones constitute 0.5–1.2‰. Serving as the main source of carbon in these carbonates was carbon dioxide dissolved in marine water.

In carbonaceous shales, the values of $\delta^{13}C$ of carbonates are somewhat lower: from −0.7‰ to −4.4‰. The decrease in $\delta^{13}C$ values is due to the fact that in the given situation the carbon dioxide of carbonates contained in the host shales has a twofold origin. Besides the carbon dioxide of marine water, carbon dioxide released with the decomposition of organic matter buried in the sediment also participates in their formation.

Carbonates of manganese ores are characterized by lighter carbon: $\delta^{13}C$ varies from −17.1‰ to −8.9‰. This range is characteristic for carbonates from many manganese deposits and corresponds to ores of diagenetic and catagenetic origin. The bulk of the carbon dioxide necessary for their formation was generated in the sediment by microbial destruction of organic matter.

The values of $\delta^{18}O$ of calcite and rhodochrosite from the Parnok deposit (from 17.3‰ to 23.9‰) are lower than those found in most diagenetic carbonates but fairly typical for minerals of metamorphic

					Host Rocks						
K-25b/5 beige	K-25b/6	K-25b/7	K-18d/10-1	K-29/1	M1-1/17	M1-1/18	M1-1/20	B2-2	B2-3	B2-4	B2-7
6.53	8.20	43.4	6.02	5.75	9.51	4.80	7.18	62.9	48.6		20.96
0.05	0.09	0.24	0.05	0.02	0.08	0.02	0.10	0.43	0.83		0.13
0.51	0.75	3.11	0.78	0.34	1.23	0.56	1.91	10.2	13.4		2.85
3.71	2.68	4.47	3.27	16.3	0.70	0.15	0.38	4.75	6.56		5.34
56.7	51.6	32.2	48.0	51.3	1.90	3.06	2.61	0.34	0.48		5.78
1.34	1.75	2.12	1.04	0.22	0.80	4.72	3.81	1.44	3.45		0.83
3.74	6.07	7.40	10.7	3.30	47.81	47.8	47.0	7.95	12.7		36.6
0.34	0.29	0.26	0.25	<0.05	<0.05	<0.05	<0.05	<0.05	0.51		0.42
0.10	0.08	0.08	0.13	0.07	0.73	0.03	0.76	3.36	4.63		0.48
0.11	0.12	0.06	0.05	0.07	<0.05	<0.05	<0.05	0.2	0.11		1.05
26.4	28.1	7.06	29.6	22.6	37.2	38.9	36.2	8.49	9.41		26.2
99.5	99.7	100.4	99.9	100.1	100.1	100.1	100.1	100.1	100.6		100.6
0.13	0.08		<0.03			0.56		3.54	0.17		1.75
−14.4	−11.5	−9.0	−8.9	−11.7	1.2	0.5	0.6	−4.4	−1.3	−0.7	−3.6
21.4	21	20.7	21.7	19.2	20.3	19.7	20.3	17.3	18.4	17.6	19.4

and metasomatic origin. These data closely match the estimates for the parameters of regional metamorphism of rocks of the deposit ($T \approx 250°C$, $P \approx 2.5$ kbar) (Brusnitsyn et al., 2009).

Our obtained values of $\delta^{13}C$ and $\delta^{18}O$ for carbonates of the Parnok deposit are close to the results of the research of other authors (Shishkin and Gerasimov, 1995; Zykin, 2002; Silaev, 2008).

One of the distinctive features of rocks of the Parnok deposit, as for many sedimentary-diagenetic manganese deposits globally, is the direct correlation between the isotopic composition of carbon and the content of manganese (Fig. 3.84), and the inverse correlation of the isotopic composition of carbon and the content of calcite (Fig. 3.85). This provides evidence of participation by oxidized carbon of organic origin in the concentration of manganese in the initial sediment evidently predominantly at the early diagenesis stage. However, analogous relationships between the isotopic composition of carbon and the contents of manganese and calcite have also been noted in deposits of metasomatic genesis (Kuleshov and Bych, 2002).

In rocks of the Parnok deposit, the constant presence of *organic carbon* has been established. The content of C_{org} in host rocks (0.56–3.54% by weight) and in manganese ores (0.08–1.42%) corresponds to the level of weakly metamorphosed "black shales."

Table 3.22 Chemical (Mass%) and Isotopic (‰) Composition of Rhodochrosite of Manganese Ores of the Parnok Deposit, Magnitnyi-1 Section (Compiled by A.I. Brusnitsyn)

	Pr-31-97	Pr-33-97	Pr-51-98	Pr-52-98	Pr-53-98	Pr-54-98	Pr-59-98
Sample	**5349**	**5350**	**5351**	**5352**	**5354**	**5355**	**5356**
SiO_2	20.8	35.2	21.3	19.9	21.5	15.9	13.8
TiO_2	0.165	0.152	0.433	0.437	0.204	0.216	0.171
Al_2O_3	0.52	1.13	4.32	5.01	1.02	1.44	0.139
Fe_2O_3 total	0.854	1.15	2.81	1.93	0.882	0.894	0.393
MnO total	64.5	55.5	39.7	53.4	64.7	61.9	72.00
MgO	<0.05	0.83	1.54	0.772	0.125	0.126	<0.05
CaO	0.654	0.56	13.6	4.7	2.5	2.14	1.02
Na_2O	0.15	0.135	0.115	0.115	11.4	0.12	0.223
K_2O	0.087	0.097	0.104	0.084	0.894	0.121	0.082
P_2O_5	<0.05	<0.05	0.068	0.557	<0.05	<0.05	<0.05
LOI	12.2	5.21	15.9	13.5	8.77	16.8	12.2
Total	100	99.9	99.8	99.9	99.8	99.6	100
$\delta^{13}C$, ‰ PDB	−11.6	−13.0	−11.3	−10.4	−12.1	−12.6	−17.0
$\delta^{18}O$, ‰ SMOW	20.5	19.6	19.8	20.8	20.1	19.3	20.5

Chemical analyses were conducted in the VSEGEI laboratory of chemical and spectral analysis using the X-ray-spectral fluorescence method under ARL-9800 spectrometer (Switzerland); analyst: *B.A. Tsimoshenko.*

On the genesis of the deposit

Detailed study of the chemical and mineral composition, textural particularities, and geological structure of ore bodies and host deposits has enabled the identification of several stages of formation of the deposit (Shishkin and Gerasimov, 1995). During the first stage—hydrothermal-sedimentary—occurred the formation of the initial manganese-containing carbonate deposits of the Parnok strata. The cited authors have accepted hydrothermal solutions as the source of ore matter, despite the absence of hydrothermal features in the sequence of the rocks of the deposit.

With the second stage of the formation of the deposit, the cited authors link the transformation of initial carbonate ores with the impact of hydrothermal-metamorphic processes, which led not only to mechanical (boudinage, cleavage, slickensides, etc.), but also to recrystallization and formation of new mineral associations (silicates and aluminosilicates of manganese and iron, carbonates, quartz, pyrite, etc.).

The third stage is connected with the establishment of a weathering crust including rich ores of manganese oxides.

On the whole, the summary of our obtained data matches well with the hydrothermal-sedimentary model for the genesis of the deposit (Brusnitsyn and Kuleshov, 2011; Brusnitsyn et al., 2013). The specifics of the studied object include the realization of the hydrothermal process in the environment of a sedimentary basin with the absence of clear indicators of volcanism. A discharge of ore-bearing solutions occurred in a relatively closed depression-trap, within the boundaries of which has periodically arisen an anaerobic atmosphere.

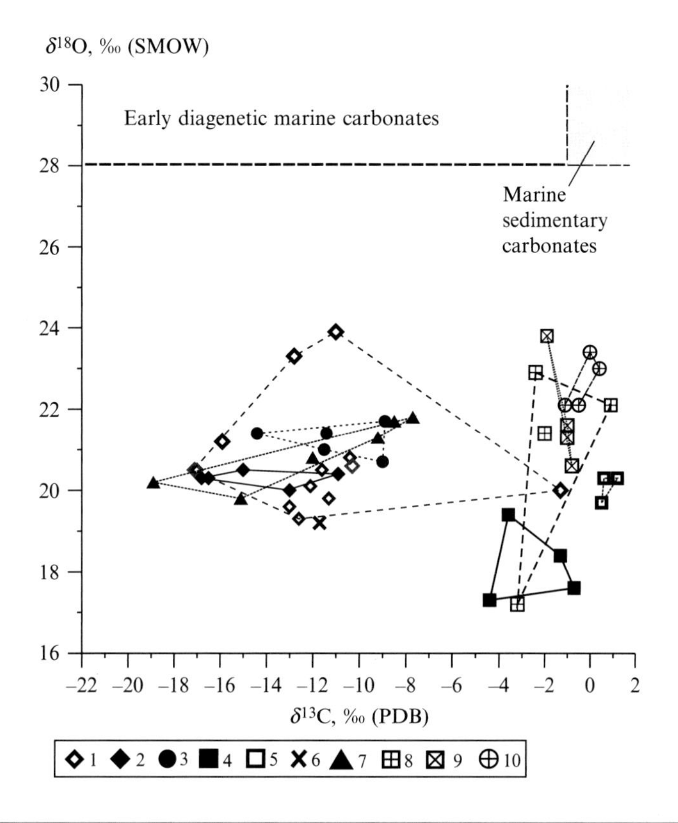

FIG. 3.83

$\delta^{13}C$ versus $\delta^{18}O$ in rocks of Parnok manganese deposits. 1–4—Manganese-ore deposit, sectors: 1—Magnetic-1, 2,—Magnetic-2, 3,—East-2, 4,—East-4; 5–6—host rocks: 5—limestones, 6—carbonaceous-siliceous-calcareous shales; 7–10—data (Shishkin and Gerasimov, 1995): 7—manganese ore; 8–10—host rocks: 8—gray pelitomorphic manganese limestones (member C-1), 9—carbonaceous clay-calcareous schists (member C-2), and 10—dark-gray loopy and lenticular-layered silty limestones (member C-3).

Hydrothermal solutions supplied the surface of the marine bed with iron, manganese, calcium, and certain rare elements. Here as a result of the changes of Eh-pH parameters occurred a differentiation and precipitation of ore matter. The mechanism of these processes is well studied in modern basins as well as in ancient sedimentary and volcanogenic-sedimentary strata (Maynard, 2003; Brusnitsyn, 2009).

Under prevailing "normal" aerobic conditions, iron is accumulated near the hydrothermal source, and manganese at a certain distance from it, where the solutions are maximally diluted by marine water. In the periods of development of anaerobic atmosphere, the values of Eh-pH of the benthic waters were conducive to the precipitation of iron, while manganese was retained in dissolved form. The longer the stagnant period was maintained, the greater the quantity of manganese was accumulated in the water mass. With a change in the hydrodynamic regime and an enrichment of benthic waters by oxygen occurred the oxidation of the manganese concentrated in them and its transfer in a solid state into the sediment. Regardless

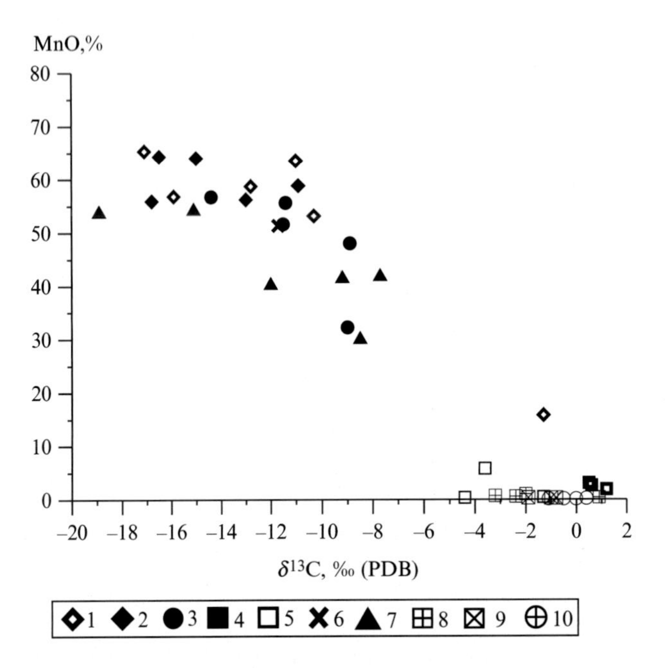

FIG. 3.84

Relationship between $\delta^{13}C$ values and MnO content in the ores and host rocks of the Parnok deposit. Legend see Fig. 3.83.

of which of the mentioned variants (or both) were realized in the Parnok deposit, a change (in space or in time) in the oxidizing-reduction characteristics of the water mass should have occurred fairly slowly. On this account, a maximum differentiation of metals will occur. Additionally, a tranquil hydrodynamic regime prevented the mechanical remixing of metal-bearing oozes of different composition.

The calcium supplied by hydrothermals was diffused in the marine water and, reacting with the carbon dioxide dissolved in the water, gave rise to calcite. Gradually settling, thinly dispersed grains of calcite formed carbonate oozes that served as protolith for pelitoform limestones. Characteristically, as for other sediments of the deposit, the limestones are rich in elements that mark the development of a hydrothermal process (Mn, As, Ba, Cu, Pb, Sb, Sr, Zn).

Processes of post-sedimentational transformation of ore-bearing deposits took course in reductive conditions, created on account of the decomposition of organic matter buried in the sediment.

In the initial sediment, iron and manganese were accumulated predominantly in oxide form. With subsequent dehydration and recrystallization of iron hydroxides, magnetite was formed. Oxides and hydroxides of manganese by means of a chain reaction with participation by organic matter were transformed into rhodochrosite. Relicts of manganese oxides were preserved in the form of interbeds of hausmannite and pyrochroite. Additionally, the initially oxide form of accumulation of metal is indicated by the presence in the ores of a positive ceric anomaly. The formation of manganese silicates is a result of later processes of cata- and metagenesis of the ores.

3.2.2.3 *Deposits of hydrothermal-sedimentary genesis of other regions*

Also numbered among the model examples of formation of ores of hydrothermal-sedimentary type are undoubtedly the deposits of the *Atasui region of central Kazakhstan* (Karazhal, Ktai, Kamys, Zhumart,

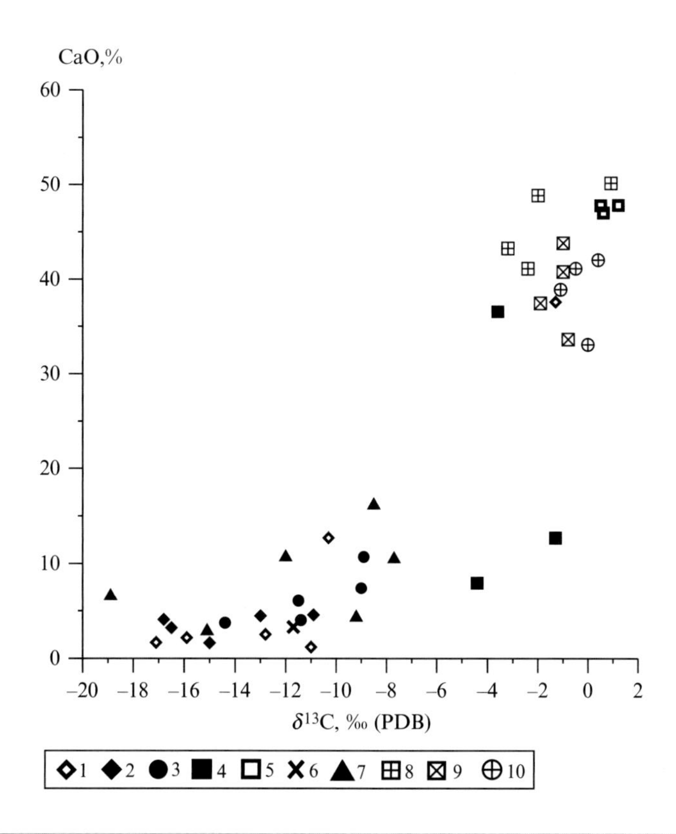

FIG. 3.85

Relationship between $\delta^{13}C$ values and CaO content in ores and host rocks of the Parnok deposit. Legend: see Fig. 3.83.

Zhairem, Ushkatyn, and others), complex deposits with which, besides ferromanganese, are genetically closely connected as well barite-lead-zinc deposits (eg, at Zhairem and Ushkatyn). However, in the isotope regard, they remain little studied.

At present, the principal regularities of the formation of deposits of the indicated region are explained fairly fully and are well elucidated in the works of a range of researchers (F.V. Chukhrov, Iu.A. Khodak, E.M. Gribov, A.A. Rozhnov, E.I. Buzmakov, G.N. Shcherba, A.B. Veimarn, V.V. Kalinin, D.G. Sapozhnikov, A.A. Gurevich, and others).

Deposits of the Atasui region in the tectonic regard are associated with the Dzhailmin graben-syncline (trough), the bed and framing of which are represented by volcanogenic-sedimentary continental and shallow-water marine strata of Givetian and Frasnian stages of the Devonian, and the trough itself is composed of a complex of Middle Paleozoic marine sedimentary rocks. Mineralization is connected principally with the Late Devonian (Famennian stage) phase of ore development and consists of marine siliciclastic-carbonate deposits of the Upper Devonian containing in some quantity carbonaceous material and an impurity of volcanogenic material.

In this region are concentrated the principal reserves of iron and manganese ores of the Famennian (approximately 20 deposits and as many ore-occurrences) (Fig. 3.86). Almost all deposits coincide with siliceous-clayey-carbonate deposits of the stagnant depressions of the marine bed. Observed is a cyclical

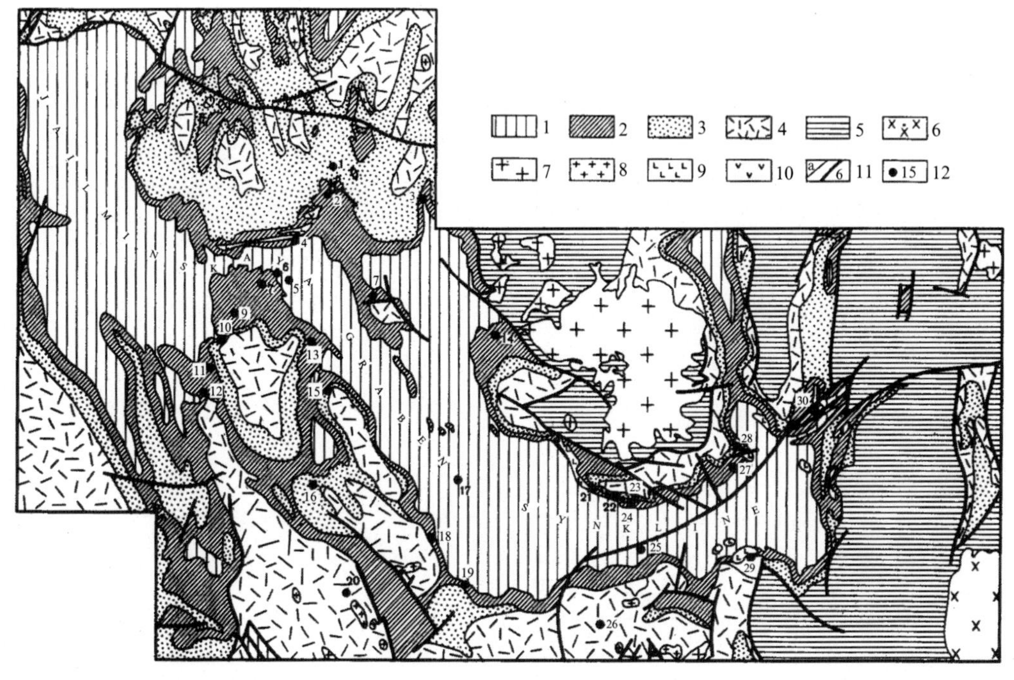

FIG. 3.86

Schematic geological map of Atasuysk ore district (Roznov et al., 1980). 1—Lower Carboniferous (C_1)—limestones, mudstones, sandstones; 2—Famennian (D_3fm)—limestone, clay-siliceous-carbonate rocks with layers of iron-manganese ores, tuff, tuffites; 3—Dayrinsk suite (D_3dr)—red conglomerates, siltstones, sandstones, lenses trachydatic porphyrites; 4—Lower—Middle Devonian ($D_{1\text{-}2}$)—terrigenous-volcanogenic deposits; 5—Lower Paleozoic (Pz_1)—metamorphosed volcanoclastic sediments; 6—granites (C_2); 7—granitoids ($D_{2\text{-}3}$); 8—subvolcanic bodies of quartz porphyry; 9—diabase dykes of porphyry; 10—gabbro-diabase; 11—geological boundaries (a) and faults (b); 12—deposits and ore occurrences: 1—Ushkatyn-I, 2—Ushkatyn-III, 3—Karsadyr, 4—North Zhairem, 5—Eastern Zhairem, 6—West Zhairem 7—Dal'nezapadny Zhairem, 8—Veernyi, 10—mt. Zhomart, 11—Middle Zhomart, 12—Zhomart, 14—Damidovsky, 15—Akkuduk, 17—Taskuduk, 18—North Bestau, 19—Bestau, 21—Lal'nezapadny Karazhal, 22—West Karazhal, 23—East Karazhal, 24—South Karazhal, 25—Aschily, 27—South Ktay, 28—Big Ktay, 29—South Klych, and 30—Bast'yube.

occurrence in sediment accumulation, expressed in the alternation of horizons of flysch structure composing the base of large rhythms with horizons of nodular-layered siliceous limestones, which contain ferromanganese ores and are deposited in the upper parts of the rhythms. The latter were formed in a more tranquil tectonic regime with intensive manifestation of exhalational-hydrothermal activity.

Industrial ferromanganese mineralization is localized predominantly in five horizons (at maximum up to 12 are counted) of varying (1–20 m) thickness. The maximum mineralization is developed in a member of red-colored limestones (D_3fm_2) of nodular-layered texture. Beds of ferromanganese ores are commonly interstratified with horizons of siliceous limestones and represented by thin alternations of layers of ore minerals (main): hematite, magnetite, stilpnomelane, braunite, hausmannite, and jacobsite with mineralized siliceous limestones and ferruginous jaspers.

It is recognized that the Atasui deposit was formed during the process of activity of long-lived deep hydrothermal systems, discharged onto the surface of the lithosphere along local weakened tectonic zones. Ore deposition occurred in several stages. During the sedimentational-diagenetic stage, the hydrothermals seep onto the bed of the marine basin. Near the gryphons were deposited ore sediments (sulfides of Fe, Pb, and Zn in conditions of low Eh and oxides of Fe and Mn in conditions of high Eh) forming concentric zones around the mouths of the sources. The composition of the ore sediments reflected the geochemical mobility of the elements (SiO_2, Fe, and Mn) composing a given zone; here is developed a concentric or semi-concentric type of zonality, expressed in the subsequent replacement of zones of ferruginous jaspers, iron ores, and manganese ores by measure of distance from a point or linear-fracture channel of entry into the basin of sedimentation of ore-bearing solutions.

In the Zhairem deposits, quantitative analysis of the distribution of the absolute masses of iron and manganese demonstrates that for all ore bodies the maximum of the masses forms ringed zones around the deposits' common center. Maximums of manganese are situated further from the center than are maximums of iron. In the mineralogical regard, a tendency has been established toward replacement from the center of zonality toward the periphery of ferruginous jaspers by hematite-magnetite ores and further by low-grade magnetite-silicate and siderite-silicate ores (chlorite and ferristilpnomelane), and subsequently by nodular-layered light-gray siliceous limestones.

By measure of burial of gryphons under strata of supraore deposits, the hydrothermal systems acquired a closed character with the metasomatic and vein ores of secondary and tertiary stages particular to them.

The concentric zonality in the Zhairem deposits is inherent as well to superimposed hydrothermal-metasomatic barite-lead-zinc mineralization. From the center to the periphery, the degree of hydrothermal restructuring of the rocks fades, and quartz metasomatites are subsequently replaced by barite, galena-barite, sphalerite-galena-barite, and non-barite sphalerite-galena metasomatites (Fig. 3.87). The focus of mineralization for both sedimentary ferromanganese and superimposed barite-sulfide

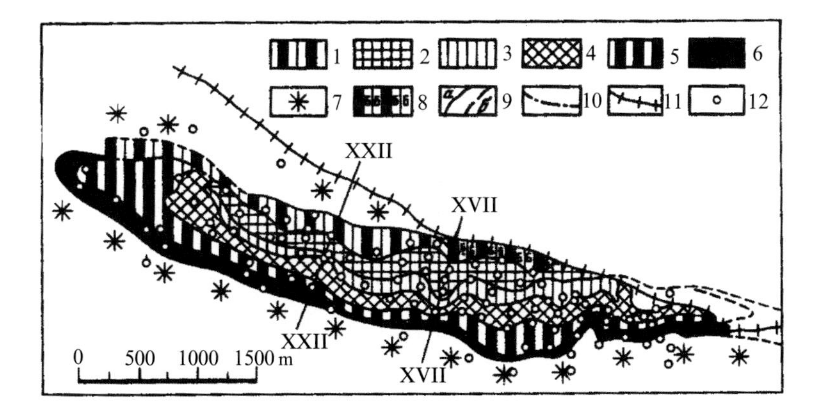

FIG. 3.87

Distribution (in terms) of mineral types of iron and iron-ore deposits in the West Karazhal (central Kazakhstan) (Roznov et al., 1980). 1—Ferruginous jasper; 2—magnetite ores; 3—magnetite-hematite ores; 4—hematite ores; 5—ferromanganese ores; 6—oxide-manganese ores; 7—red-mineralized cherty limestones; 8—barite-bearing ferruginous jasper; 9—border mineral ore types: a—reliable, b—estimated; 10—tectonic faults; 11—erosional surfaces of ore-bearing deposits; 12—drilling wells.

mineralization coincides. In the Karazhal deposit, a jasper zone situated in the center of mineral zoning also coincides in a significant plane with an area of superimposed barite mineralization.

The source of the mineral hydrothermals has by many researchers (A.A. Rozhnov, E.I. Buzmakov, V.I. Shchibrik, G.N. Shcherba, and others) been linked with magmatic chambers that feed the volcanism, synchronous to ore-genesis, of a contrast-differentiated subalkaline-basalt-alkaline-liparite formation. A volcanogenic-hydrothermal (endogenous) source of ore matter is supported by the paragenesis of ores of iron, manganese, lead, zinc, barium, and in part copper. Evidence for this can also be seen in the isotope data (Table 3.23).

As noted, the formation of the deposit bores a multi-act (polychronous) character—beginning with accumulation of ore-bearing oozes at the stage of sedimentogenesis, redistributions of matter in diagenesis, the transformation of the formed sedimentary and sedimentary-diagenetic rocks, and subsequent mineral precipitation (ore and non-ore) subjected to hydrothermal processes after burial by younger

Table 3.23 Isotopic Composition of Carbon and Oxygen of Carbonate Rocks of the Manganese Deposits Ushkatyn-III and Eastern Zhairem (Central Kazakhstan)

Number of Analysis	Number of Sample	Sample Location	Depth	Sample Characteristics	$\delta^{13}C$, ‰ (PDB)	$\delta^{18}O$, ‰ (SMOW)
Ushkatyn-III deposit						
3275	74-88	Well 9234, first shaft, profile XIX	Depth 654.3 m	Overlapping ore member variegated limy siltstones	4.2	20.7
3276	75-88		Depth 661.0 m	Ore member. Cross vein of pink calcite in hausmannite ore	−6.4	20.9
3277	76-88		Depth 663.1 m	Horizon of oxides of Fe and Mn, cross-vein of calcite	−7.6	19.2
3278	77-88		Depth 663.5 m		−10.4	18.8
3280	79-88		Depth 673.0 m	Interbed of pink limestone	4.0	18.6
3281	80-88		Depth 692.0 m	Horizon of weakly carbonatized oxides of Fe and Mn	0.1	21.4
3282	81-88		Depth 700.5 m	Gray silty limestone	1.7	20.7
3284	83-88		Depth 708.8 m	Strongly hematitized limestone	0.4	19.5
3285	84-88		Depth 721.0 m	Clastic limestone	1.6	19.9
3286	85-88		Depth 753.5 m	Gray limestone with dendrites of hematite	1.8	21.7
3287	86-88		Depth 757.0 m	Carbonatized oxide Fe-Mn ore	−7.1	19.1
3288	87-88		Depth 774.0 m	Hausmannite ore with cross-veins of pink calcite on the contact with host gray limestone	2.0	18.1
PI	88-88			(a) limestone	−5.6	19.1
				(b) pink calcite	−7.2	19.2

Table 3.23 Isotopic Composition of Carbon and Oxygen of Carbonate Rocks of the Manganese Deposits Ushkatyn-III and Eastern Zhairem (Central Kazakhstan)—Cont'd

Number of Analysis	Number of Sample	Sample Location	Depth	Sample Characteristics	$\delta^{13}C$, ‰ (PDB)	$\delta^{18}O$, ‰ (SMOW)
3291	90-88	Well 9234, first shaft, profile XIX	Depth 778.0 m	Banded oxidized Fe-Mn ore	1.8	19.5
3292	91-88		Depth 782.0 m	Hematite-carbonate rock of concretionary structure	2.0	20.6
3293	92-88		Depth 788.8 m	Limestone dark-gray	1.3	21.6
3294	93-88		Depth 791.3 m	Jacobsite ore. Calcite cross vein	−9.8	20.9
3295	94-88		Depth 797.8 m	Manganese oxides, weakly carbonatized	−1.6	21.9
3296	95-88		Depth 798.3 m	Carbonatized manganese-oxide ore. Calcite	1.2	20.8
3297	96-88		Depth 799.1 m		3.6	21.2
3308	105-88		Depth 811.0 m	Dark-gray silty limestone	−4.6	17.2
Ushkatyn-II central part						
3309	106-88	Well 9605. Profile II	Depth 934.5 m	Braunite ore with cross vein of calcite. Calcite	−7.0	17.3
3320	117-88		Depth 947.0 m	Braunite ore with cross vein of calcite. Calcite from cross vein	0.3	20.1
Eastern Zhairem deposit						
		Well 3417	Depth 573.8 m	Black and dark-gray "knotty" carbonate with cross vein of calcite. Host carbonate	−6.4	18.9
		Cross vein of calcite			−1.6	15.3
3321	118-88		Depth 557.5 m	Black and dark-gray "knotty" carbonate with cross vein of calcite. Host carbonate	−0.4	8.4
3322	119-88		Depth 559.0 m	Strongly carbonaceous limestone with cross-veins of calcite. Limestone	0.1	13.0

deposits of the chambers of subaqueous discharge of hydrothermals. In this regard, the deposits of the Atasu group of Kazakhstan are close to the above-examined Bikkulovo deposit in the South Urals.

A substantial role in the subsequent transformations is played by the supplying non-ore fluids that do not contain carbon dioxide, which led to the full alteration of the isotopic composition of the oxygen of the initial sedimentary carbonates of the host sequence (of the overlapping carbonates as well as those interstratified with the intra-ore member).

A significant group of deposits of hydrothermal-sedimentary genesis is distributed within the boundaries of *China*. Chinese geologists have identified a volcanogenic-sedimentary genetic type of deposits (Ye et al., 1988; Editorial Committee of Mineral Deposits of China, 1995; Fan and Yang, 1999). Deposits of this type were formed predominantly in eugeosynclinal basins such as South Tsinling, Tsilian, and Tianshan. In the latter two are enclosed medium and small deposits. Depending upon associations of ore and non-ore matter, three suptypes have been identified: spilite-mudstone (phyllite)-carbonate (Lijian deposit, Early-Middle Proterozoic), sandstone-mudstone-jaspilite (Motuoshala Fe-Mn deposit, Early Carboniferous), and siliceous-clayey-volcanogenic (Heixiakou deposit, Middle Cambrian). Thus can also be classified the deposits of Xialey (Late Devonian) and Longtou (Early Carboniferous).

The most studied and well known in the literature is the Motuoshala deposit. It is situated 60 km west of Balguntay city, in the Hejing region of the Xinjiang metallogenic province; in the tectonic regard, it coincides with the Motuoshala syncline (Tianshan fold belt) and is enclosed in the volcanogenic-sedimentary deposits of the Early Carboniferous that contain jaspilite associations. In terms of reserves it is classified as a medium deposit.

The ore member has a stratiform and lensoidal structure, underlain by siliciclastics and argillitic siliciclastics with interbeds of fine-grained sandstones, and is composed of three horizons: in the base (2–46 m) by an iron-ore horizon represented by hematite with interbeds of jaspilites and chert with a thickness of 2–24 m; above this is deposited a horizon (2–62 m) of interstratification of ferruginized sandstones, siliciclastics, hematite jaspilites with interbeds of rhodochrosite in the floor; the upper horizon (1–51 m) is represented by interstratified Mn-containing interbeds in sandstones and argillitic siliciclastics with hematite jaspilites. Above are deposited non-ore argillite sandstones with interbeds of chert and clays.

Iron ores are represented by hematite and layered jaspilite hematite rocks with interstratified siliceous shales and barite-containing rocks. Manganese horizons are deposited above the iron-ore deposits, forming three lensoidal bodies with a length of 0.8–1 km and a width of 0.3–0.5 km and a general thickness of 0.7–30 m (with a mean of 5.8 m). The ore mineral is represented by microspheroidal rhodochrosite, in the upper part of the ore member by oxides of manganese. Hydrothermally altered oxide ores are represented predominantly by micro- and fine-grained braunite (spherulites of rhodochrosite are replaced by braunite with cross-veins of hausmannite, braunite, specularite, magnetite, kanoite, bementite, barite, and sulfides). Hypergenic manganese oxides are represented by psilomelane, pyrolusite, goethite, and hydrogoethite.

Developed ores are represented by manganese oxides (mean content of Mn—18.77) and iron (mean content of Fe—7.17%) and contain high contents of Pb, Zn, and S.

The deposit was formed, according to Chinese researchers (Editorial Committee of Mineral Deposits of China, 1995), in the subtidal clastic facies of a shallow-water depression at a distance of 3–5 km to the east from the volcanic center. In this direction has been established a sequential replacement of rocks: basalt lavas and lava breccias—interbedding of tuffs with jaspilites and basalts—volcanogenic-sedimentary series: Fe-Mn-ore member, tuffites, and mineralized siliceous shales.

The study of the isotopic composition of oxygen in hematites (Wang Youbiao, 1985; cited in Editorial Committee of Mineral Deposits of China, 1995) has led to the conclusion that hematites of steel-gray color ($\delta^{18}O = -0.46$ to $-1.25‰$) were formed by sedimentary means in conditions of equilibrium with marine water ($\delta^{18}O = -0.5‰$ to $0.5‰$) in the usual conditions of a marine basin. The isotopic composition of red hematites is heavier: $\delta^{18}O = 4.86–8.40‰$. This serves as evidence for higher temperatures of formation of jaspilites and hematites—134–158°C. Vein quartz and specularite ($\delta^{18}O = 12.71‰$) were formed at a temperature of approximately 314°C.

Thus, a characteristic particularity of deposits of the hydrothermal(volcanogenic)-sedimentary genetic type is a supply to the basin of sedimentation of sedimentary and volcanogenic-sedimentary rocks of thermal (low- and mid-temperature) solutions rich in ore components. These solutions can have different origins and could have been connected with different geological processes (effusive and intrusive volcanism; tectonic activity of the region leading to the movement of mineralized deep catagenic waters of elisional genesis; petroleum waters; ore-bearing brines; etc.). All of the variations in geological structure of the deposits of this type and of associations of host rocks and ore components are determined by the specifics of the geological development of the (volcanogenic)-sedimentary-rock basin itself.

In many cases, hydrothermal-sedimentary ore deposition (the formation of ore-bearing oozes) is replaced by specifically hydrothermal ore-formation. This occurs in those cases where the site of discharge of subaqueous hydrothermals (gryphons) is overlapped by younger sediments and where the hydrothermal ore-bearing solutions entering the sediment are discharged inside the ooze sediment (of the examples considered, these include the Bikkulovo deposit and certain deposits of the Atasu region of Kazakhstan). The morphology of the zones of mineralization can be the most diverse—from stratiform bodies to vein and stockwork bodies.

3.2.3 HYDROTHERMAL DEPOSITS OF MANGANESE

In nature, processes of hydrothermal formation of manganese-containing and manganese rocks, rarely ores, are very widely distributed. They are known in Chile, Morocco, Korea, Japan, China, United States, Armenia, Kazakhstan, in many fold belt regions of Russia, and in a range of other countries. Deposits and ore-occurrences of this type form principally vein, rarely stratiform accumulations. By dimensions and reserves, they are with rare exception insignificant and in the economic regard are of low profitability.

Manganese minerals in rocks and ores of the examined type can form particular ore accumulations (oxides, carbonates, sulfides, silicates) and also can be included in the composition of other minerals of hydrothermal association. Manganese oxides of hydrothermal deposits are frequently rich in Ba, Sr, Ag, Pb, W, and Ti, and more rarely Be, As, and Sb. Manganese minerals are spatially and genetically closely linked with barite, fluorite, calcite, sulfides of nonferrous metals, and gold-silver mineralization and are associated with low- and mid-temperature phases of hydrothermal activity.

Manganese carbonates are commonly associated with sulfide ores of nonferrous metals and are represented predominantly by rhodochrosite, more rarely by manganocalcite and kutnohorite, frequently in association with manganese silicates (rhodonite, bustamite, tephroite, friedelite, alleghanyite) and manganese sulfide (alabandite).

A detailed survey of known deposits of manganese of hydrothermal genesis and various aspects of the mechanism of their formation are reported in the work of S. Roy (Roy, 1981). The essential information on the deposits and ore-occurrences of countries of the former USSR are contained, as noted, in a range of scientific anthologies and monographs (Margantsevye et al., 1966, 1978, 1982; Gavrilov, 1967, 1972; Margantsevo-zhelezisto-kremnistaya et al., 1981; Avaliani, 1982; Geologiia i geokhemiia et al., 1982; Sokolova, 1982; Margantsevoe rudoobrazovanie et al., 1984; Vulkanogenno-osadochnye et al., 1985; Zheleznye et al., 1985; Usloviia obrazovanie et al., 1984; and others).

The hydrothermal genesis of the deposits is fairly easily established. Thus, S. Roy (Roy, 1981) has identified a range of geological and geochemical indicators characteristic for deposits of this type: (a) the vein character of ore bodies transecting the structure of the host rocks; (b) in veins composed of oxides, silicates, carbonates, and sulfides is noted the constant presence of minerals such as barite, fluorite, and various polymorphic modifications of silica and calcite characteristic for hydrothermal deposits; (c) in vein deposits, carbonates, silicates, and sulfides of manganese are associated with hydrothermal deposits of Pb, Zn, Ag, and Au; (d) a spatial and temporal proximity has been established for the formation of veins composed of oxides of manganese oxides, barite, and fluorite, and of veins of sulfides of nonferrous metals with notable quantities of gold and silver. Furthermore, concentrations of W, Pb, Cu, Mo, Tl, As, and Sb in minerals of manganese oxides and gold- and silver-containing sulfides are very close; (e) host rocks in the majority of deposits are represented by layered volcanogenic rocks of various composition, from rhyolites to basalts; (f) in altered wall rocks in vein deposits are present adularia, calcite, epidote, sericite, clayey minerals, and quartz; (g) research on liquid inclusions in rhodochrosite and other syngenetic minerals provides evidence of temperatures characteristic for hydrothermal systems; and (h) characteristic for hypogenic manganese oxides is an enrichment in wolframite and thallium.

To all appearances, to this list can be added yet another indicator—j, the isotopic characteristics of a range of elements (carbon, oxygen, sulfur, hydrogen, and others) of vein minerals. It is well known that in the isotope regard the ore-forming fluids of hydrothermal systems that lead to the formation of certain minerals are substantially distinguished from those of exogenic systems. The isotope data for hydrothermal deposits are widely used in establishing the nature (the source) of ore and non-ore matter, the stadiality and temperatures of mineral precipitation, the source of the water of hydrothermal solutions, and other characteristics.

The isotope characteristics of hydrothermal systems for a range of elements (sulfur, carbon, oxygen, and hydrogen) have been well laid out in the scientific literature, far exceeding the framework of the present monograph. Therefore, as an example of manganese deposits we will report data on the isotopic composition of carbon and oxygen in calcites and limestones of the manganese-ore strata of the Ushkatyn-III deposit (well 9234) and the Eastern Zhairem deposit (well 3417) of the Atasu region of central Kazakhstan (see Table 3.23; Fig. 3.88). From these data, it follows that values of $\delta^{13}C$ and $\delta^{18}O$ on the whole support a hydrothermal genesis of secondary calcites and an intensive hydrothermal-metasomatic transformation of the host sedimentary carbonates. Thus, the multicolored limy siltstones overlapping the ore member are characterized by heavy values of the isotopic composition of carbon ($\delta^{13}C=4.2‰$) characteristic as a rule for carbonates of shallow-water marine basins that were formed in arid or semiarid conditions (close to the accumulation of evaporite formations). Values of $\delta^{13}C$ in calcites from cross-veins of the ore horizons and those associated with oxides of iron and manganese are characterized by low values (−4‰ to −10‰) particular to carbonates of deep genesis (Galimov, 1968; Kuleshov, 1986a). Similar C and O data have been reported from the Permian-age Guichi deposit of China (Xie et al., 2014), who favored a hydrothermal origin based on REE.

A characteristic particularity of carbonate matter of rocks of the studied wells of the Ushkatyn-III and Eastern Zhairem deposits are the very low values of $\delta^{18}O$ for host limestones as well as for secondary calcite (vein calcite and calcite associated with oxides of manganese and iron), which vary within the interval 18–21‰. Such $\delta^{18}O$ values are characteristic for carbonates of hydrothermal genesis and are not particular to sedimentary carbonate rocks of either marine or freshwater basins. This serves as evidence for intensely occurring processes of hydrothermal-metasomatic transformation of terrigenous-carbonate rocks of the host sequence under the influence of deep fluids, leading to the practically full alteration of the initial isotopic composition of the oxygen. It cannot be excluded that the nodular textures of the host carbonates are also a result of these processes.

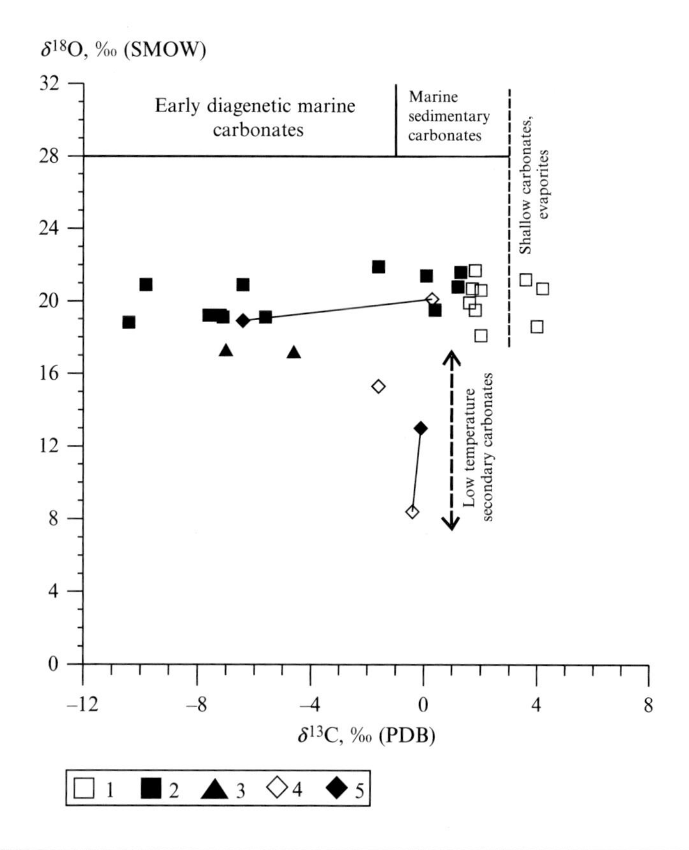

FIG. 3.88

$\delta^{13}C$ versus $\delta^{18}O$ in sedimentary carbonate rocks and hydrothermal calcite certain manganese deposits of Central Kazakhstan. 1–2—Ushkatn-III deposit, well 9234: 1—host limestones, 2—calcite from veins of iron ores; 3—Ushkatn-II deposit, well 9605, calcite from veins of braunite ores; 4–5—Eastern Zhairem deposit, well 3417: 4—host limestones, 5—calcite vein in the limestone.

Important information on the genesis of the water that participates in the hydrothermal process is provided by the data on the isotopic composition of hydrogen (δD). The results of isotope research in a range of geological objects, including manganese ore, have established that in fluids of volcanogenic and hydrothermal processes developed in the regions of modern volcanism, the principal role is played by water of meteoric genesis, with some quantity (admixture) of marine water, and possibly of endogenic (deep, mantle) genesis.

Thus, for example, the isotopic composition of hydrogen (δD) of the hydroxide group of manganite of the Tertiary deposit of Kuroko (Japan) serves as evidence for a meteoric origin of water of ore-bearing hydrothermal solutions (Hariya and Tsutsumi, 1981). The formation of manganite here occurred in equilibrium with meteoric waters at temperatures no higher than 250°C.

With plutonic processes is closely connected the formation of *contact-metasomatic and skarn deposits* of a range of elements, including manganese. Their formation is a result of a hydrothermal-metasomatic restructuring of the enclosing, frequently manganese deposits at the contact with

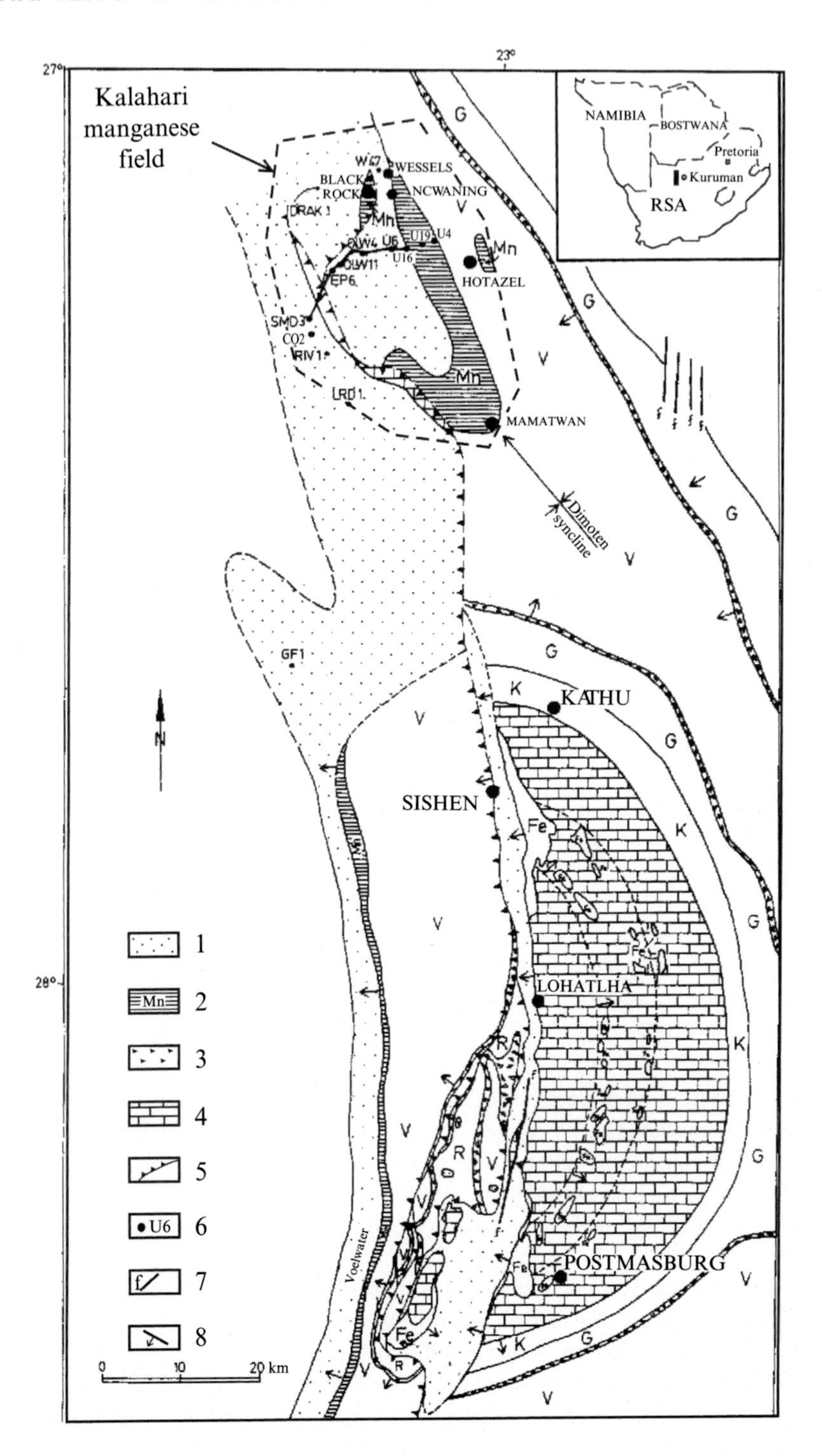

FIG. 3.89

See the legend on opposite page.

magmatic rocks. Manganese ores of this genetic type are of low economic value. The isotope characteristics of the ore-forming systems are particular to hydrothermal systems and do not fall under consideration in the present work.

3.2.4 EPIGENETIC (CATAGENETIC) DEPOSITS

In the scientific literature at present, an insignificant role is assigned to processes of authigenic manganese-ore formation in the process of lithification of sediment taking place after early diagenesis and to their role in manganese ore-genesis. However, detailed lithological and isotope research have allowed the conclusion that at the post-sedimentational stages of the evolution of sedimentary-rock basins occur substantial transformations of the initial composition of primary sediment, leading in a range of cases to the formation of manganese rocks and ores with significant concentrations of manganese up to the formation of large and extra-large (giant and supergiant) deposits.

Serving as model examples of this type of deposits are a range of deposits formed at different stages of transformation of sediments after early diagenesis and in various geological and isotope-geochemistry conditions. To these can be classified the deposits of the Kalahari manganese-ore field (Republic of South Africa), the Usa deposit (Kuznetsk Alatau) of Lower Cambrian age, the Ulutelyak deposit of Lower Permian age, a range of ore-occurrences and small deposits of Pai-Khoi, and a range of other deposits and ore-occurrences.

3.2.4.1 The supergiant deposit of the Kalahari manganese-ore field (Northern Cape, Republic of South Africa)

Deposits of manganese ores of the Northern Cape province (Republic of South Africa) are situated within the boundaries of the south-western part of the Kalahari Desert, 60 km north of the Sishen populated locale, and coincide with the Kalahari manganese-ore field (Fig. 3.89). Here are concentrated colossal reserves of manganese; in terms of resources, it is the world's largest manganese-ore basin, consisting of approximately 13.5 billion tons of manganese ore (with a manganese content in the ores from 20% to 60%). This constitutes no less than half the resources of the known deposits of the landmass (Laznika, 1992; Beukes et al., 1995). The ore-bearing strata by drill borings and mining development is traced in area up to 15 km in width and 35–40 km in the meridional direction.

Although the geological structure and material composition of the ore bodies within the boundaries of the developed deposits of the Kalahari manganese-ore field (Mamatwan, Black Rock, Wessels, Gloria, and others) have received detailed study, in the scientific literature to date there is no unified opinion regarding the origin of the initial manganese ores (braunite lutites).

FIG. 3.89

Fragment of the paleogeological sketch map of the area of deposits in the Kalahari manganese field and Postmasburg area (pro-Karoo geological time), western plunge of the Kaapvaal Craton (Beukes and Smit, 1987). Intrusive rocks are not shown. 1–4—Rocks of the Transvaal Group: 1—Lucknow Formation (basal conglomerates, shales, quartzites, and andesitic basalt lavas), 2—Voelwater Subgroup (dolomites, shales, ferromanganese formations), 3—Makganyene Formation (diamictites); 4—Campbellrand Subgroup (carbonates, shales in rare cases); 5—thrust fault; 6—boreholes; 7—faults; and 8—dip direction of layers. Letter designations for formations: V—Ongeluk (andesitic basalt lavas); R—N—iron-ore formations: R—Nelani-Rooinekke, N—Naragas-Doradale-Pannetjie, G—Griquatown, K—Kuruman.

In the present chapter, on the basis of a survey of literature data and results of microscope study of the ore matter of manganese (braunite) lutites weakly altered by superimposed metamorphic processes that were gathered from the Mamatwan deposit, an attempt has been made to elucidate certain aspects of the genesis of the initial manganese ores and rocks and to resolve the question of the source of the manganese.

Brief outline of the geological setting and the ores of the deposit of the ore field

General geological characteristics. Manganese ores of the Kalahari Desert (Northern Cape province, South Africa) coincide with deposits of the Hotazel formation, which together with the overlying carbonate rocks of the Mooidraai formation constitute the Voëlwater subgroup, which in turn is part of the Postmasburg group of the Transvaal supergroup of the Lower Proterozoic (Fig. 3.90).

The sequence of the Early Proterozoic Transvaal supergroup in the area of the deposits of the Northern Cape province is represented by the initial deposits of a carbonate platform and an iron-ore formation with a general thickness of approximately 2000 m (Beukes, 1983). These are capped by glacial diamictites with a thickness of approximately 150 m of the Makganyene formation, which, in turn, is overlapped by continental basalt-andesite sheets of the Ongeluk formation with a thickness reaching 900 m. The latter in the upper part of the section are characterized by "pillow" textures (Cornell et al., 1996).

Rocks of the Hotazel formation are deposited directly on the vulcanites of the Ongeluk formation and are overlapped by ferruginous hyaloclasts (Beukes, 1983; Tsikos et al., 2003). Higher along the section they are replaced by rhythmic-layered rocks of the BIF and by braunite lutites; their age is estimated at 2.2 and 2.4 billion years (Cornell et al., 1996; Bau et al., 1999).

Stratigraphically above the iron- and manganese-containing rocks of the Hotazel formation in the central and southern parts of the Kalahari manganese-ore basin are developed limestones and dolomites of the Mooidraai formation (Kunzmann et al., 2014).

In the northern and western parts of the Kalahari manganese-ore field deposits of the Voëlwater subgroup are discordantly overlapped by shales, quartzites, and basal conglomerates of the Mapedi Formation of the Olifantshoek group of the Paleoproterozoic. Along the zircons, their age has been established by the U-Pb method at approximately 1.9 billion years (Cornell et al., 1998).

Exposure at the surface of deposits of the Hotazel formation is observed only in the north-west of the manganese-ore field, in the area of the Black Rock deposit.

In the structural regard, the Kalahari manganese-ore field is situated to the north of the Postmasburg ferromanganese-ore area and coincides with the Dimoten syncline, which is filled predominantly by lavas of the Ongeluk formation and overlying deposits of the Hotazel and Mooidraai formations (see Fig. 3.89). In the Upper Paleozoic (pre-Karoo geological time) in the central part of the syncline the rocks have been strongly eroded and are filled by deposits of the Dwyka formation (tillites), which are considered to have filled the glacial trough valley. All rocks are overlapped by young deposits of the Kalahari formation (Fig. 3.91).

Despite the relatively weak manifestation of fold deformations, deposits of the territory are intensively dislocated by thrust-type faults; noted in sequences are numerous tectonic sheets that lead to the repetition of alternation of rocks, including those of the ore strata (Black Rock deposit, Figs. 3.91A and 3.92).

Deposits of the Hotazel formation within the boundaries of the manganese-ore field as noted are deposited on the lavas of the andesite-basalt (pillow-lava) of the Ongeluk formation and are concordantly overlapped by a complex of terrigenous-carbonate rocks of the Mooidraai formation. The latter

Super group	Group	Sub group	Formation	Major lithology	Approx. thickness (m)
			KALAHARI	Calcrete, marl, sand, clay	до 150
KAROO			DWYKA	Diamectite	до 650
KHEIS	OLIFANTSHOEK	VOLOP	VERVATER	Gray quartzite	до 3500
			GLEN LION	Brown quartzite	
			ELLIES RUS	Gray quartzite	
			FULLER	Brown quartzite	
			HARTLEY	Andesitic lava	700
TRANSVAAL			NEYLAN / LUCKNOW	Conglomerate, quartzite	450
	POSTMASBURG		MAPEDY / GAMAGARA	Shale, quartzite, lava, bazal iron-rich conglomerate	150–200
		VOEL-VATER	MOOIDRAAI	Dolomite, chert	250
			HOTAZEL	Iron-formation, manganese, lava	
			ONGELUK	Andesitic lava	500–600 (до 900)
			MAKGANYENE	Diamectite	50–150
	CHAAP	KOE-GAS	ROOINEKKE & NELANI	Iron-formation, shale	240–600
		ASBES-HEUWELS	GRIQUATOWN	Clastic-texturred iron-formation	200–300
			KURUMAN	Microbandet iron-formation	150–750
		CAMPBELLRAND	GAMOHAAN	Carbonate, shale	1500–1700
			KOGELBEEN		
			KLIPPAN		
			PAPKUIL		
			KLIPFONTEIN-HEUWEL		
			FAIRFIELD		
			REIVILO		
			MONTEVILLE		
		SCHMIDTSDRIF	LOKAMMONA	Shale, quartzite, dolomite, limestone	10–250
			BOOMPLAATS		
			VRIBURG		

FIG. 3.90

Summary stratigraphic column of the area of deposits in the Kalahari manganese field (Beukes and Smit, 1987; Gutzmer and Beukes, 1996).

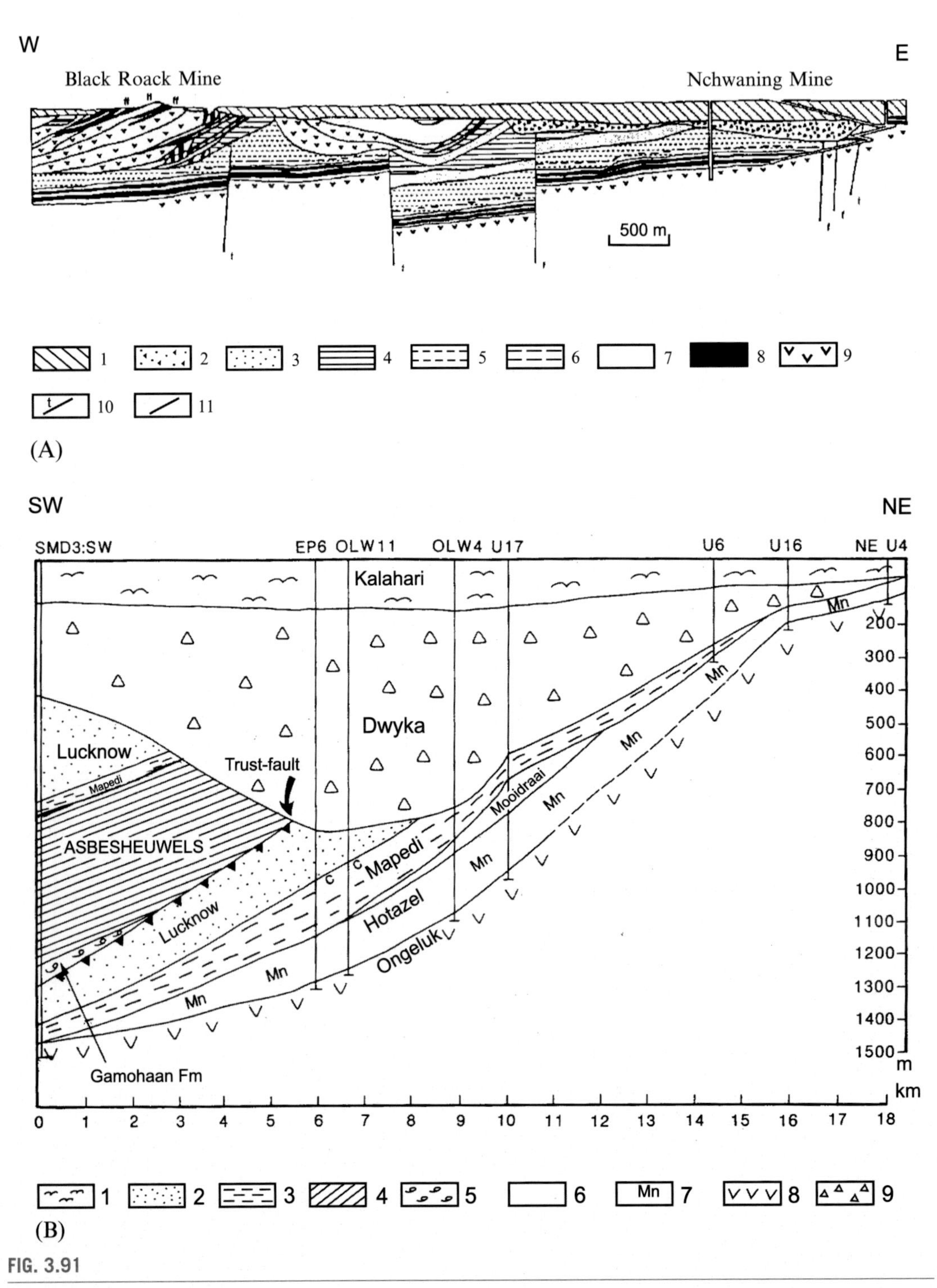

FIG. 3.91

See the legend on opposite page.

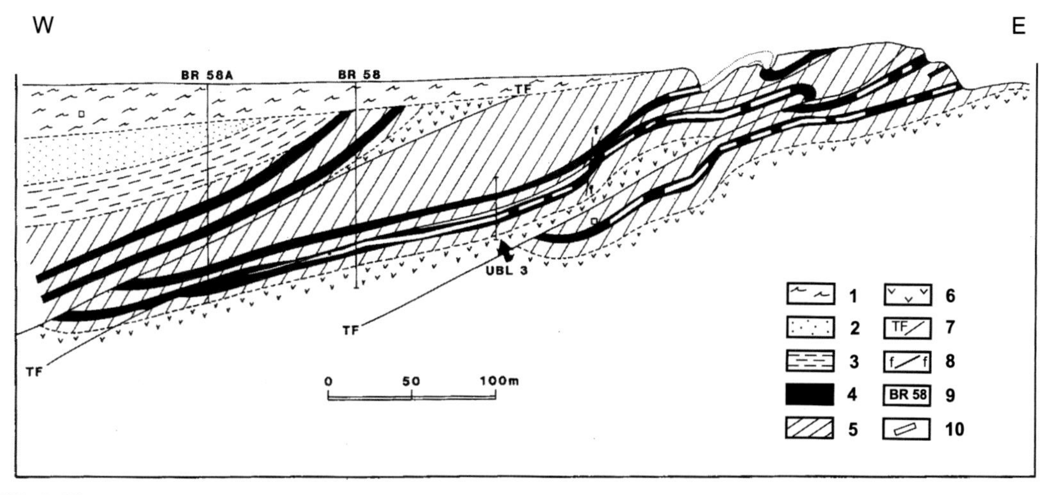

FIG. 3.92

Schematic geological setting of ore body in the Black Rock (Kalahari manganese field, South Africa) (Beukes and Smit, 1987). 1—Kalahari Formation (sand and calcretes); 2—Mapedi Formation (quartzites); 3—Mapedi Formation; 4—Hotazel Formation (manganese); 5—Hotazel Formation (ferruginous-siliceous silicites); 6—Ongeluk Formation (andesitic basalt lavas); 7—thrusts; 8—faults; 9—boreholes; 10—underground minings.

is represented by a thick strata of dolomites, dolomitic limestones, in places stromatolite limestones interstratified with carbonatized clays, siliceous shales, and in places lavas of basic and medium composition.

The structure of the rocks of the Hotazel formation has been studied most fully in deposits of manganese ores of the southern part of the manganese-ore field—the Mamatwan deposit (Fig. 3.93) and the Middelplaats deposit (Fig. 3.94) (Nel et al., 1981; Jennings, 1986). In the lower part of this formation is deposited a layered strata of hematite-quartz rocks—a ferruginous layered member of the BIF. It is represented by red- and gray-colored layered ferruginous siliciclastics with interbeds of jaspilite. Deposited above is a strata of ferromanganese lutites, which consists of an alternation of beds enriched to varying degree by different minerals of oxides of iron and manganese: hematite, jacobsite, braunite, braunite-2, hausmannite, bixbyite, and the like. Present in varying quantity in

FIG. 3.91

Schematic geological section of the (A) northern (Gutzmer and Beukes, 1995) and (B) central (Beukes and Smit, 1987) parts of the Kalahari manganese field. Legend for A: 1—Kalahari Formation (sand and calcretes); 2—Dwyka Formation (tillites); 3-4—Lucknow Formation: 3—shales; 4—quartzites; 5-6—Mapedi Formation: 5—shales; 6—quartzites; 7 and 8—Hotazel Formation: 7—banded iron formation; 8—manganese ores; and 9—Ongeluk Formation (andesitic basalt lavas); 1—thrusts; 11—faults. Legend for B: 1—Kalahari Formation (sand and calcretes); 2—Lucknow Formation (shales, quartzites); 3—Mapedi Formation (shales); 4—Asbesheuwels Subgroup (ferruginous quartzites and shales); 5—Gamohaan Formation (carbonates and shales); 6—Mooidraai Formation (dolomites and shales); 7—Hotazel Formation (braunite lutites and ferruginous shales); 8—Ongeluk Formation (andesitic basalt lavas); and 9—Dwyka Formation (tillites).

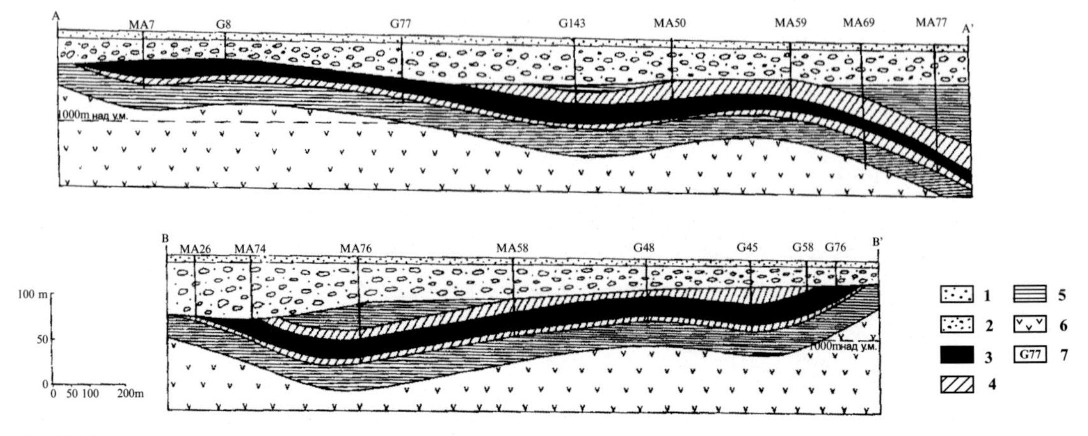

FIG. 3.93

Schematic geological sections across the Mamatwan deposit (Kalahari manganese field, South Africa) (Nel et al., 1986). A-A′—SE-NW; B-B′—SW-NE. 1—Kalahari Formation (sand); 2—Kalahari Formation (pebble, sand, and calcretes); 3–5—Hotazel Formation: 3—upper manganese-ore sequence; 4—lower manganese-ore sequence; 5—banded ferruginous silicites; 6—Ongeluk (andesitic basalt lavas); and 7—boreholes.

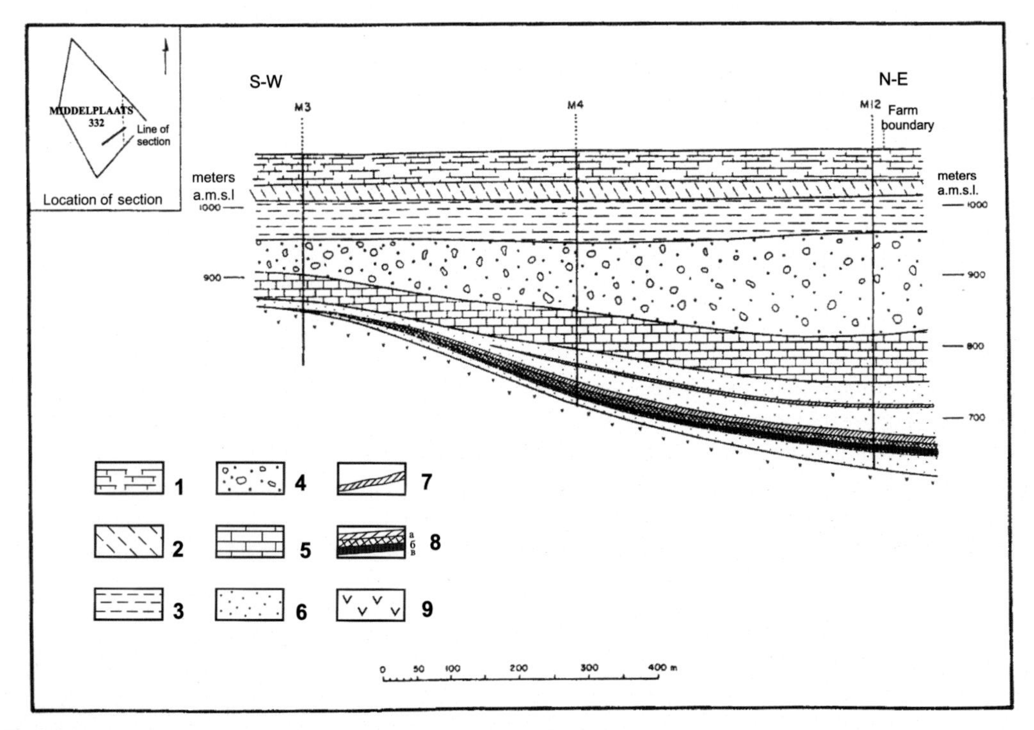

FIG. 3.94

Schematic geological section across the Middelplaats deposit (Kalahari manganese field, South Africa) (Jennings, 1986). 1–3—Kalahari Formation: 1—calcretes; 2—red clay; 3—brown clay; 4—Dwyka Formation (tillites); 5—Mooidraai Formation (dolomites); 6–8—Hotazel Formation: 6—banded ferruginous silicites; 7—middle manganese-ore horizon; 8—ores of the lower manganese-ore horizon: (a) low-grade, (b) medium-grade, (c) high-grade; and 9—Ongeluk Formation (andesitic basalt lavas).

the composition of these horizons along the cross section of the strata are carbonate minerals: rhodochrosite, manganocalcite, and kutnohorite.

In the zone of ancient hypergenesis corresponding to the pre-Karoo weathering crust, pyrolusite and psilomelane are the ore-forming minerals.

Manganese ores of the Kalahari manganese-ore field have been classified as stratiform by the majority of researchers. The ore strata within the boundaries of its eastern part is deposited under the deposits of the Kalahari formation at a depth from 8–10 to 60–70 m, sloping to the west-south-west at angles from 5–8 to 10–15 degrees; within the boundaries of the western part of the field the depth of deposition can reach the order of 800–1000 m, rarely greater. The ore bed is developed by a system of faults of north-easterly trend. In certain cases, bostonite dikes coincide with the faults.

The deposit nearest the surface of the ore strata where the output of manganese ores is produced (or was produced, in mined-out quarries) by an open method has been elucidated in several sections: the southern end of the ore field—one of the world's largest deposits—the Mamatwan deposit; along the eastern border of the manganese-ore field—the Smartt and Perth deposits; and in the north—Black Rock, Wessels, and others.

In the sequence of the ore member three ore bodies have been identified, which are most fully represented in the region of the Mamatwan deposit: lower, middle, and upper. However, only the lower ore body, which is being developed in the deposits, is economically viable (Fig. 3.95). In its structure in terms of textural, mineralogical, and chemical particularities have been identified 11 horizons (from bottom to top): B, L, N, H, C, M, Z, Y, X, W, and V (Nel et al., 1986; Gutzmer et al., 1997) (Fig. 3.96). The maximum thickness of the lower ore body reaches 45 m within the boundaries of the southern border of the field (the Mamatwan deposit) and decreases in the northerly direction up to 5 m at a distance of approximately 35 km (Wessels deposit).

The middle ore body is thin (reaching 2 m) and economically unviable. The upper ore body reaches 5 m of thickness and is under development locally (Black Rock deposit).

Rocks of the ore strata were subjected to varying degrees of metamorphism, that is to say, metamorphism was locally variable and not regional in character. This variability also determined the presence of several industrial types of rich manganese and ferromanganese ores: Wessels, Hotazel, jacobsite, Mamatwan (initial braunite lutites), and hypergenic (developed on the initial lutites). However, the principal reserves of manganese (approximately 95–97%) are present in ores of the Mamatwan type.

The least altered Mn-lutites coincide with the southern and south-eastern parts of the manganese-ore field (Middelplaats, Mamatwan, Smartt, and Rissik deposits), whereas in the northern (Gloria deposit) and particularly the north-western (Wessels and Black Rock deposits) sections of the examined manganese-ore basin are noted intensive transformations of the initial ores. These changes bear a metasomatic character and coincide as a rule with tectonic faults (Beukes et al., 1995; Gutzmer and Beukes, 1995, 1996; Gutzmer et al., 1997). Processes of metamorphism have led to the formation of rich manganese-oxide ores with mean manganese content >44% (Black Rock, Wessels, Gloria, and other deposits). However, the share of the latter, as noted, throughout the manganese-ore field constitutes no greater than 5% and most of the high-grade material has been mined out (Gutzmer and Beukes, 1995; Tsikos et al., 2003).

Processes of metamorphism of the initial braunite lutites intensively developed within the boundaries of the north-western section of the manganese-ore field have led to a decrease of the

FIG. 3.95

Mamatvan manganese deposit, general view of the ore strata. (A) General view of the south-western part of the open pit (the tower on the horizon is on the Middelplaats deposit); (B) a general view of the quarry walls, dark—stratum braunite lutites; and (C) general view of the ore strata.

initial thickness of the manganese-ore strata (reaching 4–5 m) with a simultaneous increase in content of manganese (>40%).

Types of ores. In the genetic regard, the manganese ores of the deposits of the Kalahari manganese-ore field can be divided into four types: the poor, Mamatwan type (braunite lutites); the rich, Hotazel and Wessels types (hydrothermally altered initial lutites); and the hypergenically altered type (on the initial braunite lutite Mamatwan type) (Kleyenstuber, 1984; Nel et al., 1986; Gutzmer and Beukes, 1995; Beukes et al., 1995).

Ores of the Mamatwan type, in terms of distribution, are primary sedimentational-diagenetic with low content of manganese. They represent carbonate-containing oxide ores and are composed principally by braunite, hausmannite, kutnohorite, and manganocalcite. The content of Mn varies within the boundaries of 20–40% with low concentrations of Fe (4–6%) and relatively high contents of CO_2 (12–18%) and CaO (14–16%).

Ores of the Wessels type represent a product of the hydrothermal alteration of initial ores of the Mamatwan type. They are composed predominantly of braunite-2, bixbyite, and hausmannite with insignificant content of calcite. Contents of manganese vary from 38% to 51%, iron from 9% to 20%, CaO from 4% to 6%, and CO_2 from 1% to 3%.

Ores of the Hotazel type are composed of predominantly bixbyite and hausmannite and are characterized by content of manganese from 44% to 46%, of iron 11–15%. This type of ores represents a combination of hypergenically and hydrothermally altered initial ores of the Mamatwan type.

Hypergenically altered ores represent a result of the transformation of ores of the Mamatwan type. They are developed in the zone of ancient hypergenesis below deposits of the Kalahari formation. The ores contain 54–60% Mn_3O_4, 6–9% F_2O_3, and 7–12% SiO_2.

Additionally, in the industrial regard yet another type has been identified—jacobsite ores, which are characterized by a low Mn/Fe ratio. These ores are developed in certain sections of the ore body and are characterized by a lateral alteration of initial ores of the Mamatwan type.

Particularities of microscopic structure of the ores. Within the boundaries of the manganese-ore field, to date no less than 135 minerals have been established in the composition of the ores and vein formations (Gutzmer and Beukes, 1996). The most widely distributed among them number several dozens of ore and non-ore minerals. Manganese ores are represented predominantly by minerals of manganese and iron oxides, more rarely manganese carbonates.

As has been established, the initial braunite lutites were repeatedly subjected to various regional transformations: late-diagenetic and low-temperature green shales metamorphic in facies (Kleyenstuber, 1984).

Four phases of the transformation of manganese lutites have been identified (Gutzmer et al., 1997): (a) low-temperature regional metamorphism, expressed in a parallel-layered metasomatic transformation of small lenses and ovoids of kutnohorite and manganese calcite in "grained" hausmannite with submetallic luster. Also formed at this stage were partridgeite, manganite, and calcite (Kleyenstuber, 1984). Manganite in the composition of primary ores (braunite lutites) in the deposit has not been detected; (b) discoloration and reduction of manganese ores in association with veins of pyrite-carbonate composition; (c) brecciation and streaking in fault zones in association with sparry calcite-dolomite mineralization and oxidation of surrounding manganese ores; (d) hypergenic alteration characterized by the development of goethite in fault zones and calcretization on the border of discordance with the Kalahari formation.

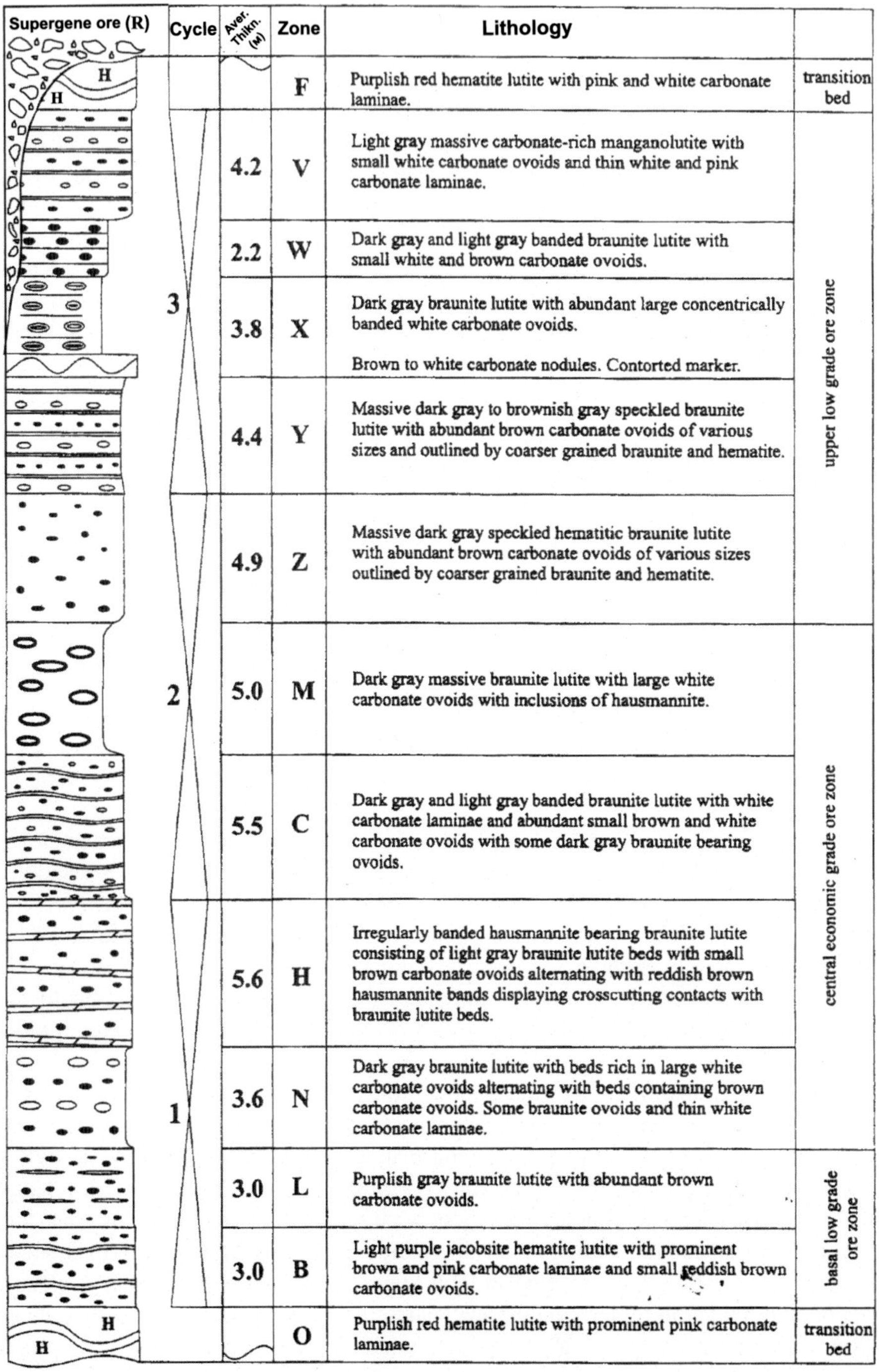

Supergene ore (R)	Cycle	Aver. Thikn. (M)	Zone	Lithology	
H			F	Purplish red hematite lutite with pink and white carbonate laminae.	transition bed
		4.2	V	Light gray massive carbonate-rich manganolutite with small white carbonate ovoids and thin white and pink carbonate laminae.	upper low grade ore zone
		2.2	W	Dark gray and light gray banded braunite lutite with small white and brown carbonate ovoids.	
	3	3.8	X	Dark gray braunite lutite with abundant large concentrically banded white carbonate ovoids. Brown to white carbonate nodules. Contorted marker.	
		4.4	Y	Massive dark gray to brownish gray speckled braunite lutite with abundant brown carbonate ovoids of various sizes and outlined by coarser grained braunite and hematite.	
		4.9	Z	Massive dark gray speckled hematitic braunite lutite with abundant brown carbonate ovoids of various sizes outlined by coarser grained braunite and hematite.	
	2	5.0	M	Dark gray massive braunite lutite with large white carbonate ovoids with inclusions of hausmannite.	central economic grade ore zone
		5.5	C	Dark gray and light gray banded braunite lutite with white carbonate laminae and abundant small brown and white carbonate ovoids with some dark gray braunite bearing ovoids.	
		5.6	H	Irregularly banded hausmannite bearing braunite lutite consisting of light gray braunite lutite beds with small brown carbonate ovoids alternating with reddish brown hausmannite bands displaying crosscutting contacts with braunite lutite beds.	
	1	3.6	N	Dark gray braunite lutite with beds rich in large white carbonate ovoids alternating with beds containing brown carbonate ovoids. Some braunite ovoids and thin white carbonate laminae.	
		3.0	L	Purplish gray braunite lutite with abundant brown carbonate ovoids.	basal low grade ore zone
		3.0	B	Light purple jacobsite hematite lutite with prominent brown and pink carbonate laminae and small reddish brown carbonate ovoids.	
H			O	Purplish red hematite lutite with prominent pink carbonate laminae.	transition bed

FIG. 3.96

Lithostratigraphic subdivision of ore body in the Mamatwan deposit (Nel et al., 1986).

Hydrothermal transformations of manganese lutites occurred multiple times; several stages (or "events") have been identified: Wessels (associations: coagulation calcite-silicate, rich hausmannite—ore and vein), Mamatwan, and Smartt, which were accompanied by formations of the corresponding mineral parageneses (Gutzmer and Beukes, 1996).

It follows to note that according to mineralogical research by Gutzmer et al. (1997), as well as that of a range of preceding authors (De Villiers, 1983; Kleyenstuber, 1984; Nel et al., 1986; and others), in many cases manganese oxides (hausmannite, partridgeite, manganite, manganomelane, todorokite, etc.) replace small lenses and ovoids of manganese carbonates. Particularly intensive transformations of this type are noted along fractures and fault zones.

The composition of kutnohorite strongly varies, from ideal $CaMn(CO_3)_2$ to ankerite and dolomite. Manganous calcite in cross-veins and ovoids of unaltered braunite lutites contains up to 13% MnO, less than 2% MgO, and maximally 1.25 mass% FeO (Gutzmer et al., 1997).

The character of relationships of carbonates that compose ovoids and laminae with minerals of manganese oxides is clearly visible under an optical microscope in thin sections under transmitted (Fig. 3.97A, C, E, G) and reflected (Fig. 3.97B, D, F, H) light. Here with various increases can be observed a gradual replacement by manganese oxides of the initial carbonate matter of ovoids and laminae.

An interesting particularity of manganese carbonates of the Mamatwan deposit is that in certain samples in the carbonate matter of the ovoids, the initial microbial structures have been preserved. In Fig. 3.98 such structures are distinctly visible in the form of chains, which represent mineralized (pseudomorphosed) purple bacteria and fibers of cyanobacteria (Zhegallo et al., 2000; Shkol'nik et al., 2004). Their presence provides evidence of an initial shallow-water formation of the sediment with subsequent mineralization of bacterial mats.

Geochemistry of isotopes

Isotope research was conducted in order to arrive at an explanation of the conditions of formation of manganese lutites of the described deposit. The principal results of this research have been laid out in the works of N. Beukes, J. Gutzmer, P. Preston, A. Tsikos, and a range of other researchers.

The most complete characteristics of the isotopic composition of carbon and oxygen of initial braunite lutites unaltered by subsequent processes of transformation were obtained by P. Preston (Preston, 2001; http://152.106.6.200:8080/dspace/bitstream/10210/1967/1/PaulaPreston.pdf) for manganese ores of the Mamatwan type in the Mamatwan deposit. This author has studied the deposit's manganese carbonates (general sample) from the core of two wells (G552 and G558), which are characteristic of braunite lutites of all horizons of the ore strata. The general scattering of $\delta^{13}C$ values for the studied samples constitutes from −13.5‰ to −6.4‰ (PDB) and that of $\delta^{18}O$ from 15.2‰ to 20.3‰ (SMOW).

The results of isotope research have allowed the author of the indicated work to educe an inverse correlation between values of $\delta^{13}C$ and those of $\delta^{18}O$: carbonates with highest values of the isotopic composition of oxygen are characterized by the lowest values of $\delta^{13}C$, and vice versa (Fig. 3.99).

It can be supposed that the noted regularity represents a line of mixing of carbonate matter from two sources bearing different isotope characteristics. Evidently, one of these groups consists of manganese carbonates of unaltered initial braunite lutites, for which are characteristic the lowest values of $\delta^{13}C$ and the highest of $\delta^{18}O$.

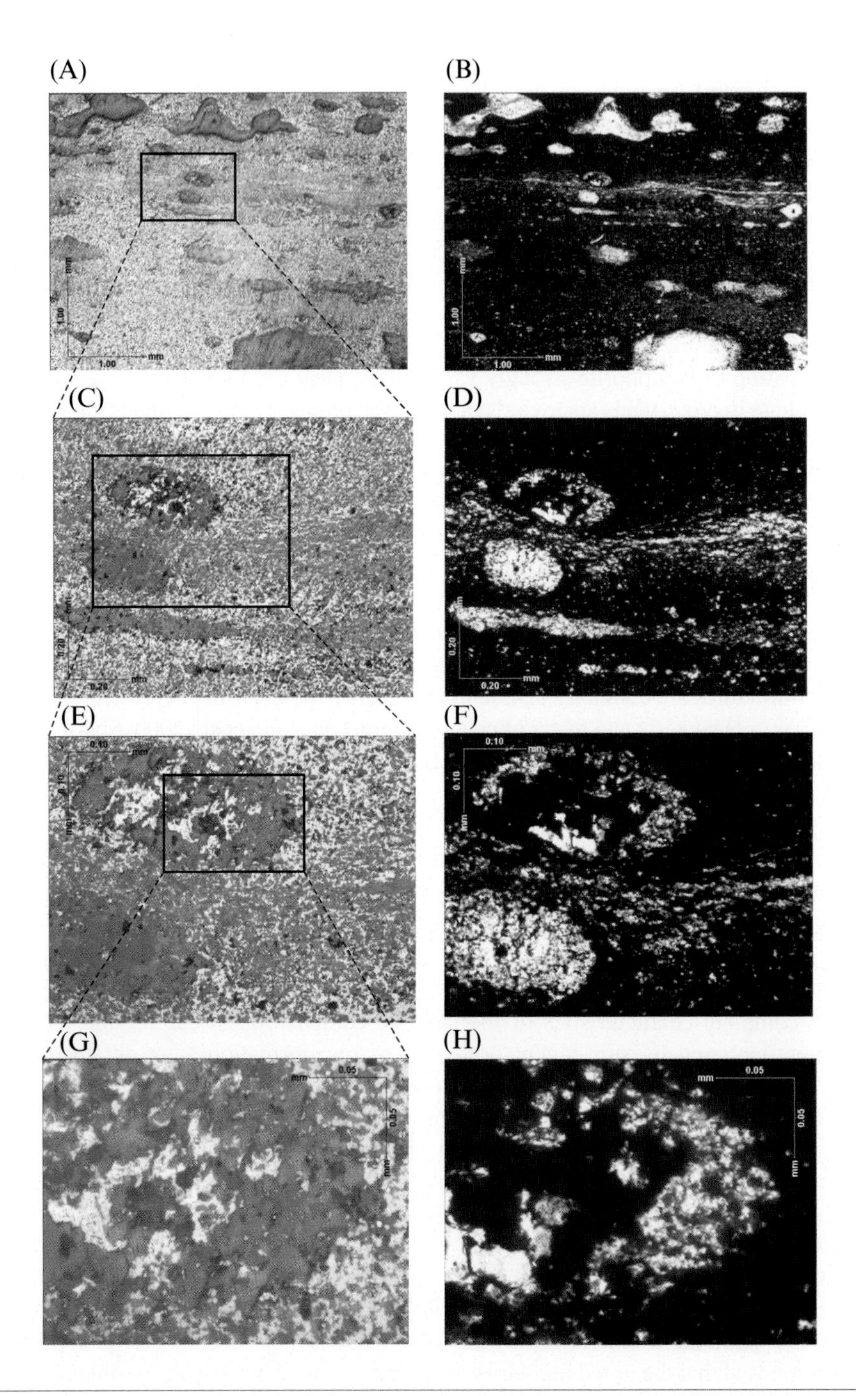

FIG. 3.97

Photomicrographs of thin sections of sample 2/04 in the reflected (A, C, E, G) and transmitted (B, D, F, H) light at different magnifications. In the reflected light, manganese oxides and carbonates are white and dark, respectively. In the transmitted light, manganese carbonates and oxides are dark and white, respectively.

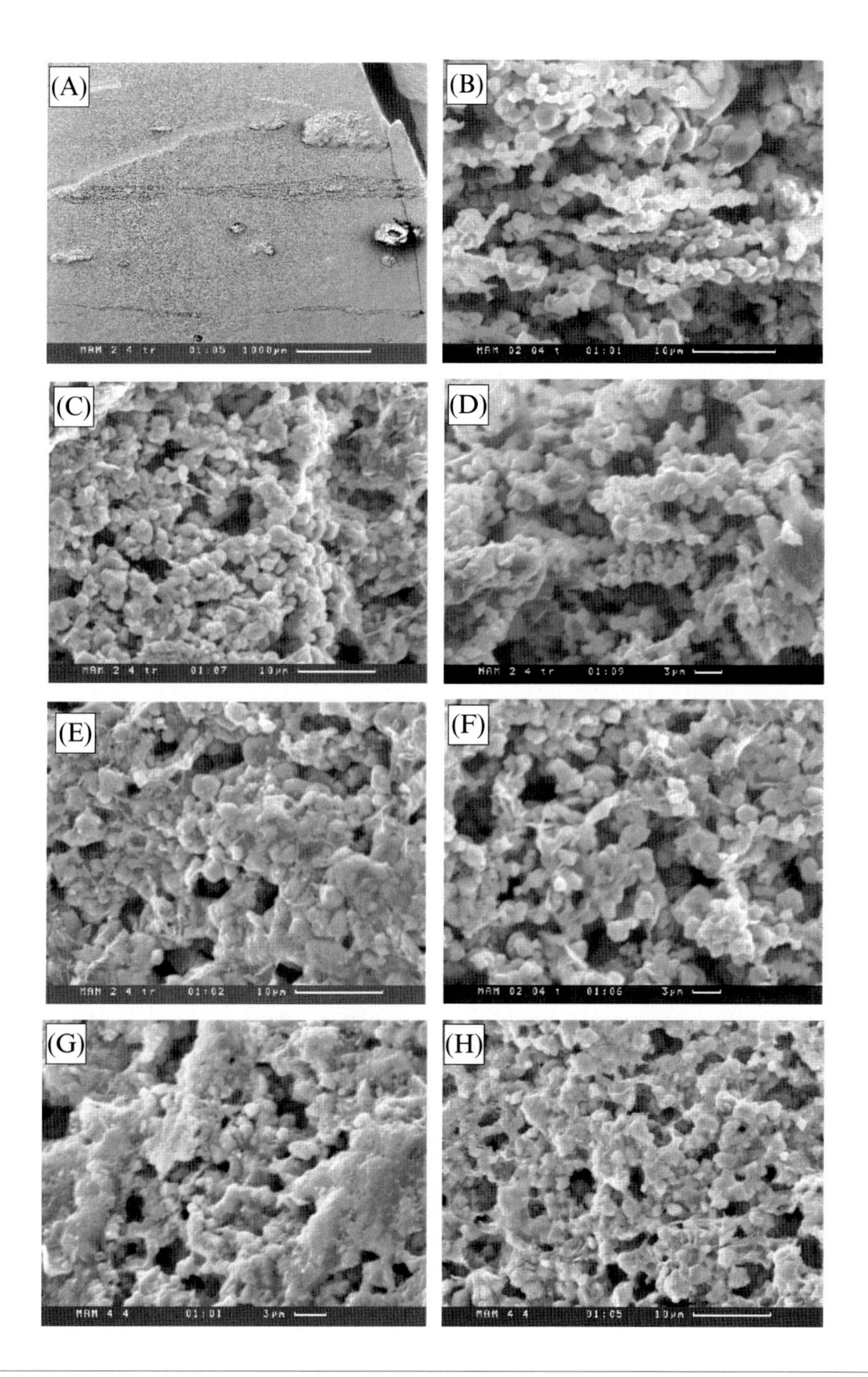

FIG. 3.98

SEM photomicrographs of sample 2/04 at different magnifications. (A) General view (oval bulges are carbonate ovoids); (B)–(H) residual microbial structures (different sectors of sample Mam-2/04, photos by E.A. Zhegallo and E.L. Shkol'nik).

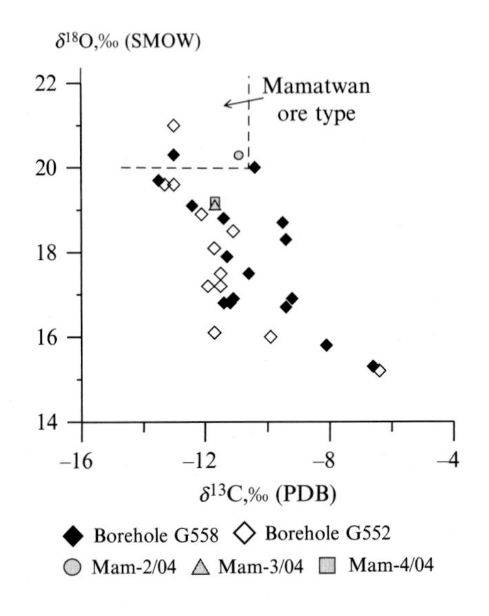

FIG. 3.99

$\delta^{13}C$ versus $\delta^{18}O$ in manganese lutites of Mamatwan deposit, Kalahari manganese-ore field, RSA [compiled on data (Preston, 2001)]. G-552 and G-558—data by Preston (2001); Mam-2/04, Mam-3/04, and Mam-4/04—our data.

Another group of carbonate matter with opposite isotope characteristics apparently represents later carbonates that compose lutites. A particularity of their isotopic composition is their high values of $\delta^{13}C$ and low values of $\delta^{18}O$. It can be supposed that such isotope characteristics are inherent to carbonate matter formed at the stage of regional metamorphism.

Preston's work also demonstrates that a characteristic particularity of ores of Mamatwan type is an inverse correlation between the isotopic composition of carbon and the content of manganese (Fig. 3.100A): the higher the manganese content in the rock, the lighter the isotopic composition of carbon. This regularity is observed in practically all known manganese deposits of the continents connected with complexes of sedimentary rocks, including deposits of Ukraine (Nikopol' and Bol'she-Tokmak) (Kuleshov and Dombrovskaya, 1988), Mexico (Molango) and China (Taojiang) (Okita and Shanks III, 1988, 1992; Okita et al., 1988), Kazakhstan (Mangyshlak) (Kuleshov and Dombrovskaya, 1990, 1993), Hungary (Úrkút) (Polgari et al., 1991), Georgia (Chiatura) (Kuleshov and Dombrovskaya, 1997a,b), Russia (Usa, Pai-Khoi, Bezmoshitsa) (Kuleshov et al., 1991; Kuleshov and Beliaev, 1999; Kuleshov and Bych, 2002), and a range of other deposits. This regularity is a result of participation by oxidized carbon of organic matter (C_{org}) in the process of manganese concentration—that is, the higher the share of oxidized carbon of organic origin in the composition of the rock, the higher the content of manganese.

It should also be noted that in zones X and M in the sequence of the ore member of the Mamatwan deposit, no correlation has been found between the isotopic composition of carbon and the content of manganese (see Fig. 3.100B). These zones are characterized by the highest

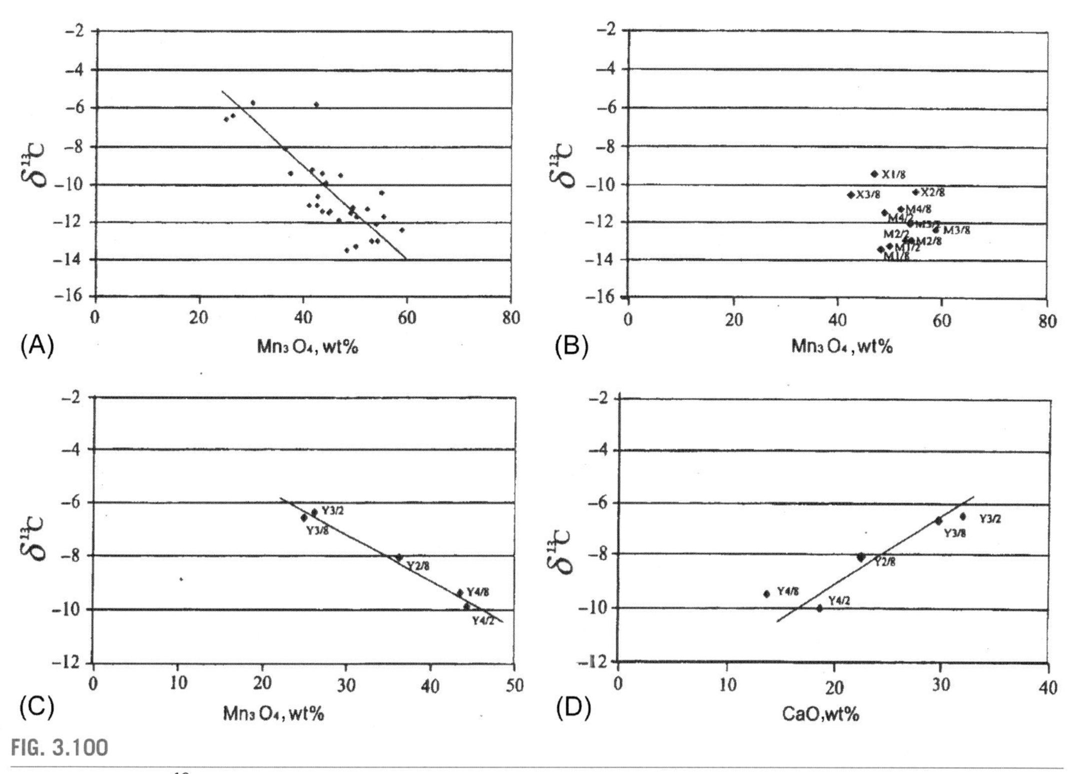

FIG. 3.100

Dependence of $\delta^{13}C$ values and Mn_3O_4 content in the ore strata of Mamatwan deposit (manganese-ore field of the Kalahari, South Africa) excluding supergene altered zone (A). (B) Manganese-rich zone in the X and M; (C) most rich in carbonates zone Y; (D) the dependence on the $\delta^{13}C$ values and content of CaO zone Y (D) (Preston, 2001).

concentrations of manganese. Moreover, for manganese ores of zone Y, which is characterized by high carbonate content, a direct correlation has been observed between $\delta^{13}C$ values and CaO content: the higher the calcium content in the rock, the lower the content of light isotope ^{12}C (see Fig. 3.100D).

In light of the explanation of the inverse correlation between the isotopic composition of carbon and the manganese content in the rock, Preston (2001) following Polgari et al. (1991) in their study of the rocks of Úrkút (Hungary), adheres to the viewpoint that the formation of manganese carbonates occurred as a result of processes of reduction of manganese oxides with participation by carbon of organic matter. This mechanism has previously been proposed by P. Okita and co-authors for deposits of Mexico and China (Okita et al., 1988; Okita and Shanks, 1988, 1992) and has been noted numerous times by the authors of the present work in the above-enumerated range of manganese deposits.

An inverse correlation between the isotopic composition of carbon and that of oxygen, as well as an analogous correlation between $\delta^{18}O$ values and the manganese content in the rock (Fig. 3.101), according to P. Preston (Preston, 2001) reflects conditions of sedimentation and early diagenesis.

However, a detailed examination of the character of the distribution of the isotopic and elemental composition in manganese ores of the Mamatwan deposit leaves room for the possibility that the established isotope regularities for the Mamatwan deposit could be due to other causes.

It is worth recalling that for ores of Oligocene (Paratethys: Nikopol', Chiatura, Mangyshlak, etc.) and Paleozoic deposits (Molango, Úrkút, Pai-Khoi, Bezmoshitsa, etc.), as indicated previously by the author of the present work in the range of publications mentioned above, no correlation has been observed between the isotopic composition of carbon and that of oxygen; $\delta^{18}O$ in manganese carbonates varies within a narrow range that corresponds to carbonates of early-diagenetic origin. Therefore, an inverse correlation between the isotopic composition of carbon and that of oxygen in ores of Mamatwan type would not be expected unless there have been subsequent (ie, after the formation of specifically these ores) processes of transformation and new formation of carbonate matter.

With considerable confidence, it can be supposed that the initial—that is, unaltered by subsequent processes—braunite lutites (ores of Mamatwan type) were characterized by values of the isotopic composition of carbon close to −14‰ to −12‰ and by those of oxygen no lower than 20–22‰ (SMOW). Subsequent (hydrothermal?)-metasomatic transformations evidently of the stage of regional metamorphism led to enrichment by heavy isotopes (^{13}C) of carbon and as a rule by light isotopes of oxygen (^{16}O). Serving as evidence for this is the inverse correlation also noted by P. Preston (Preston, 2001)

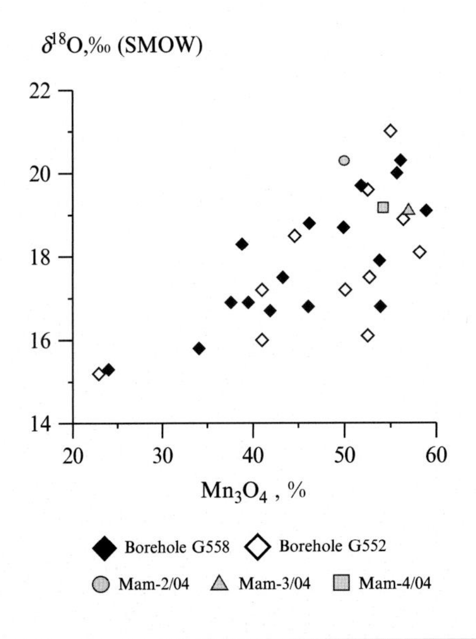

FIG. 3.101

Dependence $\delta^{18}O$ values and the content of Mn_3O_4 in manganese lutites of Mamatvan deposit, Kalahari manganese-ore field, South Africa (compiled from Preston, 2001). G-552 and G-558—data by Preston (2001); Mam-2/04, Mam-3/04, and Mam-4/04—our data.

between the content of manganese in rocks and the isotopic composition of oxygen: the lighter the isotopic composition of oxygen (ie, the more transformed the rock), the lower the manganese content (see Fig. 3.101).

Another type of secondary alteration of the isotopic composition of braunite lutites consists of transformations of the initial ores of Mamatwan type in near-fault zones. This regularity has been clearly illustrated in the work of Gutzmer et al. (1997). At the Mamatwan deposit, these authors have studied in detail the mineralogy, geochemistry, isotopic composition, and processes of transformation of the initial sedimentational braunite-kutnohorite manganese ores—lutites, along the cross sections L1 and S1, traversing the fault zones.

Unaltered braunite lutites from well M2 in terms of the isotopic composition of their carbon and oxygen were found to be similar to samples along the cross sections L1 (L2) and S1. Primary carbonates (kutnohorite) of ores of Mamatwan type here, according to the above-cited authors, are characterized by isotopic compositions from −13.9‰ to −10.8‰ for $\delta^{13}C$ and from −15.7‰ to −11.8‰ for $\delta^{18}O$, PDB (respectively, 15.1–19.0‰, SMOW).

The isotopic composition of sparry carbonate in breccia from the fault zone is heavier. For example, for sequence L2 it varies from −9.0‰ to −5.7‰ and from −6.0‰ to −4.4‰ (PDB) for carbon and oxygen, respectively (Fig. 3.102).

One of the possible sources of carbon dioxide with a heavy isotopic composition of carbon, besides carbonates of the host sequence, could have been calcretes from the overlapping deposits of the Kalahari formation, which are characterized by a heavier isotopic composition by comparison with ores of Mamatwan type: −6.0‰ and −4.3‰ for $\delta^{13}C$ and −3.1‰ and −0.8‰ (PDB) for $\delta^{18}O$ (Gutzmer et al., 1997) (Fig. 8.101).

At the Wessels deposit, Burger (1994) has established a broad range of variations the isotopic composition of carbon and oxygen in different types of ores (Mamatwan, Wessels, and secondary braunite). Here in cross-veins of high-temperature rich hausmannite ores of Wessels type are noted the lightest values of $\delta^{18}O$ (reaching 16.5‰, SMOW). Moreover, in cross-veins of ores of this type, rich in braunite-2, values of $\delta^{18}O$ are very high—up to 29.3‰ (SMOW).

It follows to note that for secondary braunite ores in this deposit a direct correlation has been observed between the isotopic composition of carbon and that of oxygen (for 4-x samples!): from the lightest (sample IMB3: −22.3‰ for $\delta^{13}C$ and 16.8‰ for $\delta^{18}O$, SMOW) to the heaviest (sample IMB2: −7.9‰ for $\delta^{13}C$ and 22.1‰ for $\delta^{18}O$, SMOW).

Analogous alterations of isotopic composition can be observed in the Mamatwan deposit along the line of cross section S1 (Gutzmer et al., 1997). These alterations occurred at higher temperatures and with participation by fluids with a lighter isotopic composition of oxygen by comparison with analogous processes of secondary transformation and new formation of minerals.

In this way, the isotope data reported above enable the conclusion that the established regularities in the distribution of the isotopic composition in manganese (braunite) lutites in many cases are the result of later processes of their transformation, which occurred evidently at various temperatures and under the influence of carbon dioxide-water solutions characterized by a different isotopic composition. This matches well with the noted observations of Gutzmer et al. (1997), who have identified four phases (or stages) of the post-sedimentational transformation of braunite lutites. It can be supposed that in one case, these were carbon dioxide-water solutions characterized by high values of $\delta^{13}C$ and low values of $\delta^{18}O$, and in the other case by a heavy isotopic composition of carbon (no lower than −6‰ to −4‰, PDB) and oxygen (no lower than 27–28‰, SMOW), having a different

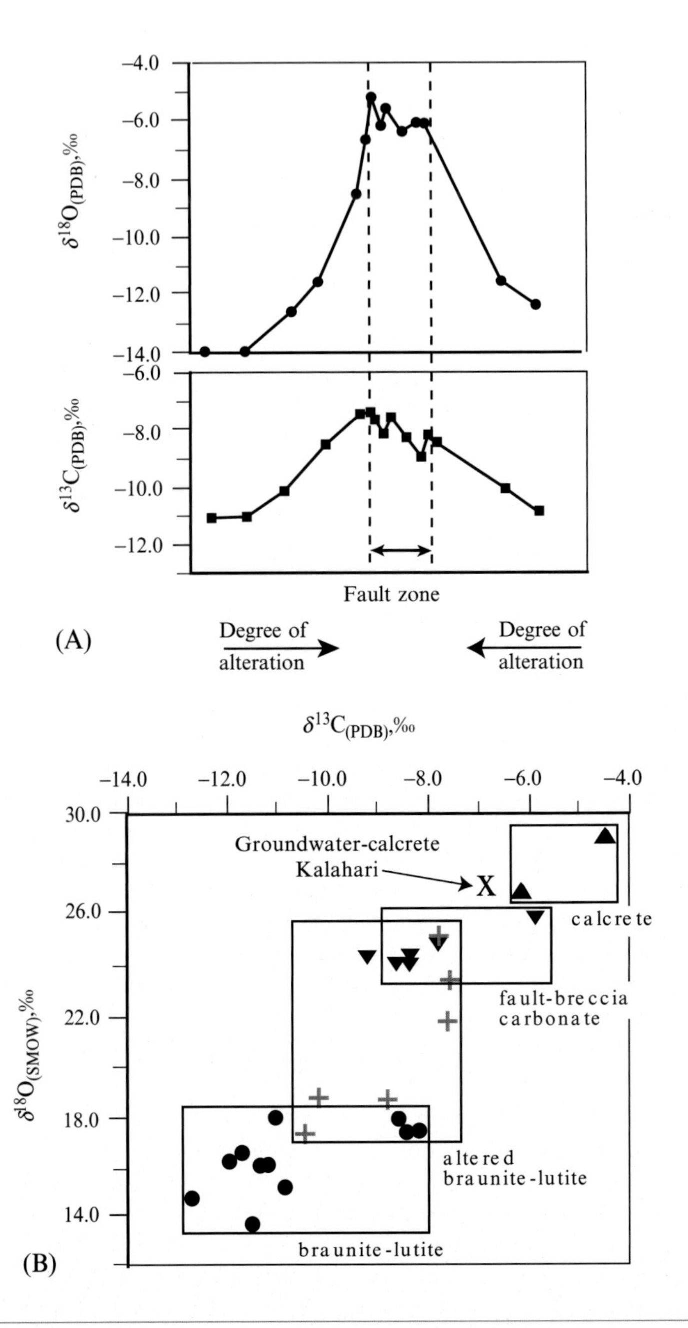

FIG. 3.102

Variations of the isotopic composition of carbon and oxygen at the intersection of L1 (mine Mamatwan, South Africa): (A) in carbonate ores matter of mamatvan-type altered rocks and secondary carbonate breccias across the strike of the fault zone and (B) $\delta^{13}C$ and $\delta^{18}O$ in different types of rocks (Gutzmer et al., 1997).

source of matter. At the Wessels deposit one of the generations of secondary carbonates in the secondary braunite ores, as noted, is characterized by low values of the isotopic composition of carbon and oxygen.

It should be emphasized that the observed distribution patterns of the isotopic composition of carbon and oxygen along the line of cross section S1 (Mamatwan deposit) (Gutzmer et al., 1997) are due to the practically full transformation of initial lutites, rather than to the presence of secondary calcite (later input into the system of carbon dioxide). This follows from the fact that the accompanying kutnohorite and calcite as a rule are close in terms of isotopic composition (in all cases, calcite is lighter in terms of oxygen by no less than 1–2‰ by comparison with kutnohorite), which can serve as evidence of their syngeneticity.

Consequently, in the isotopic regard the carbon dioxide-water system at the moment of formation and transformation of the rock of the cross section along the line S1 was practically fully homogenized. The calcite and kutnohorite were formed (or transformed) simultaneously (a single phase of formation or transformation) and are not different generations.

In this way, the transformation of initial braunite lutites apparently occurred multiple times (confirming the results of previous studies of mineral associations within the boundaries of the deposit) and at various temperatures under the influence of carbon dioxide-water solutions with different sources of matter.

Genesis of the deposit

Regarding the origin of the initial manganese ores (Mamatwan type) of the Kalahari manganese-ore field, several points of view have arisen. The earliest and most widely distributed of these holds that their origin was initially sedimentary and occurred by chemical means in shallow-water conditions (Boardman, 1964). Adherents of this viewpoint invoke various mechanisms for the explanation of the method of formation of manganese-bearing sediments, including the precipitation of dissolved Mn entering with deep waters into the shallow-water area of the basin of sedimentation.

According to the first mechanism—the *transgressive-regressive* model, developed for Phanerozoic deposits (Frakes and Bolton, 1984; Force and Cannon, 1988)—the formation of manganese-ore deposits occurred in conditions of varying water levels, which determined the setting of the boundary of distribution of manganese oxides in shelf conditions. Nel et al. (1986) have identified in the composition of manganese-ore strata of the examined basin on the example of the Mamatwan deposit three horizons corresponding to three transgressive-regressive cycles (see Fig. 8.95).

The deposition of manganese as well as iron ores occurred in conditions of a variable water level, resulting in a drifting of the “focus” of the deposition of manganese oxides and carbonates “up and down” within the boundaries of the shelf zone (Cornell and Shutte, 1995).

In the given scheme, such facies as black shales and sandstones are not found in the composition of the sequence of the ore strata of the Kalahari deposit. The observed relationships of manganese oxides and carbonates in thin sections today do not provide evidence of their synchronous (syngenetic) origin.

According to the second mechanism—the *upwelling* model—the formation of the strata of manganese oxides was due to the entry of manganese-rich deep oceanic waters into the oxidizing conditions of the shelf (Cannon and Force, 1983).

This model likewise does not preclude the contradiction issuing from the specifics of the facies composition of the deposits of the shelf, including the contradiction of the transgressive-regressive model.

Simultaneously with the above-examined models exists a theory of a *volcanogenic-exhalation* origin of manganese ores. This model in certain aspects is analogous to the upwelling mechanism. However, serving as the source of manganese in this case are hydrothermal solutions of a volcanogenic-hydrothermal system that carry this element from the strata of underlying andesites of lavas of the Ongeluk formation into the oxygen-rich bottom waters of the shelf (Cornell and Shutte, 1995). The accumulation of manganese occurred by the same means as in modern sediments of hydrothermally active oceanic regions characterized by manganese-ore specifics (with a precipitation of manganese oxides).

This model in its proposed form likewise exhibits a range of flaws and does not withstand criticism (Beukes and Gutzmer, 1996).

It would seem that the described viewpoints are based on definite geological and geochemical factors that are clearly insufficient for explaining the identified regularities and particularities of the geological structure, petrography, mineralogy, lithology, and geochemistry of manganese ores and rocks of the host sequence of the Kalahari manganese-ore basin.

Conditions of formation of manganese-containing rocks enclosed in complexes of sedimentary (volcanogenic-sedimentary) deposits in many cases are analogous or close to the conditions of formation of rocks of the host sequence. Therefore, for indirect evidence explaining the initial nature of manganese sediments (manganoprotolith, manganese lutite) of the examined basin, we should turn to the conditions of the formation of underlying and overlapping rocks, especially carbonate rocks. That is, we assign in our understanding central importance to the "background" against which the accumulation of manganese and ferromanganese rocks of the Hotazel formation occurred.

According to the research of Beukes (1983), the sequence of rocks of the Early Proterozoic Transvaal supergroup in Northern Cape province underlying rocks of the Postmasburg group (including an ore-bearing member of the Hotazel formation) is represented by initial shallow-water deposits of the carbonate platform and iron-ore formation (general thickness of approximately 2000 m) of the shelf, littoral, and supralittoral of the epicontinental and lacustrine water bodies.

Metal-bearing rocks of the Hotazel formation including members of braunite lutites from the underlying terrigenous-carbonate rocks and deposits of the Postmasburg and Ghaap iron-ore formations of the Transvaal supergroup are separated discordantly by the diamictites of the Makganyene formation and vulcanites of the Ongeluk formation deposited alongside them.

At the same time, the rocks of the Hotazel formation are concordantly overlapped by a strata of dolomites and shales of the Mooidraai formation (Kunzmann, 2014). The latter were formed, as were the carbonates of the underlying sequence, in shallow-water conditions and include fine-grained micrites with intraclasts of algal mats.

An initially microbial origin (evidently cyanobacterial mats), as we have established (see Fig. 3.98), is inherent also to carbonate inclusions in braunite lutites of the ore member. This also serves as evidence for the highly shallow-water conditions of formation of the initial rocks (evidently, supratidal).

Underlying carbonate rocks of the Gamohaan formation and the Kuruman iron-ore formation were also formed in shallow-water marine conditions. Evidence for this is found in the results of the detailed study of petrography, isotopic composition of carbon in the kerogen and its distribution along the sequence, the isotopic composition of carbon and oxygen in limestones, dolomite, and siderite, as well as the presence of microbial structures (Beukes et al., 1990).

Consequently, the reconstruction of the paleoenvironment in which the accumulation of rocks of the examined sequence occurred serves as evidence for the highly shallow-water conditions of formation of underlying and overlapping, as well as rocks specifically of the ore member. In this case, the upwelling mechanism of entry of ore matter into the ore-genesis zone is highly improbable.

It follows to note that an upwelling mechanism of entry by manganese (as a source of Mn), though accepted by many researchers as a basic explanation for the genesis of the ores, to date is not accepted unequivocally and it is not supported by any of the known deposits; it bears a strictly theoretical character. This mechanism proposes the existence of a vast, deep-water marine basin (or ocean) characterized by a stagnant, oxygen-free regime of development, featuring as a rule a zone of hydrosulfide contamination and an accumulation of organic matter (which give rise to the formation of black-shale strata). To date, conclusive geological data are lacking on the existence of such a basin to the west of the area of accumulation of iron- and manganese-ore rocks in the region of the western slope of the Kaapvaal craton.

Additionally, the upwelling mechanism of formation of manganese deposits proposes entry into the discharge zone by predominantly dissolved manganese and does not account for the significant concentrations of iron contained in rocks of the Hotazel formation nor in the ores themselves of the deposits of the Kalahari manganese-ore field enclosed in rocks of this formation.

The isotope characteristics of carbonates (limestones, dolomites) from the sequences underlying and overlapping the ore strata of the Hotazel formation are close to "common" or normal-sedimentary marine carbonates of the Phanerozoic. Thus, dolomites of the underlying Gamohaan formation are characterized by values of $\delta^{13}C$ from −0.5‰ to −0.1‰ and values of $\delta^{18}O$ from −5.9‰ to −3.9‰, PDB (accordingly, from 24.9‰ to 26.9‰, SMOW) (Beukes et al., 1990), and overlapping deposits of the Mooidraai formation—within the boundaries of 0.5–0.6‰ for $\delta^{13}C$ and −2.1‰ to 0.1‰, PDB (accordingly, 28.7–30.7‰, SMOW) (Bau et al., 1999). This also provides evidence for conditions of formation of marine carbonates close to those common for the Phanerozoic. An insignificant increase in the lightness of the isotopic composition of oxygen in carbonates of the underlying sequence can serve as evidence either for an insignificant desalinization of the shallow-water paleo-water body, or for higher temperatures of carbonate precipitation (evidently both factors have their place).

It can be supposed that the same (or close) isotope characteristics were inherent to the initial carbonate matter of the mineralized bacterial mats (to all appearances, of initially calcite composition), the residues of which we see today in the form of ovoids and laminae in the composition of our studied manganese ores of the Mamatwan type. Now they (ovoids and laminae) are composed of kutnohorite and manganese calcite (Nel et al., 1986; Gutzmer and Beukes, 1996), which evidently is the result of past processes of the metasomatic transformation of the initial carbonates (evidently aragonite, calcite, dolomite) into manganese-containing carbonates. Accordingly, the isotope characteristics of kutnohorite and Mn-calcite for unaltered braunite lutites, as noted, are not primarily sedimentational but rather are secondary and should be close to −14‰ to −12‰ for $\delta^{13}C$ and no lower than 20–22‰ (SMOW) for $\delta^{18}O$. Analogous isotope characteristics for manganese carbonates of unaltered ores of Mamatwan type, as noted, have been proposed by Gutzmer et al. (1997): from −13.9‰ to −10.8‰ for $\delta^{13}C$ and from −15.7‰ to −11.8‰ for $\delta^{18}O$, PDB (accordingly 15.1–19.0‰, SMOW).

Returning to the question of the initial composition of the manganoprotolith, their quantitative correlation of primary carbonates and oxides of manganese remains unclear. Without question, a significant volume in the initial sediment was occupied by bacterial mats, originally representing organic matter, which was subsequently mineralized (as a rule, by calcium carbonate). The presence of primary sedimentary (or hydrothermal-sedimentary) manganese oxides in such facies (or lithotypes) will be highly limited and determined by paleofacies conditions.

The environments of sedimentation in which cyanobacterial mats are developed, as is known, are characterized by extreme conditions for organic life (its "blight"), arid or semiarid conditions, and the practically full absence of terrigenous impurity (supralittoral, intertidal zone, isolated and semi-isolated shallow-water marine basins, and the like). Participation by known mechanisms of entry by manganese (in the form of oxides) into such sediment (by means of continental runoff, upwelling, discharge of subaqueous hydrothermals, a volcanic source, etc.) encounters a range of difficulties, which in our case are difficult to overcome.

The same difficulties arise when arriving at a mechanism for the supply of the precursor elements of the initial manganoprotolith as a result of the discharge of ore-bearing hydrothermals in deep-water conditions. These theories today are actively being developed (Gavrilo, 1972; Bonatti et al., 1976; Crerar et al., 1982; Roy, 1986; Flohr and Huebner, 1992; Varentsov et al, 1993; Starikova 2001; and others) and are widely applied in explaining the genesis of many deposits and ore-occurrences of manganese developed in regions with active manifestation of volcanic and hydrothermal activity (see Section 3.2).

According to these theories, treated in detail in the work (Brusnitsyn et al., 2009), the accumulation of ore-bearing sediment occurs in the zone of the discharge of hydrothermal solutions circulating in the strata of underlying rocks as well as the discharge of those issuing onto the bottom of the water body. It is proposed that the zonal paleohydrothermal deposits with jaspelite "core" and manganese deposits on the periphery, characteristic for the South Urals, were formed by this means (Brusnitsyn et al., 2000, 2009; Starikova et al., 2004; Kuleshov and Brusnitsyn, 2005).

Manganese ores of the deposits of the Kalahari ore field are substantially distinct from those of the South Urals. The presence of mineralized (kutnohorite, manganous calcite) residues of cyanobacterial fibers, which serve as evidence for shallow-water conditions of the primary sediment, together with the morphological particularities of the ore deposits (extended bed bodies of consistent thickness by comparison with "hummocky" near-hydrothermal buildups) and isotope particularities enable us to consider that the braunite lutites of the deposits of the Kalahari basin had a mechanism of formation substantially distinct from that inherent to hydrothermal-sedimentary deposits of the Franciscan (Hein and Koski, 1987) and Primagnitogorsk basins (Kuleshov and Brusnitsyn, 2005; Brusnitsyn et al., 2009).

The predominant quantity of manganese (commonly in isomorphic mixture with iron) is contained in minerals that as a rule are secondary with relation to carbonate minerals and are developed on account of the latter. It can therefore be supposed that the basic mass of silicate minerals of manganese was formed as a result of the influence of metasomatic solutions rich in manganese, silicon, and in part iron.

Of course, this proposal is connected with a range of questions requiring explanation in turn. One of these is the question regarding the source of the ore-forming metasomatic (hydrothermal-metasomatic?) fluids themselves, as well as that of their associated ore matter—manganese. The single most plausible source of such fluids could only be the solutions that supply the underlying vulcanites of andesite composition of the Ongeluk Formation. This proposal has already been expressed (Cornell and Shutte, 1995). The hydrothermal-sedimentary mechanism of formation of the ores themselves in the proposed variant has been subjected, as noted, to thorough criticism by the noted authors (Beukes and Gutzmer, 1996); as it stands, the mechanism is untenable and requires further revision.

In the present case, it can be supposed that metasomatic fluids, saturated by ore components in the underlying andesite lavas, infiltrated into still non-lithified sediments and advanced within the beds laterally (Fig. 3.103). At this stage of development of the sediment metasomatic processes led to the formation of its material composition, including to "manganization"—microbial mats mineralized by

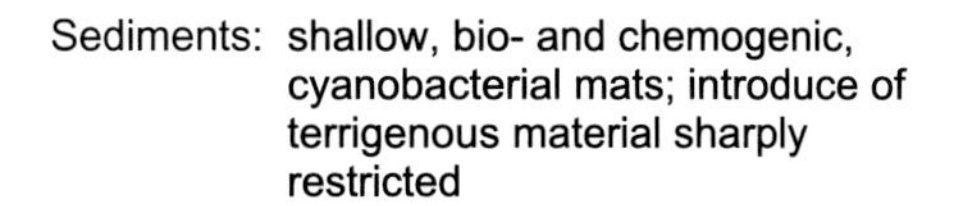

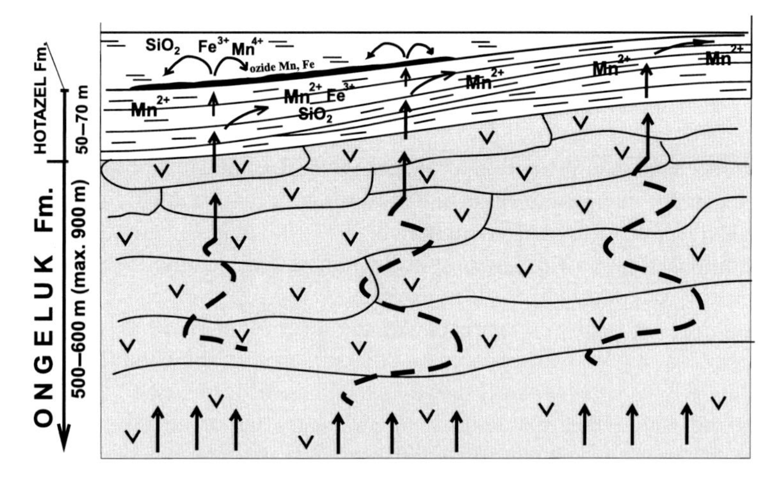

FIG. 3.103

Schematic diagram of rocks and ores formation of Kalahari manganese deposits (South Africa).

carbonates of calcium and magnesium are replaced by manganous calcite and kutnohorite. Moreover, in many cases also preserved are initial biogenic as well as structural and textural sedimentational indicators.

It cannot be excluded that part of such solutions could have reached the surface of the sediment—that is, a "seepage" of such solutions could have occurred through still non-lithified sediment and entered into the bottom waters of the shallow-water paleo-water body, bearing with them dissolved (ore and non-ore) components. In this case, a precipitation of manganese (as well as of iron and silicon) into the sediment occurred in the form of oxides. This mechanism of entry by ore matter into the sediment is substantially distinct from the above-examined example of hydrothermal discharge.

The scale of this process, its mechanism, and the physical-chemical conditions of the metasomatic restructuring of the initial rocks are still problematic and require further study. At this point, it can only be noted that examples are known to us of manganese ores that are metasomatically formed, but unsubjected to subsequent metamorphism, in the form of carbonates. One of these is the Usa manganese deposit of Kuznetsk Alatau (Russia, Keremovo oblast) (Kuleshov and Bych, 2002), the resources of which, without accounting for the section of the deposit destroyed as a result of tectonic movements (blocks sunken or destroyed by erosion), as well as the part of the deposit destroyed under hypergenic conditions (left-bank and right-bank sections), significantly exceed 100 million tons. In distinction from deposits of the Kalahari basin, highly permeable clastic carbonate rocks of near-reef formations of the Early Paleozoic here were subjected to processes of manganese-ore metasomatism.

In explaining the genesis of manganese ores of the Kalahari deposit, an important question concerns the mechanism and conditions of oxidation of the initial manganese carbonates, insofar as manganese lutites in modern form are represented predominantly by oxides (braunite). The transition of manganese carbonates into oxide forms, as is known, requires an environment with high values of Eh (Garrels and Christ, 1968; Brookins, 1988). Similar conditions could have arisen with the aeration of the sedimentary strata as well as with the entry into the system by carbonate-manganese rocks of oxygen-rich vadose (meteoric) waters. The probability is not excluded that such a mechanism prevailed for the examined sequence of rocks of the Kalahari manganese field, insofar as even insignificant variations of the water level (or eustatic movements of the landmass) could have led to the realization of such conditions for the shallow-water basin of sedimentation characteristic for the examined basin.

The proposed mechanism of formation of the ores of the examined deposit enables an understanding of the isotope data for manganese carbonates composing ovoids and laminae of unaltered braunite lutites. Lighter values of $\delta^{18}O$ (by approximately 8–10‰) inherent to carbonates of manganese ores could reflect conditions of a transformation of initial mineralized residues of microbial mats by metasomatic solutions, the waters of which (deep or meteoric) were invariably characterized by a lighter isotopic composition of oxygen by comparison with the water of the basin. It is also possible that these processes took course at higher temperatures by comparison with the conditions of formation of the overlapping and underlying shallow-water as well as with those of biogenic carbonates.

The isotopic composition of carbon undoubtedly indicates participation in this process by oxidized carbon of organic matter (as is well described in the scientific literature).

After the accumulation of primary sediment (including of manganoprotolith) of the Hotazel formation, over the course of the extended geological development of the territory of the examined basin and by measure of the accumulation of overlapping sediments and their slope into the depth, the rocks underwent processes of regional metamorphism in facies of green shales with the formation of a mineral paragenesis particular to manganese (braunite) lutites. Subsequent metamorphic changes for these rocks have been described in detail by N. Beukes, J. Gutzmer, and coauthors in a range of works mentioned above.

Finally, it should be noted that the genesis of manganese (braunite) lutites in the deposits of the Kalahari manganese-ore basin cannot be considered separately from the understanding regarding the conditions of formation of ferruginous quartzites of the Hotazel formation, which are part of the composition of the ore strata and are interstratified with braunite lutites along the entire thickness of the sequence. This question has been treated by the specialized work of Tsikos et al. (2003).

In the cited work, detailed research has been conducted within the boundaries of the deposits of the northern and southern parts of the manganese-ore field on the isotopic composition of calcite, kutnohorite, dolomite, ankerite, hematite, magnetite, and quartz from various types of rocks of the iron-ore formation (poor and rich in carbonate, containing Fe-Mn carbonates) and manganese ores. The results of the research have allowed the authors to conclude that the isotope data for calcites and ankerites of the deposits of the southern part of the manganese-ore field ($\delta^{13}C = -18$‰ to -4‰; $\delta^{18}O = 12$–20‰) constitute evidence of past processes of diagenesis with participation by carbon of oxidized organic matter, accompanied by a reduction of iron of high valence.

Subsequent processes of transformation of the rocks occurred under the influence of fluids that arose along tectonic faults as well as along the zone of discordance on the border of the Hotazel and Olifantshoek formations. A theory of the fault pattern as a controlling factor in the superimposed processes of transformation of manganese lutites has been demonstrated in detail in a range of works by J. Gutzmer and N. Beukes with coauthors (Gutzmer and Beukes, 1995, 1997; Beukes et al., 1995; Gutzmer et al., 1997; Chetty and Gutzmer, 2012; and others).

In this way, the study of the internal structure of the ore matter (petrographic, SEM), as well as the survey of publications on the Kalahari manganese-ore field, which concerns primarily the geochemistry of stable isotopes of carbon and oxygen of carbonates, quartz, and oxides of iron from various types of rocks of the ore strata of the Hotazel Lower Proterozoic formation, allows for fresh understandings of certain aspects of the primary genesis of braunite lutites (ores of Mamatwan type).

The internal structure of manganese ores (the character of interrelations of carbonate and oxide minerals of manganese, the presence of microbial textures) and the particularities of their isotopic composition enable the proposal that metasomatic solutions that "infiltrated" the initial biogenic-sedimentary (predominantly of bacterial-mat origin) deposit of the shallow-water Early Proterozoic paleobasin can serve as a source of manganese. These solutions carried ore components (primarily manganese and iron) along with silicon into the system of initial, evidently still non-lithified sediment.

The possibility cannot be excluded that a portion of these solutions could have either seeped onto the floor of the paleo-water body and/or delivered manganese oxides into the initial sediment. To all appearances, the underlying lavas of andesite composition of the Ongeluk formation served as a source of manganese.

3.2.4.2 Ulutelyak manganese formation

Particularities of geological structure

The Ulutelyak manganese deposit is situated in the eastern part of the East European platform, 95 km east-north-east from Ufa city and 14 km from Asha city (Bashkortostan), and is classified as a small deposit. Its reserves and resources have been estimated differently by various authors; the estimated reserves of ores in different conditions constitute from 11.3 to 20–25 million tons with resources and metallogenic potential from 50 to 500 million tons and greater (Magadeev et al., 1997; Cheban, Gareev, 1997; Margantsevye et al., 1999; Mikhailov, 1992; Zharikov, 2000). The deposit is represented predominantly by carbonate-manganese-containing rocks and poor manganese ores with an insignificant quantity of oxide-carbonate and oxide ores. This deposit from the moment of discovery (1936–1940) has been studied by a range of researchers (Khabakov, 1944; Betekhtin, 1946; Kheraskov, 1951; Varentsov, 1962b; Makushin, 1970; and others). At present, it has been fairly well explored by bore drilling (mapping, prospecting, and geological-exploratory, 1967–70) and quarry developments. On the basis of these works, the principal regularities have been established for the geological structure of the deposit, the mineral and chemical composition of the manganese-containing and manganese rocks and ores, their genetic particularities, and their distribution (Sarkisian, 1945; Gribov, 1972a,b, 1974, 1978, 1982; Makushin, 1970, 1972, 1975; Mizens, 1979; and others). To date, however, the origin of the deposit and the source of the ore matter itself remain unclear.

In the geological regard, the Ulutelyak manganese deposit coincides with the south-eastern end of the Upper Kama depression, divided into the Bashkir and Tatar uplifts. From the Cis-Urals trough, it is separated by the Ryazan'-Okhlebinino embankment of submeridional trend.

In the structural regard, this deposit is situated in the region of the northern termination of the Cis-Urals (White) depression near the tectonic contact with the more ancient formations of the Karatau structural complex. The central part of the manganese-ore deposit (field) is located in the core of the Kazayak anticline, which is composed of rocks of the Artinskian and Kungurian stages of the Lower Permian (Fig. 3.104).

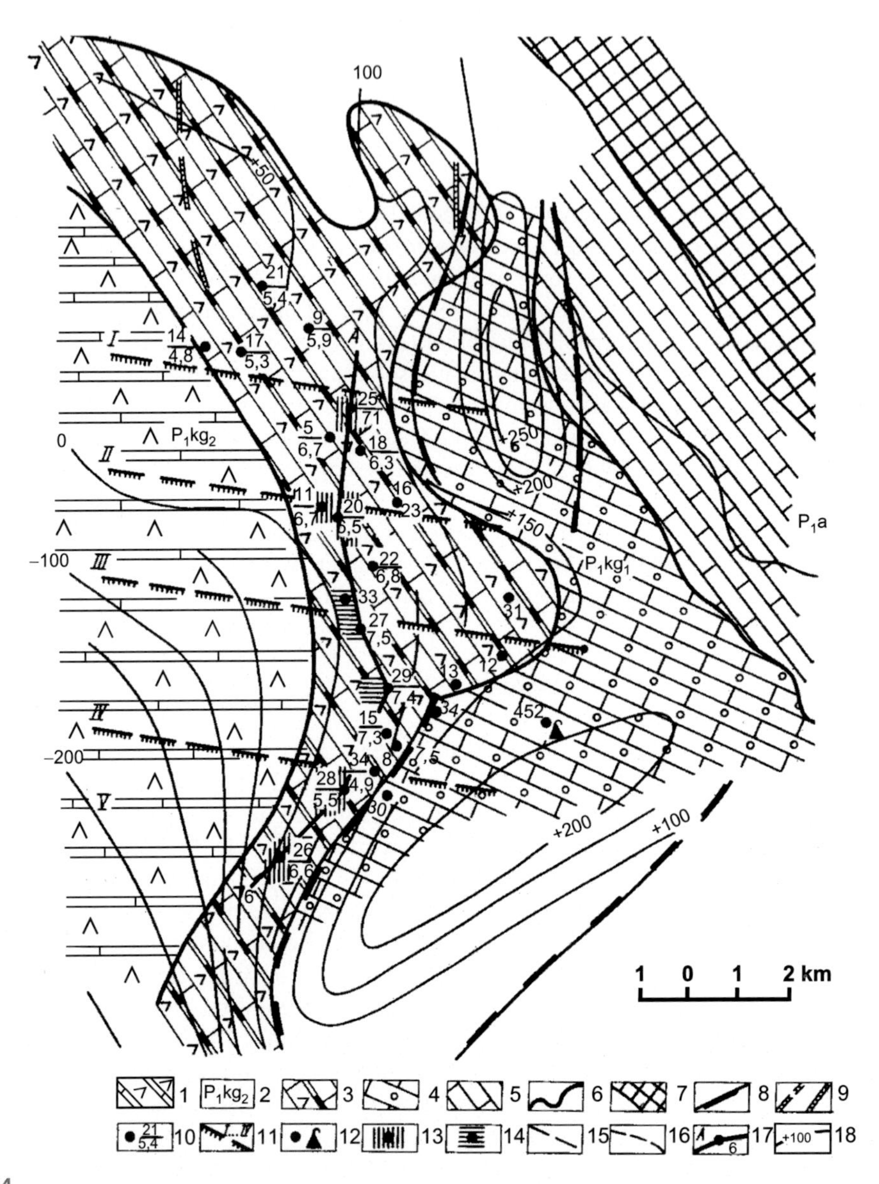

FIG. 3.104

Geostructural map of the manganese-ore zone near the Karatau ore node. Modified after Makushin (1970, 1975). 1—Terrane of the Kungurian limestone-marl-dolomite-anhydrite facies (ore-barren members); 2—rocks of the Iren suite (P1kg2); 3—manganiferous rocks of the Ulutelyak suite; 4—zone of the transgressive occurrence of carbonate members of the Ulutelyak Formation on the Artinskian rocks; 5—Artinskian limestones (P1a); 6—geological boundaries; 7—Karatau anticlinorium; 8—contours of local synsedimentary structures of the Ulutelyak time; 9—contour of the positive synsedimentary structure of the late Ulutelyak time; 10—prospecting wells in 1968–70 (numerator shows the well number; denominator, thickness of the upper ore horizon, m); 11—boundaries of the structural-facies blocks and their numbers; 12—prospecting wells with the emission of hydrocarbon gases in 1967; 13–14—wells recovering alabandine mineralization: 13—low, 14—intense; 15 and 16—zone of maximal Mn contents in the: 15—lower ore horizon, 16—upper ore horizon; 17—profile; 18—stratoisohypse of the Artinskian section roof.

Manganese mineralization coincides with the Filipov horizon of the Kungurian stage (P_1kg fl), which in the area of the deposit is deposited discordantly on underlying carbonates of the Artinskian stage (Makushin, 1970) and is represented by rocks of the Ulutelyak suite (in the composition of the Ulutelyak suite is included a complex of Upper Artinskian-Lower Kungurian deposits of the Lower Permian deposited in the base of the halogen formation and represented by rhythmic alternation of two sulfate and two carbonate members), having a gentle attitude (not exceeding 5 degrees) with a fall to the south-west (Fig. 3.105) (Makushin, 1970, 1975; Gribov, 1982).

In the base of the Filipov horizon of the Ulutelyak suite are deposited anhydritic rocks (P_1kg fl_1) with a thickness of 35–40 m, which upwards along the sequence are replaced by carbonate (limestones, marls) manganous rocks (P_1kg fl_2) with a thickness reaching 5.3 m. Above is deposited a section of anhydritic rocks (P_1kg fl_3) with a thickness reaching 50 m. The vertical section of the ore-hosting strata is terminated by a bed of manganous limestones (P_1kg fl_4) with a thickness reaching 8 m.

Carbonate-manganese-containing rocks of the Filipov horizon are bituminized to varying degree; in a range of cases, they contain fluorite mineralization (Margantsevye et al., 1999).

The Filipov deposits without visible interruption are overlapped by anhydrites and gypsums of the Iren horizon of the Kungurian stage with a thickness reaching 25 m.

Industrial manganese mineralization coincides with clayey-limy (marly) layers of the Filipov horizon (P_1kg fl_2 and P_1kg fl_4). Moreover, the greatest contents of manganese have been noted in the upper horizon (5.5–8.2%; those of the lower horizon constitute 2–4%) (Makushin, 1975).

In the structure of the manganese carbonate horizons, they are divided along three rhythms, beginning with thin-layered marls and clayey-limy dolomites. Upwards along the section, they are replaced by layered ball clayey limestones and terminate in layered oolitic limestones. The thickness of the rhythms constitutes from 1.6 to 3.0 m.

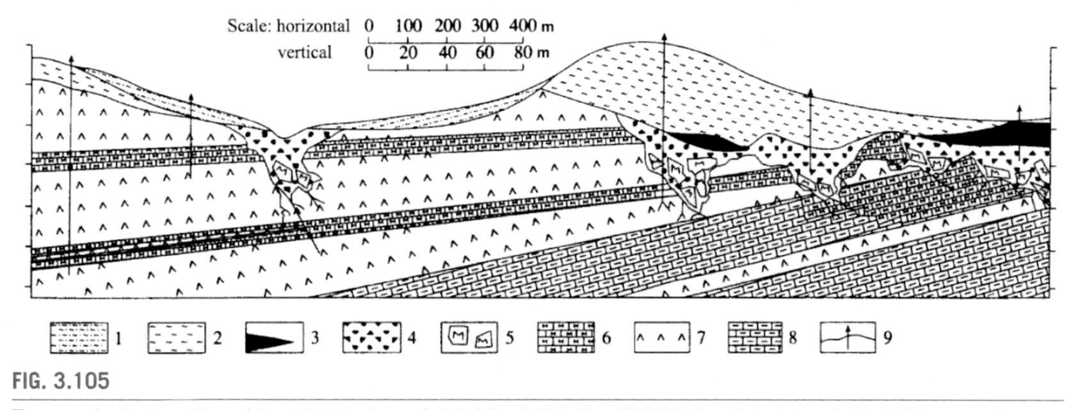

FIG. 3.105

Type geological section of the Ulutelyak ore field. After Mikhailov (2001), 1—deluvial and alluvial loam and sandy loam (Q_{2-4}); 2—loesstype loam cover and sandy loam at the base (N_2-Q); 3—powdery and dense silicified manganese ores (Pg_3-N_1); 4—deluvial-rubbly oxidized manganese ores; 5—semi-oxidized boulder-rubbly ores of the caprock and karst depressions (K-Pg); 6—manganiferous limestones; 7—gypsum and anhydrites; 8—limestones and marls; 9—wells.

Maximum manganization is characteristic for clayey limestones of the middle part of the rhythms; in the first rhythms have been noted the greatest concentrations of manganese. Manganese carbonates are represented by manganous calcite, manganous dolomite, calciorhodochrosite and magnesium-manganese-calcium carbonate (Gribov, 1972b, 1978).

Subsequent processes of karst- and crust development in the Mesozoic and Cenozoic eras led to the leaching of gypsums and carbonates and as a result to the deformation of the ore beds. Hypergenic transformation (leaching and oxidation) of manganese-containing carbonates led to the manifestation of secondary manganese oxides, which fill the karst caves and form the most industrially valuable secondary oxide ores (see Fig. 3.105).

Among secondary minerals of manganese, vernadite and psilomelane have been established (Gribov, 1972b).

Results of lithological and isotope research

With the aim of explaining the conditions of formation of manganese rocks and ores, studies were conducted on the lithological particularities and isotopic composition of carbon and oxygen of rocks from the ore-bearing strata itself, represented by interbeds of manganese-oxide ores and non-ore clayey-carbonate interlayers as well as of underlying and overlapping non-ore carbonates (the so-called flagstones).

Lithological particularities. By means of an optical microscope were studied the principally carbonate interbeds of the ore horizon, practically not containing manganese oxides and overlapping carbonates (the so-called thick plates). In the studied rocks, organogenic residues were found throughout; however, their highest content coincides with the upper oxidized carbonate horizon of the ore strata. Among these are present residues of foraminifera of the *Nodosaridae* and *Hemigordiopsidae* families (as determined by E.Ia. Leven, GIN RAS), as well as undetermined residues of large foraminifera-fusulinids, gastropods, and ostracods (Figs. 3.106 and 3.107). Also present are oncolites with characteristic micro-organogenic (algal) internal structure (Fig. 3.108).

Fine clasts of carbonate matter are commonly microbially corroded (by cyano- and archaebacteria, calcimicrobes, micromicelles, and the like). The studied samples are almost entirely devoid of impurities of clastic material (quartz, feldspars, and the like).

A characteristic particularity of rocks of the ore strata is the heterogeneity of their chemical composition: the content of iron, manganese, and calcium within a single sample vary substantially even with an insignificant distance between the studied points (data of a microprobe integrated with SEM, Fig. 3.109). The presence of silica has been noted throughout.

The results of microscope study under optical and scanning microscopes enable us to conclude that the rocks are represented by material of siltstone-pelitic and in part thin- to fine-sandstone dimensions, which to a significant degree have been subjected to recrystallization. In the composition of the rocks, besides the aforementioned rare clasts of undetermined residues of small macrofauna (evidently gastropods and bivalves), microfauna (primarily foraminifera), and other organic residues (see Figs. 3.106 and 3.107), are also present numerous microbial structures (Figs. 3.110 and 3.111A–E, G–K) representing mineralized residues of bacterial mats. Framboids of pyrite have been noted rarely (Fig. 3.111F).

It follows to note that no essential difference has been detected in the structure and composition of organic residues contained in the interbeds of oxide and carbonate rocks inside the ore strata or in the overlapping and underlying limestones.

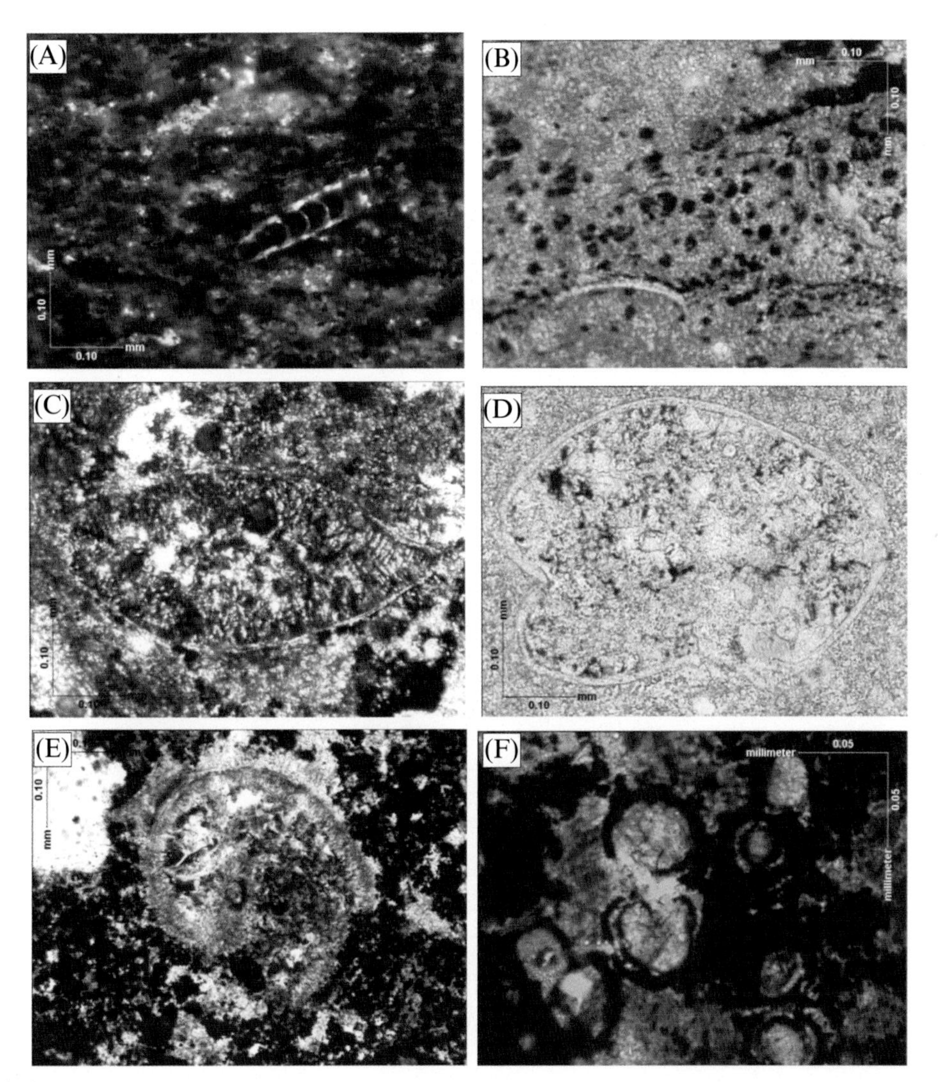

FIG. 3.106

Remnants of small Mn-bearing ostracod and foraminifer tests. Photomicrographs of thin sections (crossed nicols). (A) Foraminifera (sample 14a/03); (B) fragment of ostracodes (sample 14a/03); (C) ostracodes (sample 8/03); (D) ostracodes (sample 3/04); (E) foraminifera (sample 10/03); and (F) cluster of alabandite(?) (sample 10/03).

In this way, the substantially carbonate composition of non-ore interbeds, the small size of clasts of carbonates (from hundredths and tenths up to a few millimeters), the absence of an impurity of terrigenous material (quartz, feldspars, and the like), and the presence of isolated residues of repressed macrofauna and abundant microbial formations serve as evidence for extreme (high-stress) conditions of the development of organic life and for a regime of sediment accumulation of the ore-bearing strata

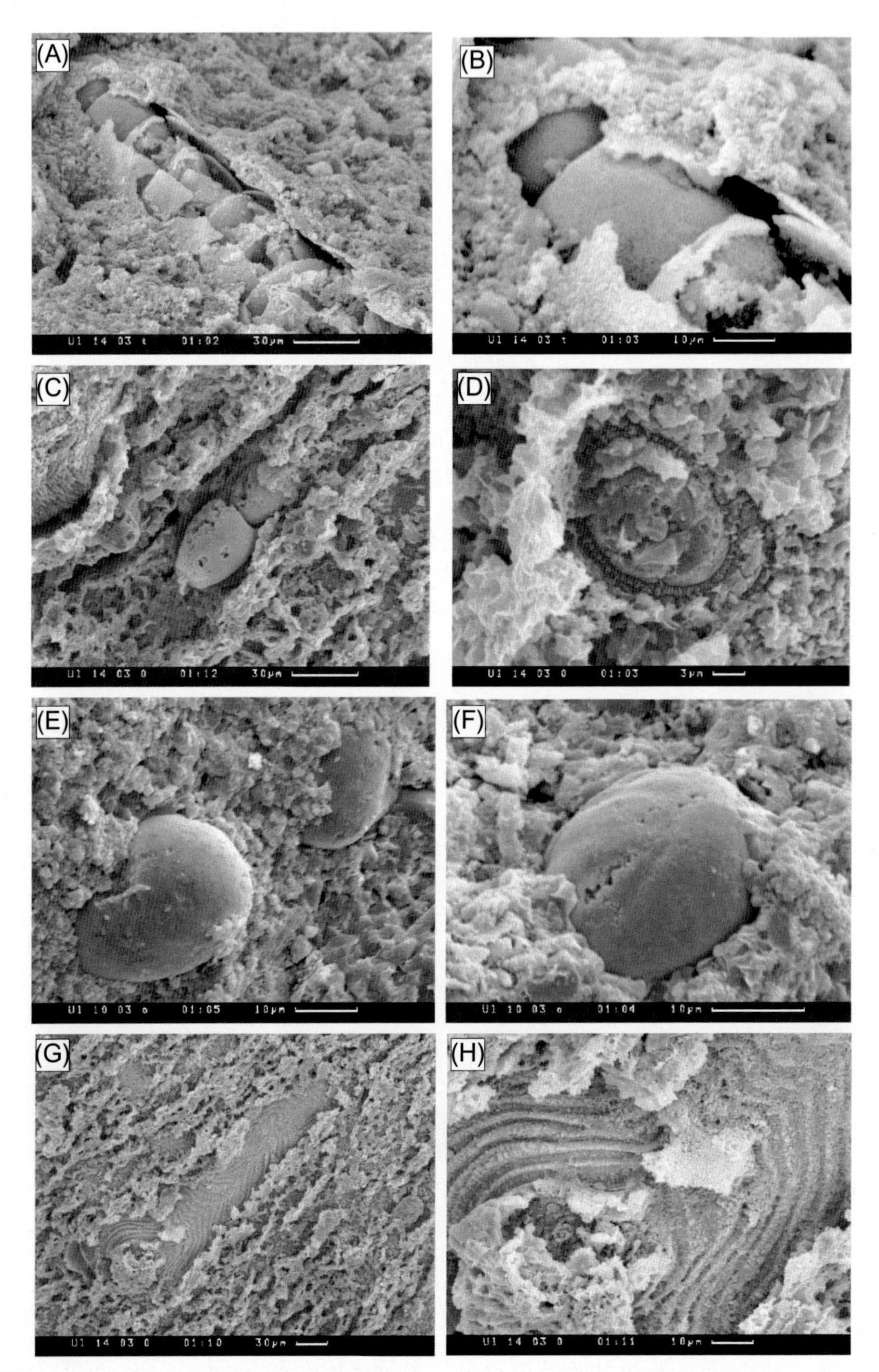

FIG. 3.107

Organogenic remnants in the manganese carbonate (A)–(D) and Mn-oxide-carbonate (E)–(K) rocks (SEM microphotographs taken by E.A. Zhegallo and E.Ya. Shkol'nik). (A)–(D) Sample 14a/03: (A) remnant of fusuline, (B) same, increased, (C) remnant of fusuline, (D) transverse section of foraminifera; (E) and (F) sample 10/03, remnants of foraminifera; (G) and (H) remnants of plant detritus at different magnifications.

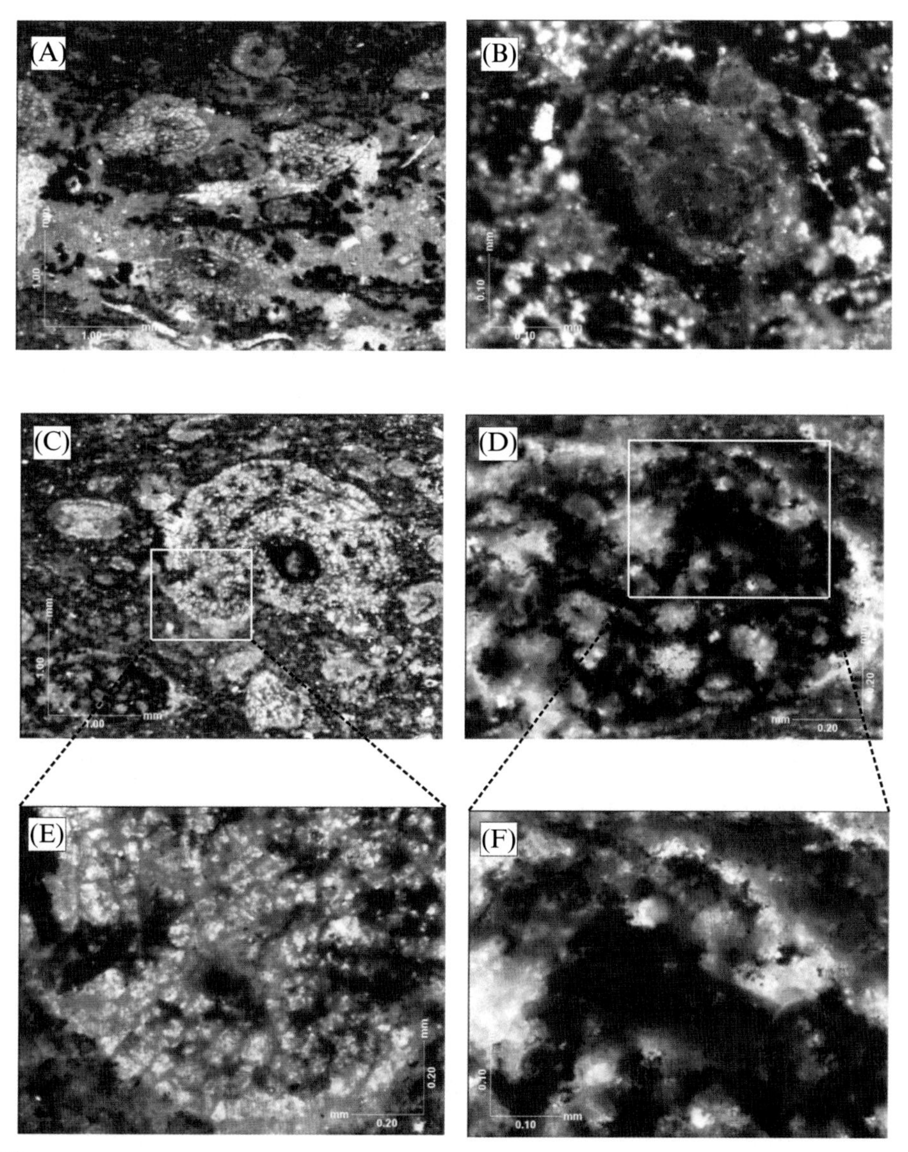

FIG. 3.108

Structure and pattern of manganization in oncolites and ooids at different magnifications, Ulutelyak deposit (sample 14a/03, crossed nicols). (A), (C), and (E) oncolites. (B), (D), and (F) ooids.

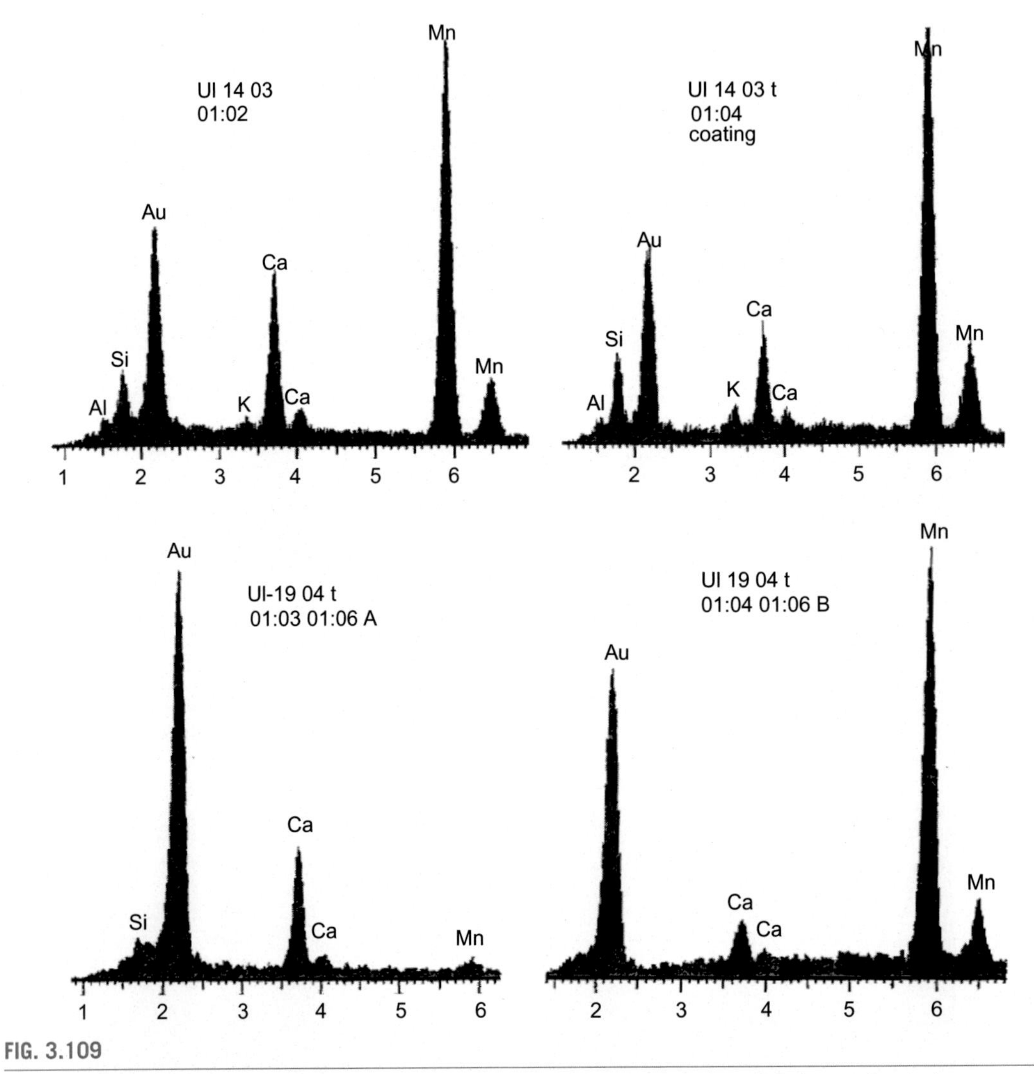

FIG. 3.109

The dot chemical composition of samples of 14/03 and 19/04, Ulutelyak deposit (microprobe analyzer combined with a scanning microscope).

in the Permian basin. Such conditions are commonly characteristic for semi-isolated basins of the sabkha type, where the biochemogenic accumulation of deposits predominates in conditions of dry and warm (semiarid and arid) climate with intensive evaporation and active development of bacterial mats. Continental runoff under such conditions is sharply limited due to the weak compartmentalization of the surrounding relief.

Oxides of manganese within the boundaries of the ore body represent the result of oxidized manganese-containing carbonates (manganese calcite and dolomite, possibly manganocalcite) under hypergenic conditions. Their distribution in the rock is uneven, due to the primary heterogeneity of

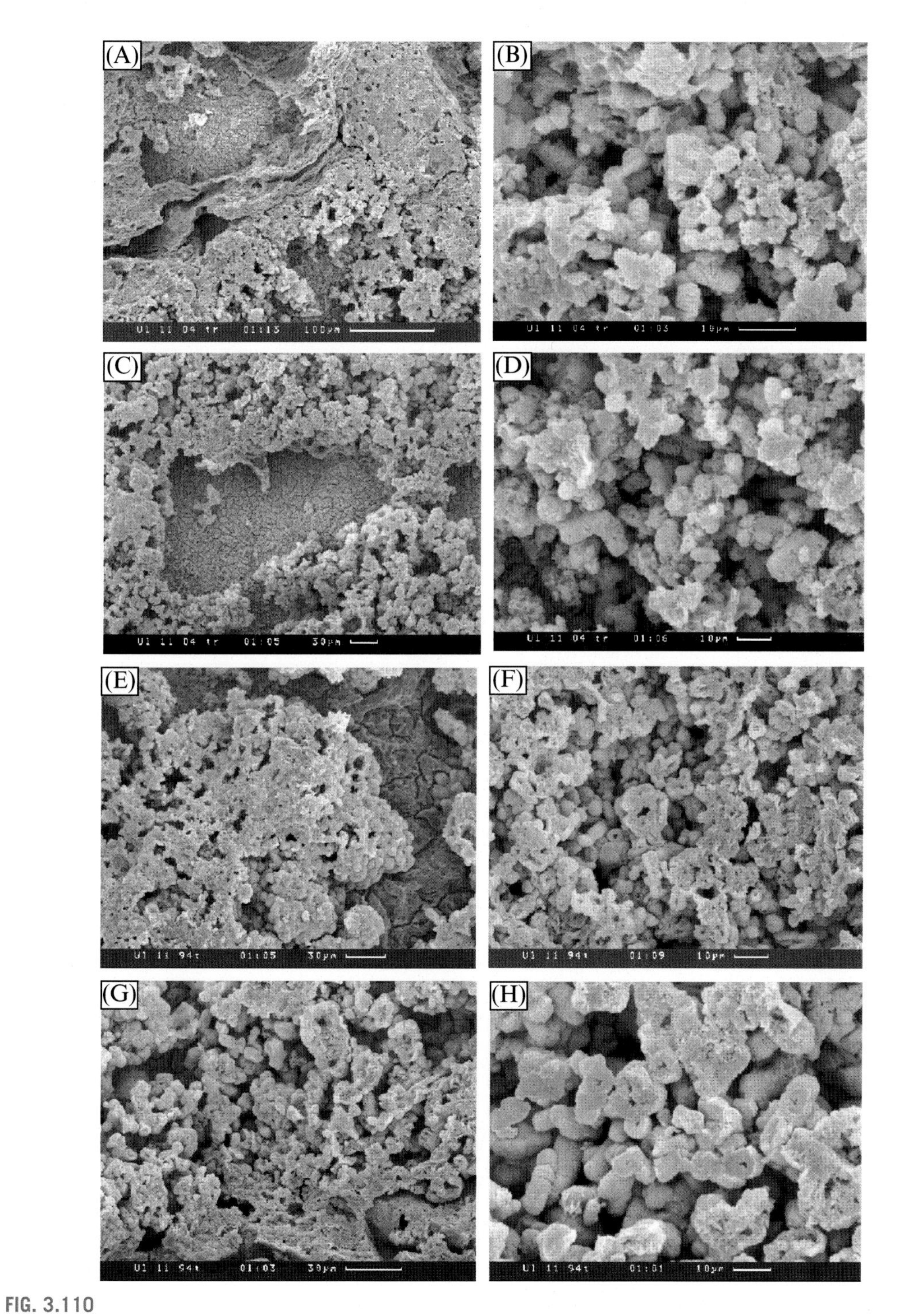

FIG. 3.110

Mineralized microbial remnants in carbonate rocks, Ulutelyak deposit (A–H—different parts of the sample 11/04 at different magnifications, SEM microphotographs taken by E.A. Zhegallo and E.Ya. Shkol'nik).

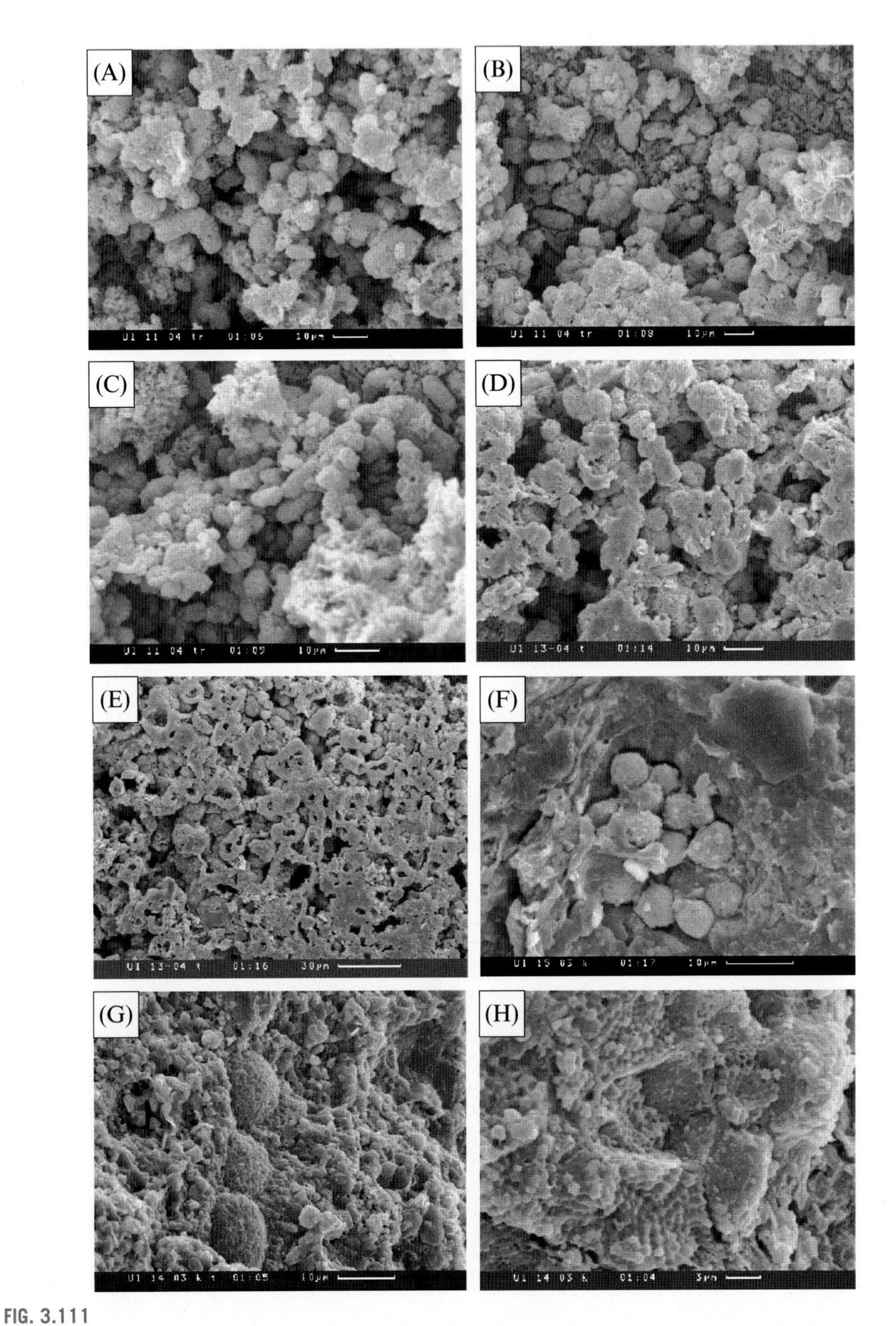

FIG. 3.111

Mineralized microbial remnants (A)–(E) and (G), (H) and segregations (framboids) of pyrite (F) in carbonate rocks, Ulutelyak deposit (SEM microphotographs taken by E.A. Zhegallo and E.Ya. Shkol'nik).

the sediment as much by the organogenic (microbial, algal, and the like) nature of initial matrix (see Figs. 3.106, 3.108, and 3.112A–F), as its "microball" texture (Fig. 3.112G–I). In many cases, alabandite is characterized by an analogous distribution in the rock (see Fig. 3.106F) (Gribov, 1972a).

It follows to note that in the overlapping limestones of the "thick plate" in sections situated close to the ore member have also been subjected to manganization and bitumization, which coincide with the "intergranular" space of the "microballs" (Fig. 3.113).

Thin cross-veins of secondary, evidently hypergenic calcite have been noted extremely rarely in thin sections (isolated observations).

Isotopic composition of carbon and oxygen. For the purposes of isotope research, samples were gathered from the northern and southern quarries of carbonate and oxide-carbonate rocks already subjected to hypergenic changes (containing manganese oxides). To that end, samples containing secondary hypergenic calcite (incrusted, joint fill, and the like) were excluded.

The obtained isotope data are recorded in Table 3.24 and displayed in Figs. 3.114 and 3.115. From these data, it follows that on the whole the values of the isotopic composition of carbon and oxygen vary within a broad interval. The highest values of $\delta^{13}C$ (2.8–4.5‰) are characteristic for carbonates of the underlying and overlapping "flagstones." Such values are substantially higher than those in common marine carbonates and are characteristic for carbonates formed in shallow-water semi-isolated continental basins under semiarid and arid conditions.

The isotopic composition of the sulfur of anhydrites of the upper strata (anhydritic quarry; values of $\delta^{34}S$ constitute 14.3‰ and 14.5‰) is characteristic for sulfates of this age.

The isotopic composition of carbon and oxygen of carbonate matter in oxide-carbonate ores is characterized by fairly broad variations: from −3.8‰ to 3.3‰ for $\delta^{13}C$ and from 22.0‰ to 29.6‰ for $\delta^{18}O$; that is, they are on the whole rich in light isotopes of carbon and oxygen by comparison with the overlapping and underlying "flagstones" and non-ore marly interbeds of the ore member (northern quarry). Furthermore, a direct correlation can be seen between values of $\delta^{13}C$ and $\delta^{18}O$—samples with the heaviest isotopic composition of carbon are characterized by the highest values of $\delta^{18}O$, and vice versa (see Figs. 3.114 and 3.115). Falling into the range of heavy isotopic composition are not only carbonates of the underlying and overlapping "flagstones" and non-ore carbonate interbeds of the northern quarry, but also a share of the oxide-carbonate samples of the ore horizon of the southern quarry.

The observed lightest values of isotopic composition of both carbon and oxygen of manganese carbonates probably result from the presence of late (authigenic) carbonate, the formation of which is connected with the entry into the system of the initial sedimentary rock by infiltrating solutions, which contained a certain quantity of CO_2 and were characterized by a lighter isotopic composition of oxygen of water as well as that of carbon of dissolved carbon dioxide (range "A" in the graph of Fig. 3.114).

What is the origin of these solutions?

Above all, an increase in the lightness of the isotopic composition of the initial carbonates could have taken place under the influence of modern soil waters and infiltrational waters of meteoric genesis. This process can be discerned today—within 1–2 years the surfaces of oxide-carbonate rocks exposed in quarries are covered by a crust of manganese oxides and incrustations of hypergenic calcite. An impurity of such calcite in the composition of the sample could also be caused by our observed regularity in the distribution of $\delta^{13}C$ and $\delta^{18}O$ values.

With an explanation of the established regularity, however, this mechanism in the given case can be excluded for the following reasons. Firstly, our studied samples, according to visual and microscope observations, contain practically no impurity of secondary calcite. Secondly, calcite (the first

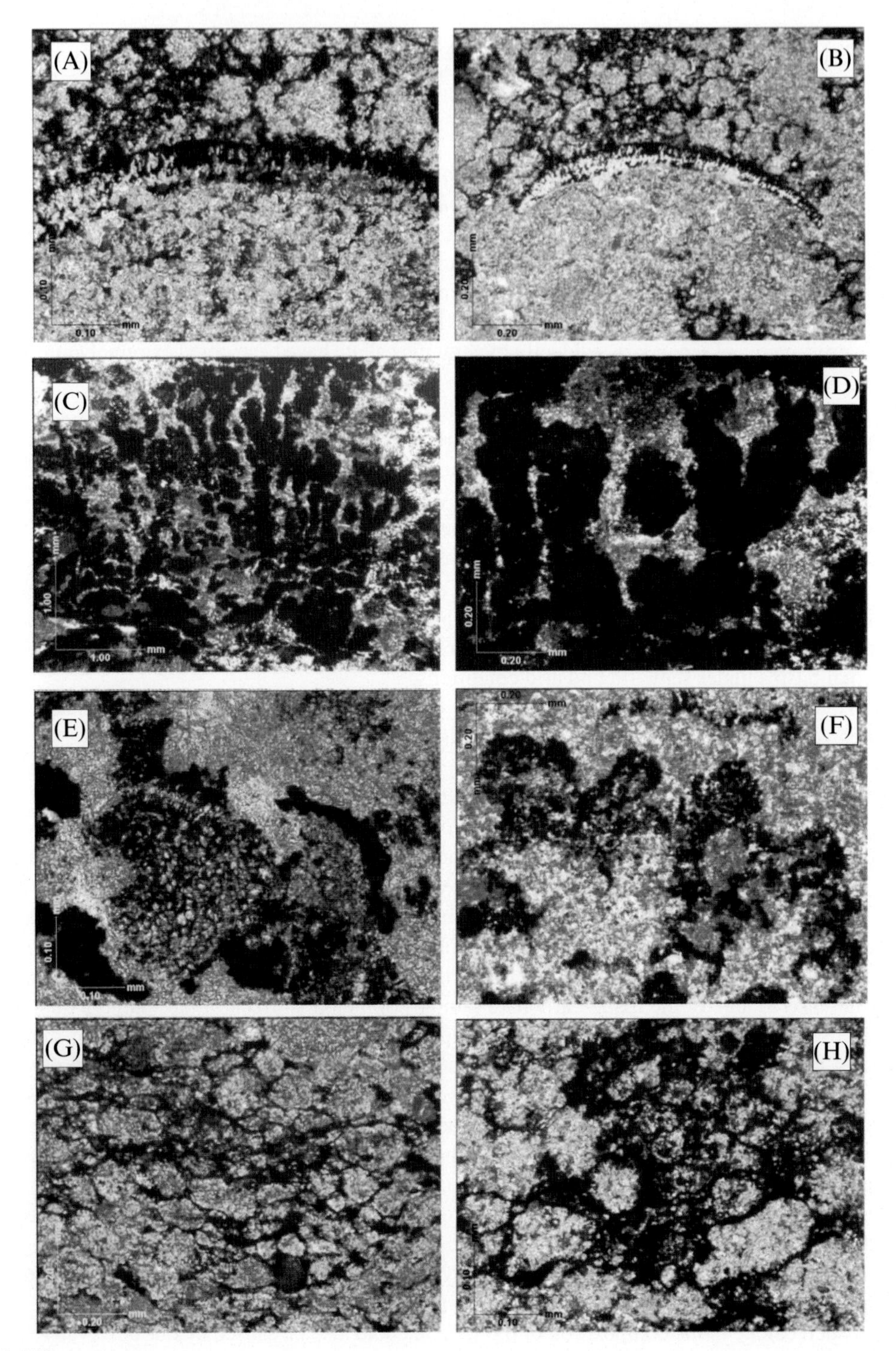

FIG. 3.112

Manganiferous organogenic remnants and carbonate matrix of the Ulutelyak deposit. (A) ostracod test, sample 15/04 (crossed nicols); (B) ostracod test, sample 15/04 (parallel nicols); (C) algal texture, sample 8/04 (parallel nicols); (D) algal texture, sample 8/04 (parallel nicols); (E) algal texture, sample 12/04 (crossed nicols); (F) algal (?) texture, sample 15/04 (crossed nicols); (G) manganized cyanobacterial (?) matrix, sample 3/04 (crossed nicols); and (H) manganized cyanobacterial (?) matrix, sample 5/04 (crossed nicols).

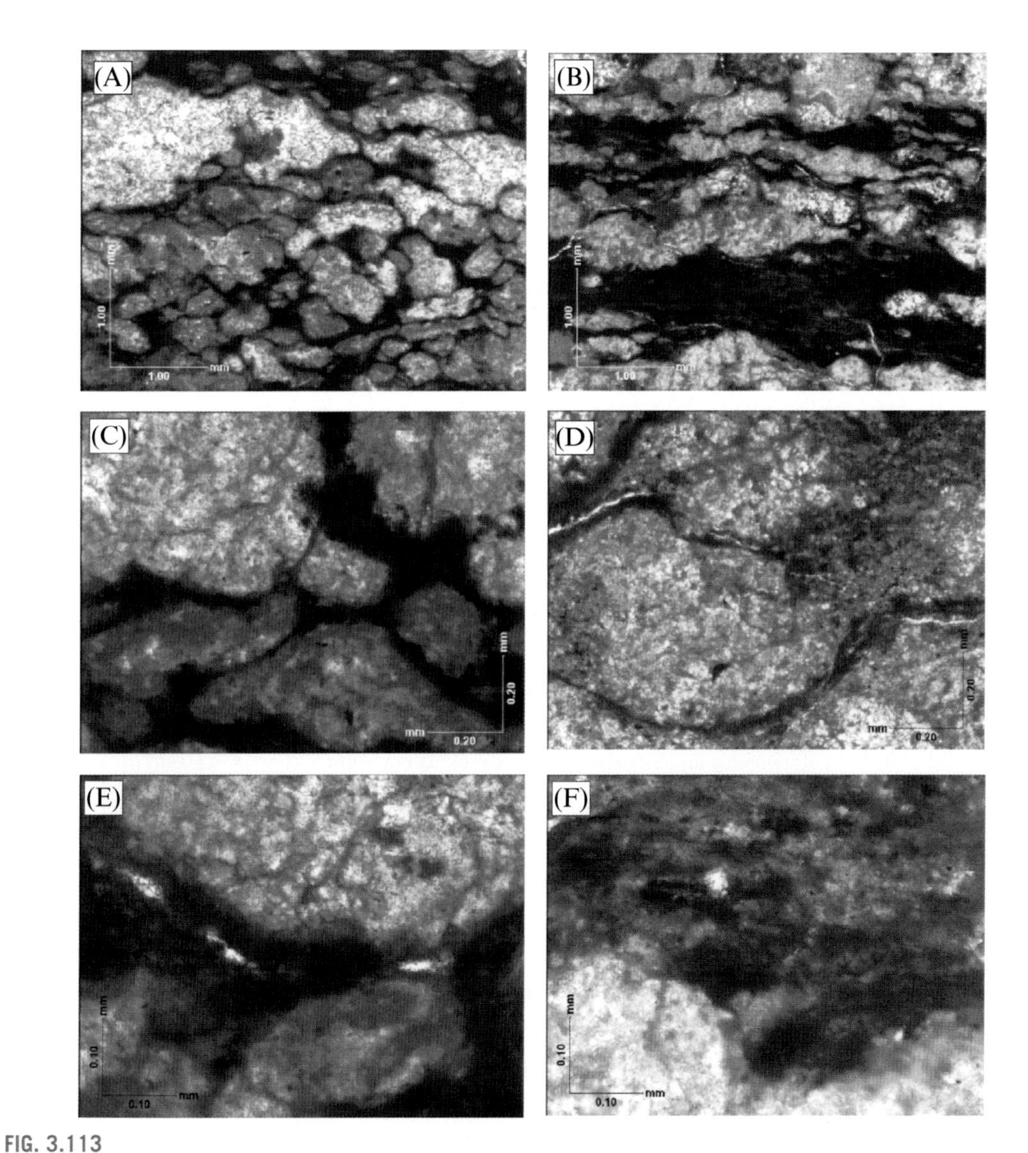

FIG. 3.113

Character of manganization of the carbonate matrix in the overlying carbonate rocks ("thick slab"), Ulutelyak deposit. (A)–(C) Sample 15/03 at different magnifications (crossed nicols) and (D)–(F) sample 18/03 at different magnifications (crossed nicols).

portion of CO_2 with a two-stage decomposition of the sample in orthophosphoric acid) in all samples in terms of isotopic composition is close to (or located in isotopic equilibrium with) the weakly soluble carbonate (manganocalcite, manganous dolomite) of this sample. Thirdly, the ratio of calcite/manganous carbonate in the studied samples varies from 1:1 to 1:8; moreover, there is no correlation between the distribution of $\delta^{13}C$ and $\delta^{18}O$ values and the value of this ratio. If the observed increase

Table 3.24 Isotopic Composition of Carbon, Oxygen, and Sulfur in Rocks of the Ulutelyak Manganese Deposit

Number of Analysis	Number of Sample	Characteristics of Sample	$\delta^{13}C$, ‰ PDB	$\delta^{18}O$, ‰ SMOW
"Northern" quarry				
5218	9_03	Underriding carbonates ("lower plate"); calcite	2.8	29.2
Block of carbonate-oxide rock (ore)				
5220	11_03	Interbed of thin-layered oxide-carbonate rock 0.30 m higher than the foundation of the block; calcite	3.7	28.8
5221	12_03	Higher by 0.55 m, clayey-marl interbed; calcite	2.7	28.0
5222	13_03	Higher by 0.25 m, carbonate-oxide rock; calcite	2.9	29.7
5223	14_03	Higher by 0.50 m, oolitic oxide-carbonate rock; calcite	−3.8	22.3
5224	15_03	Overlapping carbonate ("thick plate"); calcite	3.1	29.5
5225	16_03		4.5	29.9
5227	18_03		4.4	29.7
5228	19_03		4.4	30.0
"Southern" quarry				
Separate block of oxide-carbonate-manganese ore				
5411	2_04	Rare-oolitic layered oxide-carbonate rock, 20 cm below the upper edge of the "block"; manganocalcite	−0.7	23.1
5412	3_04	Rare-oolitic layered oxide-carbonate rock, lower by 10 cm; manganocalcite	2.5	28.7
5414	5_04	Rare-oolitic layered oxide-carbonate rock, with sections of friable manganese oxides and crystals of secondary calcite; manganocalcite	3.3	29.4
5416	7_04	Substantially clayey thin-layered interbed with manganese oxides, lower by 25 cm; manganocalcite	1.2	26.9
5417	8_04	Thin-layered oxide-carbonate rock, lower by 15 cm; manganocalcite	2.3	27.8
5418	9_04	Clayey interbed, lower by 30 cm; manganocalcite	−2.4	24.2
5419	10_04	Horizon of thin-layered oxide-carbonate rocks, lower by 20 cm; manganocalcite	−0.1	26.1
5420	11a_04	Layered fine-oolitic oxide-carbonate rock; lower by 30 cm, oxidized interbed, manganocalcite	3.0	29.7
	11b_04	Layered fine-oolitic oxide-carbonate rock, non-oxidized interbed; manganocalcite	1.6	28.2
5421	12_04	Oxidized fine-oolitic carbonate rock; lower by 35 cm; manganocalcite	−1.1	27.6
5422	13_04	Coarse-layered oxide-carbonate rock, lower by 30 cm, foundation of edge; manganocalcite	−0.3	22.5
5426	17_04	Carbonate of separate edge of oxide-carbonate ore, lower by 20 m; manganocalcite	4.1	29.6
5428	19_04	Carbonate of separate edge of oxide-carbonate ore, lower by 25 m lower; manganocalcite	−1.9	22.0
Anhydritic quarry				$\delta^{34}S$, ‰
	1_03	Anhydrite, quarry bottom, by watermark		14.5
	5_03	Anhydrite, quarry wall, higher by 10.5 m		14.3

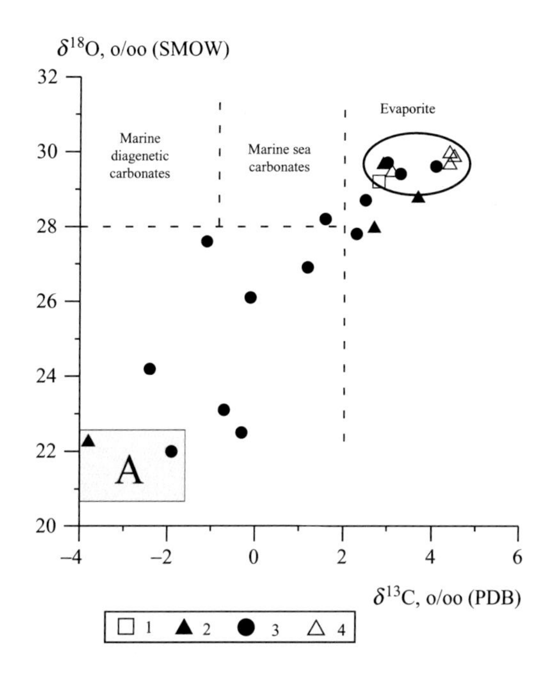

FIG. 3.114

Distribution of $\delta^{13}C$ and $\delta^{18}O$ values in rocks of the Ulutelyak manganese deposit. 1—Underlying flagstone (northern quarry); 2—block of manganese-oxide-carbonate ores in the northern quarry; 3—block of manganese-oxide-carbonate ores in the southern quarry; 4—overlying flagstone (northern quarry). Zone A is the inferred source of the isotopically light carbon dioxide.

in the lightness of the isotopic composition of the studied manganese and manganous carbonates were connected with the presence of secondary (hypergenic) calcite, then we would be justified in expecting a direct correlation between the degree of the increasing lightness of isotopic composition and the quantity of such calcite.

To some degree, evidence for the priority of the established regularity can be found in the unevenness of the distribution of $\delta^{13}C$ and $\delta^{18}O$ values along the sequence (see Fig. 3.115).

Consequently, the observed increase in lightness of the isotopic composition of carbon and oxygen in the studied manganese and manganous carbonates is due to another process and occurred prior to the modern (Paleogene-Quaternary, possibly Mesozoic) transformation under hypergenic conditions (although it is highly complicated to completely rule out the presence of hypergenic calcite).

The nature of the transforming metasomatic solutions and the mechanism of this transformation remain unclear. It can be supposed that such waters are closely connected with petroleum waters characterized by elevated concentrations of manganese, a light isotopic composition of carbon and oxygen, and an elisional origin. This transformation evidently occurred at the nano-level, without recrystallization of the initial rocks, and consequently without alteration of the isotope-oxygen background of the initial rocks. In thin sections, we do not observe isolated phases of secondary carbonate.

The geological data do not accommodate mechanisms of kinetic division of isotopes with decarbonatization as an explanation for lighter values of the isotopic composition of carbon of

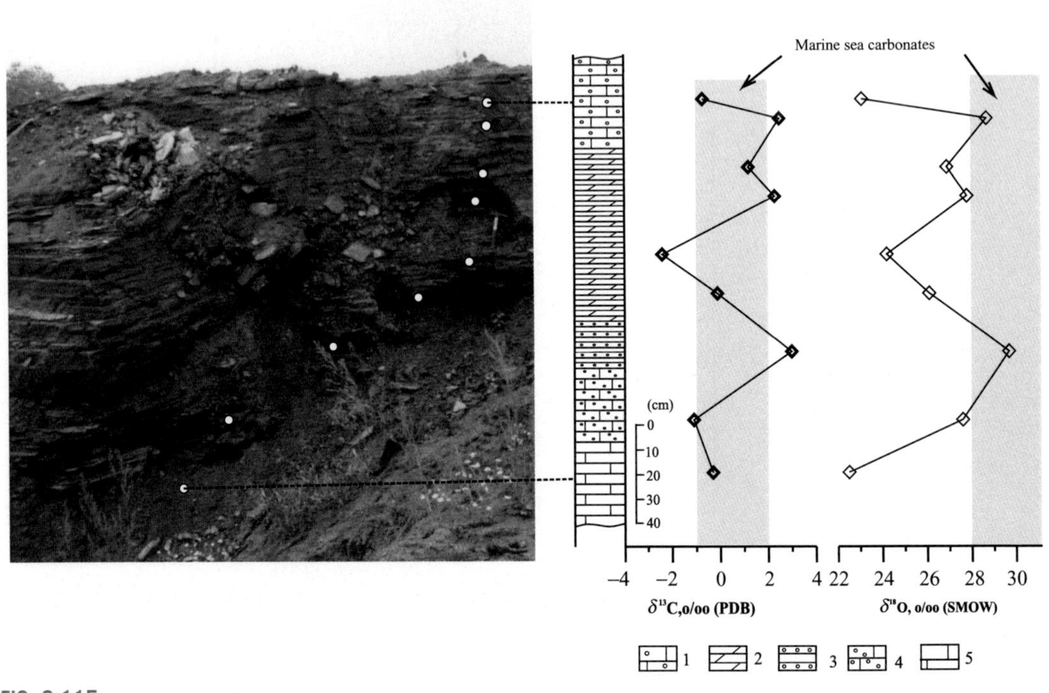

FIG. 3.115

Distribution of δ^{13}C and δ^{18}O in manganese-oxide-carbonate rocks from a separate boulder in the southern quarry at the Ulutelyak deposit. 1—Sparsely oolitic bedded, 2—thin bedded with clayey admixture, 3—bedded fine-oolitic, 4—fine-oolitic, and 5—coarse bedded.

manganese-containing carbonates supposing a subtraction from the system of the heavy isotope carbon ^{13}C together with carbon dioxide; hence, their discussion is not given further consideration.

On the genesis of manganese rocks and ores

Various viewpoints purport to explain the origin of primary manganese and manganese-containing rocks of the Ulutelyak deposit. The most widespread are theories of a sedimentational-diagenetic genesis. In the opinion of Makushin (1970), the formation of manganese deposits occurred in a salt- and brackish-water marine lagoon during periods of small local transgressions against the background of a general (progressive) degradation of the sea [a regressive stage of development of the region, according to (Gribov, 1982)]. A source of manganese here was terrigenous material entering from the paleo-land mass, situated within the boundaries of the modern Karatau massif.

As theorized by Gribov (1972b), the formation of rocks of the Ulutelyak suite, in which are localized manganese-containing carbonates of the Ulutelyak deposit, occurred during a regressive stage of the development of the region in a semi-closed marine basin under conditions of arid climate, with a limited supply of terrigenous material and fresh waters from the Urals. The formation of interbeds of manganese-containing carbonates occurred during periods of intermittent marine transgressions from the north, which led to a certain decrease in the salinity of the waters. A more drastic desalinization of the waters of the basin in the area of the Ulutelyak deposit is connected with a strengthened tributary of fresh waters from the Urals landmass, probably caused by a certain humidification of the climate of the collecting area, which led to an accumulation following after sulfate rocks by limestones and limy-dolomite marls, including manganese-containing varieties.

It follows to note that within the framework of a sedimentational-diagenetic theory, it is impossible to account for all presently available geological, mineralogical, and geochemical data on the Ulutelyak deposit. Hence, other mechanisms of formation of manganese rocks have been advocated, particularly in connection with the discharge of manganese-rich anoxic waters of the zones of hydrosulfuric acid concentration (Makushin, 1975; Ivanov and Petrovskii, 1998; Golota, 2000). The discharge mechanism of such waters likewise cannot fully explain the regularities of distribution established at the deposit for initial manganese deposits in the sequence of rocks of the evaporite formation.

The elucidation of the genesis of manganese ores is closely connected with the resolution of the question as to the source of manganese itself. Based on the information available in the literature in tandem with our obtained data, it is difficult to agree with the prevailing theories of a sedimentary-diagenetic origin of the ores and the source of manganese resulting from terrigenous runoff from the paleo-landmass situated within the boundaries of the modern Karatau massif (Makushin, 1970; Gribov, 1972b). One of the essential facts in contradiction to these theories is the absence of a terrigenous impurity (siltstone, sandstone, and larger fractions) in the composition of the ore strata and of its overlapping deposits.

Furthermore, given what we know about the conditions of carbonate-accumulation in the Permian basin, initially only calcium carbonate could have been accumulated by microbial means (mineralized bacterial mats). The development of organic life occurred in extreme (high-stress) conditions in which macrofauna had sharply repressed (small) forms; their residues are represented by calcium carbonate. Initial sedimentary-diagenetic oxides of manganese and iron, which are characteristic for known deposits of manganese such as Chiatura, Nikopol', Mangyshlak, and others, and which are commonly present in the form of ooliths and cement of clasts of terrigenous rocks, to date remain unknown in the structure of the Ulutelyak deposit.

A fundamentally different point of view regarding the genesis of the manganese deposits coinciding with the strata of sedimentary rocks is connected with catagenetic processes. The possibility of catagenetic manganese ore-formation in elisional basins in connection with the Mangyshlak and Laba deposits has been expressed by V.N. Kholodovyi, V.I. Dvorovyi, and E.A. Sokolova. However, common catagenic, including elisional processes, that are not connected with phases of petroleum genesis are evidently incapable of ensuring sufficient concentrations of the components necessary for the formation of ore (in part manganese-ore) deposits. According to the prevailing theory (Paragenesis et al., 1990), only petroleum waters containing large quantities of carbon dioxide, methane, hydrosulfide, and other aggressive components were likely to leach from the host rocks and carry over large masses of ore elements into the discharge zone. This mechanism of entry by manganese is indicated in the example of certain manganese deposits (Pavlov and Dombrovskaya, 1993).

A theory involving participation by petroleum waters in the formation of manganese carbonates connected to the Ulutelyak deposit has been proposed by B.M. Mikhailov. According to Mikhailov's theory, the mechanism of formation of manganese-containing carbonates was fairly complex: the formation of manganocalcite and manganodolomite, as well as alabandite, possibly "took place during late diagenesis in oozes of anoxic sinkholes of the Permian basin, into which petroleum waters could have infiltrated from the underlying petroleum deposits" (Mikhailov, 2001, p. 10). Further, he notes that "toward the surface in the composition of the petroleum waters have entered divalent manganese, evidently initially contained in the waters of the Early Permian basin". Also noted is the occurrence during the Mesozoic and Paleogene of the formation of a carbonate cap rock in the rocks of the carbonate-sulfate strata with participation by petroleum waters with "desulfurization of anhydrites and gypsums."

Geological, lithological, mineralogical, geochemical, and isotope data, at our disposal, allow us to accept the theory of a catagenic (post-early-diagenetic) origin of manganese and manganese-containing carbonates of the Ulutelyak deposit. Within this theoretical framework, manganese entered into the initial carbonate and carbonate-clayey sediment together with petroleum waters of elisional genesis, as well as isotopically light dissolved carbon dioxide; the latter in many cases led to a change in the isotopic composition of the initial carbonate matrix. The migration paths of such waters were in our opinion determined by the hypsometric setting of the roof of the sulfate members of the Ulutelyak suite, on the whole repeating the contour lines of the roof the Artinskian deposits (see Fig. 3.104).

It should be noted that the formation of alabandite in relation to manganese carbonates occurred later—manganese sulfide replaces grains of the most manganese-rich carbonates at the stage of epigenesis (Gribov, 1972a). A source of sulfur in this process is hydrosulfide formed as a result of sulfate-reduction of anhydrites and gypsums of the host strata. The hydrocarbons necessary for the reduction of sulfates evidently entered the sediment along with the petroleum waters.

Thus, it can be supposed that the initial carbonate sediment (predominantly mineralized bacterial mats of various textures, including oolitic and oncolitic, with insignificant impurity of terrigenous material and rare small residues of macrofauna) underwent repeated processes of transformation during early diagenesis, when the connection between bottom and ooze waters had not yet been lost, as well as during late diagenesis (at different stages of catagenesis or epigenesis). Post-early-diagenetic transformations occurred evidently as a result of the entry into the system by petroleum waters (solutions). Furthermore, metasomatic transformations of the initial sediment occurred on the nano-level and did not lead to the recrystallization of matter.

The injection of such waters (solutions) occurred in multiple acts; during the early stages, these waters were manganese-containing, which led to an enrichment of initial calcites and dolomites by manganese. Connected with the supply of hydrocarbons into the system, predominantly during the concluding stages, are processes of sulfate-reduction and the formation of one of the latest epigenetic minerals—alabandite.

The multi-act character of the formation of mineral phases in rocks of the Ulutelyak deposit has been mentioned numerous times by E.M. Gribov, A.A. Makushin, and B.M. Mikhailov in the above-cited works.

Manganese oxides in rocks of the Ulutelyak deposit are secondary. With their insignificant content in the rock, they occupy an "intergranular" space and are represented by incrustational and earthy "powder" formations. In oxide ores can be noted a layered manganization that inherits the initial microbial-organogenic texture. Moreover, in thin sections have been noted sections where organogenic residues are unaffected by these processes.

In this way, the results of microscope and isotope-geochemistry research enable the explanation of certain characteristic particularities of the conditions of formation of the carbonate matter hosting the manganese oxides.

Above all, as researchers have previously established (Makushin, 1970; Gribov, 1972a,b, and others) and is also confirmed by the results of our works, carbonate rocks hosting the ore matter were accumulated in a shallow-water semi-isolated basin under conditions of a semiarid and arid climate. This has allowed a range of researchers (Khmelevskii, 1968; Makushin, 1970; Gribov, 1972a,b; Khmelevskii and Ianchuk, 1990) to identify an independent manganese carbonate-gypsum formation of arid type.

The formation of "initial" carbonate-manganese-containing and manganese minerals is secondary and connected evidently with the discharge of petroleum waters of elisional genesis injected into the sulfate-carbonate strata of the evaporite formation of the Lower Permian. Divalent manganese was also supplied along with these solutions. Originally, the ore-hosting deposits were represented by organogenic (predominantly mineralized microbial) carbonates of calcium and to a lesser degree of magnesium, with an insignificant terrigenous (possibly aeolian) impurity.

The proposed existence of an anoxic zone with hydrosulfuric contamination and elevated concentrations of manganese carried in with the surrounding peneplaned landmass in an extremely shallow-water Kungurian paleobasin, at least for the studied rocks, is unsupported by factual data.

3.2.4.3 Manganese carbonates of the Pai-Khoi ore-occurrence

Geological setting

In the region of the distribution of ore-occurrences and small deposits of manganese rocks and ores of Pai-Khoi have been identified three large structures—the Kara and Korotaikha depressions divided by the Pai-Khoi uplift (Termal'nyi et al., 1989). The borders of these structures pass along high-amplitude thrusts, the displacement planes of which fall principally to the north-east (Fig. 3.116). At the same time, the given region is part of a structure of higher rank, occupying the northern part of the Cis-Urals marginal trough and the southern part of the Pai-Khoi-Novaya Zemlya fold region. Within the boundaries of the Pai-Khoi uplift are developed Paleozoic deposits of the Kara structural-facies zone, represented by deep-water associations of rocks and forming the Kara allochthon within the modern structural plan.

Carbonate-manganese ores in Pai-Khoi coincide with Upper Paleozoic deposits of the Kara shale structural-facies zone (Yudovich and Ketris, 1981; Yudovich et al., 1981, 1987). Deposits of the Paleozoic, beginning from the Lower Ordovician and ending with the Permian, form an uninterrupted cross section (Eliseev et al., 1984, 1986; Osadochnye et al., 1984) and include numerous horizons of carbonate concretions (Beliaev and Yudovich, 1983).

Manganese mineralization principally coincides with three stratigraphic levels: C_2-P_1, D_3^2-C_1^1, and D_2 (Yudovich and Ketris, 1981; Yudovich et al., 1981, 1987; Rogov and Galitskaya, 1983, 1984; Rogov et al., 1988) and is manifested in the form of concretions and concretional (or concretoid), more rarely bedded formations.

The carbonate-manganese ores, according to many scholars, are classified as sedimentary-diagenetic (Yudovich and Ketris, 1981; Yudovich et al., 1981; Eliseev et al., 1984). This is the prevailing viewpoint at present and matches well with modern theories in which carbonate concretions in the sedimentary basins are formed primarily in strata of ooze during the stage of early and significantly more rarely late diagenesis (Zaritskii, 1959, 1970; Seibold, 1962; Makedonov, 1970; Makedonov and Zaritskii, 1977; Zaritskii and Makedonov, 1985). These authors suppose that during later, post-diagenetic (catagenetic) stages of the geological life of the sediment, concretion-formation with rare exception practically does not occur.

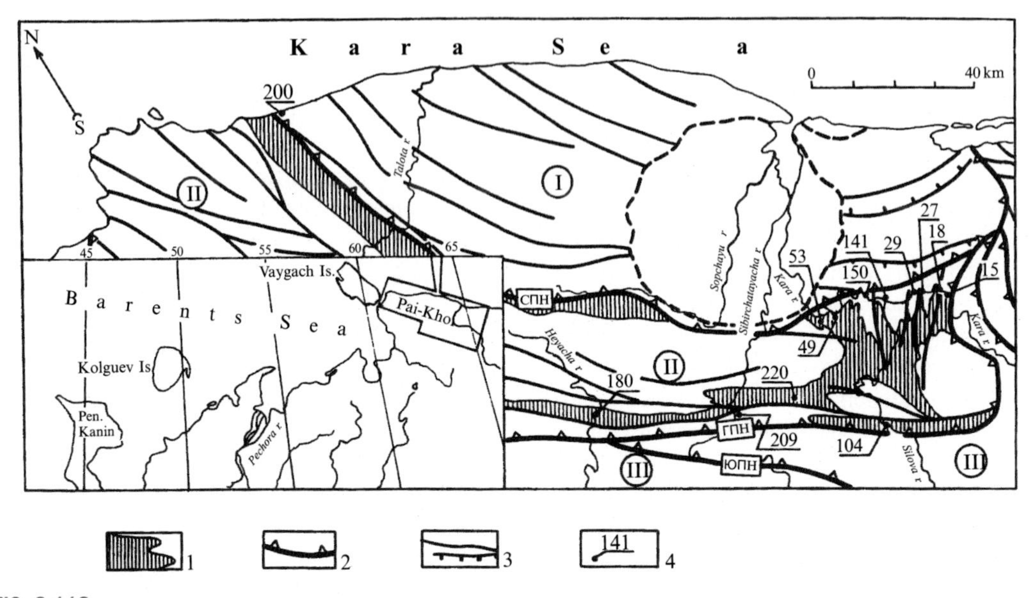

FIG. 3.116

Schematic structural-tectonic map of the study area of Pay-Khoi. Inset—the geographical position of Pai-Khoi. Tectonic zoning given by V.V. Yudin (Thermal, 1989). I—Kara depression; II—Pai-Khoi uplift, Kara shale allochthon; III—Pripaykhoyskaya area of Korotaikha depression. Main overthrust sheets and thrusts: GPN—Main Pay-Khoi overthrust, YPN—South Pay-Khoi overthrust, and SPN—North Pay-Khoi overthrust. 1—Area of the Devonian deposits; 2 and 3—overthrust and thrusts: 2—main, 3—secondary; 4—numbers of studied outcrops (in number of the sample in Table 3.25 correspond to the first digit of the numerator of the sample).

Rogov and Galitskii (1983) have proposed a hydrothermal-sedimentary origin for the manganese carbonates of the Pai-Khoi concretions. Our research (Beliaev and Kuleshov, 1994) has permitted us to draw conclusions regarding their polygenic nature, accommodating both points of view.

We conducted a study of specifically Mn-containing carbonate concretions and nodules gathered from the D_2-P_1 deposits as well as from later hydrothermal-metasomatic formations—carbonates from the transverse veins, secondary carbonate of superimposed carbonatization, and calcite from amygdules in basalts. Engaged as well is the available material in the literature on the isotopic composition of non-ore post-sedimentational carbonate rocks of Pai-Khoi.

Concretional and bedded manganese-containing carbonates in the C_2-P_1 deposits. The ore-hosting strata is represented by two types of sequences. The first is represented by a strata of carbonaceous clayey and siliceous shales with sparse thin beds of manganous dolomitic limestones and inclusions of manganous concretions.

The second type of sequence is represented by a flysch-type strata of clayey and limy-clayey shales. In the lower part, this strata is composed of a rhythmic alternation of clayey and limy-clayey shales with numerous horizons of manganous carbonates (Rogov and Galitskaya, 1984; Rogov et al., 1988; Yudovich et al., 1987). Contact between these two types of deposits is tectonic throughout.

Concretional and bedded ores of D_3fm-C_1t. This manganese-bearing level has been known since the 1930s. All known occurrences of manganese-oxide ores here are connected with this level. It is represented by an alternation of ferruginous jasperoid siliciclastics, manganous limestones, and ferruginous micaceous-limy-siliceous shales, frequently with amphibole. Manganese mineralization is connected with thin- to micro-layered carbonate-siliceous rhythms, forming members with a thickness from 0.1 to 5 m. But more frequently, manganese carbonates form dense lensoidal or spherical concretions 2–5 mm "thick," which inside the layer (along the strike) reach 2.5–3 m in diameter.

Concretional ores of D_2. This ore-bearing level coincides principally with the upper part of the section of the Padeiskii suite of the Kara type; it is represented by a diverse set of concretions and forms several well-expressed horizons.

Besides manganese carbonates from concretions, for purposes of comparison were studied separate samples of calcite from intersecting veins of various composition, as well as carbonates of carbonate- and carbonate-siliceous rocks altered by later processes and calcite of schistic amygdaloidal basalts of the Madagavozhskii complex bearing a hypothetical Late Devonian age.

The mineral composition of manganese carbonates. Complex (chemical, X-ray, infrared-spectroanalysis) study of carbonate phases of carbonate rocks and ores permits diagnosis of a large set of ore-forming minerals (Yudovich et al., 1981, 1998; Yudovich and Ketris, 1981).

Rhodochrosite is represented by manganese carbonate, with certain quantities of Ca and Fe isomorphically replacing Mn. Accordingly, ferruginous-calcium and ferruginous varieties can be identified that are characterized by specific reflexes in diffractograms and absorption bands in infrared-spectrograms. In many cases, zinc-containing rhodochrosites have been established. For example, in sample Ka57/1026, representing a large zonal concretion from black siliceous shales (Yudovich et al., 1998), ferruginous rhodochrosite dominates with an insignificant impurity of Mn-calcite (the basic reflection in X-rays corresponds to an interplanar distance of 2.82 Å). The calculated formula is close to $(Mn_{0.76}Fe_{0.14}Ca_{0.09}Mg_{0.01})CO_3$.

Oligonites, more rarely manganosiderites, almost always occur with notable contents of Mg and an impurity of Ca. For example, the following formula was found for a sample of siliceous-carbonate concretion No. 1487 from the carbonaceous argillites (Yudovich et al., 1998): $(Mn_{0.36}Fe_{0.32}Ca_{0.17}Mg_{0.15})CO_3$. The most intensive reflections in X-ray diffractograms correspond to interplanar distances of 2.80 and 2.90 Å.

Manganodolomites are characterized by dolomitic structure with an impurity of Mn and insignificantly of Fe isomorphically replacing Mg.

Widely distributed as well are manganocalcites and manganous calcites.

Isotope data and their discussion

All obtained isotope data are recorded in Table 3.25 and displayed in Fig. 3.117. From these data, it follows that carbonate Mn-containing nodules are characterized by wide general boundaries of variation in the isotopic composition of carbon and oxygen, which constitutes from −12.1‰ to 0.8‰ for $\delta^{13}C$ and from 19.5‰ to 27.8‰ for $\delta^{18}O$. Moreover, against the background of such a broad spectrum of isotopic ratios, separate groups of Mn carbonates coinciding with deposits of different stratigraphic levels bear particular isotope characteristics and in the graph of Fig. 3.117 form independent fields. In certain cases, these fields are partially overlapping. Thus, manganese carbonates coinciding with deposits of $C_{2\text{-}3}$-P_1 age are characterized by values of the isotopic composition of carbon from −12.1‰ to −4.8‰ and of oxygen from 22.3‰ to 26.1‰; those coinciding with deposits of D_3-C_1 age vary from −4.8‰

Table 3.25 Isotopic and Chemical Compositions of Manganese-Containing Carbonate Concretions of the Kara Zone of Pai-Khoi

Number of Analysis	Number of Sample	IR	CO_2	CaO	MgO	MnO	FeO	Fe_2O_3	$\delta^{13}C$, ‰ CPDB	$\delta^{18}O$, ‰ OSMOW
1	2	3	4	5	6	7	8	9	10	11
Sedimentational and manganized carbonates $C_{2\text{-}3}$-P_1										
2301	141/807	43.18	23.78	26.92	1.49	1.26	1.44	0.55	−4.8	23.1
2302	141/808a	–	3.36	2.37	2.17	1.88	3.52	0.94	−6.8	22.3
2303	41/813	5.44	43.47	29.75	17.39	1.91	1.36	0.25	−8.6	24.7
2304	150/877	64.1	14.1	14.39	1.5	1.6	0.63	–	−7	23.9
2305	150/887a	–	28.6	21.56	7.61	4.88	3.3	4.52	−8.2	22.9
2306	150/889	29.74	26.86	20.47	2.77	5.08	5.26	0.57	−8	24.2
2307	150/898	4.52	40.58	29.47	16.06	2.11	1.52	0.15	−11.3	26.1
2308	150/900	6.04	39.65	28.35	16.01	1.86	1.25	0.32	−11.7	25.5
2309	150/901	7.56	38.53	27.79	14.48	1.84	2.61	0.41	−9.4	23.1
2311	150/885a	–	10.4	7.98	3.33	2.7	4.31	0.93	−7	23.1
Sedimentational and concretional manganese-containing carbonates D_3fm-C_1t										
2313	29/145	11.9	37.55	39.93	2.3	5.47	0.09	–	−1.8	27.5
2314	27/128	22.36	33.53	30.22	5.14	5.84	1.55	–	−0.4	26.5
2315	49/326	85.28	4.73	0.73	0.99	3.51	0.29	4.91	−4.8	24.4
2316	49/327a	–	8.45	6.75	1.21	4.7	6.78	2.22	−4.4	23.1
2317	104/393	9.84	36.83	32.15	2.98	12.92	1.95	1.12	−2.4	25.6
2318	104/396a	–	29.83	28.39	1.81	8.76	1.54	0.99	0	26.4
2319	104/402	15.48	36.83	40.68	0.94	5.24	0.19	0.07	−0.5	25.5
2320	180/1086	8.24	37.75	29.18	5.67	16.24	–	–	−0.7	24.3
2322	180/1089	48.14	21.03	17.71	2.65	6.08	–	–	−0.4	25.9
2323	180/1097a	–	18.36	16.91	2.93	5.18	4.3	–	0.8	25.5
2324	180/1096	50.02	19.22	14.45	2.18	7.7	0.06	–	−0.5	23.7
2325	180/1092	35.26	27.5	28.09	1.5	6.56	–	–	−0.8	25
Concretional manganese carbonates D_2ef-gv										
2342	15/75	66.84	0.41	0.74	–	0.04	0.5	0.05	−7.1	19.7
2343	15/75-2	21.44	29.55	31.75	0.71	8.5	3.52	0.51	−9.7	22.8
2344	15/75-3	41.34	22.48	24.12	1.06	5.3	1.87	0.41	−8.3	19.5
2345	15/75-4	19.16	29.88	28.79	1.42	11.3	2.95	1.41	−8.4	21.1
2346	15/75-5	36.82	24.7	18.46	1.59	15.2	1.8	0.05	−5.7	22.7
2347	15/75-6	47.86	17.25	16.24	2.83	4.7	2.73	2.02	−6.6	19.8
2348	15/75-7	55.94	14.43	14.52	1.77	3.1	2.59	1.33	−6.8	19.9
2349	15/75-8	35.42	24.9	25.59	1.42	6	1.58	1.4	−6.8	19.8
Carbonates of late generations										
	220/1492a	44.9	22.7	25.87	0.78	0.03	1.98	–	−0.2	22.8
2326a	220/1501	30.58	30.23	36.34	0.83	0.11	0.78	0.16	−1.9	22.3
2328	18/53	32.42	16.92	1.66	0.39	16.5	13.16	13.07	−9	21.1

Table 3.25 Isotopic and Chemical Compositions of Manganese-Containing Carbonate Concretions and Concretoids of the Kara Zone of Pai-Khoi—Cont'd

Number of Analysis	Number of Sample	IR	CO_2	CaO	MgO	MnO	FeO	Fe_2O_3	$\delta^{13}C$, ‰ CPDB	$\delta^{18}O$, ‰ OSMOW
1	2	3	4	5	6	7	8	9	10	11
2330	27/1273	25.9	30.95	20.47	8.74	8.24	0.03	3.66	1.9	20.8
2336	53/370	1.12	39.85	53.11	0.31	0.04	–	0.09	−10.4	15.2
2337	18/54a	27.18	24.6	2.33	–	16.4	24.31	1.1	−4.2	17.4
2338	53/367	93.76	2.5	2.83	–	0.01	0.37	0.07	−10.8	18.1
2339	53/369	20.74	32.3	41.49	0.68	0.09	1.02	0.03	−5.3	20.3
2341	209/1470	0.46	43.91	55.48	0.08	0.02	0.03	–	0.2	25.5

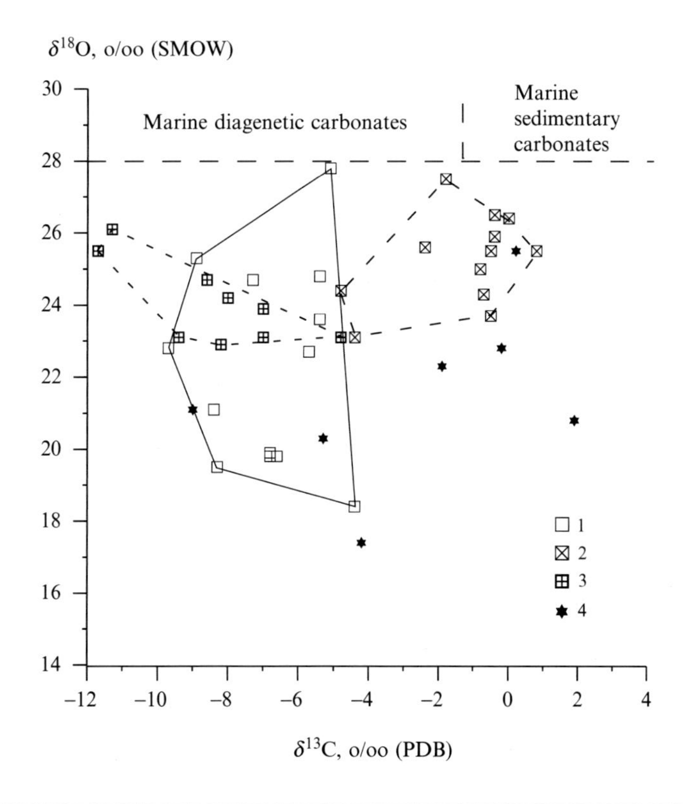

FIG. 3.117

$\delta^{13}C$ versus $\delta^{18}O$ in carbonate-manganese concretions of Kara structural-facies zone of Pai-Khoi. 1—Concretions from deposits of D_2 ages; 2—concretions from deposits of D_3-C_1 ages; 3—concretions and layers occurring in the sediments of C_{2-3}-P_1 ages; and 4—carbonates of later generations.

to 0.8‰ and from 23.1‰ to 27.5‰, respectively; and finally, Mn carbonates deposited in D_2 deposits vary from −9.7‰ to −5.1‰ and from 19.5‰ to 27.8‰, respectively.

Established distribution patterns of isotope ratios in the studied nodules enable us to see that the greatest difference in the isotopic composition of carbon and oxygen is inherent to Mn carbonates from the D_3-C_1 deposits. These occupy the range of the highest values of isotopic composition of carbon and oxygen. Mn carbonates from the Middle Devonian and Middle-Upper Carboniferous-Lower Permian deposits in terms of isotope characteristics of carbon and oxygen on the whole are close to each other.

Obtained isotope data for the examined manganese and manganese-containing carbonates of Pai-Khoi on the whole are close to those for the above-examined Oligocene ores of prominent deposits of Ukraine (Nikopol', Bol'she-Tokmak), Georgia (Chiatura, Kvirila), and Kazakhstan (Mangyshlak), and are also analogous to ores of the Bezmoshitsa occurrence in Cis-Timan. At the same time, in Pai-Khoi Mn-rocks in terms of the isotopic composition of carbon as well as oxygen exhibit certain differences along genetic lines.

The greatest scattering of isotope ratios is observed in crosscutting veins and carbonates of later (epigenetic) generations. Thus, the isotopic composition of carbon in carbonate-quartz veins varies within the boundaries −10.8‰ to −5.3‰ and corresponds on the whole to Mn carbonates from the deposits of C_{2-3}-P_1 and D_2 ages. Carbonate of amygdule-stone basalts and carbonatized phthanite are characterized by the highest values of $\delta^{13}C$: from −1.9‰ to −0.3‰ for the former and 1.9‰ for the latter.

In this way, it can be seen that non-ore and Mn carbonates of the Kara structural-facies zone of Pai-Khoi are characterized by a wide scattering of isotope ratios of carbon and oxygen. This serves as evidence for different conditions of formation and different sources of their matter. In Fig. 3.117, it is visible that on the whole in terms of isotope characteristics they occupy a range of lower values for the isotopic composition of carbon as well as oxygen by comparison with carbonates of marine normal-sedimentary and diagenetic origin.

Carbon. Manganese-containing and manganese carbonates of Pai-Khoi in terms of their isotopic composition of carbon cover a fairly broad interval characteristic for carbonates of various genetic groups of carbonaceous matter of the Earth's crust. Samples with heavier isotopic composition fall into the range of $\delta^{13}C$ values particular to carbonates of normal-sedimentary marine origin. Such isotopic composition is characteristic of ores from D_3-C_1 deposits and secondary carbonate of phthanite.

Concretions from deposits of Middle Devonian and Upper Carboniferous-Lower Permian age, as well as calcites from amygdule-stone basalts and transversing veins, are characterized by a lighter isotopic composition and fall into the range of values particular to carbonates of diagenetic and catagenetic (epigenetic) origin. Low values of $\delta^{13}C$ are caused by the presence in their composition of isotopically light oxidized carbon of organic origin.

A light isotopic composition of carbon ($\delta^{13}C$ values vary from −14% to −12 to −8% to −6‰) also characterizes carbon dioxide of deep origin (Galimov, 1968; Kuleshov, 1986a), which in one variant could have entered the basin of sedimentation along deep faults from the zones of the upper mantle during periods of tectonic and magmatic activity in the region. The author excludes these untenable viewpoints from further discussion.

The isotopic composition of carbon in studied ores of Pai-Khoi by comparison with Oligocene Mn ores of Georgia, Ukraine, and Kazakhstan and other examined examples on the whole is characterized by narrower boundaries of variation in the range of higher values of the isotopic composition of carbon. Thus, as shown above, for Nikopol' ores, values of $\delta^{13}C$ vary within the range −24.6% to −4.9‰, in part from −34.5% to −7.3‰, and in Mangyshlak ores from −32.9% to −4.7‰. These data serve as evidence that in the process of ore development at Pai-Khoi the role of organic carbon was substantially lower by comparison with its role in deposits of the Oligocene.

Oxygen. In studied concretions and later carbonates of the Pai-Khoi formations, oxygen is characterized by a light isotopic composition, and in its basic mass is substantially lighter by comparison with marine sedimentary and carbonates of marine diagenetic origin (see Fig. 3.117). This enables confirmation that a significant proportion of examined Mn-containing carbonates at the moment of formation were not located in isotope-oxygen equilibrium with dissolved bicarbonate of marine water (DIC) and were apparently deposited in conditions of late diagenesis or during the stage of catagenesis, when the isotope-oxygen exchange with the bottom waters of the paleobasin was disrupted.

The studied Mn-containing concretions in terms of their isotopic composition of oxygen on the whole are distinguished as well from the above-examined manganese ores of Oligocene age, which can serve as evidence for their different conditions of formation.

Substantial differences have not been noted in the isotopic composition of oxygen among the studied groups of carbonate matter of the Kara zone of Pai-Khoi. However, it can be noted that carbonates of Mn coinciding with D_2 deposits are characterized by a broader spectrum of variations in values of $\delta^{18}O$.

If it is supposed that the formation of Mn ores occurred with participation by waters of marine genesis ($\delta^{18}O=0$) and was oriented to the middle boundary of variations in $\delta^{18}O$ in ores equal to 22–26‰, then the temperatures of carbonate precipitation should be elevated and constitute 40–70°C. Moreover, carbonates with lower values of $\delta^{18}O$ should have been deposited at higher temperatures.

Carbonate matter of carbonate-quartz veins by comparison with ore nodules are on the whole rich in light isotopes of ^{16}O, which could be the result of their higher temperatures of formation. At the same time, carbonate of phthanite and scattered carbonate of amygdules-stone basalts in terms of the isotopic composition of their oxygen on the whole are not distinguished from carbonate-manganese nodules.

The enrichment of the studied carbonate ore nodules and late carbonates by light isotopes of oxygen could also be a result of the desalinization of the basin of sedimentation or of participation in the process of the formation of ore concretions by water of meteoric genesis. In our view, this theory is not supported by geological observations, and we exclude fully this possibility from further discussion.

Origin of carbonate-manganese rocks and ores of Pai-Khoi

The measured data on the isotopic composition of carbon and oxygen of Mn carbonates from the Middle-Upper Paleozoic deposits of the Kara zone enable confirmation that the process of manganese ore-genesis in the Upper Paleozoic Kara paleobasin took a completely different course than did the analogous process in sedimentary-rock basins with the formation of sedimentary-diagenetic manganese carbonates.

The available isotope data confirm the fact that to date, specifically sedimentary manganese carbonates have not been detected in Pai-Khoi, or for that matter in any of our earlier studied manganese deposits. Therefore, the hypothesis proposed by Sapozhnikov (1967) of a chemogenic precipitation of manganese carbonates from the marine water in conditions of discharge into the shallow waters by rising streams from deep waters (upwelling) is in need of correctives.

The viewpoint espoused by Platonov et al. (1992) purporting an initially sedimentary origin for the deposits of Mn ores of Pai-Khoi and Novaya Zemlya is to all appearances likewise untenable. On the basis of an analysis of the geological development of the Pai-Khoi-Novaya Zemlya region in the

Middle-Late Paleozoic, an analysis of facies zonality of the Upper Carboniferous-Sakmarian deposits of the southern island of Novaya Zemlya, Kolguev island, and the Timan-Pechora region, and a comparative analysis of the Pai-Khoi-Novaya Zemlya and Black Sea basins, Platonov and coauthors arrived at the bold conclusion of a single mechanism of formation of the Mn ores of these two manganese-bearing basins. The genetic results of these authors find support neither in our isotope data, nor in geological fieldwork. On the other hand, the concretional morphology of Mn carbonates, as rightly noted by Ia. Yudovich et al. (1981) serves as evidence affirming their post-sedimentational formation.

Before to launching into the discussion of the questions as to which stage of the geological life of the sediment featured the formation of ore concentrations of Mn in the form of various types of concretions and nodules and which of these were the sources of manganese, we turn first to the available lithological and geochemical data for ore-hosting deposits of this region.

The results of the detailed study of the lithology, geochemistry, and mineralogy of the Paleozoic deposits of the Kara structural-facies zone of Pai-Khoi, in conjunction with results of the determination of the index of color of conodonts, the degree of graphitization of organic matter, and the dimensions of ultracrystallites of quartz in siliciclastics, have enabled a conclusion regarding the intensive high-temperature (reaching 350–400°C) epigenetic restructuring of these deposits and the substantial role of processes of folding and dynamic metamorphism in the formation of mineral associations in the studied rocks (Termal'nyi epigenez et al., 1989; Osadochnye formatsii et al., 1984).

No less informative in this regard are also the isotope data on carbon and oxygen obtained previously by N.V. Sukhanov and coauthors (Beliaev et al., 1983; Yudovich and Sukhanov, 1984). These authors have conducted studies of various carbonate rocks of the shale (Kara) zone of Pai-Khoi, along with carbonate matter from various carbonaceous siliciclastics (phthanites, chert, jasperoids, siliceous shales, and the like) and concretions of various composition (carbonate, siliceous-carbonate, pyrite-calcite-siliceous, etc.) of a broad age spectrum—from the Silurian to the Lower Permian.

The isotope data are in good agreement with the results of subsequent lithological-geochemical research. As has been demonstrated, in siliceous weakly dolomitic limestones of Lower Carboniferous age, values of $\delta^{13}C$ (−1.5% to 1.7‰) and $\delta^{18}O$ (27.1–29.8‰) correspond to those of normal-sedimentary marine carbonates. The isotopic composition of calcite non-ore deposits in the studied sections of the Paleozoic is characterized throughout by low values of $\delta^{13}C$ (−10.5% to −6.6‰) and $\delta^{18}O$ (19.0–23.3‰), which constitutes evidence of their post-sedimentational and post-diagenetic origin.

One of the characteristic features of carbonate deposits of the Paleozoic of the Kara structural-facies zone of Pai-Khoi is the presence, alongside the bedded normal-sedimentary marine carbonates, of isotopically anomalous carbonate strata represented by different lithological-petrographical types: layered, spotted, and ocellar, containing different quantities of calcite, dolomite, and siliceous matter. Low values of the isotopic composition of carbon (up to −30‰) and oxygen (up to 21.4‰) constitute evidence of their later origin and of their formation (in certain cases possibly transformation) during the stage of the catagenesis of sedimentary rocks of the basin. Moreover, their formation evidently occurred at elevated temperatures and active participation by isotopically light carbon dioxide formed in the process of oxidation of organic matter, including hydrocarbons of the petroleum series.

Thus, practically all genetic types and varieties of carbonate matter, including both Mn-containing and Mn-ore nodules, within the boundaries of the Kara structural-facies zone in Paleozoic deposits were subjected to isotope research. The isotope data allow us to consider that the forming of carbonate matter occurred under the most diverse conditions throughout the duration of the geological life of the sedimentary-rock basin—beginning from the moment of formation of specifically sedimentary carbonates (ie, those isotopically balanced with dissolved bicarbonate (DIC) of the seawater), the redistribution and new formation of concretional carbonate formations evidently during stages of insignificantly diagenesis and primarily catagenesis, and ending with superimposed carbonatization as a result of the subsequent hydrothermal restructuring of the Paleozoic depositions. A substantial role in the formation of carbonate bedded rocks may belong to metasomatic processes by formation of marbled limestones and "ocellar" dolomitic limestones of the Famennian and Serpukhovian stages.

The above conclusions are well illustrated by the graph of Fig. 3.118, constituted according to isotope data of the cited works (Beliaev et al., 1983; Yudovich and Sukhanov, 1984). In the graph, it is apparent that the studied non-ore carbonates have different natures and form independent fields within the coordinates $\delta^{13}C$-$\delta^{18}O$. Moreover, the principal quantity of studied carbonates has a post-sedimentational origin.

The results of geochemical research have enabled an explanation for the generally elevated contents of Mn in Paleozoic depositions. Manganese content along the sequence and the lithological types of rocks is distributed unevenly. The greatest concentrations of Mn have been identified particularly in concretions and siliciclastics, more rarely in carbonates of the Upper Paleozoic,

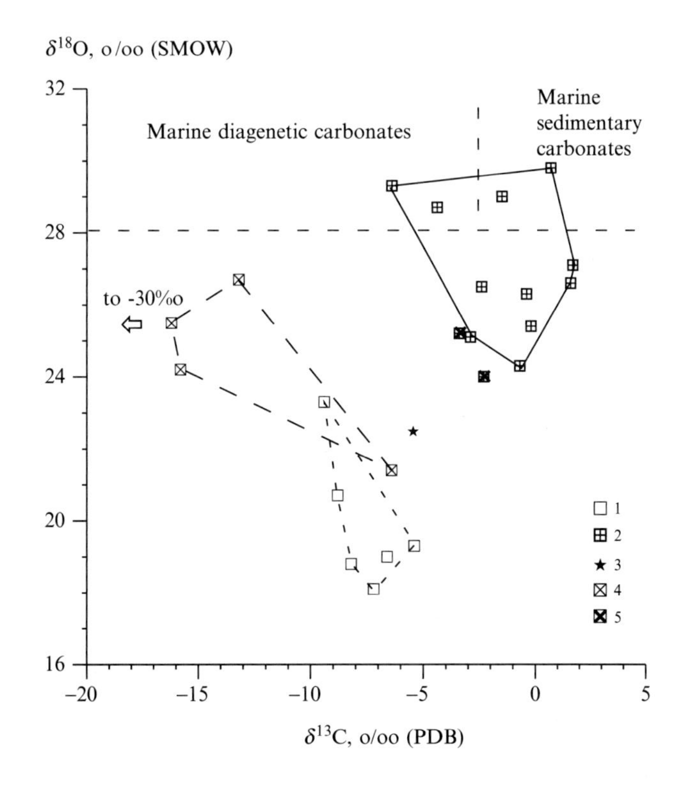

FIG. 3.118

$\delta^{13}C$ versus $\delta^{18}O$ in barren carbonate rocks of Kara area, Pay-Khoy [compiled on (Beliaev et al., 1983; Yudovich and Sukhanov, 1984)]. 1—Concretions of different composition; 2—bedded carbonate rocks; 3—carbonaceous shales; 4—marble and "spotty" carbonates; and 5—cherty carbonates.

and coincide with defined stratigraphic levels (Yudovich et al., 1981). The noted circumstance and the available isotope data allow us to consider that the process of manganese-ore formation in Paleozoic depositions of Pai-Khoi is connected not only with concretion-formation, but rather occurred much more widely owing to processes of a different type of carbonate development under conditions of catagenesis and coincides with defined stages (or phases) of carbonatization of Paleozoic depositions bearing a cyclical (pulsation) character. It is evident that the incoming catagenic solutions were characterized by different chemical compositions, were distinguished in terms of content of ore and non-ore components, and contained carbon dioxide of different geneses and naturally of different isotopic composition. All of this was reflected in the chemical and isotopic composition of the studied carbonates. However, manganese mineralization coincides only with certain stages of entry by catagenic solutions.

The formation of manganese-containing carbonate nodules was connected only with certain (no fewer than three) stages of concretion development. This was reflected in the character of the distribution of $\delta^{13}C$ and $\delta^{18}O$ values. In the graph of Fig. 3.118, the carbon dioxide of these formations forms corresponding fields. This can be seen also in Fig. 3.119, where a correlation is shown between the distribution of values of the isotopic composition of carbon and the contents of Mn (see Table 3.25) in the studied samples. From this graph, it follows that manganese-ore carbonates that coincide with deposits

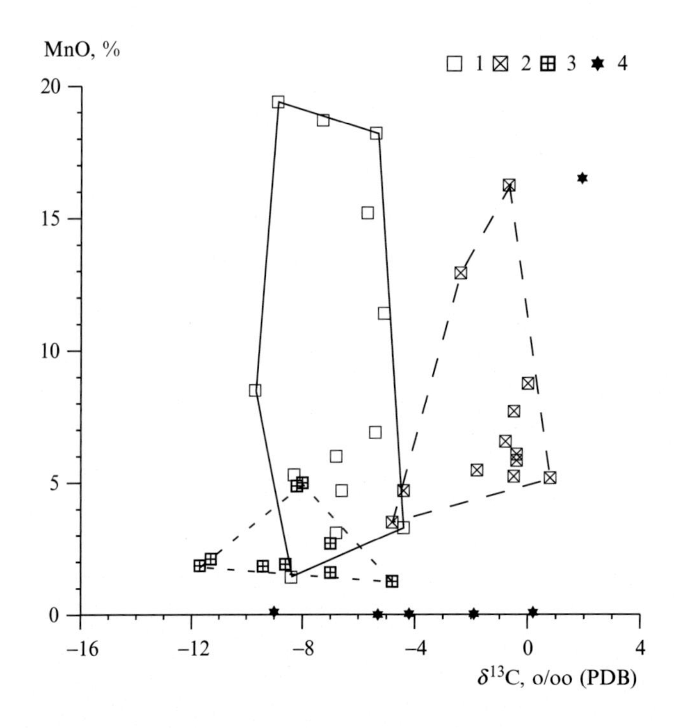

FIG. 3.119

Dependence of $\delta^{13}C$ value and MnO content in the carbonate-manganese ores of Pai-Khoi. 1—Concretions from deposits of D_2 ages; 2—concretions from deposits of D_3-C_1 ages; 3—concretions and layers occurring in the sediments of $C_{2\text{-}3}$-P_1 ages; 4—carbonates of later generations.

of C_{2-3}-P_1 age were formed from solutions characterized by the lowest contents of Mn. Ore-forming solutions that led to the formation of ore carbonates in Middle Devonian and Upper Devonian-Lower Carboniferous deposits carried with them significant concentrations of manganese but were distinguished from one another in terms of their sources of carbon dioxide. For the first, carbon dioxide was borrowed entirely from marine sedimentary carbonates. In carbon dioxide of carbonates from Middle Devonian deposits, a notable role was played by oxidized organic matter, although the principal share is also constituted by redeposited carbon dioxide of sedimentary genesis.

In the study of the genesis of Mn carbonates of this region, the most complicated task is the resolution of the question as to the source of Mn and the nature of the ore-bearing solutions. On the basis of the obtained isotope data and by analogy with certain generations of manganese ores of the Chiatura and Kvirila deposits, it can be supposed that the principal part of the studied manganese carbonates at Pai-Khoi (particularly D_2 and C_{2-3}-P_1) was formed during the catagenesis stage as a result of the discharge of elisional waters in shallow-water parts of the basin.

A distinctive particularity of Pai-Khoi ores separating them from those of Chiatura and Kvirila is the large degree of participation by carbon dioxide of sedimentary carbonates in the forming of the isotopic composition of the CO_2 of catagenic solutions. Today, we do not notice typical early-diagenetic concretional carbonates in Paleozoic depositions of Pai-Khoi. If they did exist, then they apparently underwent intensive later transformation during the catagenesis stage and lost the isotopic markers of their initial early-diagenetic nature.

Carbonate concretions from D_3-C_1 depositions in terms of their origin are evidently distinguished from the above-mentioned and could represent an independent type of Mn-ore carbonates. In terms of the isotopic composition of their carbon, they are distinguished substantially from all studied ores of other deposits examined above and are characterized by the heaviest isotopic composition particular to marine sedimentary carbonates. Carbon dioxide of oxidized organic matter is practically absent from their composition. The coincidence of these ores to the most deep-water facies of the paleobasin, the wide distribution in the host deposits of various chemogenic siliciclastics, as well as the heavy isotopic composition of carbon enables us to consider that their formation is not connected with elisional waters of the zones of oil-and-gas-generation but rather represents a special stage of carbonate concretional mineral precipitation during the stage of catagenesis of sedimentary rocks of the basin. To all appearances, their formation occurred during the initial stages of the formation of the elisional waters and represents the earliest and highest-temperature stage of elisional catagenesis (according to Kholodov, 1982a,b).

One of the distinctive particularities of Paleozoic deposits of Pai-Khoi is also manifest in the elevated contents in them of many ore and non-ore elements: Cu, Zn, Ni, As, Pb, Ba, Sr, and the like, up to the formation of independent deposits of barites, phosphorites, and other economic minerals (Osadochnye et al., 1984; Yudovich et al., 1981; and others). Apparently connected with these processes are also industrial concentrations of many ore and non-ore economic minerals (Mn, Al, Ba, phosphorites, etc.).

Significant concentrations of many lesser elements, as noted, are also included in the crystal lattice of carbonate minerals, as reflected in their names (eg, zinc rhodochrosite). Elevated contents of these elements provide a solid basis for considering deep fluids as a source of the studied manganese ores (Rogov and Galitskaya, 1983, 1984). We adhere to a different viewpoint and consider that the enrichment of transforming solutions by indicated elements can occur in the strata of the sediment itself and, in particular, under the influence of aggressive petroleum waters (Paragenezis et al., 1990).

In this way, the process of manganese ore-genesis in the deposits of Pai-Khoi occurred completely differently than in manganese-ore basins featuring the formation of sedimentary-diagenetic carbonates of manganese (Ukraine, Georgia, Kazakhstan). The study of the isotopic composition of carbon and oxygen in various types of carbonate rocks, including manganese-ore rocks (beds, concretions, nodules, and the like) (Yudovich, 1983; Beliaev and Yudovich, 1983), conducted on the basis of our data as well as the data in the literature, enables the conclusion that the Paleozoic terrigenous-carbonate and carbonaceous-siliceous-clayey deposits of Pai-Khoi featured the wide occurrence of processes of post-sedimentational carbonate precipitation. The greatest intensity of this process is manifested during the post-early-diagenetic (catagenetic) stage of the geological life of the depositions of the basin of sedimentation, when the connection was already lost between the ooze sediment and the bottom waters of the paleobasin. New formation of carbonate took course under the influence of catagenic waters and products of the metamorphism of organic matter. Specifically marine sedimentary and diagenetic carbonates have a sharply subordinate significance in the composition of Paleozoic rocks of the studied region.

Post-sedimentational (especially catagenetic) concretional carbonate development occurred pulsationally (cyclically) on account of the discharge of catagenic carbon dioxide solutions of various generations set in motion as a result of the region's tectonic activity. This was reflected in the isotopic composition of the carbon and oxygen of the deposited carbonates. Within the $\delta^{13}C$-$\delta^{18}O$ coordinates different generations of carbonates form specifically "isotopic" fields. Certain stages of concretional carbonate development were characterized by elevated manganese content, which in certain cases was responsible for the formation of ore concentrations of manganese.

3.2.4.4 Deposits of reef facies

A separate group among epigenetic (catagenetic) deposits of manganese is occupied by deposits that formed as a result of the metasomatic replacement of calcite by manganese in carbonate deposits of reef and near-reef facies. Serving as typical representatives of such deposits are the Usa deposit (Kuznetsk Alatau, Russia), a series of deposits of China, and certain points of manganese-ore mineralization in other regions where these processes are distinctly manifested.

The *Usa manganese deposit* is situated within the boundaries of the western slope of the Kuznetsk Alatau in the region of the middle course of the Usa River. In this region has been discovered a large quantity of small deposits and occurrences of manganese ores, which coincide with various manganese-bearing stratigraphic levels (Bych et al., 1975; Mirtov and Tarasova, 1980). The Usa deposit is enclosed in rocks of Early Cambrian age.

General information on the geological setting and material composition of the ores. The geological structure, mineralogy, and geochemistry of the ore matter and its host deposits of the Usa deposit at present have been well studied (Radugin, 1941; Mukhin, 1940; Betekhtin, 1946; Dodin, 1947; Getseva, 1947; Sokolova, 1960, 1961; Varentsov, 1961, 1962a; Khodak et al., 1966; Rakhmanov and Eroshchev-Shak, 1966; and others).

In the structural regard, the region of the deposit coincides with a large synclinal fold of submeridional strike complicated by depositions of the Riphean and Vendian stages. In the central part are developed Cambrian and Devonian deposits. Coinciding with the axial part of the structure is the zone of the Belaya Usa fracture zone, which in the modern erosional section is manifested in a series of large subparallel disturbances of north-westerly trend and their accompanying transverse and diagonal disjunctives of north-easterly trend. From the south, the synclinal structure is truncated by the Kibras fault, along which Lower Cambrian and Devonian deposits are interlinked with Lower-Middle Riphean strata. In the eastern part of the region are situated the Murtinsiy gabbro-diabase and Tumuiassky syenitic intrusives.

The deposit is divided according to conditions of deposition and orographic setting into three sections (ore fields) separated by faults: right-bank (Pravoberezhnyi), left-bank (Levoberezhnyi), and southern (Azhigolsky). The southern part of the deposit in relation to the deposits of the right-bank section is embedded in a downthrown tectonic block.

Within the boundaries of the entire deposit, manganese-bearing deposits are complicated by folds of small amplitude, which are broken by a series of tectonic faults.

The most studied in the geological regard is the northern part of the right-bank section. Here has been identified the western flank of the large synclinal structure, complicated by a series of flexure-type bends and disturbances of various characters and amplitudes (Fig. 3.120). Its eastern flank is truncated by the eastern zone of the faults, along which deposits of the ore-bearing strata make contact with limestones of the Usa suite of the Lower Cambrian.

In the western part of the ore field has been observed a synclinal fold thrust over the western flank of the above-described syncline. The slope is normal, at an angle of 75–80 degrees.

Within the boundaries of the deposit have been observed numerous zones of thrust faults with slickenlines, zones of mylonites and cataclastites, and tectonic breccias. A survey of the north-western part of the right-bank section revealed thrust faults in the form of small sheets (Fig. 3.121).

Ore-hosting rocks of the Usa deposit are of Lower Cambrian age and are classified to the Ust'-Kundat suite of Kuznetsk Alatau. Stratigraphically this interval falls into the Aldanian Stage in almost its entire volume (Tarasova and Mirtov, 1971; Tarasova et al., 1973, 1974; Mirtova, 1981).

The ore-hosting strata is concordantly overlaid on deposits of the Upper Vendian, which are represented by thin-layered stromatolitic limestones with interbeds of clastic marbled dolomites with inclusions of numerous laminae and small lenses of black cherts.

Lower Cambrian deposits in the region of the deposit are divided into two members: lower and upper. The lower stratum is ore-bearing and is subdivided into two submembers, lower subore and upper ore (Mirtova, 1983).

The subore portion of the lower member begins with mixed dolomite-limestone transitional from Vendian deposits—dark-gray thinly rhythmically layered fine-clastic limestones (fine-grained sandstones and siltstones) that contain an admixture of clasts of pelitoform dolomite and an interbed of black micro-grained, occasionally dolomitic limestone. Frequently noted are interbeds (up to 2 cm) of black chert. The thickness of the rhythms is 3–7 cm. Occasionally occurring are interbeds of phosphorites and fine-clastic dolomite-limestone gritstones. Above are dark-gray rhythmically layered clastic limestones (sandstones, siltstones) with an insignificant (1–5%) impurity of pyroclastic material (plagioclase, volcanic glass) with interbeds of black spongolites. The thickness of these transitional deposits constitutes approximately 9 m.

The subore horizon itself is represented by thin rhythmically layered clastic weakly manganous (1–5% Mn) limestones (rhodochrosite-limestone breccia with basal dark-gray siliceous-limy cement, black limy-rhodochrosite siltstones and sandstones with interbeds of black cherts, gray clastic limestones with thin interbeds of spongolites, and pyroclasts) with interbeds of spongolites and poor carbonate-manganese ores (5–13% Mn). The general thickness of the subore horizon constitutes >140 m.

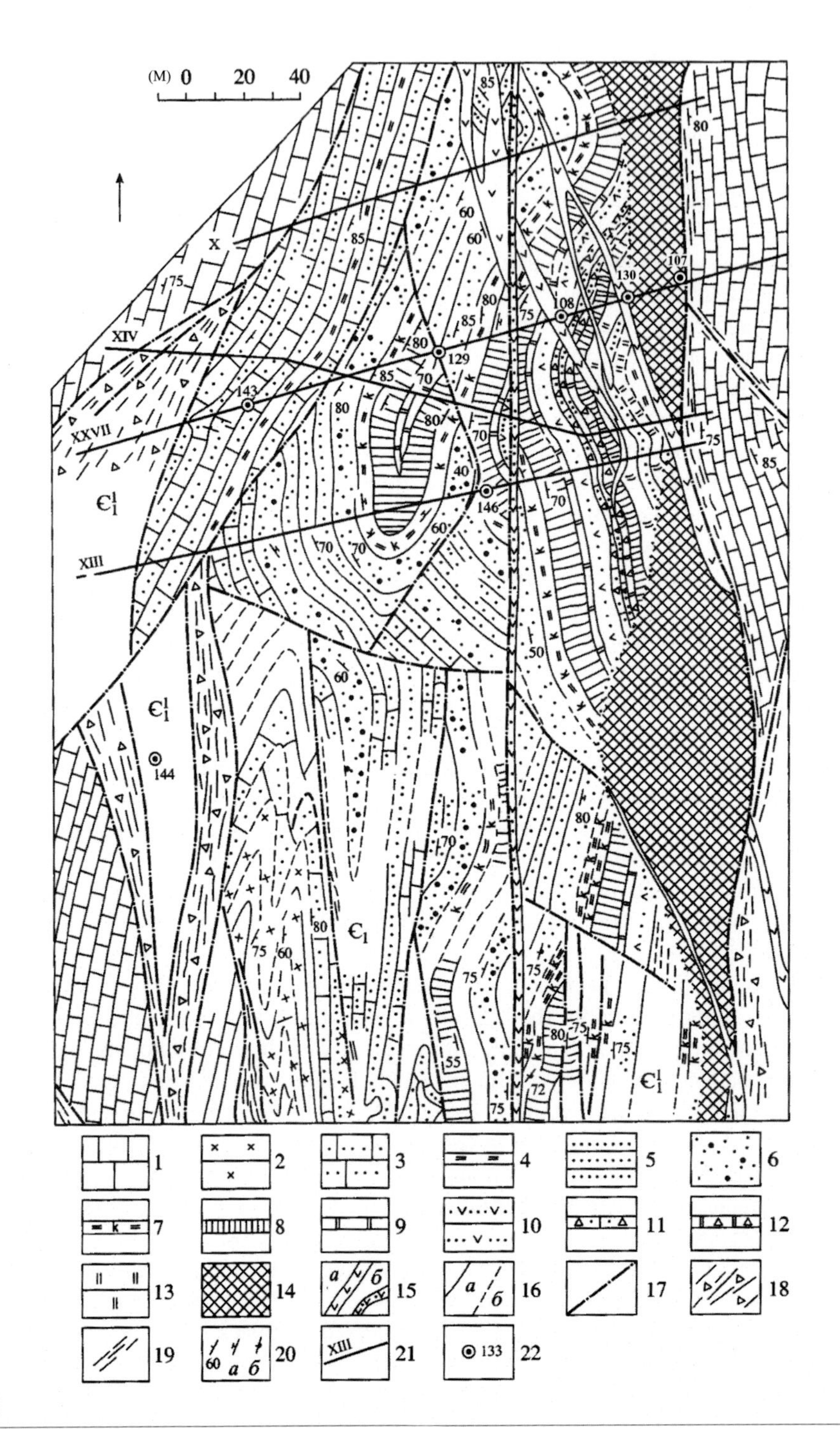

FIG. 3.120

See the legend on opposite page.

The ore submember encloses 10 beds of carbonate and chlorite-carbonate-manganese ores with a thickness from 2.5 to 14 m, divided by inter-ore interbeds with a thickness from 2 to 5–6 m. The ores are rhythmically layered, predominantly clastic and clastic-spherulitic, more rarely spherulitic and pelitoform. In terms of composition, they have been delineated as rhodochrosite, limestone-rhodochrosite, and chlorite-rhodochrosite varieties (Mirtova, 1983). Data on the chemical composition of the ores are recorded in Table 3.26.

Rhodochrosite ores are represented by gray and dark-gray (up to black) rhodochrosite siltstones and sandstones (with rare interbeds of gritstones in the base of the rhythms) of various structures (pelitoform, micro-grained, spherulitic, microspherulitic, and clastic). Occasionally the ores are entirely composed of rhodochrosite spherulites and their clasts. Rarely occurring are clasts of phosphorite of micro-coagulation and ball texture and an impurity of pyroclastic material. Occasionally phosphorites form thin laminae and small lenses. Rhodochrosite ores compose independent beds but also occur in the form of interbeds in seams of chlorite-carbonate ores. Noted is a thin impurity of sulfides. Content of Mn in rhodochrosite ores exceeds 20%.

Calcareous-rhodochrosite ores are dark-gray up to black as well as clastic (clasts of the sandstone-siltstone variety); in terms of textural particularities are analogous to rhodochrosite ores. Present in a significant quantity in the composition of the clasts are limestones, frequently phytogenic (Fig. 3.122A and B) and containing fragments of archaeocyathids (Fig. 3.122C and D). The cement is manganocalcite or calcite, more rarely rhodochrosite or siliceous. The content of manganese in the rock is dependent upon the quantity of limestone clasts and varies from 10% to 20%.

Chlorite-rhodochrosite ores are represented by rhythmic interbedding of dark-gray, black, and pinkish-gray rhodochrosite ores of various structures (clastic, clastic-spherulitic, microspherulitic, pelitoform) with chlorite and rhodochrosite-chlorite laminae of green and greenish-gray color. These ores

FIG. 3.120

Geological scheme of the northern part of the Pravoberezhny sector of Usa manganese deposit (composed by S.M. Tarasova, Yu.V. Myrtov, A.F. Bych etc.). 1—Usa suite; light-gray arhaeocyate and epiphitonic limestones, with rare interbedded onkolitic layers and limestone breccia. 2–13—ore-bearing stratum (Ust-Kundat suite): 2—laminate spongolites with clastic limestone layers and admixture of pyroclastic material; 3—calcareous and calcareous-rhodochrosite sandstones and rhythmic-badding siltstones with interlayers of spongolite; 4—calcareous-rhodochrosite siltstones with interlayers of rhodochrisite-calcareous sandstones; 5—rhythmic-badding calcareous-rhodochrosite siltstones with interlayers of rhodochrisite-calcareous its variety (10–15% Mn); 6—spherulite-clastic rhodochrosite ores (rhythmic-badding rhodochrosite sandstones, grits, siltstones) with interbedded of micro-laminated chlorite-rhodochrosite ores (more 20% Mn); 7—thin interstratification of motley siltstones, calcareous grits and breccia, rhodochrosite tuff-sandstones and tuff-alevrolites; 8—green-gray chlorite-rhodochrosite ores with layers and lenses of rhodochrosite grits (20% Mn); 9—interbedding of clastic limestones and mica-sulfide-chlorite schists (over psammitic and ash tuffs); 10—interstratification of rhodochrosite tuff-alevrolites, tuff-sandstones and chlorite-rhodochrosite ores; 11—greenish-gray fine-clastic limestones, with interbedded calcareous breccia and tuff-shists; 12—light-gray limestone grits with interlayers of green and cherry tuff-shists; 13—black rhodochrosite-calcareous siltstones black with interbeds of light-gray clastic limestones; 14—psilomelan-pyrolusite ores of oxidation zone; 15—dike rocks: (a) basic and (b) alkaline composition; 16—border of patterns: (a) tracked (b) supposed; 17—faults; 18—tectonic breccias; 19—zone of crush; 20—elements of occurrence: (a) normal, (b) overturned, 21—exploration line and its number; 22—exploration well and its number.

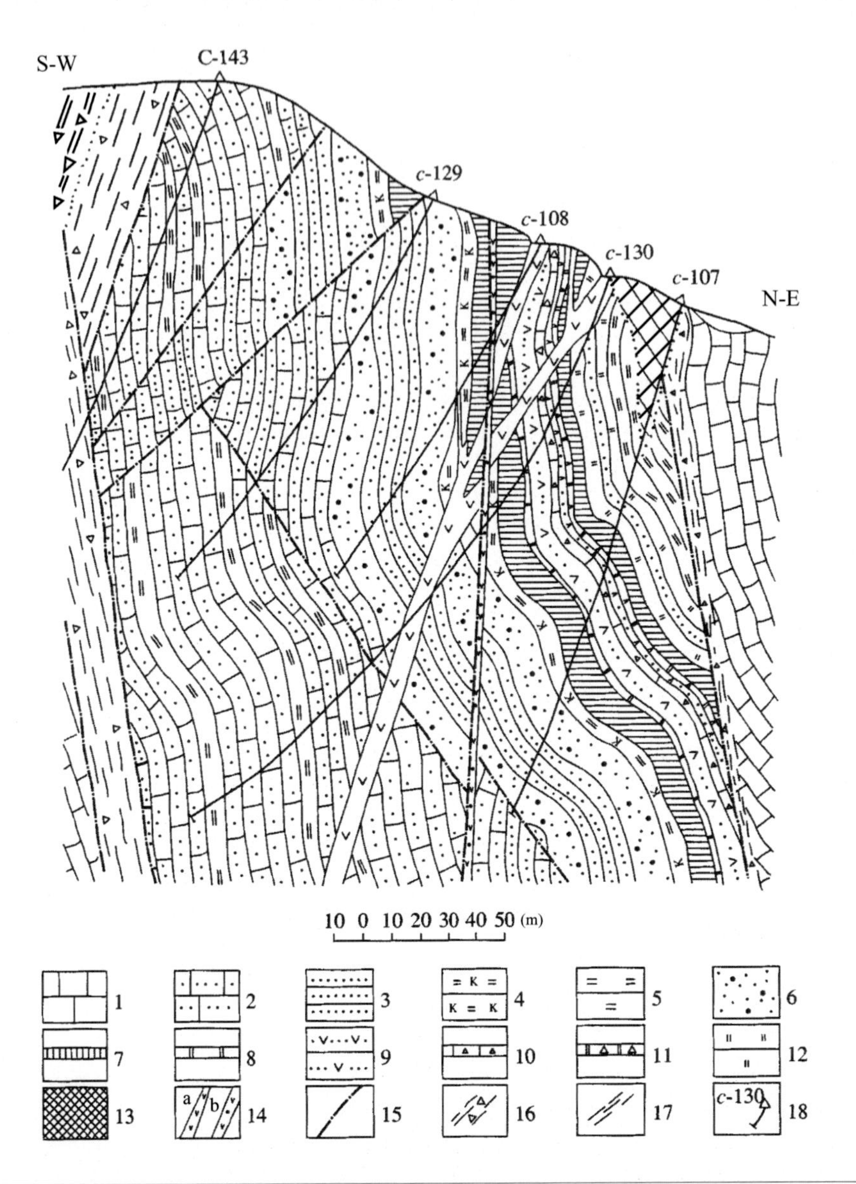

FIG. 3.121

Geological cross section along the line XVII of Pravoberezhny exploration seqtor of Usa manganese deposit (composed by S.M. Tarasova, Yu.V. Myrtov, A.F. Bych). 1—Usa suite; light-gray arhaeocyate and epiphitonic limestones, with rare interbedded onkolitic layers and limestone breccia; 2–12—ore-bearing stratum: 2—rhythmic-badding calcareous sandstones and aleurolites with spongolitic interlayers; 3—rhythmically layered limestone-rhodochrosite sandstones (10–20% Mn); 4—thin interstratification motley cherts, calcareousgrits and breccia, tuff-sandstones and rhodochrosite tuff-aleurolites; 5—calcareous-rhodochrosite siltstones with rhodochrosite-calcareous sandstones (10–15% Mn); 6—ritmichnosloistye spherulites-clastic rhodochrosite-ore interlayers of micro-chlorite-rhodochrosite ores (less than 20% Mn); 7—micro-layered chlorite-rhodochrosite ores with interlayers of rhodochrosite grits; 8—interstratification of clastic limestones and mica-sulfide-chlorite schists (over psammitic and ash tuffs); 9—interstratification of rhodochrosite tuff aleurolite, tuff-sandstones and chlorite-rhodochrosite ores; 10—fine-clastic greenish-gray limestones with calcareous breccias and tuff schists; 11—light-gray calcareous gritstones with green grits and cherry colored tuff schists; 12—black rhodochrosite-calcareous siltstones interbedded with light-gray clastic limestones; 13—oxidation zone, psilomelan- pyrolusite ores; 14—dikes: (a) diabase, (b) syenite; 15—faults; 16—tectonic breccia; 17—zone of mylonitization; 18—exploration well and its number.

Table 3.26 Chemical Composition of Rocks and Manganese Ores of the Right-Bank Section of the Usa Deposit (Kuznetsk Alatau)

Analysis	SiO_2	TiO_2	Al_2O_3	Fe_2O_3	FeO	MnO	MnO_2	MgO	CaO	P_2O_5	Na_2O	K_2O	BaO	Cr_2O_3	NiO	H_2O^+	H_2O^-	CO_2	Total
2060	6.06	0.05	1.60	0.24	0.13	2.19	0.32	1.04	46.3	0.03	0.08	0.22	0.08	0.27	0.08	0.14		40.8	99.6
2065	2.48	0.03	1.08	0.07	0.26	2.81	0.65	2.26	47.3	0.09	0.02	0.18	0.07	0.29	0.08	0.01	0.13	42.2	100
2066	4.62	0.02	0.99	0.76	0.12	5.81	0.65	1.96	43.5	0.09	0.03	0.13	0.04	0.29	0.07	0.28	0.08	40.8	100.2
2067	5.47	0.02	1.16	0.32	0.85	4.14	0.54	1.65	45.4	0.07	0.05	0.13	0.09	0.28	0.05	0.85	0.14	39.6	100.1
2068	11.1	0.02	0.83	0.75	0.23	4.94	0.43	10.1	32.6	0.07	0.01	0.12	0.10	0.20	0.06	0.01	0.14	38.4	100.1
2069	8.26	0.09	2.31	0.37	0.44	1.28	0.43	2.27	45.5	0.09	0.03	0.35	0.34	0.32	0.07	0.65	0.10	37.2	100.0
2070	5.24	0.03	1.01	0.24	0.26	7.15	0.65	0.83	42.8	0.07	0.02	0.11	0.06	0.48	0.14	0.32	0.09	40.2	99.7
2071	4.75	0.02	0.77	1.86	0.62	30.4		2.31	21.2	0.15	0.01	0.09	0.03	0.38	0.13	0.32	0.20	37.1	100.3
2072	17.0	0.07	1.65	2.86	7.10	27.0		6.75	12.2	0.23	0.03	0.12	0.03		0.04	2.88	0.22	22.8	101.0
2073	12.1	0.04	1.20	2.27	2.65	35.0		4.34	11.0	0.22	0.02	0.11	0.02	0.35	0.22	2.03	0.21	27.1	99.0
2074	7.15	0.03	1.13	1.59	0.14	18.3	0.97	2.77	32.2	0.11	0.02	0.16	0.07	0.15	0.04	1.35	0.28	34.2	100.6
2075	91.9	0.01	0.40	0.48		1.57	1.08	0.23	2.28	0.01	0.15	0.06	0.03		0.03		0.21	2.25	100.7
2076	3.79	0.02	0.92	0.99	0.12	9.02	0.87	2.70	40.2	0.07	0.09	0.13	0.08	0.13	0.03	0.40	0.14	39.0	98.6
2077	6.44	0.02	0.64	1.72	1.65	45.2		1.70	7.98	0.13	0.09	0.13	0.04	0.05	0.10	0.61	0.18	33.4	100.1
2078	12.3	0.08	2.26	1.10	1.13	12.0		5.41	29.3	0.09	0.05	0.17	0.30	0.17	0.09	0.66	0.21	34.8	100.2
2079	11.0	0.26	4.43	1.04	0.57	5.19	0.65	2.59	36.2	0.29	0.13	0.70	1.51	0.23	0.07	0.95	0.22	33.6	99.5
2080	34.5	0.34	5.98	2.55	3.45	7.00		3.59	18.8	0.57	0.20	1.35	2.33	0.03	0.07	3.07	0.80	15.8	100.5
2083	34.5	0.33	5.13	2.40	7.56	19.1		9.20	3.24	0.34	0.37	0.68	0.85		0.11	5.16	1.34	9.25	99.5
2084	27.3	0.59	5.14	2.19	7.00	9.51		11.2	15.3	0.38	0.09	0.37	0.43	0.01	0.07	4.21	0.80	15.1	99.6
2085	12.7	0.04	0.80	0.99	4.02	50.8		2.42	3.29	0.23	0.05	0.08	0.01		0.05	1.64		22.9	100
2086	4.19	0.05	1.37	0.14	1.23	4.75	0.97	1.55	44.7	0.07	0.08	0.14	0.12	0.24	0.07		0.09	38.8	98.5
2089	1.86	0.05	1.09	1.21	3.60	32.1		4.39	13.0	0.26	0.07	0.13	0.04	0.01	0.07	2.02	0.08	28.7	98.7
2090	13.4	0.04	1.13	1.65	2.36	40.0		3.51	6.08	0.19	0.06	0.17	0.18		0.08	1.09	0.47	28.5	98.9
2091	14.2	0.05	1.20	2.11	4.39	39.7		4.18	4.88	0.25	0.05	0.17	0.09		0.06	1.17	0.28	26.5	99.3

Analyses conducted in the GIN RAN chemical laboratory. Analyst: *M.I. Stepanets.*

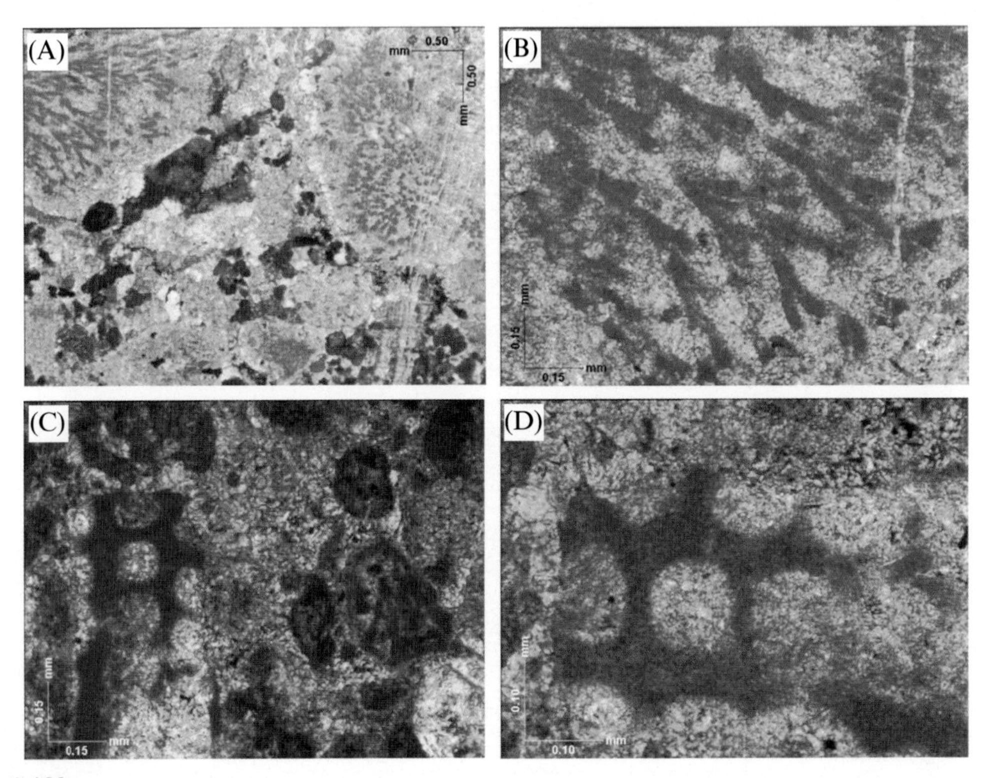

FIG. 3.122

Microphotographs of manganese-ore thin sections from Usa deposit (crossed nicols). (A) and (B) sample 2068: (A) magnification 2.5, (B) magnification 10; (C) and (D) sample 2071: (C) magnification 2.5, and (D) magnification 10.

are the most phosphate- and sulfide-rich, present in the form of thin laminae. Sulfides also form nest-type clumps and abundant thinly scattered impurity. Content of manganese constitutes 17–25%.

Silicate ores occur fairly rarely and in small quantity. They form narrow (0.2–0.5 m) zones along the exocontacts of dikes. The rocks are reddish-brown, fine-grained, frequently of porphyritic structure. Their basic mass is composed of fine-grained aggregates of rhodonite, bustamite with abundant impurity of stilpnomelane. Nodular structures (up to 1 cm in diameter) are filled by coarse grains of axinite and bustamite.

Oxidized ores are also widely distributed within the boundaries of the deposit. These are developed principally in zones of oxidation and coincide with large tectonic faults. Ores are represented by psilomelane, pyrolusite, and vernadite varieties.

The inter-ore interbeds are composed of rhythmic-layered rhodochrosite-limestone sandstones and siltstones, breccia of phytogenic and archaeocyathidic limestones with carbonate-siliceous cement, jasperoid multicolored siliciclastics, and psammitic and ash tuffs of basic composition.

The fragmentary (observed) thickness of the ore horizon, according to the data of geological-exploratory surveys, constitutes 102.5 m. The total thickness of the subore and ore horizons within the boundaries of the right-bank ore field constitutes approximately 250 m.

In this way, characteristic for the ore-bearing strata on the whole is a dark-gray (up to black) color of the rocks, a thin rhythmic layering, a fine-clastic, substantially carbonate composition of the rocks, an abundance of interbeds of cherts and spongolites, and a consistent tuffogenic impurity in rocks and ores. Attention is drawn to the absence of even an insignificant impurity of aluminosilicate terrigenous material and the abundance of organogenic residues (planktonic forms, archaeocyathids).

Deposits of the ore-bearing member are concordantly overlapped by strata of light-gray and gray limestones of the Usa suite. Within the boundaries of the deposit, occurrences of this suite are represented by a thick strata of lenticular-layered algal (epiphytic) and archaeocyathid-bearing limestones with interbeds of clastic limestones. The suite contains a transitional horizon represented by an alternation of algal-archaeocyathidic and stromatolite-oncolitic limestones, limy gritstones, and breccia with interbeds of phosphorite-sulfide shales and tuff-shales. The transitional horizon is characterized by the presence of laminae and clasts of phosphorite, the presence of an impurity of pyroclastic material, and rare interlayers of manganous rocks.

For the purposes of isotope research, stone material of all varieties of carbonate-manganese ores and manganese-containing carbonate rocks developed was gathered within the boundaries of the most studied part of the deposit—directly from the trenches of exploration profiles X, XIII, and XIV of the northern part of the right-bank section, as well as from the core material of certain wells from the left-bank section of the deposit. Direct contacts of the ore strata with carbonate rocks of the Usa suite are absent from the collected samples. For this reason, samples of limestones of the Usa suite were gathered from a range of trenches of profile X and of a natural outcrop (rock exposures) in the continuation of profile XIV.

According to the results of previously conducted mineralogical research (Getseva, 1947; Varentsov, 1962a; Rakhmanov and Eroshchev-Shak, 1966) as well as our data (Figs. 3.123 and 3.124), ore carbonates are represented primarily by an isomorphic mixture of minerals of the manganocalcite-rhodochrosite series. Evidently also present is calcite-magnesian kutnohorite (samples 2068, 2073, and others).

Isotope data. The results of isotope research are recorded in Table 3.27 and displayed in Fig. 3.125. From these it follows that the studied carbonate ores and manganese-containing rocks on the whole are substantially distinguished from marine carbonates of sedimentary origin and are characterized by a broad interval of values of the isotopic composition of carbon: values of $\delta^{13}C$ vary within the boundaries −18.4‰ to 0.7‰; moreover, the isotopic composition of oxygen varies within a relatively narrow range—values of $\delta^{18}O$ vary from 18.4‰ to 23.0‰. Limestones of the Usa suite are characterized by the heaviest isotopic composition of carbon ($\delta^{13}C = -1.9$‰ to 1.0‰) and oxygen ($\delta^{18}O = 21.2$–24.3‰).

The obtained isotope data provide evidence for the fact that carbon of the studied rocks has a dual origin—on the one hand it was derived from carbonates of initially sedimentary origin (−2‰ to 2‰), and on the other hand it is represented by carbon of oxidized organic matter (−24‰ to −22‰).

According to the employed method of decomposition of Mn carbonates (a $PbCl_2$ reagent was used), the obtained isotope data characterize the general (average) carbon of the sample. Therefore, it can be assumed that the scattering of $\delta^{13}C$ values is determined by the primary isotopic heterogeneity of the rock, caused by the physical mixture of carbonate generations of various geneses. The latter could have been characterized by a different initial isotopic composition analogous to that of initially sedimentary (with a heavy isotopic composition of carbon) as well as to that of post-sedimentational (with a light isotopic composition of carbon) carbonates. However, our conducted methodological research (Erokhin and Kuleshov, 1998; Kuleshov, 2013) indicates that the calcite component in the samples in terms of isotopic composition is analogous (or close) to that of manganese-containing carbonates of the same sample and is distinguished from initially sedimentary carbonates by lower values of $\delta^{13}C$. It can

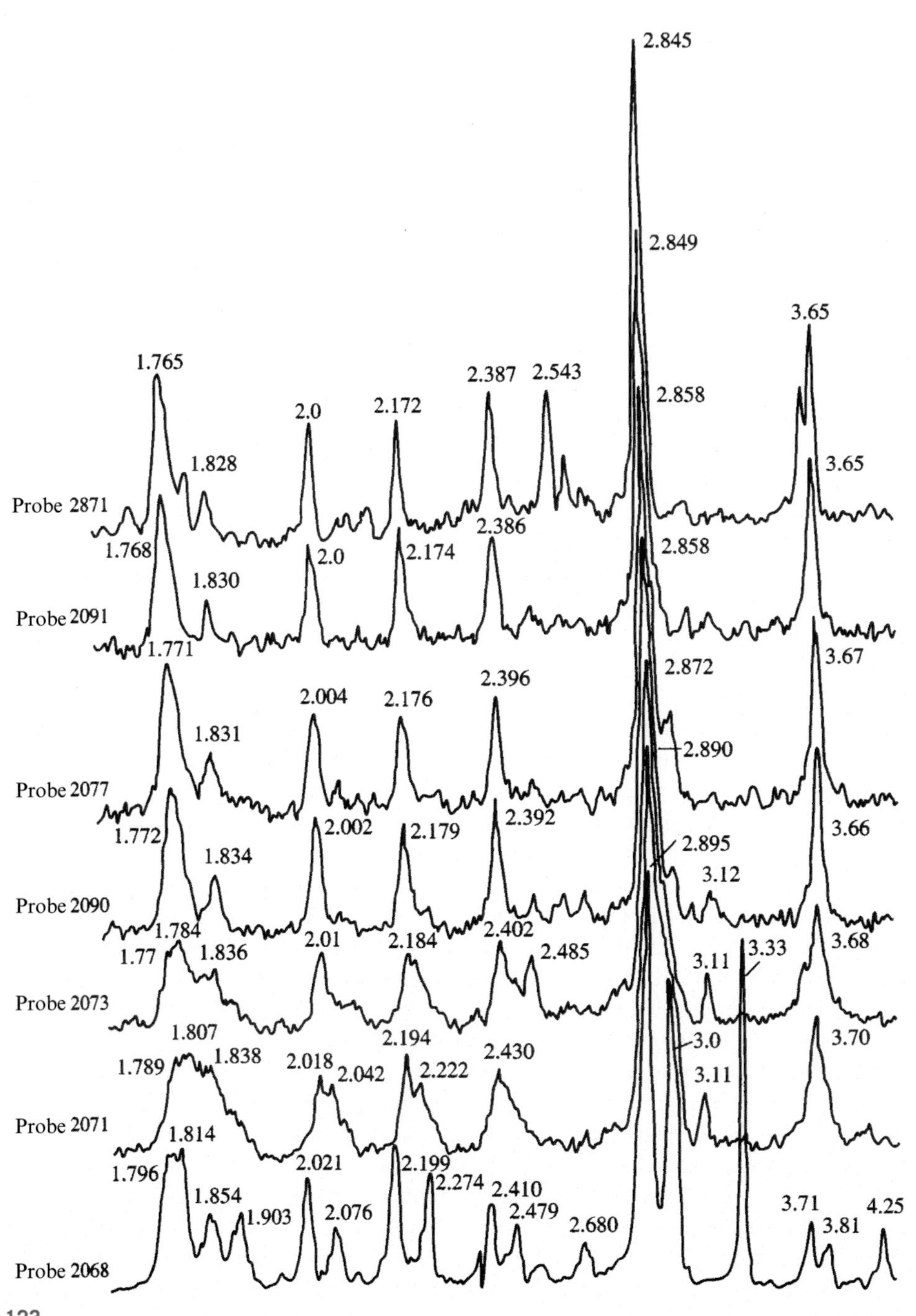

FIG. 3.123

X-ray diffractograms of manganese carbonate ores of Usa deposit.

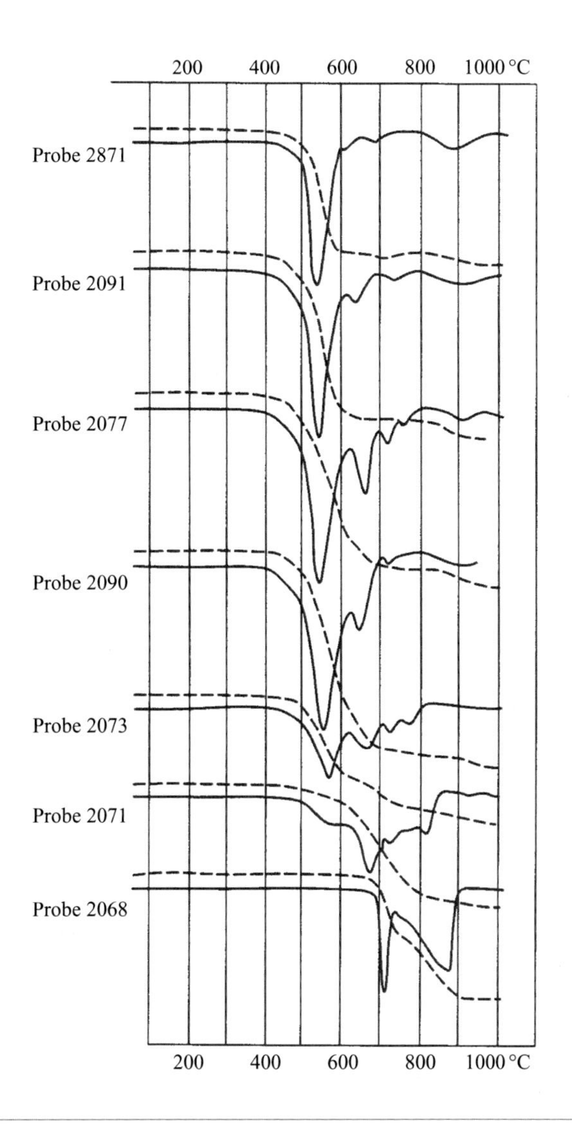

FIG. 3.124

Thermogram curves of carbonate-manganese carbonate ores of Usa deposit.

therefore be concluded that the forming of the isotopic composition of carbon (and that of oxygen in the same manner) in the rock occurred as a result of a unitary physical-chemical process of formation of the manganese-containing rock.

The isotopic composition of oxygen by comparison with carbon in the genetic regard has turned out to be less informative. On the whole the increasing lightness of $\delta^{18}O$ values of the initial clastic carbonate rocks and ores of the ore-bearing strata occurred under the influence of metasomatic (or metamorphic) fluids isotopically light in terms of oxygen. This took place either at the stage of the formation of ore matter or later.

It follows to note that the transformation of initially sedimentary carbonates within the boundaries of the deposit occurred so intensively that even in limestones of the Usa suite situated close to the ore strata, the initial isotope markers were not preserved in the isotopic composition of oxygen. For example, in sample 2061 (light-gray marbleized limestone, MnO=0.27%), the value of $\delta^{18}O$ is close to that of ore carbonates and constitutes 23.6‰. In the zone of the eastern contact of the ore body, limestones

Table 3.27 Isotopic Composition of Carbon and Oxygen and MnO Content in Rocks and Manganese Ores of the Right-Bank Section of the Usa Deposit (Kuznetsk Alatau)

Number of Analysis	Number of Sample	Location of Gathering and Characteristics of Sample			$\delta^{13}C$, ‰ PDB	$\delta^{18}O$, ‰ SMOW	MnO, %
Right-bank section							
2066	1/86	Trench X	Gray marble		−1.9	21.2	2.6
2061	2/86		Light-gray marble		0.5	23.6	0.3
2062	3/86	Trench XIV	Ore, fine-grained		−3.3	19.8	15.8
2063	4/86			With sulfides	−1.9	22.5	10.4
2064	5/86			Without sulfides	−1.3	21.3	7.4
2065	6/86			Light-gray, medium-grained, silicified	−2.1	21.8	3.3
2066	7/86			Dark-gray, fine-grained	−0.3	22.0	6.3
2067	8/86				0.7	22.9	4.6
2068	9/86		Fine-clastic ore breccia		−1.9	22.8	5.3
2069	10/86		Gray ore gritstone		−2.5	21.8	1.6
2070	11/86		Ore breccia with spherulites		−1.6	20.3	7.7
2071	12/86		Ore breccia, dark-gray, thick	Thin-layered	−7.3	21.5	30.4
2072	13/86		Ore dark-gray, thin-grained		−7.7	23.0	26.9
2073	14/86				−11.8	20.3	35.0
2074	15/86		Layered ore tuff gritstone		−5.6	21.5	19.1
2075	16/86		Ore fine-grained, silicified		−7.3	18.9	–
2076	16a/86		Chloritic ore tuff gritstone		−1.9	19.9	9.7
2077	17/86		Ore dark-gray, fine-grained		−11.2	22.3	45.2
2078	18/86		Ore gray, layered, fine-grained		−6.9	21.0	12.0
2079	19/86		Silicified carbonate ore breccia		−6.1	19.8	5.7
2080	20/86		Ore dark-gray, brecciated		−8.0	19.6	7.0
2083	23/86		Ore dark-gray, layered		−11.5	18.5	19.1
2084	24/86a		Ore breccia, light-gray clast		−6.8	19.0	9.5
2084	24/86b		Dark-gray cement		−9.6	18.9	–
2085	25/86		Ore greenish-gray, thin-banded		−18.4	20.2	50.8
2086	26/86		Ore greenish-gray, fine-grained, silicified		−5.5	20.6	5.5
2089	28/86		Ore dark-gray, crypto-grained		−13.0	20.4	32.1
2090	29/86		Ore gray, thin-layered		−15.9	20.4	40.0
2091	30/86		Ore dark-gray, banded, oxidized		−15.9	22.6	39.6

Table 3.27 Isotopic Composition of Carbon and Oxygen and MnO Content in Rocks and Manganese Ores of the Right-Bank Section of the Usa Deposit (Kuznetsk Alatau)—Cont'd

Number of Analysis	Number of Sample	Location of Gathering and Characteristics of Sample			$\delta^{13}C$, ‰ PDB	$\delta^{18}O$, ‰ SMOW	MnO, %
Outcrop on the eastern contact of the ore body (continuation of trench XIV)							
2093	32/86	Trench XIV	Gray marble		0.9	24.2	–
2094	33/86				1.0	24.3	–
2095	34/86	Trench XIII	Gray ore siltstone		−12.4	21.6	–
2096	35/86		Dark-gray ore sandstone		−4.5	20.5	–
2097	36/86		Ore sandstone with sulfides		−6.0	23.0	–
2098	37/86		Ore gritstone		−1.7	19.1	–
2099	38/86		Carbonate rock with chlorite and sulfides, calcite		−0.4	20.0	–
2100	39/86		Ore tuff sandstone		−8.3	20.7	–
2101	40/86		Ore gritstone		−8.0	20.2	–
2102	41/86		Ore breccia		−8.5	20.4	–
2103	42/86		Ore thin-layered, fine-grained		−13.6	20.4	–
2104	43/86		Ore sandstone		−10.8	20.7	–
Left-bank section							
2105	44/86	Well 409, depth 104.8–107.0m, ore siltstone			-13.7	20.9	–
2106	45/86	Trench XIII	Depth 192.0–195.5 m	Ore siltstone with chlorite	−10.3	20.6	–
2107	46/86		Depth 242.2–242.4 m	Ore tuff siltstone	−9.7	20.9	–
2108	47/86		Depth 246.2 m	Ore tuff siltstone silicified	−4.2	20.0	–
2109	48/86		Depth 283.0 m	Ore carbonaceous-siliceous siltstone	−4.6	19.1	–
2110	49/86		Depth 306.8 m	Ore siliceous siltstone	−3.5	18.4	–
2111	50/86		Depth 330.9 m	Ore carbonaceous siltstone with interbeds of carbonaceous shales	−5.7	19.2	–
2112	51/86		Depth 338.3 m	Quartz-carbonate vein with clast of Mn ore	−8.5	19.8	–

Dash: *no data.*

of the Usa suite are also characterized by a light isotopic composition of oxygen (24.2‰ and 24.3‰). Their isotopic composition of carbon has evidently remained static (0.9–1.0‰).

On the genesis of ores of the deposit. At present the literature is host to opposing viewpoints on the question of the origin and source of matter of the Usa deposit. The most widespread theory proposes an initially sedimentary formation of carbonate-manganese ores (Betekhtin, 1946; Pushkina, 1960). Detailed geochemical research of manganese ores and host rocks conducted by Varentsov (1962a) has enabled the elucidation of the principal distribution patterns of the chemical elements and has led to the conclusion that the precipitation and formation during diagenesis of the basic mass of iron and of manganese carbonates occurred in the shallow-water basin of sedimentation of the Usa eugeosyncline. Serving as a source of manganese, according to Varentsov, were the products of the weathering crusts of the paleo-landmass situated south-west of the deposit (at the location of the Kuznetsk Alatau anticlinorium).

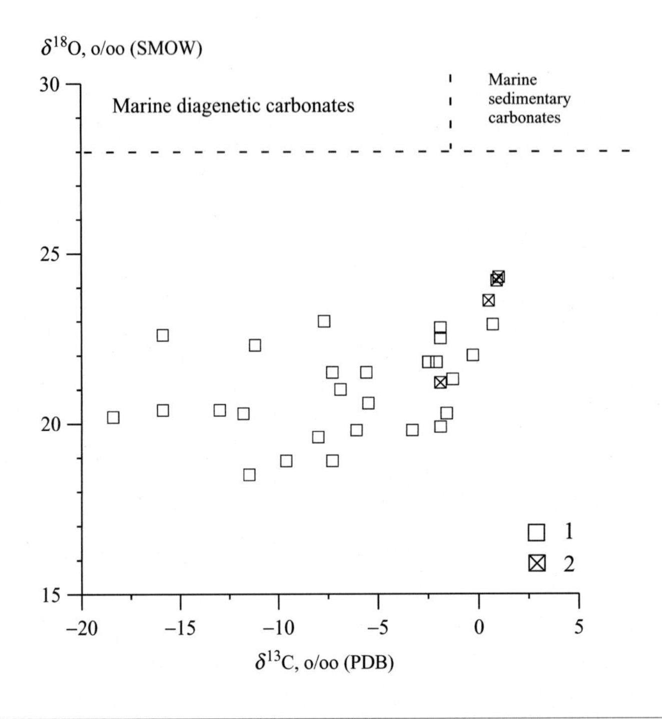

FIG. 3.125

$\delta^{13}C$ versus $\delta^{18}O$ in manganese carbonate ores of Usa deposit. 1—ores and 2—host limestones.

An analogous viewpoint is maintained by Iu. Khodak et al. (1966). However, these authors do not exclude the possibility of a deep source of manganese connected with fumarolic activity.

Sokolova (1960, 1961) has proposed a volcanogenic-sedimentary origin for the manganese ores. This theory is based on the fact that immediately in the district of the deposit the ore strata is facially replaced by volcanogenic depositions, taking as an analogue of this the volcanogenic rocks traced in the watershed of the White Usa and Upper Kibras rivers. However, subsequent works (Mirtov et al., 1973) have clarified the stratigraphic situation of these deposits, which were carried to the Kotlas suite of the Lower-Middle Riphean. Consequently, they could not be facies analogues of the Lower Cambrian ore-bearing strata.

Mirtova (1983) has refined the theory of a volcanogenic-sedimentary origin of the manganese ores of the deposit, considering it rather to be a hydrothermal-sedimentary origin. In distinction from E.A. Sokolova, she proposes that the connection with volcanism here is more distant. This is indicated in her opinion by the character of the volcanogenic material in the ore strata, where are distributed fine-clastic, predominantly siltstone and ash tuffs, as well as by the lack of a single age for all volcanogenic occurrences in the district of the deposit.

The obtained data in terms of the isotopic composition of carbon and oxygen in manganese-containing carbonates and manganese ores are unable to provide an unequivocal answer to the question of the origin of the Usa deposit. However, they do include information on the source of carbon dioxide and conditions of formation of manganese carbonates. Thus, for example, a characteristic particularity of the studied rocks of this deposit is the notable inverse correlation between the

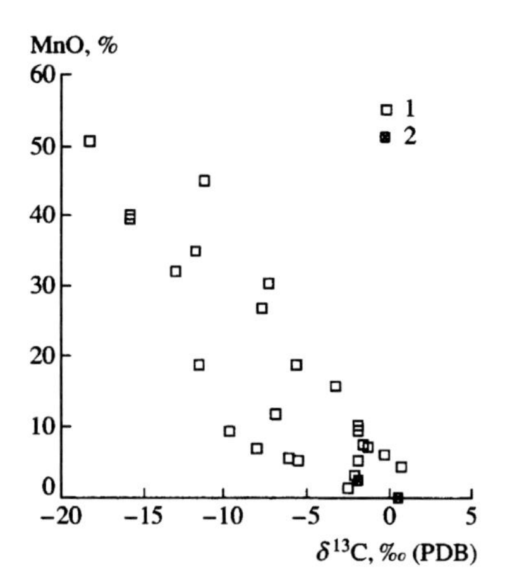

FIG. 3.126

Dependence of $\delta^{13}C$ values and MnO content in ores and rocks of Usa deposit. 1—Ores, and 2—host limestones.

ratios of the isotopic composition of carbon and the content of manganese (Fig. 3.126): the higher the content of manganese in a sample, the lower the values of $\delta^{13}C$ and consequently the greater quantity of rocks rich in light isotope of carbon ^{12}C, and vice versa. That is, rocks with the greatest content of manganese in large degree are rich in carbon of organic origin. This circumstance is a characteristic feature of carbonate-manganese ore-genesis and is manifested, as noted several times above, in practically all deposits. It can be supposed that the ores of the Usa deposit were formed in the same way as in the deposits examined above—namely, with active participation by oxidized carbon of organic matter.

The question of which stage of the geological life of the initial sediment represented by clastic organogenic and phytogenic carbonates featured the formation of carbonate-manganese ores at the Usa deposit is difficult to judge in terms of the isotope data.

The isotopic composition of oxygen in manganese carbonates of the ore strata, as well as in a range of limestones of the Usa suite, does not bear indicators of an initially sedimentary nature. The obtained data provide evidence for the transformation of the isotopic composition of oxygen of initial sedimentary rocks during late metasomatic (or metamorphic) processes as a result of isotopic exchange with the supplying fluids. It cannot be excluded that carbonate-manganese ores were formed at this stage.

The results of isotope research compel us to revisit the discussion of a metasomatic origin of manganese ores. The possibility that their formation occurred by such means has been examined in detail previously by Radugin (1941). The isotope data on oxygen match well with this point of view.

The situation is complicated in light of the interpretation of the distribution patterns of the isotopic composition of carbon. If it is considered that carbon in the process of ore-genesis was taken from two sources—sedimentary carbonates and oxidized organic carbon—then the metasomatic fluids should have carried with them carbon dioxide with very low values of $\delta^{13}C$ (no higher than the lowest obtained values) and should be close to −25‰ to −22‰. It is understood that the ore-forming solutions in the given situation could not be connected with processes of volcanism and with an endogenic (mantle) source of carbon dioxide [the $\delta^{13}C$ of the carbon of mantle origin should be close to −8‰ to −6‰ (Galimov, 1968)].

Clear evidence for this is also seen in the direct correlation between the contents of manganese and the isotopic composition of carbon in the rock (see Fig. 3.126).

It is possible to accept the proposal of Varentsov (1962a) of an initially diagenetic origin for manganese ores of the Usa deposit, along with their transformation by subsequent processes of metamorphism and metasomatism (Rakhmanov and Eroshchev-Shak, 1966). The studied data generally do not contradict this theory, which, however, is unable to explain completely the observed correlation between the isotopic composition of carbon and the content of manganese, also characteristic for limestones of the Usa suite. Additionally, samples of clastic limestone, initially phyto- and organogenic—that is, its non-manganous part—should be distinguished in this case by the high values of the isotopic composition of carbon particular to sedimentary carbonates, while rhodochrosites and manganocalcites should be characterized by lower values of $\delta^{13}C$. However, our methodological research (Erokhin and Kuleshov, 1998; Kuleshov, 2013) has shown that the isotopic compositions of the carbon in these groups of carbonate matter are approximately identical.

Furthermore, with an examination of the correlation between the content of Ca, Mn, and Fe in the rock and the quantity of CO_2 (Fig. 3.127; see Table 3.26), certain regularities can be noted. For example, the content of iron in the rock is found in inverse proportion to the content of CO_2 (see Fig. 3.127C). This serves as evidence that iron is not connected with carbonate matter. At the same time, for a significant number of samples is noted a direct linear correlation between the content of calcium and carbon dioxide, complicated by an insignificant quantity of measured points on account of the presence in the sample of manganese carbonates (see Fig. 3.127A). This same regularity can be observed as well in the distribution of measured points within the coordinates MnO-CO_2 (see Fig. 3.127B). Only in this case is a concentration of points observed in the district of the setting of limestones.

In ores of the Usa deposit, a direct correlation has been noted between the contents of iron, manganese, and phosphorus (Fig. 3.128). Such a correlation has previously been described in detail by Varentsov (1962a) for rocks of the right-bank section of the deposit. However, in our case (see Figs. 3.127D and 3.128B), it bears a more complex character.

Evidently, the observed distribution of chemical components in the rocks of the Usa deposit was determined by the input into the system of initially sedimentary carbonate strata by manganese-containing carbon dioxide metasomatic fluids. This was expressed in the distribution patterns of chemical elements in the rocks, as well as in the interdependence of the isotopic composition of carbon and oxygen with certain basic components. Thus, for example, a linear correlation is noted between the isotopic composition of carbon and the content of CO_2 in the rock (Fig. 3.128A), which also confirms the mixing of carbon dioxide of initially sedimentary genesis (heavy isotopic composition and high contents of carbon dioxide particular to sedimentary carbonates) with externally supplied transforming solutions (light isotopic composition and low concentrations of CO_2). The forming of rocks of the deposit evidently occurred as a result of the two-stage action of metasomatic fluids characterized by different isotope-geochemistry parameters. This, probably, is caused by the presence of two "regions" in the graphs of distribution of components such as iron, manganese, calcium, and phosphorus and their respective isotopic composition of carbon and oxygen (see Figs. 3.127 and 3.128).

On the basis of isotope data, it is difficult to judge the initial nature of manganese carbonates of the Usa deposit, insofar as these rocks have undergone post-sedimentational transformation. However, the established isotope-geochemistry regularities and microscope observations have lent support to the theory as previously proposed by Radugin (1941) of a metasomatic origin for these ores.

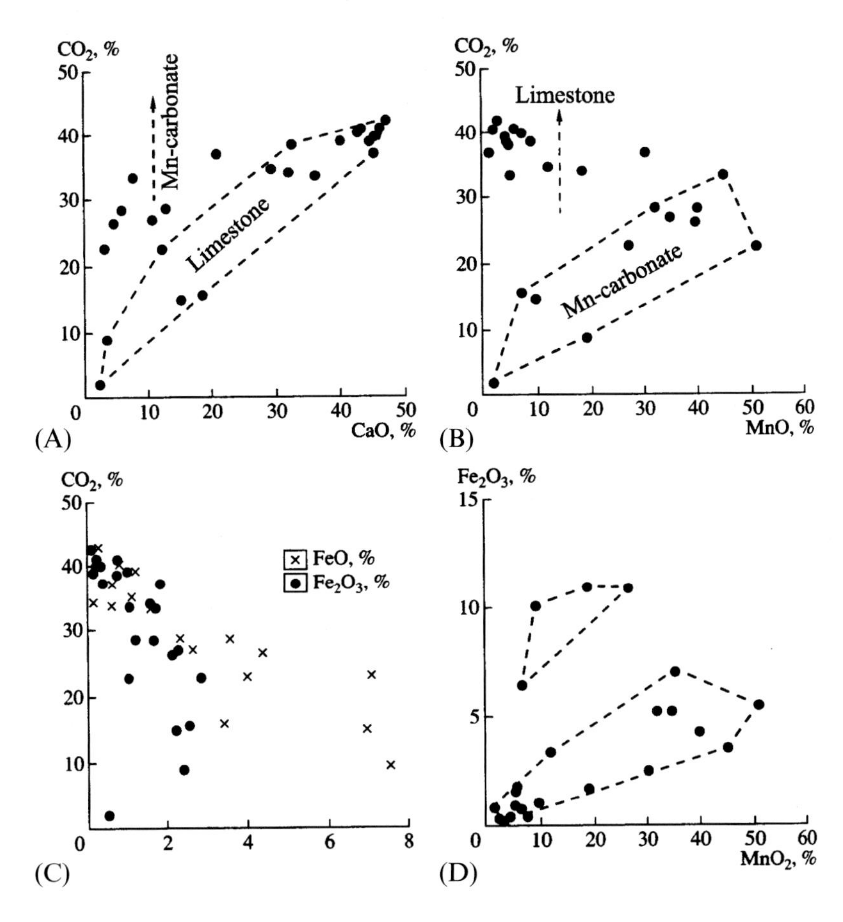

FIG. 3.127

Relationship between contents of CO_2 and CaO (A), CO_2 and MnO (B), CO_2 and FeO, Fe_2O_3 (C), Fe_2O_3 and MnO_2 (D) content in ores and rocks of Usa deposit.

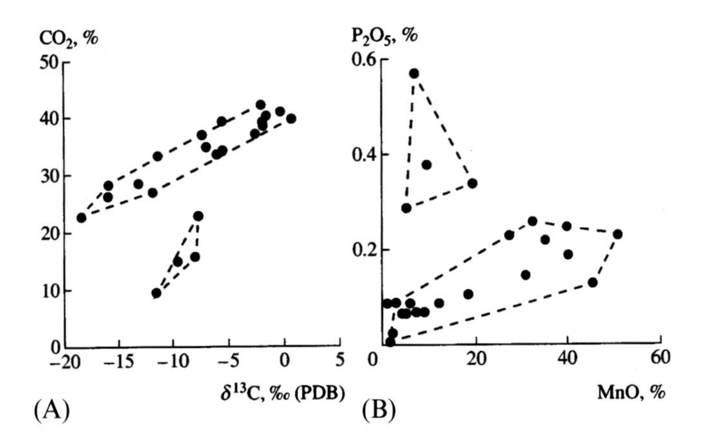

FIG. 3.128

Relationship between contents of CO_2 and $\delta^{13}C$ values (A), contents of P_2O_5 and MnO (B) in ores and rocks of Usa deposit.

Serving as evidence in favor of a metasomatic genesis of the ores, according to K.V. Radugin, is the great diversity of the composition of the rocks along with the particularities of the geological structure of the deposit. For example, the layers of one of the exploratory trenches tended not to occur in parallel with the horizons of another neighboring trench. This could be explained by the unevenness of the metasomatic process, rather than by a rapid mobility of the initial (lagoonal) sediments.

Among indicators of metasomatism, K.V. Radugin has observed diffuse contacts of the ores of variously colored hues and the presence of inclusions of marbles partially subjected to metasomatism ore matter. Noted as well was a replacement of tectonic breccia in the district of the deposit by ore, which itself was nowhere transformed into breccia and was metasomatically formed after the formation of the plication and folding of rocks of the ore-bearing strata. Consequently, these structures in the ores were inherited.

One of the particularities of the studied ores is their clastic texture (Fig. 3.129). It can be supposed that the enrichment of the rock by ore components (manganese) could have occurred only after accumulation of the strata of clastic carbonate (eg, limestone—sample 2079: content of MnO—5.19%, MnO_2—0.65%, and CaO—36.17%; see Table 8.26). Manganese carbonates could not have served as an initial product of failure, insofar as under hypergenesis conditions they are extremely unstable and rapidly oxidize (in quarries of deposits under development, eg, at Chiatura and Kvirila, it can be seen that carbonate-manganese rock from the surface is covered by a rind of manganese oxides in the course of the first 1–2 years).

Indicators of the metasomatic origin of the ore matter evidently could also be spherulitic (Fig. 3.129B) and microspherulitic (Fig. 3.129C) textures of rhodochrosite ores. Moreover, in certain cases (sample 2073, Fig. 8.128C) have been observed sections of varying degrees of recrystallization.

In this way, the results of isotope research of carbonate-manganese ores of the Usa deposit enable a refinement of the understanding of the nature of the carbon dioxide of the ore matter and certain sides of the geochemistry of the manganese-mineralization process itself. Unarguable is the fact of active participation by oxidized carbon of organic matter (hydrocarbons) in the formation of manganese carbonates. Furthermore, the degree of participation by isotopically light organic carbon is a deciding factor in the concentration of the ore element itself—manganese.

Many questions regarding the conditions and the mechanism of formation of the deposit remain unresolved. However, it can be supposed that carbonate-manganese ores of the Usa deposit were formed as a result of the metasomatism of carbon dioxide—ore (Mn, Fe, P) solutions of the initial strata of clastic carbonate rocks (siltstones, sandstones, breccia) of the lagoonal near-reef facies. For this reason, the observed textures in different types of ores particular to sedimentary rocks (layering, rhythm, the dimensions of the grains, and the like) were evidently inherited from the initial sediments.

The question of the origin of carbon dioxide fluids acting on the rocks is complicated. Low values of $\delta^{13}C$ serve as evidence of participation in their composition by oxidized carbon of organic origin. Such carbon dioxide is characteristic for the products of oxidation and destruction of organic matter and is generated in deep zones of sedimentary-rock basins (the zone of oil and gas generation).

Classified as a typical representative of manganese deposits formed with (hydrothermal) metasomatic restructuring of carbonate reef facies could be the Jiaodingshan Co-Mn-ore deposit and the Dawashan manganese-ore deposit, as well as a range of small occurrences situated alongside (Sichuan province, China). These deposits are embedded in rocks that compose organogenic (algal) reef build-ups of the Late Ordovician (Wufengian strata, Ashgillian Epoch) and coincide predominantly with zones of intersections of tectonic faults of north-easterly and north-westerly strike (Fan et al., 1999).

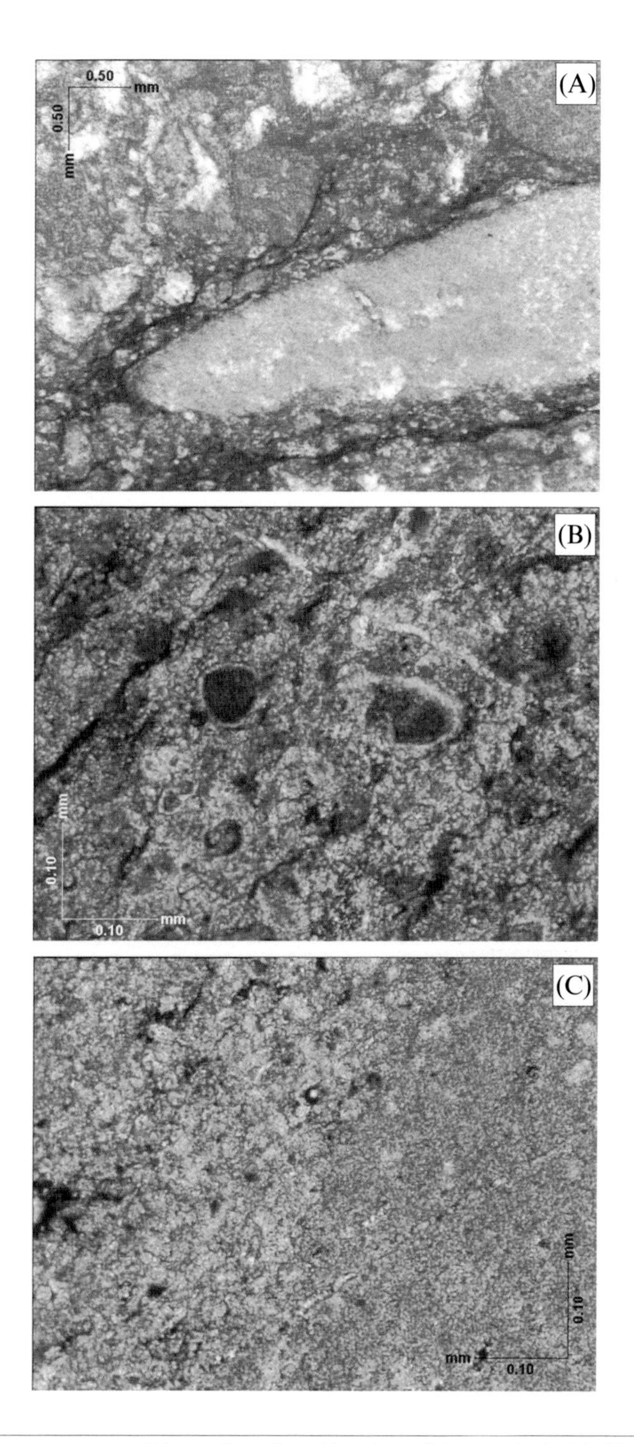

FIG. 3.129

Microphotographs of manganese-ore thin sections from Usa deposit (crossed nicols). (A) Sample 2079, magnification 2.5; (B) sample 2077, magnification 20; and (C) sample 2073, magnification 20.

The *Jiaodingshan* deposit represents an atoll-like reef buildup with a diameter of approximately 1000 m, in which three facies have been identified: a reef complex, the products of its failure, and lagoonal formations. The content of manganese in the rocks constitutes 8.25–32.2%. The ore bodies are composed of rhodochrosite, forming stratiform and lensoidal bodies concordantly deposited with non-ore host marls, limestones, black shales, and sandstones. The ores are massive, layered, spherulitic, and laminated. Layered carbonate-manganese ores are high-quality, composed of dark-red and yellowish-brown seams consisting of varying degrees of crystallization (from micro- to coarse-crystalline) of grains and spheroid formations of rhodochrosite and Ca-Mg-rhodochrosite, in places interstratified with black alternations of hausmannite composition. Massive ores are of predominantly dark-gray color and are composed of micrite and recrystallized rhodochrosite and Ca-Mg-rhodochrosite. Spheroidal ores of gray color are composed of spheroids of rhodochrosite and Ca-Mg-rhodochrosite of various dimensions, in places forming oolite-like texture. In the ores is present a significant quantity of algal residues.

Besides rhodochrosite in the ores are present kutnohorite, manganosite, hausmannite, bementite, clinochlore, magnetite, goethite, and barite. In ores and host rocks are present significant contents of cobalt, reaching ore concentrations.

The *Dawashan* deposit is represented by several ridged reef buildups, the length of which varies from 20 to several hundred meters. For example, ore body No. 3 represents a reef buildup with a height of 14 m (Fig. 3.130).

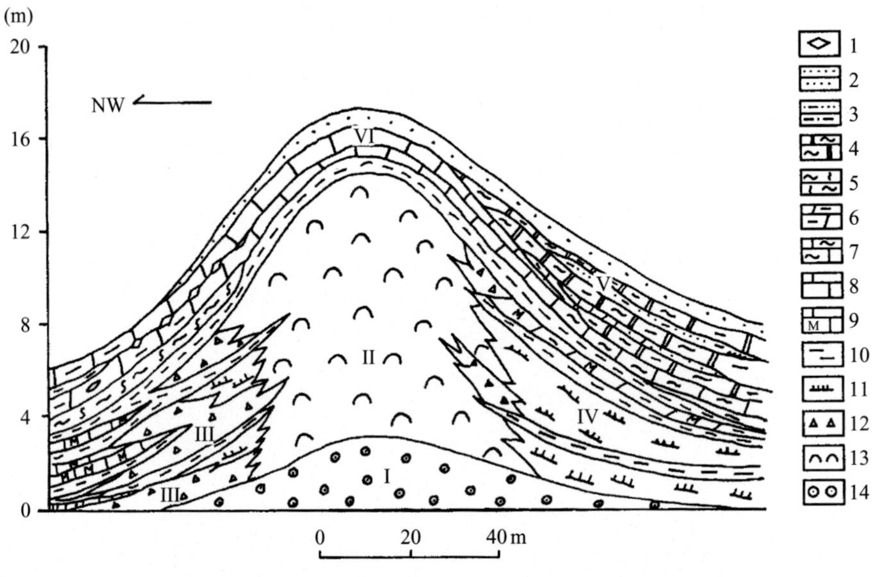

FIG. 3.130

Schematic cross section and lithofacies of algal reef complex of the Davashan manganese deposit (Fan et al., 1999). 1—Chert lens; 2—quartz sandstone; 3—carbon-rich argillaceous sandstone; 4—siliceous dolostone; 5—chert; 6—marl; 7—silicious sandstone; 8—limestone; 9—Mn-bearing limestone; 10—black-shale; 11—laminated-bandet rhodochrosite ore; 12—brecciated rhodochrosite ore; 13—massive algal rhodochrosite ore; 14—spheroidal iron-manganese ore. Roman numerals indicate: I—reef-base bank, II—reef core, III—reef-front algal flat, IV—back-reef algal flat, V—back-reef fill, and VI—reef cap.

The ore body and host deposits are represented by several facies composing different parts of the buildup: the base and core of the reef, frontal clastic facies, interior algal shallow-water deposits, interior reef facies of the filling, and the crowning reef deposits. Rocks composing the base of the reef are commonly represented by spheroidal manganese ores, simultaneously as the facies composing the core of the reef are composed predominantly by algal rhodochrosite ores. The frontal clastic facies are represented by breccia ores, the internal facies by layered varieties.

Manganese ores of this deposit are composed of rhodochrosite, Ca-rhodochrosite, kutnohorite, hausmannite, manganite, and bementite; the textures are massive, layered, laminar, spheroidal, clastic, and algal. The massive ores are composed of rhodochrosite and are the most high-quality. Layered ores are represented by two varieties: (1) interstratified dark-red and dark-gray, occasionally yellowish-pink and black interbeds with a thickness of centimeters, composed of rhodochrosite and Ca-rhodochrosite with hematite and hausmannite; (2) interstratified black and pale-yellow-gray interbeds, composed predominantly by Ca-rhodochrosite of various granularity. Laminar ores are represented by an alternation of pale-yellow-gray and black microlenses consisting of Ca-rhodochrosite and kutnohorite. Spheroidal ores are composed of spheroids predominantly of Ca-rhodochrosite with a diameter of 0.1–2 mm.

Organogenic structures in the ores are represented by oncolites, stromatolites, and residues of blue, green, and red algae.

For an explanation of the genesis of the deposit, Fan et al. (1999) have studied the isotopic composition of carbon and oxygen in various types of rocks of the Jiaodingshan and Dawashan deposits. The obtained data are recorded in Fig. 3.131. From these data, it follows that the isotopic composition of the initial carbonates was fully transformed and the isotope indicators of their initially sedimentary nature were not preserved; all values of $\delta^{13}C$ and $\delta^{18}O$ are characterized by the low ratios particular to epigenetic carbonates. Moreover, the lightened isotopic composition of carbon, as for practically all above-examined deposits, is determined by the participation in the formation of secondary carbonates and the transformation of primary carbonates by isotopically light carbon dioxide of biogenic origin that formed as a result of oxidized carbon of organic matter. Low values of $\delta^{18}O$ are contingent upon elevated temperatures of transformation and the participation by transforming fluids with a lighter isotopic composition by comparison with that of the medium of sedimentation of carbonate rocks of the initial reef facies. The range of $\delta^{13}C$ and $\delta^{18}O$ values for ores of the Chinese deposits fully coincides with that of ores of the Usa deposit (Kuznetsk Alatau, Russia) (see Fig. 3.125).

Processes of metasomatic replacement of calcium carbonates by manganese can be observed as well in points of mineralization of various districts of development of organogenic buildups. Serving as one of the demonstrative examples are the *bryozoan bioherms of the Neogene of the Taman Peninsula*. As described by D.I. Golovina (Issledovanie et al., 2012), Neogene deposits of the Taman Peninsula are represented almost entirely by clayey rocks. These determine in essence the little-dissected modern relief and the tranquil character of the shoreline—its lack of deep gulfs and bays.

In terms of morphology, buildups of bryozoan limestone (Andrusov, 1961) have been clearly identified, which form ridges of hills in the lowland part of the peninsula, as well as capes (Tuzla, Popov Kamen, Panagia) on the shore and crescent chains of reefs, extending the capes into the Black Sea. The Taman bioherms do not form unitary bodies that are consistent along the strike and

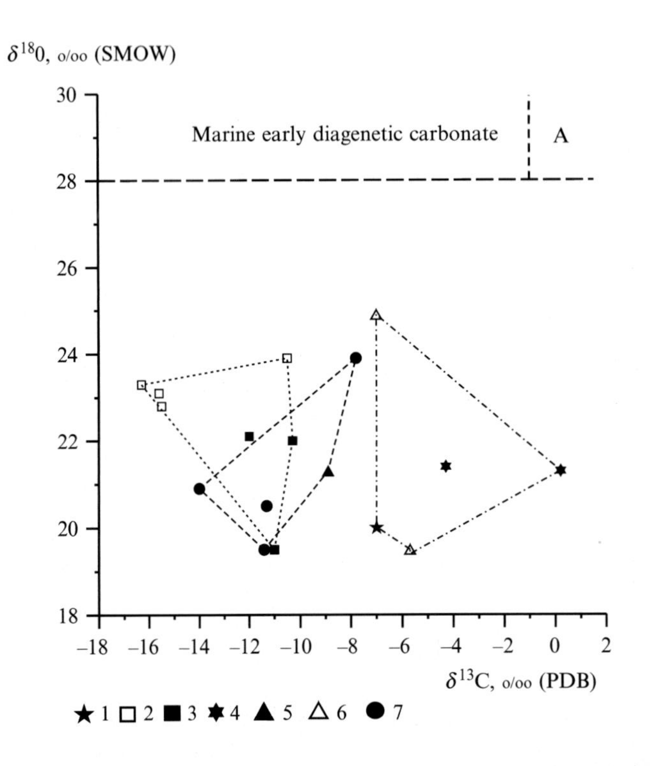

FIG. 3.131

$\delta^{13}C$ versus $\delta^{18}O$ in manganese carbonate ores of the Jiaodingshan and Dawashan manganese deposits [compiled after Fan et al. (1999)]. A—marine sedimentary carbonates; 1–3—Dawashan deposit: 1—limestone, 2—Ca-rhodochrosite, 3—rhodochrosite ores; 4–7—Jiaodingshan deposit: 4—limestone, 5—Ca-Mg-rhodochrosite, 6—Mn-dolomite, 7—rhodochrosite ores.

thickness. These are separate blocks reaching 20–30 m in height and are on the order of 50–150 m along the strike.

Bryozoan limestone represents a bioherm; the share of directly bryozoan and other sessile organisms similar to sponges, algae, and the like varies within wide boundaries—from a few to 80%. A sample of bioherm was gathered from a building-stone quarry near Cape Panagia and represents dense limy rock with an incrustation of dark-colored formations.

The stratigraphic setting of the bioherms of the Taman Peninsula is speculative; at present, they are accepted in the capacity of a lithological benchmark for the border between the Sarmatian and Meotian stages of the Eastern Paratethys, with an age estimated at 9 million years (Chumakov et al., 1996).

The chemical composition and structure of carbonate rocks of the bioherm can be seen in Figs. 3.132–3.134, where residues of various organisms, including stromatolites (see Fig. 3.134) consisting of carbonates of calcium and manganese, are distinctly visible.

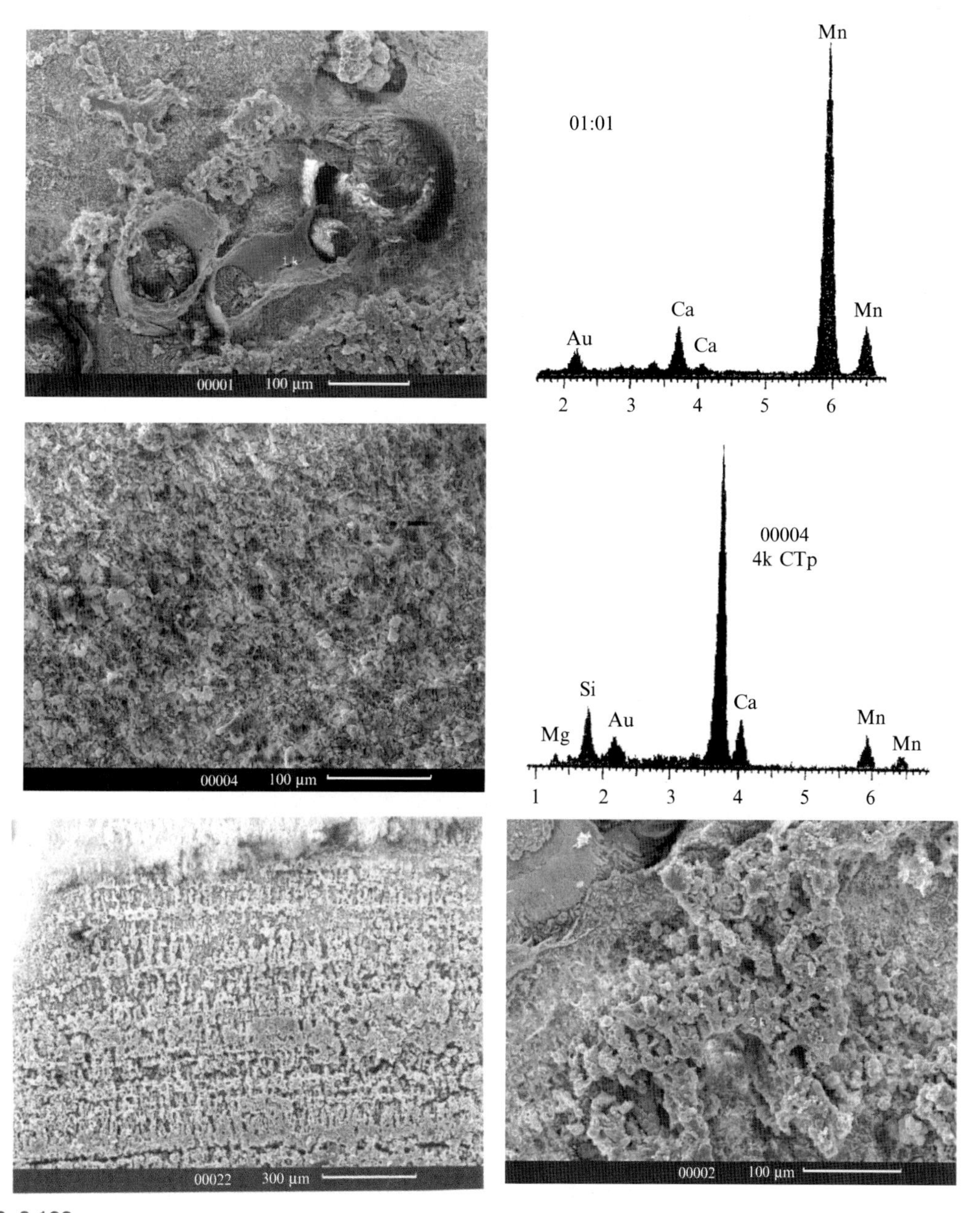

FIG. 3.132

The structure (samples 1, 2, 4, 22) and chemical composition (samples 2, 4) of the bryozoan bioherm, Taman peninsula (photos by E.A. Zhegallo and E.L. Shkolnik).

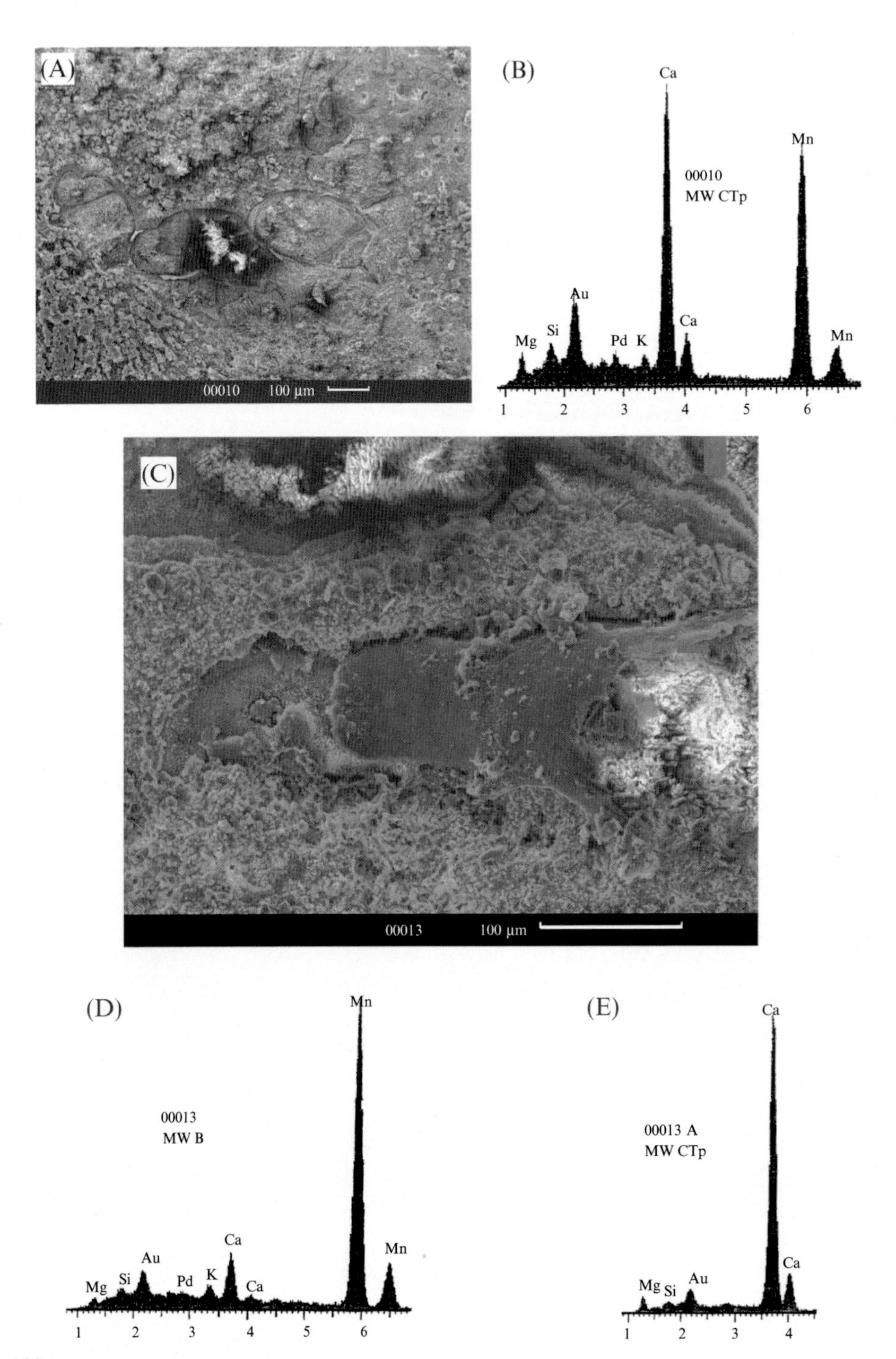

FIG. 3.133

The structure (A—sample 10, C—sample 13) and chemical composition (B—sample 10, matrix; D—sample 13, matrix; E—sample 13, stromatolite) of the bryozoan bioherm, Taman peninsula (photos by E.A. Zhegallo and E.L. Shkolnik).

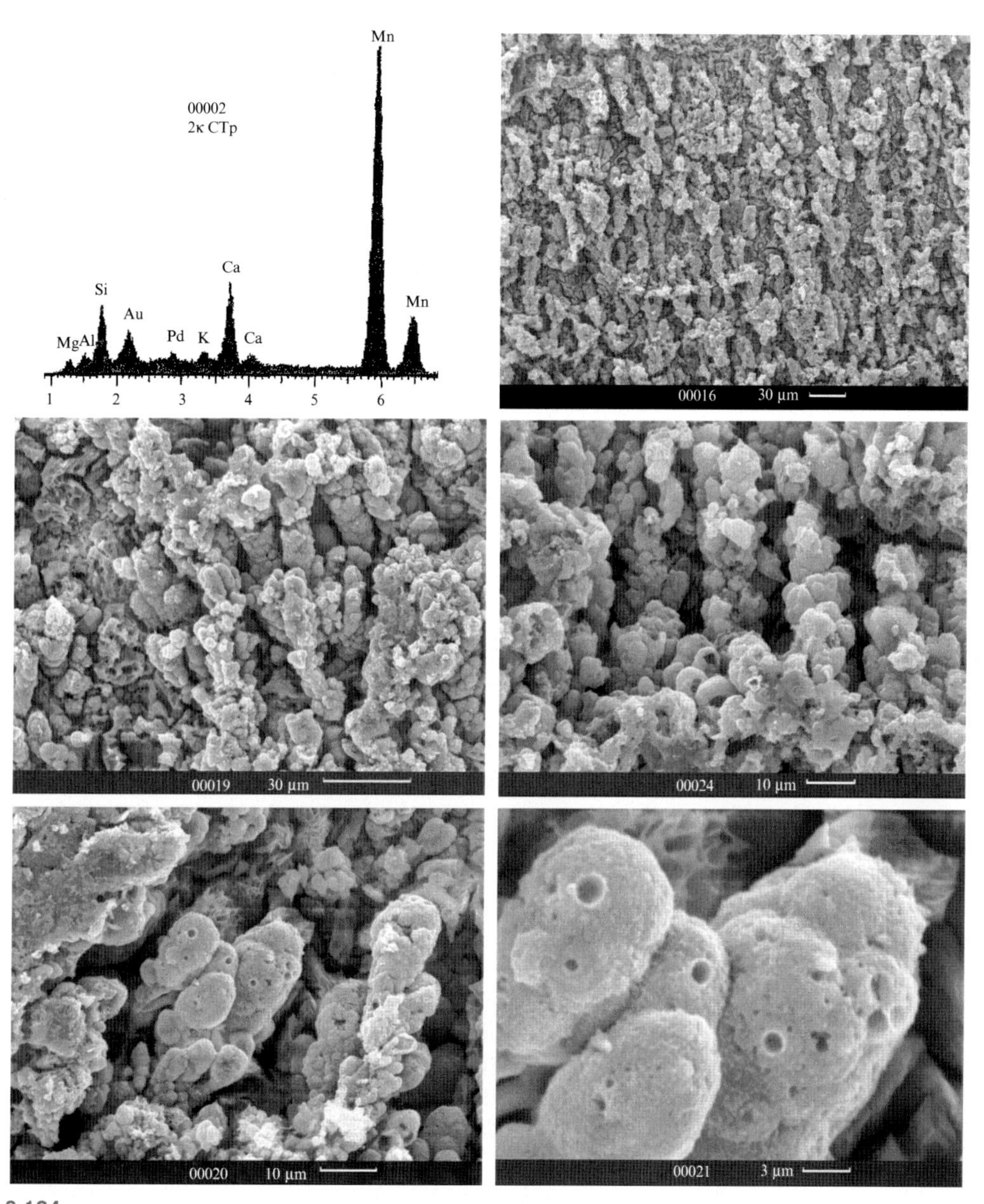

FIG. 3.134

The structure and chemical composition of the bryozoan bioherm by different magnifications, Taman peninsula (photos by E.A. Zhegallo and E.L. Shkolnik).

3.2.5 FERROMANGANESE DEPOSITS OF FERRUGINOUS-SILICEOUS FORMATIONS

It follows to classify to a separate genetic group manganese rocks and ores that are spatially and genetically connected with ferruginous-siliceous rocks that vary in terms of composition and degree of metamorphism [itabirites, jaspilites, banded ferruginous siliciclastics, ie, the BIF, etc.] of a broad age range—from Early (Archean) to Late (Neoproterozoic) Precambrian, occasionally even up to the Middle Paleozoic. The quantity of manganese in ferruginous-siliceous rocks strongly varies—from traces up to the formation of specifically ferromanganese ores (with a low ratio of Mn/Fe). In certain cases, deposits of this type are characterized by high metallogenic potential (reserves and resources).

Such deposits are known in certain regions of development of ferruginous-siliceous formations, for example in the Republic of South Africa (Griqualand West and elsewhere), Brazil (Morro do Urucum region, and elsewhere), Kazakhstan (south-western Karatau and elsewhere), and the Far East (South Khingan group of deposits).

However, it is necessary to note that despite the wide distribution of banded ferruginous siliciclastics in the composition of rocks of Precambrian consolidated blocks of various continents, deposits of ferromanganese (and manganese-iron) ores have a limited distribution. In many districts of development of Precambrian ferruginous quartzites to date, industrial concentrations of manganese have not been detected. For example, within the boundaries of the distribution of ferruginous quartzites (jaspilites) composing the crystalline foundation of the East European platform in the district of deposits of the Kursk Magnetic Anomaly and Krivoy Rog (Ukrainian crystal shield), no sections with industrial concentrations of manganese have yet been discovered.

Deposits of manganese (and ferromanganese) ores connected with ferruginous siliciclastics coincide in many cases with the weathering crusts on them. In the capacity of an example can serve small deposits and occurrences of ferromanganese ores in one of the world's largest iron-ore basins—Hamersley (Western Australia). In the weathering crusts on the iron-ore formations of the Archean and Lower-Middle Proterozoic of Pilbara province (Australia), representing zones of hematite-goethite mineralization of layered iron-ore formations (BIF, Marra Mamba Iron Formation, Hamersley Basin), within the boundaries of the Christmas Creek, Cloud Break, and Mt. Levin deposits are present blocks of ferromanganese ores with high manganese content. These deposits at present are not under development (the ratio of Mn/Fe constitutes 0.5–2).

Opinions differ regarding the genesis of rocks of ferruginous-siliceous formations; they are classified as (chemogenic) sedimentary as well as volcanogenic- and hydrothermal-sedimentary, hydrothermal, and metasomatic types and their varieties (Geologiia i genezis et al., 1972; Formozova, 1973; Voitkevich and Lebed'ko, 1975; Zhelezorudnye mestorozhdeniia et al., 1981; Precambrian et al., 1981; Iron formations, 1983; Mikhailov, 1983; Roy, 1985; Klein and Beukes, 1989; Klemm, 2000; and others). Still others are of the opinion that a substantial role in their formation is played by organic life (Zhabin and Sirotin, 2009).

With the elucidation of the genesis of ferruginous siliciclastics attempts have been made to employ also isotope data on carbon as well as oxygen of carbonates (siderites, dolomites, calcites), SiO_2, magnetite, and hematite (Perry and Tan, 1972; Backer and Clayton, 1976; Beukes and Klein, 1990; Yapp, 1990; Hoefs, 1992; Hren et al., 2006; and others).

Such research has also been conducted, as noted above (Section 3.2.4), for quartz-hematite rocks of the Mamatwan deposit. The available isotope data in many cases provide evidence of processes of transformation and new mineral formation during post-sedimentational stages of lithogenesis.

The elucidation of the genesis of rocks of ferruginous-siliceous formations, despite the fact that they are frequently connected with Fe-Mn deposits, in the present work is not examined. An interpretation of the isotope data on this question available in the scientific literature lies beyond the framework of the present monograph.

3.2.6 METAMORPHOSED DEPOSITS OF MANGANESE

To this group of manganese rocks and ores can be classified metamorphosed varieties of many genetic types (sedimentary-diagenetic, hydrothermal-sedimentary, etc.), the primary nature of which in many cases cannot be accommodated by a straightforward definition. We do not allocate specifically metamorphic ores of manganese to a separate group—despite the high content of manganese in metamorphic rocks embedded in a range of metamorphic minerals (eg, spessartine in the gondites of India, Ghana, and Brazil; rhodochrosite, caryopilite, tephroite, pumpellyite, piemontite, rhodonite, parsettensite, and other mineral associations in the South Urals deposits, etc.), industrial concentrations of manganese are not reached; deposits of manganese among these metamorphic rocks are formed only in weathering crusts (zone of hypergenesis). It appears that to this group can be classified many deposits of manganese that are embedded in metamorphic rocks of the Archean and in many cases of the Early Proterozoic.

A separate genetic group of deposits could be represented by concentrations of manganese rocks that coincide with near-contact zones of magmatic bodies infiltrating the manganese-containing rocks of various composition (near-contact, skarn, and the like). In terms of dimensions and reserves of manganese, they are small and do not present practical interest.

Serving as a classic example of deposits of the metamorphosed type is *the Nsuta rhodochrosite-ore deposit* (*Ghana*).

It follows to note that this deposit until recently in the scientific literature has been classified to the oxide hypergenic type. However, the principal reserves of manganese at this deposit in the present time are embedded in carbonates (rhodochrosite). This allows us to classify the manganese ores (rhodochrosite) to the group of metamorphosed type.

The Nsuta manganese-ore deposit is one of the world's largest. It was discovered in 1913 and has been intensively developed from 1916. The geological structure of the district and of the deposit itself, the material and chemical composition of the ores themselves are well studied and fairly completely discussed in scientific publications (Service, 1943; Sorem and Cameron, 1960; Thienhaus, 1967; Ntiamoan-Agiakwa, 1979; Kleinschrot et al., 1991, 1993; Melcher, 1995; Yen et al., 1995; Mucke, 1999; Nyame et al., 2002; Nyame 2008, and others). For this reason, we limit ourselves to only a brief characterization of the deposit.

In the scientific literature, in describing the rocks and ores of the Nsuta deposit principal attention has been paid to manganese oxides (predominantly nsutite and cryptomelane, secondarily to pyrolusite, psilomelane, braunite, hausmannite, lithiophorite, manganite, etc.). These represent the product of the oxidation of initial manganese-containing rocks—gondites, carbonates, and other more manganese-poor formations (phyllites, tuffs). Manganese oxides within the boundaries of the deposit at present possess a limited distribution (abandoned) and are not characteristic for the given deposit. Currently, mainly carbonate-manganese ores are being extracted.

Primary (metamorphosed) manganese ores (carbonates) and manganese-containing rocks (gondites, phyllites) in the stratigraphic regard are included in the composition of the sequence of deposits of the Birimian series (supergroup) of the Lower Proterozoic (Fig. 3.135), within the boundaries of

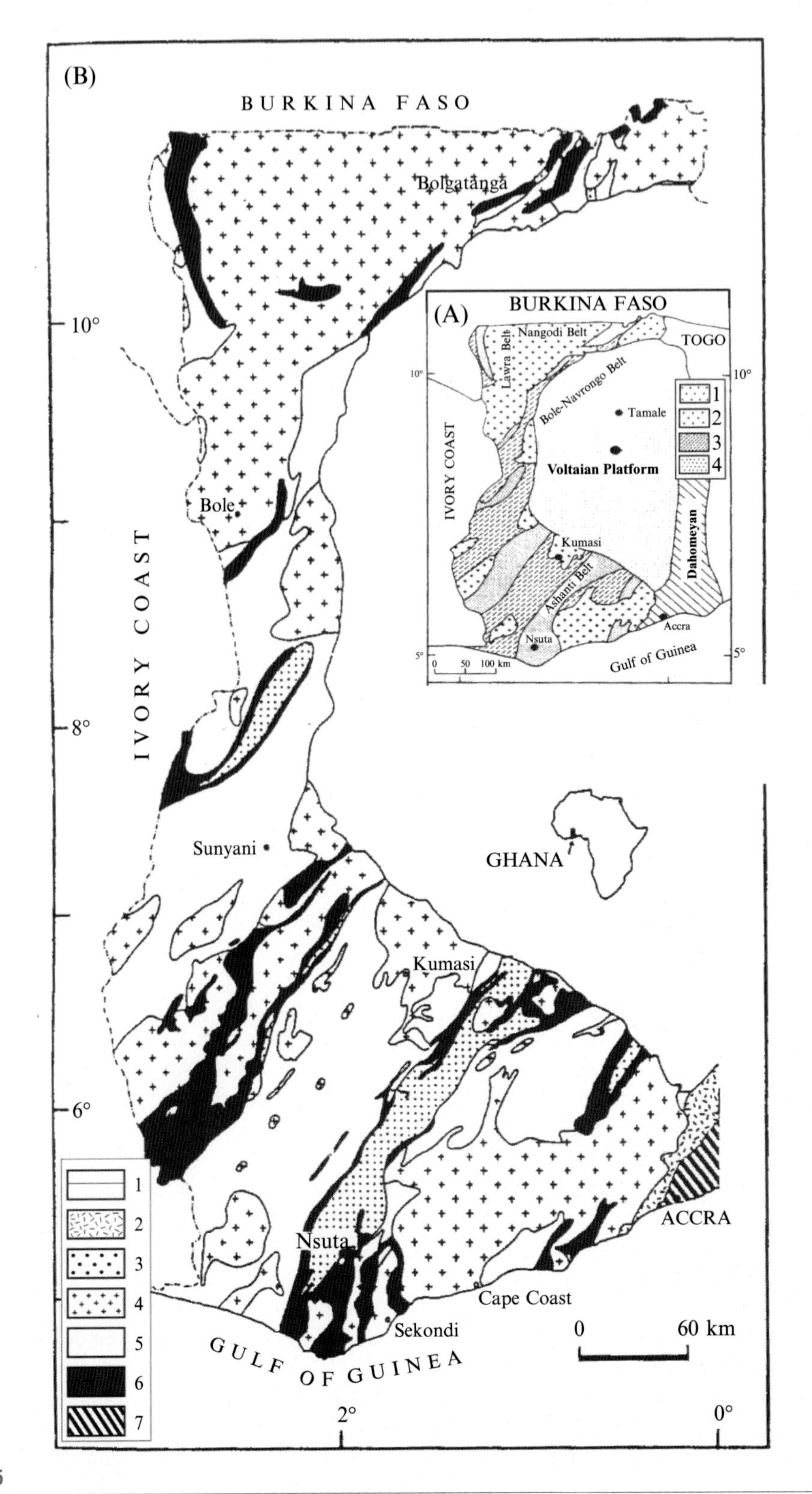

FIG. 3.135

See the legend on opposite page.

which a manganese-containing horizon (thickness reaching 50–60 m) occupies an intermediary position—underlain (Lower Greenstones) and overlain (Upper Greenstones) by greenstone rocks. In the tectonic regard, these rocks coincide with the greenstone rocks of the Eburnean province of the Ashanti greenstone belt, which in turn are included in the composition of the southern segment of the West African Craton (see Fig. 3.135A).

In the structural regard, rocks of the ore horizon are intensively contorted concordant with the host deposits in folds characterized by steep angles of dip (60–80 degrees to the east-north-east). Within the boundaries of the deposit five sections have been identified—former hills—A, B, C, D, and E, respectively, and five oxide and carbonate-manganese-ore quarries of the same names (Hills A, B, C, D, and E). At present, the mining of the ore horizon (manganese carbonates) is conducted only in certain quarries.

According to modern understandings, the primary ore body is composed of manganese carbonate (rhodochrosite) in an isoclinal fold with steep angles of dip (50–70 degrees) to the east-north-east. The thickness of the ore body in the flanks (eg, quarry Hill D; Fig. 3.136) constitutes in mean 30 m. In the hinge of the fold according to drill core data there is a "bulge" in thickness (no less than 100–120 m). The lower mark of the ore body descends up to 560–600 m below sea level (quarry Hill E). The contents of manganese in ores generally vary from 32% to 44%, and in near-contact thin zones with host rocks constitute approximately 18–20%.

For elucidation of the genesis of primary carbonate-manganese ores X.-V. Yeh and coauthors conducted a study of the mineralogy and the isotopic composition of carbon and oxygen in carbonates of the ore bodies and host rocks (24 samples) (Yeh et al., 1995). It was demonstrated that the ore bodies consist exclusively of Ca-rhodochrosite and rhodochrosite. Also, in the host rocks kutnohorite, dolomite, calcite, rhodochrosite, siderite, ankerite, and magnesite are present. In the primary rocks rhodochrosite predominates. $\delta^{13}C_{PDB}$ in carbonates varies from −15.9‰ to −5.0‰, and $\delta^{18}O_{SMOW}$ from 15.9‰ to 18.2‰ in Mn ores; and from −4.8‰ to 1.1‰ and from 13.6‰ to 16.4‰ in host rocks, respectively.

The author of the present work has studied the exposed section of the ore body in quarry Hill D (see Fig. 3.136). The values of $\delta^{13}C$ and $\delta^{18}O$ in the rhodochrosite were found to be close to those obtained by previous researchers. The distribution of ratios of the isotopic composition along the cross section is uneven; no regularity in distribution could be established.

A detailed study of C and O isotopic composition of Mn carbonates of Nsuta deposit was conducted by Nyame and Beukes (2006). They examined different genetic and mineralogical types of carbonate from host rocks and ore bodies of Hills A, B, C, D, and E (79 samples): rhodochrosite, kutnohorite, and dolomite of different morphological types and textures (microcrystalline, crystalline, granular, concretionary, veins).

FIG. 3.135

Geological position of Nsutite manganese deposit, Ghana [after Kleinschrot et al. (1993) and Melcher (1995)]. (A) 1—Voltaian system: sand and mudstones; 2—Eburnean system: granitoids; 3 and 4—Birimian system: 3—sedimentary facies, 4—volcanic facies, partly covered by Tarkwain sedimentary formation. (B) 1—Volaian system: quartzite, shale, arkose and mudstone; 2—Togo series: quartzite, shale, phyllite; 3 and 4—Tarkwaian system: 3—quartzite, phyllite, grit, konglomarate 4—granitoids; 5 and 6—Birimian system: 5—volcanic-sedimentary facies: tuff, pyroclast, argillite; 6—volcanic facies: basaltic lava, pyroclast; 7—Dahomeyan system: acidic and basic gneiss and schist.

FIG. 3.136

General view of the quarry "D" (A) and place of sampling on isotope analysis (B) (upper numeral—the isotopic composition of carbon, $\delta^{13}C$, ‰; the lower digit—the isotopic composition of oxygen, $\delta^{18}O$, ‰).

$\delta^{13}C_{PDB}$ and $\delta^{18}O_{SMOW}$ values in the analyzed carbonates showed wide variations from −20.8‰ to −1.5‰ and from 12.4‰ to 18.2‰, respectively (Fig 3.137). Based on the isotope data two distinct groups were defined. The first, isotopically light, showed small variation in $\delta^{13}C$ and $\delta^{18}O$ values (−7‰ to −2‰ and 13.4–14.7‰, respectively) and was represented by ores and carbonates of microcrystalline and microconcretionary textures from host rocks of Hills A, B, C, and E. Such carbonates, in the opinion of this authors, should be referred to as least altered. Carbonates of the second group record significant shifts in $\delta^{13}C$ and $\delta^{18}O$ values and are typical for crystalline and granular Mn carbonate, predominantly rhodochrosite ores of Hill D. Mn carbonates of this group show an isotope shift toward lighter $\delta^{13}C$ and heavier $\delta^{18}O$ by as much 13‰ and 5‰, respectively, compared to carbonates of the first group and can clearly be referred to as altered.

Note, $\delta^{13}C$ values become slightly more negative and $\delta^{18}O$ more positive from host Mn-bearing phyllite toward the Mn-carbonate ore body.

F.K. Nyamo and N.J. Beukes on the base of lithological features, mineralogy and isotope data defined a restricted set of $\delta^{13}C$ and $\delta^{18}O$ values commonly associated with least altered carbonate-bearing rocks.

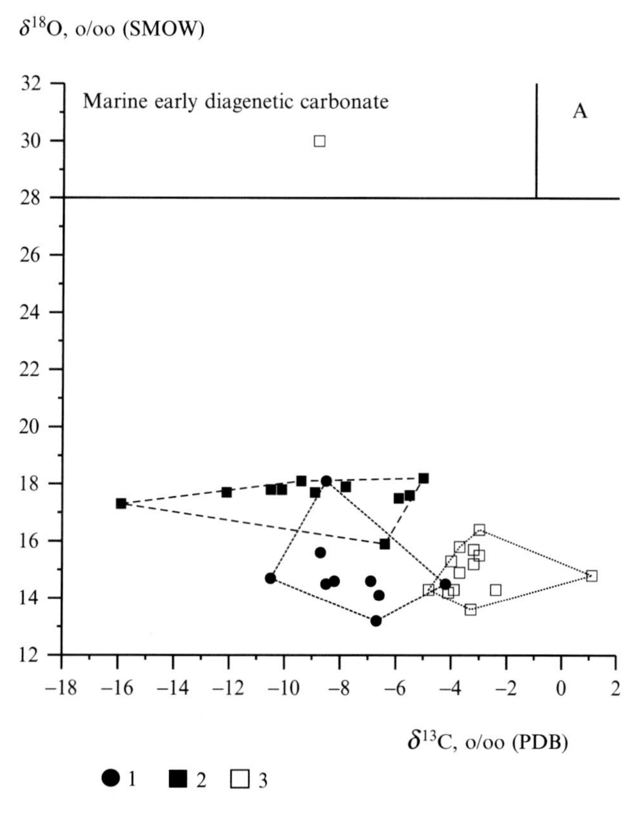

FIG. 3.137

$\delta^{13}C$ versus $\delta^{18}O$ in carbonate-manganese ores of Nsuta deposit (Gana) 1—our data, 2 and 3—data of Yeh et al. (1995): 2—manganese carbonate (ore), 3—host carbonate rocks.

They suggested a "primary" origin for carbonates of the first group by deposition and/or early-diagenetic conditions of carbonate formation. The isotope ratios were likely inherited from sedimentary-diagenetic conditions and remained practically unaffected by the metamorphic overprint.

On the other hand, greater variability $\delta^{13}C$ and $\delta^{18}O$ values typically recorded in coarse and granular carbonates are suggested to be "secondary" in origin. The isotope values' obvious in shift toward lighter $\delta^{13}C$ and heavier $\delta^{18}O$ are the result of isotope exchange between carbonate minerals and a later (mainly aqueous?) fluid. The mechanism of exchange according to F.K. Nyame and N.J. Beukes is unclear.

Thereby, F.K. Nyame and N.J. Beukes conclude that Mn-carbonate bearing rocks concentrated through an initial sedimentary process which was later modified by local, possibly structure-related processes.

Note, mentioned authors do not discuss a source of the manganese.

Literature (Yeh et al., 1995; Nyame and Beukes, 2006) and our (see Fig. 3.126) isotope data allow us to be in agreement with the opinion of authors of cited works about initial (hydrothermal?)-sedimentary-diagenetic origin of Mn carbonates and host rocks of Nsuta deposit. But, opposite to F.K. Nyame and N.J. Beukes, we agree with Yeh et al. (1995) that very low values of $\delta^{18}O$ in manganese ores support a post-sedimentation hydrothermal transformation of initial Mn-carbonate minerals. Herewith, the host rocks including Mn carbonates underwent a greater transformation in comparison to carbonates of ore body.

In Fig. 3.137 compiled on data (Nyame and Beukes, 2006), we can see the distribution of C- and O-isotope composition of different Mn-carbonate minerals of Nsuta deposit, and position of Mn carbonates of Moanda (Gabon) deposit. In all cases, Mn carbonates of host rocks and some part of Mn ores are characterized by isotopically light $\delta^{18}O$ and heavier $\delta^{13}C$ values. Mn ores of ore body Hill D (Fig. 3.137C) show, opposite, isotopically heavier $\delta^{18}O$ and lighter $\delta^{13}C$ values. Primary, not altered Mn carbonates of Moanda deposit are characterized by isotopically heavier $\delta^{18}O$ values and located in the field of Early Proterozoic carbonates of marine diagenetic origin. (Note, limestones and dolomite of Transvaal Supergroup of South Africa are characterized by $\delta^{13}C_{PDB}$ and $\delta^{18}O_{SMOW}$ values from −2.8‰ to 0.6‰ and from 20.3‰ to 31.0‰, respectively, Beukes et al., 1990; Bau et al., 1999; roughly correspond to box "A," Fig. 3.137.)

The measured isotope data and geological features of the Nsuta deposit lead to the supposition that Mn carbonates ore bodies of the Nsuta deposit were formed in volcanic-influenced sediments, like Mn carbonate of ampelites of the Moanda deposit, at the early-diagenetic stage by active participation of oxidized organic carbon. During subsequent transformations of Mn-carbonate ore bodies in the geological development of deposits of the Birimian series, initial isotope ratios were changed. At one of the earliest stages of metamorphism, probably amphibolite facies (John et al., 1999), Mn carbonates were recrystallized to coarse and granular textures with change of O-isotope composition (box "B"). It can be supposed that alteration took place by influence of CO_2-poor hydrothermal solutions and carbon isotope composition of carbonates were not significantly changed (preserving the diagenetic signature). Transformations in carbonates with retention of initial C-isotope ratios often take place by regional metamorphism to amphibolite and granulite facies. For example, Archean metamorphic marbles of the Aldanian crystalline shield (Russia) have retained their C-isotope composition and changed O-isotope ratios toward lower $\delta^{18}O$ values (Kuleshov, 1986).

John et al. (1999) have argued that Birimian rocks of amphibolite facies, Nsuta included, have undergone retrograde metamorphism to greenschist facies conditions. According to Nyame et al. (2002), available geological evidence indicates pervasive low-grade greenschist facies metamorphism. It can be supposed that CO_2 and H_2O hydrothermal solution of this episode of metamorphism (box "C," Fig. 3.137) were characterized by isotopically light $\delta^{18}O$ and heavier $\delta^{13}C$ values. Clearly, light $\delta^{18}O$ is a result of isotope exchange between water and siliceous, mainly volcanogenic host rocks, which as is known, are characterized by light isotope ratios ($\delta^{18}O_{SMOW}=5–10‰$). Carbon isotope composition of dissolved CO_2 in introduced fluids was represented mainly by dissolution of initial isotopic heavy marine carbonates of the host rock sequence. The observed scatter of $\delta^{13}C$ and $\delta^{18}O$ values in Mn carbonate are a result of mixing of CO_2 fluids (box "C") and metamorphosed Mn-carbonate ores (box "B").

We can assume that by the transformation of Mn-ore bodies took place remobilization and migration of ore carbonate matter in an adjacent strata of sequence Nsuta deposit. Such Mn carbonate (ores and scattered in host rocks) will be characterized by isotopically light $\delta^{18}O$ and heavier $\delta^{13}C$ values.

As sources of manganese in carbonates of the ore bodies according to Yeh et al. (1995) could be volcanic exhalations and reduced manganese oxides of the host rocks.

Thus, the data obtained by previous researchers (Yeh et al., 1995; Nyame and Beukes, 2006) and our data (see Figs. 3.136 and 3.137) enable us to consider that carbonate-manganese ores of the Nsuta deposit were formed with active participation by oxidized carbon of organic origin evidently in the zone of early diagenesis. At present, we do not have at our disposal data on the type of formation of primary rocks—Are they sedimentary-diagenetic or hydrothermal-sedimentary? For this reason, we classify the Nsuta deposit to the metamorphosed group.

3.2.7 DEPOSITS OF MANGANESE OF WEATHERING CRUSTS

The manganese ores of weathering crusts are the richest and accordingly the most valuable in the practical regard. High-priority industrial significance is accorded to precisely these ores. They represent oxide-manganese (frequently ferromanganese) rocks of the hypergenesis zone developed on initial manganese-containing rocks and poor manganese ores.

Manganese-rich zones in the weathering crust can be developed on any initial manganese-containing rocks: gondites, ampelites, manganese and manganous carbonates, metamorphic and other manganese-containing rocks in various climatic zones. But the most valuable in the industrial regard apparently coincide with humid regions; the most developed laterite weathering crusts are found in the equatorial latitudes of the countries of Africa and South America, as well as widely distributed in Australia, India, China, and elsewhere.

Among deposits of this type, depending upon the type of accumulation of ore matter, various subtypes can be identified—infiltrational, karst, pisolite, and residual weathering crusts.

Serving as model examples of exogenic manganese deposits are the known deposits of India, Brazil, Ghana, Russia, and other countries. These have been described in detail in the scientific literature (Symposium, 1956; Roy, 1981; Varentsov, 1996; and others), but we will consider here only the pisolite ores of Gabon, Australia, and Brazil.

3.2.7.1 Deposits of the Francevillian formation, Gabon

One of the examples well known in the scientific literature and fairly complex in the genetic regard are the deposits of oxide-manganese and ferromanganese ores of the *Francevillian formation* (*Gabon*) of Lower Proterozoic age.

Located in Gabon is a series of the largest deposits of oxide-manganese ores situated in the district of populated points of Moanda-Mounana-Franceville-Okondja, with total reserves of manganese (and ferromanganese) ores among the deposits of the basin no less than 300–350 million tons (Fig. 3.138). These coincide with Lower Proterozoic deposits of the Francevillian formation, which fill the intracratonic Franceville basin. The largest of the deposits now under development and correspondingly the most studied is the Moanda deposit, the largest in West Africa.

The deposits of the Francevillian formation in the given area are represented predominantly by terrigenous marine rocks; carbonate deposits, vulcanites, and ferruginous quartzites play a sharply subordinate role in the succession. Rocks are gently folded but are broken into separate blocks by a series of faults. The general thickness of the series constitutes approximately 4000 m.

Deposits of the Francevillian formation have been divided into several lithological-stratigraphic isolations—Fa, Fb, Fc, Fd, and Fe strata. In the southern part of the basin, the deposits of the formation are discordantly superimposed on gneisses of the Archean and are represented by red-colored sandstones with horizons of conglomerates of continental (delta alluvia) origin (maximum thickness reaches 1000 m). Manganese content is connected with deposits of the strata from Fa to Fd, but the most important deposits are those coinciding with Fb.

The manganese-bearing horizon is deposited stratigraphically above the non-ore or manganese-poor carbonaceous black shales (ampelites) and dolomites of the Francevillian horizon Fb_1. Within the boundaries of the manganese-rich horizon has been established a thin (0.2–0.5 m) basal bed of massive ores of Mn, represented by oxides and hydroxides (pyrolusite, manganite, groutite, lithiophorite, and nsutite) and containing residual rhodochrosite.

To all appearances, the manganese deposits of the Moanda-Mounana-Franceville district (also possibly the Okondja district situated to the north-east) represent the residues of a once-unitary deposit that has been eroded and at present is preserved in the form of separate deposits coinciding with the still-extant plateau (see Fig. 3.138).

An ore strata with a horizon of pisolites as a rule coincides with the upper parts of the plateau. Manganese-bearing deposits are widely developed also on the plateau: the Franceville district—Beniomi, Bordeaux, Lafobe, Menai, Yéyé, Mvona, and Papa; and within the boundaries of the Okondja district—on the Lebai plateau.

The full stratigraphic section of the deposit was recorded in the work of G. Leclerc and F. Weber for the Bangombé and Okouma plateaus (Fig. 3.139) (Leclerc and Weber, 1980). From above, the ore horizon is overlapped by argillites and clays of brown, reddish-brown, and yellowish-brown color, referred to as cover (COV). Below follows an ore stratum, within the boundaries of which at the deposit (on all plateaus) have been identified five types of ore horizons (correspondingly, types of ores): ores of the weathering crust (CRO), the zone of pisolites (PIS), the transitional zone (ZTR), platy ores (PLA), block ores (BLO), and massive ores (MAS), which underlie the platy and BLO. On the Beniomi plateau, they are composed of cryptomelane.

The thickness of the ore strata constitutes from 0.5 to 4–5 m. Horizons of different types of ores throughout the area of the plateau are distributed unevenly—they can be absent or compose horizons up to 2 m in thickness.

At all deposits within the boundaries of the enumerated plateaus has been observed a typical sequence of the ore strata analogous to that reported for the Bangombé and Okouma plateaus (Beniomi plateau, Fig. 3.140).

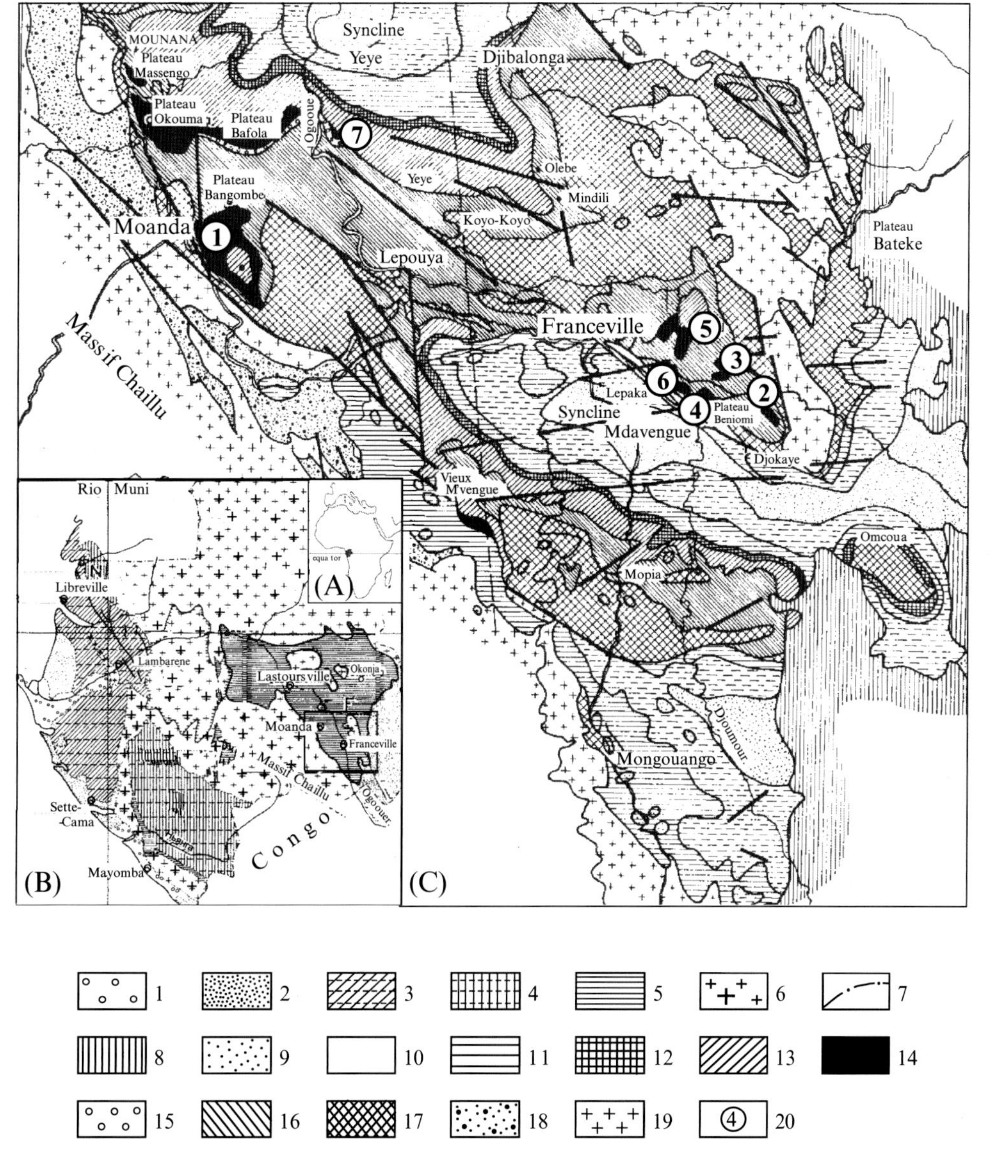

FIG. 3.138

Schematic geographic (A) and geological (B) locations and positions of manganiferous rocks of the Franceville Formation in the Moanda-Franceville area (C) (modified after Weber (1970) and Leclerc and Weber (1980)). 1–6—Rocks: 1—Quaternary, 2—Tertiary, 3—Paleozoic and Mesozoic, 4—Neoproterozoic, 5—Paleoproterozoic, 6—Archean; 7—Gabon state boundary; 8–17—Lower Proterozoic Franceville Group: 8—Fe Formation: sandstones with slate and rare rhyolitic tuff intercalations, 9—Fd Formation: carbonitized slates with acid tuff intercalations, 10—Fc Formation: carbonitized slates, ampelites, jasperites, 11–15—Fb Formation: 11–12—Fb2 Subformation: 11—terrigenous-carbonate rocks with a conglomerate member at the base, 12—quartz sandstones, pelites, ampelites, 13—manganese-ore sequence, 14–15—Fb1 Subformation: 14—ampelites, dolomites, 15—pelites, dolomitic sandstones with a conglomerate member at the base, 16–17—Fa Formation: 16—fine-grained sandstones, black shales, pelites, 17—alluvial-deltaic sandstones, conglomerates; 18—Tertiary gravelstones and sands; 19—Archean quartzites, amphibolites, itabirites, and granitoids; 20—manganese and ferromanganese-ore deposits: 1—Moanda, 2—Beniomi, 3—Mvouna, 4—Bordo, 5—Menai, 6—Lebai, 7—Yeye.

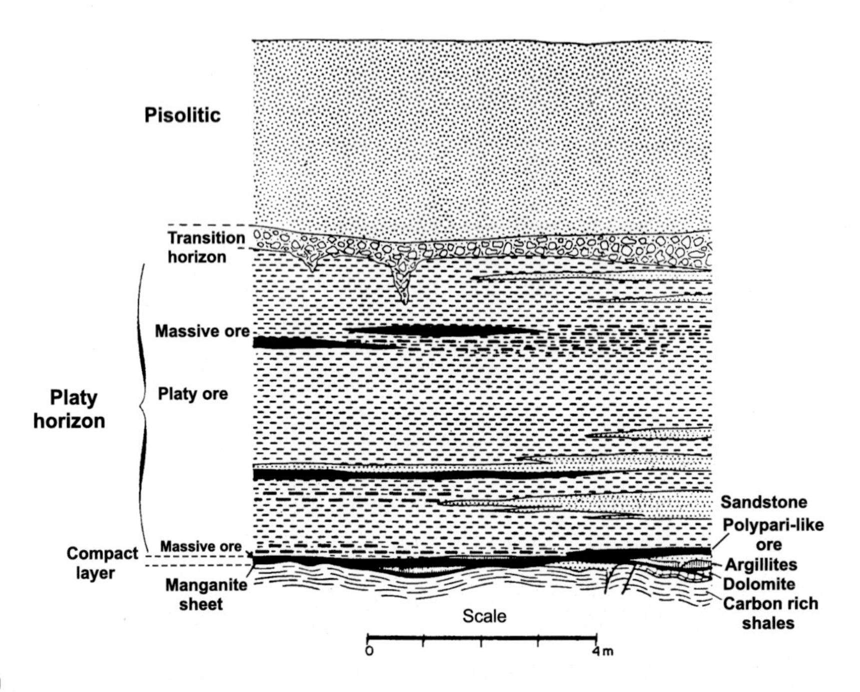

FIG. 3.139

Schematic structure of the manganese-ore sequence of the Franceville Formation (Weber, 1970).

Deposits of various types of manganese ores represent formations of weathering crusts developed on the initial manganese-containing carbonate-manganese carbonaceous shales—ampelites (the content of manganese in them according to the data of drill borings at the Moanda deposit reaches 15% and higher). This viewpoint today prevails and is undisputed for the Moanda deposit and a range of smaller deposits situated nearby (Bangombé, Okouma, Massengo, Bafoula, and others).

An exogenous origin for various types of manganese oxides of the hypergenesis zone is supported as well by the character of the deposition of the ore strata, which repeats the form of the relief—practically all types of manganese ores are developed on the top of the plateau where are noted the greatest thicknesses of manganese oxides, as well as on the slopes of the indicated highlands, descending almost always toward their base, where has generally been noted a contraction in the thickness of the ore strata and of the hypergenesis zone (Fig. 3.141).

At the same time, in certain plateaus among manganese-ore pisolites occur spherical, "well-rounded" clasts of various composition—granites, arenites (sandstones), quartz, volcanic rocks (evidently of acidic composition), and other strongly altered rocks, the primary nature of which has not yet been established (Fig. 3.142). It may seem that the "sphericity" is the result of mechanical processes (rounding) in aquatic (shore-marine) conditions. In this case, it could be supposed that the spherical ("rounded") friable arenites and altered vulcanites once were dense and subjected to processes of mechanical processing in an aquatic medium. In the opposite scenario, which apparently took place, we should accept that such forms were formed in the hypergenesis zone independent of the hardness of the initial rocks.

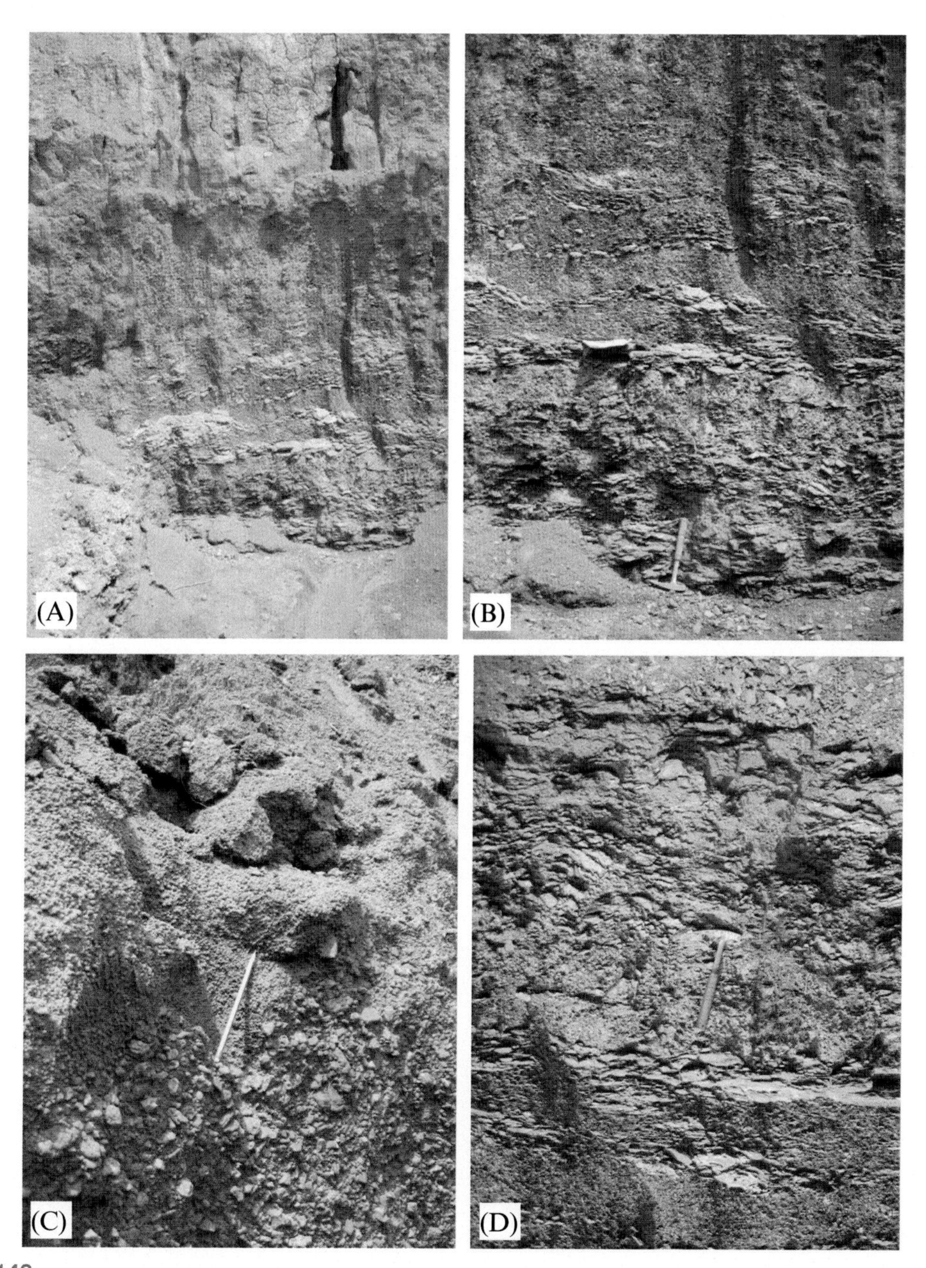

FIG. 3.140

A typical sequence of the weathering crust with oxides of iron-manganese ores Franceville formation, Beniomi plateau; Horizon Fb_1. (A) general view, (B) pizolitic and platy ores; (C) pizolitic ores; and (D) massive ores.

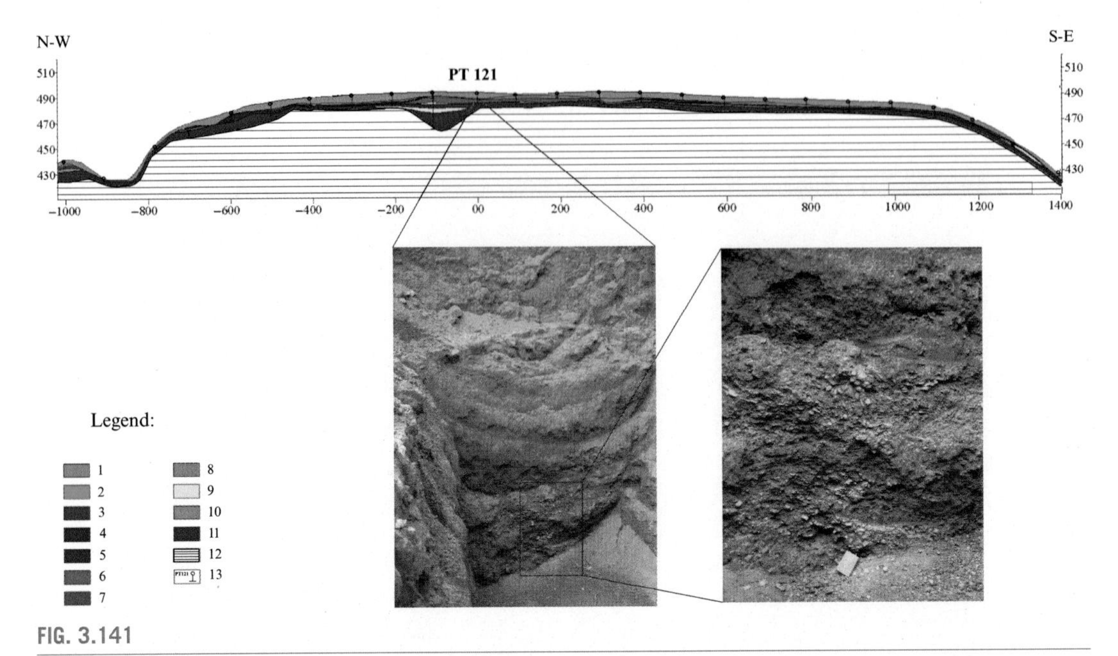

FIG. 3.141

Schematic structure of the weathering crust of the iron and manganese-oxide pizolitic deposits, Ffransvilian basin, plateau Menai (compiled by geologists CVRD, with the changes of the author). 1–12—Sections of the ore sequense (horizons): 1—overlapping loose horizon (COV), 2—pizolites (PIS), 3—the transition zone (ZTR), 4—laterites (LAT), 5—dolerite (DOL), 6—platy ores (PLA), 7—block ores (BLO), 8—jasperites (JAS), 9—arenites (GRE), 10—massive ores (MAS), 11—manganese-bearing ampelites (APB), 12—black ampelites (APN), and 13—exploration wells.

FIG. 3.142

Rounded inclusions of granite and loose arenites, volcanic rocks and rounded pebble of quartz among the ferromanganese pizolites. (A) Eye plateau, trench PT-444, granite pebbles and (B) Eye plateau, trench PT-444, pebbles of quartz and arenite.

3.2.7.2 Deposits of weathering crusts of Australia

In the genetic regard, it follows to classify to the hypergenic type the oolitic oxide ores of the *Carpentaria Basin*, especially the Groote Eylandt deposit. The latter is situated in the western coastal part of the island of Groote Eylandt (Gulf of Carpentaria, Arafura Sea, Australia). Manganese mineralization here is traced in an area of approximately 150 km^2 (Fig. 3.143).

The Groote Eylandt deposit is one of the world's largest (reserves of over 200 million tons); it is of important economic value for Australia and plays a large role in the global economics of manganese-ore raw material.

The geological structure of the deposit and surrounding territory at present has been fairly well studied (Slee, 1980; Ostwald, 1980, 1988; Bolton et al., 1988; Varentsov, 1982, 1996; and others). The principal part of manganese mineralization coincides with the Mullamen layers in the western and south-western part of the island. The bed of manganese oxides is developed primarily in relatively depressed sections of the western coastal lowlands and forms an almost solid ore body extending in the submeridional direction for 22 km with a width of approximately 6 km.

The ore body consists of various types of oxide-manganese ores intricately interbedded irregularly along the lateral: oolitic, pisolitic, concretional, and coarse-lump of sinter formations with an alternating quantity of light, cream, and reddish predominantly kaolinite clays, to a lesser degree quartz sand. The thickness of the ore bed varies from 0.5–2 to 15–20 m (on average, approximately 3–4 m), while that of the overlapping friable deposits in the district of development of manganese ores varies from several centimeters up to 30 m.

Ores are divided into seven principal types according to their physical characteristics, ratio of ore/barren rock, degree of lateritization and cementation, and mineral composition: friable pisolitic, cemented pisolitic, concretional, structureless, sandy and siliceous, ferruginous, and disseminated ores (Fig. 3.144).

Occurrences and insignificant deposits of manganese have been identified also to the west and south-west of Groote Eylandt, in the surrounding sedimentary basins. Thus, for example, in the Arafura basin are known small occurrences and deposits such as Masterton, Robinson River, Camp, Manganese, Photo, and others. The largest of them is Masterton No. 2 (also known under the name Calvert Hills). Manganese-ore mineralization in these as a rule is secondary and surface; ore beds reach thicknesses of 2–4 m. Manganese content in the ores constitutes 41–51% and higher.

In the MacArthur basin (Arnhem Land) has also been discovered a series of small deposits and occurrences (more than 70) represented by secondary hypergenic ores, including pisolitic (oolitic?). However, only two of these—Caledon 1 and Peter John—have practical value.

According to the widely accepted theories of Australian geologists (J. Ostwald, B.R. Bolton, K.J. Slee, J.L. MacIntosh, J.S. Farag, and others), oxide ores of the Groote Eylandt deposit have a shallow-water-marine sedimentary-diagenetic origin. Serving as a source of manganese are manganese-containing marls of the Mullamen layers, which contain up to 20% rhodochrosite. In the southern part of the deposit, as shown by the results of drilling, manganese-containing marls are deposited stratigraphically below the horizon of oxide ores. Friable pisolitic ores were formed, according to Australian geologists, in a tranquil marine environment, and were cemented as a result of the remobilization of manganese during the restructuring of friable pisolites under hypergenic conditions. Structureless ores could also have formed from manganese-containing oozes, whereas less distributed concretional ores were crystallized from colloidal gels.

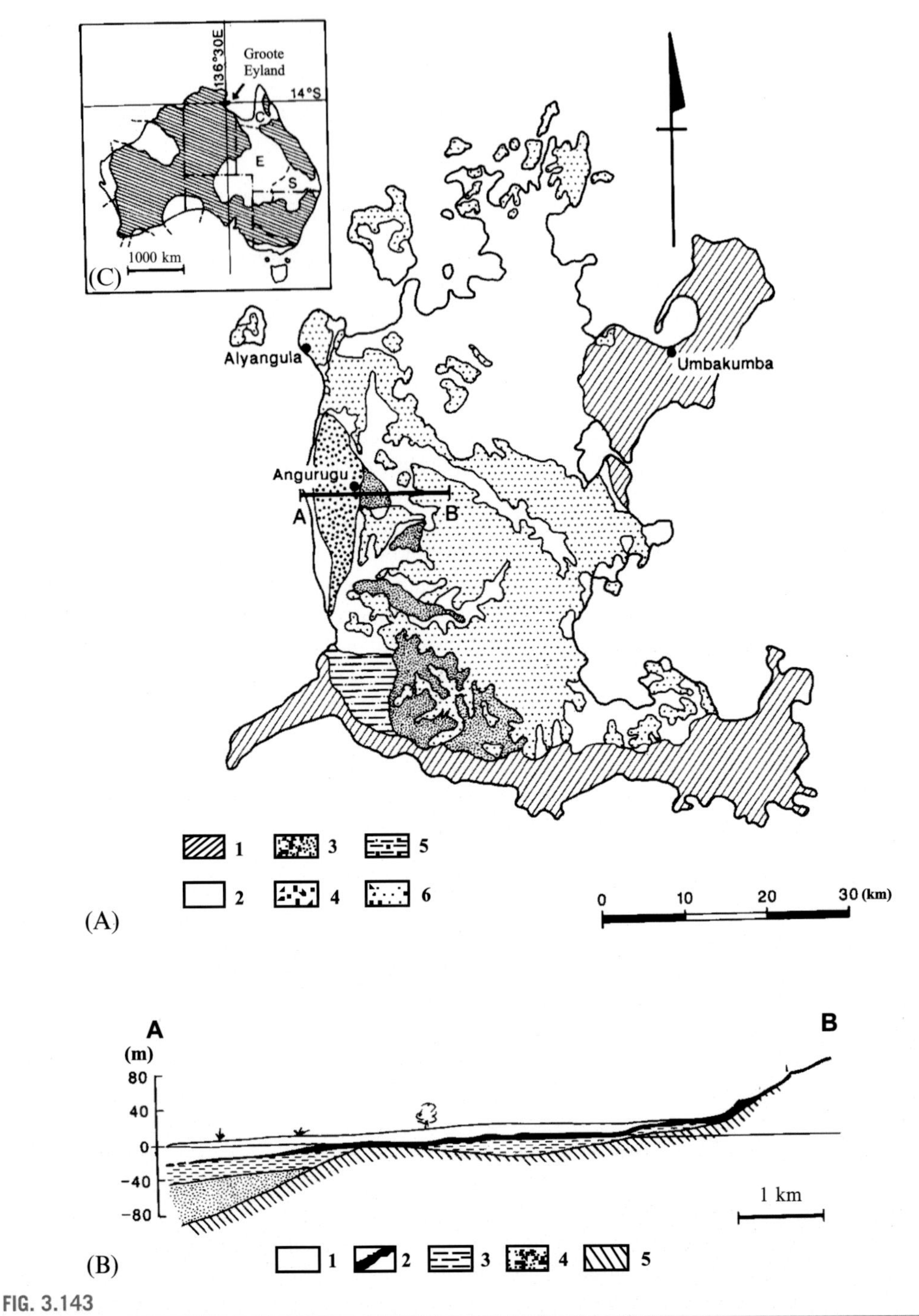

FIG. 3.143

Location and geological structure of the Groote Eylandt deposit (after Bolton et al., 1988). (A) Location and schematic geological map of the Groote Eylandt area with the position of manganese ores: 1 and 2—recent sediments: 1—sandy dunes, 2—sand, clay; 3–5—Lower Cretaceous rocks: 3—disseminated MnO_2, 4—MnO_2 pisolites, 5—carbonatized manganese siliciliths; 6—Middle Proterozoic quartzites. In the inset: (C) Carpentaria basin, (E) Eromanga basin, (S) Surat basin. (B) The geological cross section of the Groote Eylandt manganese-ore deposit: 1—post-ore clays and sols, 2–4—Lower Cretaceous rocks: 2—manganese-oxide ores, 3—clays, 4—sands, 5—Proterozoic quartzites.

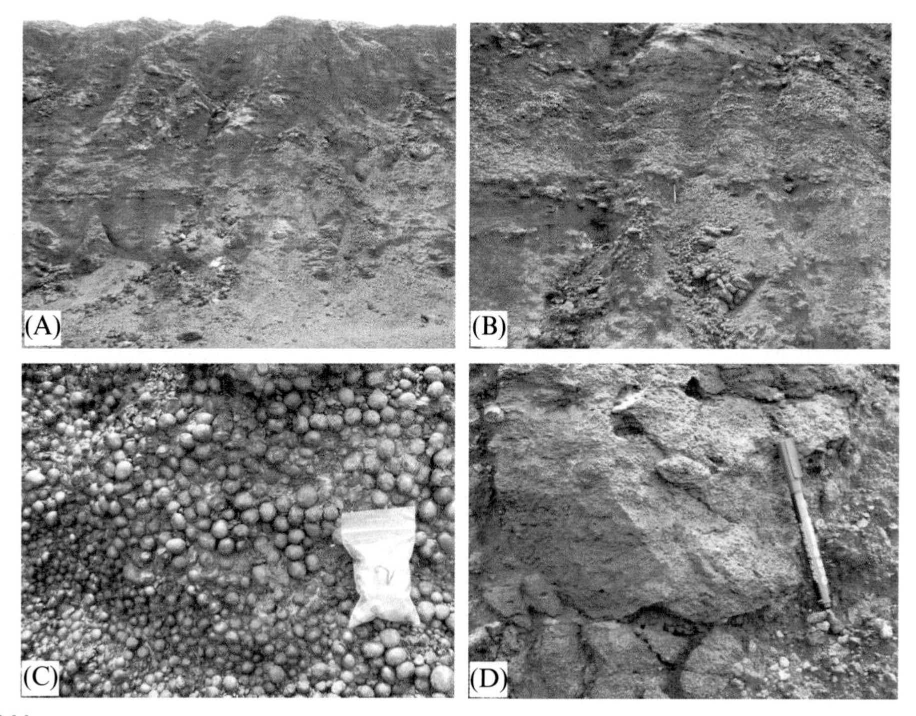

FIG. 3.144

The structure of the manganese-ore horizon in the Groote Eylandt manganese deposit (pit wall). (A) A general view of the quarry wall, (B) middle part of the wall of the quarry, (C) large pizolitic ores, and (D) small pizolitic ores.

The presence of a residual pisolitic structure in the structureless ores can serve as evidence that these two types of ore were formed by remobilization of manganese in the process of diagenesis as well as later, under hypergenic conditions. The processes of transformation of initial sedimentational-diagenetic ores in hypergenic conditions are widely developed (weathering crusts of the Upper Cretaceous and younger age) and have led in many cases to an increase in the concentration of manganese in the ores.

Detailed studies of the mineralogy and geochemistry of oxide ores of manganese at the Groote Eylandt deposit have allowed I.M. Varentsov to consider them as a product of local redeposition of weathering crusts developed in the Late Oligocene-Early Miocene that were formed on the Lower Cretaceous manganese-bearing deposits (Mullamen layers) (Varentsov, 1982; Varentsov and Golovin, 1987; Varentsov, 1996).

The persuasive conclusions of I.M. Varentsov have likewise allowed the author to classify the examined ores to deposits of weathering crusts. Processes of formation of pisolitic ores of manganese at this deposit were apparently analogous to the processes characteristic for the above-examined deposits of the Franceville basin (Gabon)—Moanda and others where the weathering crust was developed on the manganese-containing ampelites (at the Groote Eylandt deposit, as noted, in the manganese-rich Mullamen layers).

3.2.7.3 Deposits of weathering crusts of various countries

Laterite weathering crusts are widely distributed in equatorial and near-equatorial *countries of Latin America*, especially in Brazil, where a significant quantity of deposits with significant resources of manganese has been discovered (Symposium, 1956).

Oolitic ores of the weathering crust analogous to ores of examined deposits of Gabon (the group of deposits of the Franceville basin), and Australia (Groot Eylandt and others) are widely developed in the northern equatorial part of Brazil, in part in Goiás state. Information on manganese ores of the weathering crust of this district in the scientific literature is extremely scanty. For this reason, we report their general characteristics according to the factual material available to the author of the present monograph.

In the geological regard, in this district are developed rocks of Proterozoic age, which are represented by metamorphic complexes—various gneisses, metatuffs (Lower Proterozoic)—as well as by weakly metamorphosed clastic complexes—conglomerates, various siliciclastics, volcanogenic and volcanogenic-sedimentary rocks, metavulcanites (Middle Proterozoic), and a complex of granite rocks (Upper Proterozoic); they include numerous hydrothermal veins of quartz composition.

Manganese oxides in the weathering crust of the examined district are developed very widely and are represented by oolitic and pisolitic formations as well as by sinter and nodular formations. Oolitic (and pisolitic) rocks occupy an area no less than a few square kilometers (possibly, dozens of kilometers) (Fig. 3.145). From the surface they are strongly eroded, containing a large quantity of iron (see Fig. 3.145D; Mn—16.26%, Fe—18.03%, SiO_2—18.2%), but in places sections of sinter and nodular manganese oxides representing high-quality manganese ores have been observed here (see Fig. 3.145F; Mn—60.40%, Fe—0.010%, SiO_2—1.40%) (Table 3.28).

It follows to note that, besides oolitic ores, also observed in the district are exposures of sinter and nodular manganese oxides, frequently with high contents of iron. Examples of ore-occurrences (and undeveloped deposits) are shown in Fig. 3.146.

Today we lack at our disposal any data on the geological setting and structure of the ore horizon(s), thickness, and distribution. Also unclear is the composition of the continental rocks on which were formed manganese-oxide ores of two morphological types: (1) oolitic (and pisolitic) and (2) sinter formations in the form of rinds, which in many places sheath the surface of the outcrop.

It can be supposed that manganese oxides of the oolitic type have a wide areal distribution in the form of either a stratified bed or sheet horizon in the composition of a laterite weathering crust with a thickness hypothetically of no less than a few meters. Their areas of development can occupy significant spaces and correspond to medium and even large deposits.

Important in the practical regard is the second type of ores—manganese oxides in sinter forms, which frequently sheath the surface. The ores are characterized by high content of manganese and low content of iron.

At present, it is not possible to estimate the quantity and ratio of the described types of ores at the ore-occurrences (deposits?). For this task, geological-exploration surveys must be conducted.

A separate group of deposits of weathering crusts is potentially constituted by manganese-ore bodies of the filling of karst sinkholes. The process of accumulation of manganese oxides in karst bands is developed in many regions of the Earth's landmass. However, industrial concentrations are fairly rare for manganese ores of this type in the hypergenesis zone. One of the regions of development of deposits of this genesis could be the deposits of Western Australia (Woodie-Woodie district). Here the ore

FIG. 3.145

Outcrops of pizolitic and oolitic iron-manganese ores, State of Goiás (Brasilia). (A) General view, (B) ore outcrops on the road, (C)–(F) ore: (C) oolitic; (D) large-oolitic and pizolitic; (E) block of weathered oolitic ores; (F) sinteric nodular high-quality ores (Mn content—60.57%).

bodies have insignificant dimensions (a few hundred meters), in form are close to isometric, composed of high-quality (in the industrial regard) manganese oxides (Fig. 3.147).

Another well-known district of the distribution of manganese ores that fills karst bands is without a doubt the series deposits of the Postmasburg manganese-ore district (Republic of South Africa). I.M. Varentsov considers their formation as a model example (Varentsov, 1996). See also recent work on Chinese deposits by Deng et al. (2016).

Table 3.28 Chemical Composition (%) of Manganese-Oxide Ores of the Occurrences of Brazil (X-Ray-Fluorescence Analysis, Phase Composition of Iron and Manganese Omitted)

Number of Sample	SiO_2	TiO_2	Al_2O_3	Fe, tot	Mn, tot	CaO	MgO	K_2O	Na_2O	P_2O_5	S	Cl
14\07	10.8	0.180	4.59	6.05	40.4	0.71	0.62	4.77	0.19	0.77	0.015	0.050
14\07	10.8	0.180	4.11	6.07	40.4	0.70	0.55	4.73	0.18	0.78	0.016	0.032
15\07	0.70	0.014	0.80	20.6	35.4	0.75	0.29	1.98	0.15	0.75	0.070	<0.002
16\07	11.6	0.120	3.89	8.95	38.9	0.71	0.59	4.08	0.20	1.07	0.026	<0.002
17\07	6.60	0.100	3.02	10.3	41.1	0.71	0.46	3.58	0.16	1.00	0.033	0.014
18\07	4.50	0.083	2.61	9.32	43.5	0.70	0.45	3.43	0.16	0.92	0.029	0.015
19\07	4.60	0.064	2.49	5.05	48.7	0.70	0.44	2.25	0.15	0.69	0.015	<0.002
21\07	4.50	0.072	3.82	59.9	0.29	0.67	0.52	0.45	0.11	1.43	0.190	<0.002
22\07	6.30	0.160	2.69	12.2	41.7	0.73	0.56	2.13	0.25	0.87	0.043	0.067
23\07	6.00	0.130	3.39	6.44	49.0	0.69	0.53	1.44	0.19	0.86	0.017	0.120
24\07	23.0	0.140	5.73	1.83	40.1	0.78	0.71	3.53	0.25	0.84	<0.005	0.044
24/07-OB	22.1	0.140	6.13	1.85	40.2	0.77	0.76	3.58	0.28	0.86	<0.005	0.072
25\07	16.0	0.100	3.71	3.46	43.7	0.82	0.59	2.87	0.24	0.72	0.010	<0.002
26\07	18.2	0.630	16.3	18.0	16.3	0.66	0.84	2.18	0.12	0.12	0.049	<0.002
27\07	30.7	0.530	15.0	15.6	11.5	0.65	0.65	1.55	0.11	0.09	0.041	<0.002
28\07	1.40	0.028	0.73	0.01	60.4	0.65	0.27	0.13	0.10	0.17	<0.005	0.260
28/07-OB	1.31	0.032	0.74	0.08	60.6	0.65	0.27	0.13	0.11	0.17	<0.005	0.330
29\07	10.3	0.740	23.5	33.1	4.90	0.64	0.45	0.73	0.12	0.14	0.110	<0.002
30\07	8.70	0.270	3.41	1.44	30.2	0.68	0.38	1.01	0.12	0.22	<0.005	0.088
31\07	3.20	0.120	1.25	0.32	31.6	0.69	0.31	0.37	0.09	0.30	<0.005	0.190
Standard												
44-G	10.3	0.089	1.68	1.24	44.2	3.15	1.99	1.10	0.68	0.48	0.034	
Reference values	10.9	0.100	1.67	1.32	44.1	3.22	2.08			0.50	0.034	

3.3 ISOTOPIC PARTICULARITIES OF THE FORMATION OF MANGANESE ROCKS AND ORES

The above-examined principal models of the formation of initial (non-metamorphosed and non-transformed in hypergenic conditions) manganese ores constitute evidence that manganese accumulation occurred in sedimentary-rock basins with different regimes of sedimentation in various paleotectonic and paleoclimatic environments coinciding with sequences of terrigenous-, chemogenic-, and biogenic-sedimentary and volcanogenic-sedimentary rocks, and that manganese-containing rocks (and ores) were formed at different stages of sediment- and lithogenesis.

Primary manganese ores are represented by a broad spectrum of minerals of manganese carbonates and oxides, in which iron is frequently present in isomorphic mixture with manganese. The carbonate form of manganese in the ores (primary manganese rocks) in the composition of known deposits (not excluding

As	Ba	Co	Cr	Cu	Ni	Th	V	Zn	CO_2	H_2O^-	H_2O^+
0.0032	0.168	0.165	0.016	0.0031	0.015	0.001	0.005	0.130	<0.2	0.79	2.49
0.0028	0.164	0.165	0.015	0.0033	0.013	0.002	0.005	0.130	<0.2	0.80	2.40
0.0063	0.348	0.206	0.020	0.0114	0.041	<0.0005	0.004	0.230	<0.2	0.64	5.19
0.0066	0.213	0.167	0.014	0.0054	0.027	0.001	0.009	0.240	<0.2	0.88	3.58
0.0010	0.170	0.181	0.015	0.0022	0.018	0.001	0.002	0.170	<0.2	0.87	3.77
0.0042	0.176	0.201	0.014	0.0035	0.024	0.002	0.003	0.200	<0.2	0.68	3.36
0.0074	0.344	0.243	0.002	0.0115	0.026	0.001	0.048	0.210	<0.2	1.00	3.37
0.0400	0.124	0.013	0.010	0.0017	0.044	0.001	0.007	0.021	<0.2	0.39	9.76
0.0047	0.168	0.088	0.019	0.0033	0.019	<0.0005	0.003	0.230	<0.2	0.49	3.48
0.0055	0.145	0.035	0.019	0.0023	0.012	0.001	0.003	0.160	<0.2	0.39	2.52
0.0055	0.118	0.130	0.017	0.0015	0.029	0.002	0.005	0.350	<0.2	0.39	2.13
0.0010	0.119	0.133	0.019	0.0024	0.032	<0.0005	0.004	0.360	<0.2	0.40	2.10
0.0003	0.113	0.189	0.015	0.0028	0.079	0.001	0.006	0.300	<0.2	0.89	2.97
0.0160	0.259	0.057	0.018	0.0095	0.018	0.002	0.020	0.022	1.12	2.49	8.03
0.0091	0.256	0.029	0.025	0.0129	0.021	0.002	0.020	0.031	0.58	1.22	7.01
0.0120	0.408	0.066	0.017	0.0163	0.039	0.001	0.011	0.130	0.37	0.14	0.68
0.0078	0.406	0.066	0.013	0.0159	0.037	0.002	0.011	0.130	0.38	0.15	0.70
0.0280	0.163	0.063	0.035	0.0117	0.011	0.003	0.023	0.004	0.86	1.82	12.6
0.0170	4.710	0.130	0.002	0.0308	0.033	0.001	0.064	0.082	0.68	1.16	4.38
0.0110	6.930	0.182	<0.001	0.0303	0.036	0.001	0.097	0.104	0.38	0.54	3.40
0.0046	0.711	0.006	0.009	0.0060	0.029	0.001	0.022	0.008			
0.0040		0.010			0.033						

those classified to the Mamatwan deposit, Republic of South Africa) unarguably predominates. In the genetic regard, practically all natural manganese carbonates, including manganese ores, are authigenic—that is, they were formed during a post-sedimentational stage of lithogenesis (in the broad sense of the term).

A model for the formation of carbonate-manganese ores that coincide with sedimentary rocks has been developed in detail by Strakhov et al. (1968) and his followers on the example of the Oligocene deposits of the Eastern Paratethys. It has been demonstrated that processes of authigenic formation of manganese carbonates in the diagenesis zone took place with active participation by carbon dioxide that formed with the microbial oxidation of organic matter.

It follows to note that exceptions to this rule can probably be found in manganese and manganese-containing rocks of certain ore-occurrences of Pai-Khoi (Kuleshov and Beliaev, 1999; Starikova et al., 2004; Starikova and Kuleshov, 2009; Starikova and Zavileiskii, 2010; Starikova, 2012) and possibly the Kvirila deposit (Georgia), where the role of oxidized carbon of organic matter in their formation is insignificant.

FIG. 3.146

Sinteric and nodular forms of manganese and iron oxides at different occurrences, State of Goiás (Brasilia). (A) outputs of manganese ore, and the bed of a dried boards (unnamed) stream; (B) manganese-ore sinter outputs in line with the same stream; (C) quality manganese-ore outcrop in the board of the same stream; (D) the output of the ore body quality manganese ore; (E) outcropping iron-ore sinter; (F) the field of iron ore.

A distinct spatial and in many cases genetic link between the deposits of manganese and carbonaceous formations is noted already beginning from the Precambrian: in the Archean Anabar shield (Khapchan series), the Madagascar graphite series, the Proterozoic KMA (Tim suite, Tim-Yastrebovskaya structure), ampelites of Gabon (Moanda deposit), Birimian series of Ghana (Nsuta), the Amapa series of Brazil (Serra do Navio deposit), and a range of other Precambrian formations. Carbon-containing

FIG. 3.147

General views of the host rocks of manganese-ore body, Woodie-Woodie district (Australia). (A) Development of the ore body; it is clearly isometric form, (B) the structure of the wall of the quarry mined ore body; clearly visible vertical bedding of the host rocks, typical of the area of karst cavities; and (C) the structure of the upper part of the wall of the quarry filled with water; clearly visible vertical nature of occurrence of the host rocks, typical of the area of karst cavities.

rocks (carbonaceous shales) are present in stratigraphic sections of many deposits of manganese of the Phanerozoic. This has allowed a range of researchers to propose models of formation of manganese ores associated with carbonaceous facies of sediments (Gurvich, 1980; Polgari et al., 2009; and others).

Isotopic composition of carbon and oxygen

The results of isotope research of manganese carbonates of ancient deposits (South Urals basin, Mexico, Hungary, China, and elsewhere) (Kuleshov and Dombrovskaya, 1988; Okita and Shanks, 1988; Okita et al., 1988; Polgari et al., 1991; Kuleshov, 2001; and others) have fully supported the theory of N.M. Strakhov and his followers positing a mechanism of formation of this type of ores in conditions of early diagenesis.

One of the characteristic particularities of the examined examples of manganese carbonates is that they all without exception are characterized by a lighter isotopic composition of carbon by comparison with that of sedimentary marine carbonates. Values of $\delta^{13}C$ for these decline to isotope ratios close (or in some cases even analogous) to those of carbon of organic matter, and in a range of cases to those even lower, characteristic for carbon of hydrocarbons of the petroleum series and of methane (Kuleshov and Brusnitsyn, 2004).

The abundance of carbonaceous matter and sulfides along with the presence of rhodochrosite and siderite in manganese-containing rocks serves as evidence for the starkly expressed reductive conditions of formation of the initial sediment. Primary carbonaceous oozes could be accumulated in conditions of a normally aerated basin as well as in basins with hydrosulfide contamination. The latter are powerful "accumulators" of the manganese supplied into a basin of sedimentation. In the case of the entry of such waters into the shallow-water zones of the basin in the aeration zone can be expected a precipitation into the sediment by dissolved manganese in the form of oxides. This viewpoint today is widely supported by many researchers in connection with various manganese deposits.

It follows to note that in nature, specifically sedimentary carbonates of manganese—that is, those that have precipitated directly from the mass of lacustrine or marine water (by chemical means or with participation by organisms)—are not distributed. Practically all of them are authigenic, and their formation occurs within the sediment during early stages of diagenesis and (or) later during catagenesis (late diagenesis), as a result of either the redistribution of manganese already embedded in the sediment itself (probably in the form of oxides) or of the supply of new portions of manganese (characteristic for catagenesis).

However, the study of the isotope geochemistry of manganese carbonates in such deposits has revealed facts that are difficult to explain by means of accepting the mechanism of only a diagenetic formation of manganese ores. This primarily relates to the regularities of formation of the isotopic composition of oxygen. Thus, a significant proportion of manganese carbonate ores of our studied deposits as well as in certain cases even predominating (Pai-Khoi, Chiatura, Kvirila) are characterized by low values of $\delta^{18}O$ not particular to diagenetic carbonates of normal-sedimentary marine genesis. This serves as evidence that they were formed under other conditions. Such isotope ratios could be the result of either elevated temperatures or a lighter isotopic composition of oxygen of the system (medium) of ore-formation. In many cases, these are the two determining factors, but in concrete geological objects they can be manifested to varying degrees.

This signifies that a significant part of manganese carbonates was formed under conditions when the connection had already been lost between the interstitial waters of manganese-containing sediment and the bottom water—that is, their formation occurred during stages of late diagenesis (catagenesis). At this stage of the geological life of the sediment the exchange processes by the water of the mass of the sediment itself with the bottom waters of the basin of sedimentation had already terminated. Here the process of authigenic carbonate precipitation evidently takes its course at higher temperatures by drawing in new sources of carbon dioxide and water, which in terms of the isotope characteristics of carbon and oxygen are distinguished from those that predominate in the zone of early diagenesis.

Carbon dioxide-water solutions of the catagenesis zone can be of various natures (Kholodov, 1982a,b; Makhnach, 1989). However, participating in the manganese-mineralization process in many cases are solutions of elisional origin genetically linked with zones of oil and gas generation. The fundamental possibility of this process has previously been fairly fully examined in the scientific literature on the example of various geological objects (Kholodov, 1982b; Pavlov, 1989; Paragenezis et al., 1990; Pavlov, Dombrovskaya, 1993; Kuleshov and Dombrovskaya, 1997b).

It should be emphasized that although a paragenetic association of deposits of manganese (in certain cases also of iron) with oil- and gas-bearing basins has been noted, a direct genetic link between them is not evident. Apparently, a genetic connection between manganese deposits and petroleum waters (more precisely, solutions rich in water-soluble hydrocarbons and products of the metamorphism of organic matter) bears a more complex character and consists of a combined geochemical evolution of organic and mineral (manganese-ore) matter in a defined stage in the history of the development of the sedimentary-rock elisional basins themselves.

The link between the geochemical history of manganese and organic matter on the example of the studied deposits is fairly complex. The enrichment of manganese in the initial sediment in many cases begins already at the stage of sedimentation and early diagenesis and actively continues later during catagenesis. For this reason, manganese-ore deposits that formed at different stages of lithogenesis are distinguished by geochemical (and isotope) particularities and an interrelation between Mn and organic matter characteristic of the given stage (phase).

The prevailing opinion posits a genetic connection between manganese deposits and petroleum waters (Paragenezis et al., 1990). In certain cases, this is evidently plausible (Kuleshov and Dombrovskaya, 1997b). However, if this connection is a determining factor for the formation of manganese deposits, then against the background of the wide distribution of oil and gas deposits of various scales (oil- and gas-bearing basins, provinces, regions, districts, deposits) we should be discovering with greater frequency (at least the spatial limitations) deposits and ore-occurrences of manganese within oil- and gas-bearing districts. In reality, deposits of manganese in such basins are a fairly rare phenomenon. For this reason, the origin of the combined geochemical history (evolution) of organic CO_2 and dissolved manganese should be sought evidently at earlier stages of the metamorphism of organic matter—that is, before the commencement of oil generation.

It should also be noted that participation by catagenic (elisional, petroleum, etc.) waters in the manganese-mineralization process on the level of modern geological knowledge remains a fundamental possibility. The mechanism of enrichment of such waters by ore components (Fe, Mn, P, Pb, Zn, Cu, Ba, etc.) is not always clear. Also complicated is the question of during which stage of catagenesis this occurs. Still to be clarified are the principal regularities of the migration of ore solutions and the "expulsion" of ore matter into the host deposits. For all intents and purposes, in many cases elisional processes (eg, those developed in oil- and gas-bearing basins) represent a complex multicomponent geochemical system consisting of the chemical particularities of clayey, carbonate, and halogenic deposits and the organic matter embedded within them. In conditions of deep submersion (>3–4 km) at elevated temperatures and pressure, depending upon the geological setting (thickness and composition of the sedimentary strata, depth of submersion, structural-tectonic setting of the elisional oil- and gas-bearing basin, tectonic activity of the district, and the like), will occur with varying intensity processes of transformation of the chemical composition of the initial sedimentary matter with the formation of new authigenic minerals and the forming of the composition of catagenic waters, including those rich in manganese and in other ore components. All of these questions require further specialized study, given their importance to prospecting surveys for manganese-ore raw material.

Many questions regarding the evolution of the mobile carbon dioxide-water system, the composition and means of migration of catagenetic fluids within the boundaries of sedimentary-rock basins for various stages of lithogenesis to date have yet to be studied. Enormous reserves of hydrocarbons and manganese within the boundaries of elisional basins and their combined geochemical history could not be reflected in the geochemical characteristics and balance of matter within the boundaries of the deposits. As an example of this, we can consider the interrelation of manganese resources and the isotopic composition of carbon of ore carbonates in the largest deposits of the CIS countries (Fig. 3.148). From this figure, it follows that the role of oxidized carbon of hydrocarbons is in many cases the determining factor for the scale of the manganese deposits.

In this way, the isotope data on carbon and oxygen for bed deposits of manganese embedded in sedimentary strata enable the identification of two genetic types of primary carbonate-manganese ores and correspondingly two types of deposits (in terms of the predominant type of ores). To the first of these are classified specifically diagenetic (or sedimentary-diagenetic) ores. Representative could be those ores developed within the boundaries of deposits of the Nikopol' manganese-ore basin and Mangyshlak deposit. To this group can also be classified the ores of the Bezmoshitsa occurrence, as well evidently as deposits of the North Urals group, which in the isotope regard to date have yet to

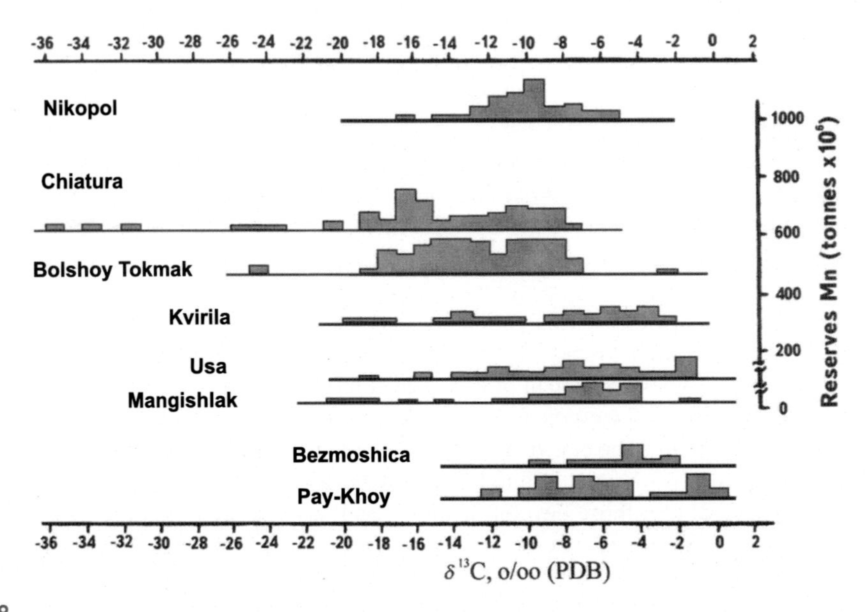

FIG. 3.148

Carbon isotope composition (δ^{13}C) versus resources of several manganese deposits in Russia and CIS countries.

be studied. The mechanism of formation of these ores is classic, having been studied in detail in the works of N.M. Strakhov and his followers.

Carbonate-manganese ores of the second type are distinguished by their isotopic composition of oxygen from the diagenetic ores and were formed later, evidently during the stage of catagenesis. These ores represent either concretional (Pai-Khoi) or metasomatic formations (the predominant share of ores of the Kvirila and Chiatura and evidently Usa deposits and several Chinese deposits). In the genetic regard, these ores as a rule represent the result of repeated supply into the zone of the deposit of ore carbon dioxide-water solutions with the formation of some quantity of generations of carbonates (including non-ore). As a result, carbonates of various stages (or phases) are differentiated from one another by isotope as well as chemical characteristics. For example, for Pai-Khoi and Chiatura ores several such generations can be identified.

The forming of the second type of ores allows a significant broadening of the prospecting criteria for manganese-ore raw material and of the prospects of uncovering new deposits of manganese (predominantly in the form of carbonates).

One of the principal questions for the study of the origin of deposits of manganese ores, as is known, is the clarification of the source of manganese. Isotope data on carbon and oxygen in resolving this question can provide only indirect, auxiliary information connected with the nature of carbon dioxide-water ore-depositing systems—the solutions and the matter dissolved in them.

As demonstrated above, in nature specifically sedimentary manganese carbonates—that is, those that would have been precipitated directly from the mass of lacustrine or marine water—do not exist. All of them are authigenic, and their formation occurs within the sediment at the stage of early diagenesis as a result of the redistribution of the manganese already embedded in the sediment itself (probably in the form of oxides) or later during catagenesis. In the latter case, manganese is commonly supplied from (hydrothermal?) metasomatic solutions.

If we accept an elisional origin of catagenetic waters participating in manganese-ore carbonate precipitation, taking into account the fact that this process occurs with a determining role played by carbon dioxide of oxidized hydrocarbons (C_{org}), then the genetic connection of these waters with the zones of gas and oil generation can be revealed. In this case, there would be no cause for doubting the catagenetic nature of the ore matter. The question is: how great is the influence on the balance of the ore matter of discharged ore-bearing (more reliably, enriched by ores and non-ore components) catagenetic solutions in the zone of diagenesis and catagenesis by comparison with continental runoff? In our opinion, for many large deposits of manganese, such as Chiatura, Kvirila, Mangyshlak, Usa, and to some degree Nikopol', it unquestionably follows to give preference to a catagenetic source.

The predominant mode of formation of primary oxide-manganese (and ferromanganese) ores are ooliths and pisoliths (deposits: Chiatura, Georgia; Nikopol', Ukraine; Mangyshlak, Kazakhstan; and elsewhere). These coincide with well-aerated shallow-water coastal-marine sections and were formed in an oxidizing environment. Entering into the coastal zone of the water bodies (marine and freshwater) dissolved compounds of manganese under these conditions form insoluble compounds of oxides and hydroxides of Mn^{4+}, which precipitate from the sediments evidently by microbial means (with participation of microbes such as *Metallogenium*).

It can likewise be supposed that in the case of submarine discharge of metal-bearing hydrothermals the formation of hydrothermal-sedimentary deposits of manganese likewise takes course with active participation by microorganisms.

Isotopic composition of strontium

The isotopic ratios of $^{87}Sr/^{86}Sr$ in manganese carbonates can serve as an additional criterion for the conditions and source of matter with their formation. We (V.I. Vinogradov, M.I. Buiakaite, GIN RAS Moscow) have conducted an investigation of isotope-strontium ratio in carbonate-manganese ores of Oligocene deposits of Eastern Paratethys: Nikopol' (Ukraine), Bol'she-Tokmak (Ukraine), Chiatura (Georgia), and Kvirila (Georgia) (Table 3.29). The general scatter of these ratios was found to be significant—from 0.78073 to 0.7109.

If we consider that the ratio of $^{87}Sr/^{86}Sr$ in authigenic carbonates of sediments of the zone of early diagenesis of basins of Oligocene age should correspond to this ratio in marine water, then this ratio for the studied manganese formations should be situated within the interval 0.7075–0.7081 (Veizer and Compston, 1974; Palmer and Elderfield, 1985; Koepnick et al., 1985; The Geologic Time, 2012). The measured ratios in many cases exceed the boundaries of this interval. The lowest were found in rocks of the Kvirila deposit, and the highest were characteristic for rocks of Nikopol', Bol'she-Tokmak, and in one case for Chiatura.

Table 3.29 Isotopic Composition of Strontium in Manganese Carbonate Rocks

Number of Sample	Characteristics of Sample	$^{87}Sr/^{86}Sr$
Nikopol' deposit, Aleksandrovskii quarry		
1619	Oxide-carbonate ore; carbonate	0.7080
1622		0.7084
1622a		0.7089
1623	Friable brown manganese carbonate	0.7080
1624	Dense carbonate of dark-gray color	0.7086
1625	Low-grade manganese carbonate rock	0.7097
Bol'she-Tokmak deposit; northern section		
1628	Well 5512, depth 93.5 m. Lump carbonate ore	0.7086
1629	Well 5512; depth 94.6 m. Ore cavernous-cellular	0.7084
1630	Well 5512; depth 95.0 m. Cavernous manganese carbonate rock	0.7109
1631	Well 5512; depth 95.4 m. Lump ore	0.7092
1632	Well 5512; depth 95.65 m. Lump carbonate cavernous ore	0.7096
1632a	Well 5512; depth 95.65 m	0.7079
1633	Well 5512; depth 106.9 m. Lump ore	0.7082
Kvirila deposit, Rodinauli section (Terjola GRP)		
1645	Well 519, depth 587.95 m. Ferromanganese oolites	0.7073
1646	Well 519; depth 587.7 m. Cemented pink manganese carbonate	0.7077
Chiatura deposit, Darkveti uplands, section No. 4		
1703	Large-oolitic limestone, oolites	0.7080
1727	Large-oolitic limestone, black-grown oxide ore	0.7079
1733	Large-oolitic limestone, carbonate ore	0.7082
1736	Large-oolitic limestone, manganese oxides	0.7077

b, geological-exploration crew.

Low ratios of isotopes of strontium are due possibly to subsequent elaboration of host and ore rocks of hydrothermal solutions having a connection with magmatic (deep) rocks and which were characterized by a ratio of $^{87}Sr/^{86}Sr$ lower than that in the rocks of Oligocene age.

High ratios of $^{87}Sr/^{86}Sr$, exceeding values of 0.7081, serve as evidence for the presence in the composition of manganese rocks of radiogenic strontium. It can be supposed that this is a result of the supply of ^{87}Sr into the system of terrigenous sediments of the Oligocene of ore-depositing solutions during the process of the formation and transformation of manganese carbonates.

The wide scattering of ratios of $^{87}Sr/^{86}Sr$ in the studied manganese and manganese-containing rocks supports the multistadiality of the formation of ores of the deposits and generally agrees with the data on the isotopic composition of carbon and oxygen.

For a conclusive explanation of the reasons for the enrichment by the radiogenic isotope of strontium of manganese carbonates and their particularities, it is necessary to ascertain the concentrations of rubidium and strontium in the rocks (which goes beyond the framework of our prior works).

In this way, the above-examined examples of the formation of deposits of manganese ores of various genetic types can serve, in the author's opinion, as models for the interpretation of geological data for other already known deposits and for the prospecting of new deposits of manganese. The following chief regularities of the conditions of formation of manganese ores and rocks can be identified.

1. The primary manganese rocks and ores of the deposits most important in the practical (economic) regard are represented by carbonates and hydroxides of manganese that were formed in sedimentary-rock basins and are paragenetically (in many cases also genetically) closely connected with sedimentary and volcanogenic-sedimentary rocks. The mineral form of deposited compounds of manganese is determined by the prevailing physical-chemical conditions (Eh, pH, and the like) in various zones of the basin of sedimentation or during a given stage of lithification of the initial sediment (lithogenesis). Primary manganese oxides are formed in a medium with high oxidizing potential (Eh) predominantly in coastal zones of shallow waters or on the bottom of the basins with submarine discharge of hydrothermal fluids, while manganese carbonates are authigenic and form inside the sediment itself in reductive conditions.
2. The formation of manganese carbonates occurs inside the sediment at the stages of diagenesis and (or) catagenesis as a result either of the redistribution of manganese embedded in the sediment itself or of the supply of new portions of manganese into the medium of already lithified sediment at the stage of late diagenesis (synonymous with catagenesis, epigenesis). Specifically sedimentary carbonates of manganese—that is, those that were precipitated directly from the mass of lacustrine or marine water—are not distributed in nature.
3. A general regularity of the formation of manganese carbonates under conditions of early diagenesis is the necessary participation in this process by microbially oxidized carbon of organic matter, and for manganese carbonates of post-early-diagenetic stages (phases) of lithogenesis, isotopically light carbon dioxide, formed with the metamorphism (destruction) of organic matter under conditions of submersion of the sedimentary mass into the depths (including during various stages of oil and gas generation). Buried organic matter during post-sedimentational stages of transformation of the sediment enables the concentration of manganese up to ore concentrations.

4. The concentration of manganese approaching the commercial scale of deposits occurs primarily under exogenous conditions—on the borders of separation of the atmosphere—of rock (zone of hypergenesis) and during sedimentary-diagenetic (early diagenesis), to a lesser degree during subsequent stages of lithification of the initial sediment and transformation of sedimentary or volcanogenic-sedimentary rock—late diagenesis (diagenesis of deep burial, concretion-formation).
5. The most valuable in the industrial (economic) regard deposits of manganese today are represented by deposits of ancient and modern weathering crusts (hypergenesis zone), especially in near-equatorial zones of certain countries of South America and Africa, as well as Australia and India. Therefore, these regions are unquestionably the most prospective for the discovery of new deposits of high-grade ores (eg, the Brazilian and Guiana shields of South America, the Precambrian rocks of western Central Africa, and elsewhere). Manganese "caps" of weathering crusts are developed also in many deposits with humid climate of other districts of the Earth's landmass (Urals, Siberia, and the like). In terms of the scale of the development of zones with rich concentrations of manganese, the latter are substantially inferior to the near-equatorial.

CHAPTER

4 THE MAJOR EPOCHS AND PHASES OF MANGANESE ACCUMULATION IN THE EARTH'S HISTORY

The formation of manganese rocks and ores has occurred over the course of practically the entire geological history of the formation of the Earth's upper crust. Manganese-containing and manganese rocks in the composition of the lithosphere are known beginning from the Archean; they are characteristic for deposits of the Proterozoic and are present in many geological-stratigraphic layers of the Phanerozoic; they are also widely distributed (predominantly in scattered form) in sediments of modern water bodies (oceans, seas, lakes). However, accumulation of manganese rocks reaching industrial concentrations—that is, the formation of deposits in rocks of the lithosphere—has occurred sporadically. Moreover, for a range of regions, periods of manganese accumulation have not always been synchronous in the history of their geological development. For example, for the territories of the countries of the CIS (Commonwealth of Independent States) (Varentsov and Rakhmanov, 1974) and China (Editorial Committee of Mineral Deposits of China, 1995) manganese-ore epochs are of various ages and are characterized by their own particularities inherent to the character of the sediment accumulation of the corresponding basins of sedimentation.

The author has delineated epochs and periods of manganese accumulation based on the scale of this process manifested above all in the resources of this metal contained in rocks of a given age. Attempts at delineating such epochs have been conducted multiple times (Roy, 1981; Glasby, 1988; Laznica, 1992; Varentsov, 1996; Maynard, 2003, 2010, 2014). However, new geological data have enabled more detailed explanation of certain aspects of manganese ore-genesis and its scale in the geological history of formation of the rocks of the Earth's crust.

In the examination of a given period or epoch and their stages, we recognize that ore-hosting rocks are predominantly sedimentary and volcanogenic (hydrothermal)-sedimentary rocks that have accumulated in the sediment-rock basins (Kuleshov, 2011a,b). These basins are characterized by various regimes of sedimentation, by the presence of various ore-hosting facies and formations, and by varying duration of accumulation of manganese-containing rocks.

Also taken into account is that the formation of manganese rocks and ores occurred in different stages of basin development and during lithogenesis of sedimentary and volcanogenic-sedimentary deposits. For example, post-early-diagenetic (metasomatism, formation of concretions) manganization of initial sediments (and rocks) can occur in sedimentary basins at different stages of the geological life of the sediment rock and can be substantially 'detached' in time from the moment of their formation. In terms of age, we have formally dated such manganese rocks and ores to their host deposits.

Isotope Geochemistry. http://dx.doi.org/10.1016/B978-0-12-803165-0.00004-5

By virtue of their insignificant distribution and negligible resource value, we do not examine specifically volcanic or associated hydrothermal manganese ores nor the manganese-containing weathering crusts on these rocks.

A separate group is represented by hypergenic deposits of manganese that formed in the weathering crusts on the source rocks of different composition and age. Naturally, we have dated the manganese accumulation in deposits of this type to the geological time of formation of the initial (source) manganese-containing rocks and ores, rather than to the period of formation of their weathering crusts.

A generalization of the data available in the literature and the author's own observations enables a general characterization of the principal periods of manganese accumulation in the history of the formation of the Earth's lithosphere and of their association with ancient (principally the largest) paleocontinents (Fig. 4.1).

Based on the restriction of manganese-containing and manganese rocks and ores to given stratigraphic levels of the Earth's upper crust, and following the principle of classification accepted in the scientific literature, time intervals can be delineated—*the metallogenic (manganese-ore) periods* of manganese accumulation in sedimentary and volcanogenic-sedimentary rocks (including their metamorphosed varieties). The major divisions of Geologic Time—*Archean*, *Proterozoic*, and *Phanerozoic*—each contain major manganese metallogenic episodes unlike iron, for which large deposits only occur close in time to the Archean/Proterozoic boundary. However, within each Era, commercial-scale manganese genesis was confined to relatively brief intervals of time. For example, virtually all the commercial tonnage of Mn in the Cenozoic was deposited in the Early Oligocene, the Rupelian Stage. This segment of geologic time lasted 5.8 million years which constitutes only 8.8% of the Cenozoic (66 million years). These metallogenic episodes generally correspond to the most prominent tectonic events of the Earth's development but differ in duration as well as in quantity of metal accumulated—that is, in terms of their metallogenic significance. The degree of geological knowledge about the rocks of these metallogenic periods likewise varies.

In terms of the scale of manganese content (reserves and stocks of manganese) within the boundaries of the periods are delineated epochs and stages (the formation of gigantic and large deposits, occasionally of medium and small sequences). The term "metallogenetic epoch" is used by the author of the present work generally with the same connotation as used by the disciples of the metallogenetic school of VSEGEI (Russian Geological Research Institute) (Bilibin, 1955; Semenenko, 1962; Tvalchrelidze, 1970; Bilibina et al., 1978), although in certain cases it is also applied to manganese ore-genesis in the delineated epochs that occupy a more prolonged time interval; for example, the metallogenic epochs spanning the Early and Middle Proterozoic, Lower and Middle Paleozoic, etc. This delineation is contingent upon the quantity (reserves and stocks) of contained in the rocks and ores manganese of corresponding age as well as the history of geological development of the basins of manganese-ore sedimentation within the boundaries of the paleo(super)continents (see Fig. 4.1).

4.1 ARCHEAN METALLOGENIC PERIOD (3500–2500 MILLION YEARS)

On the whole, this period is prospective for manganese in the industrial regard. However, despite the fact that rocks of Archean age occupy around 10% of the landmass, deposits with significant supplies of manganese, other than the deposit of Morro da Mina (Brazil), at present have not been discovered.

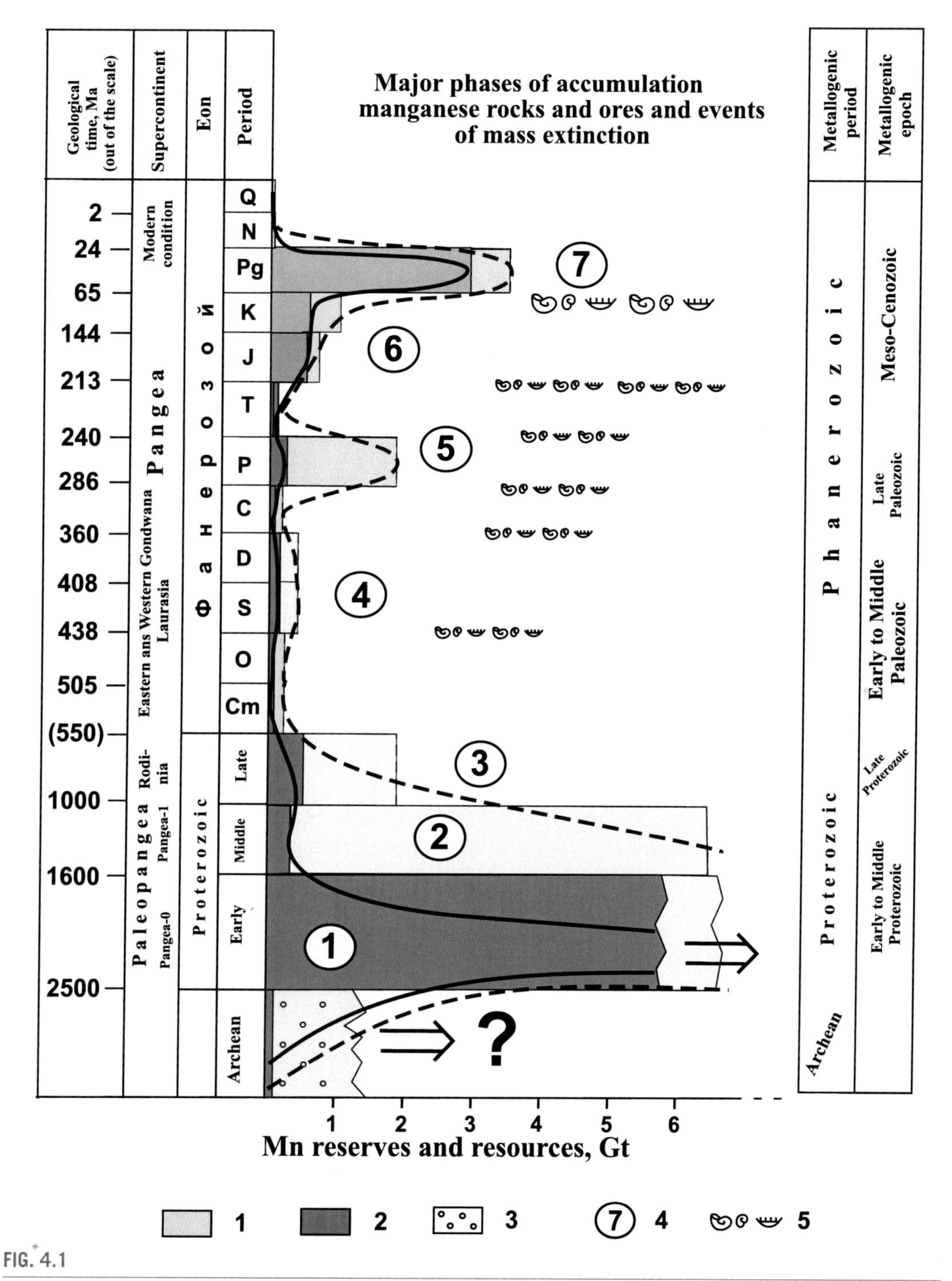

FIG. 4.1

Distribution of Mn reserves and resources in different age rocks of the Earth's lithosphere. (1) Reserves; (2) resources; (3) Archean manganiferous rocks; (4) major metallogenic phases of the accumulation of manganese rocks and ores: (1) Early Proterozoic, (2) Middle Proterozoic, (3) Late Proterozoic, (4) Early-Middle Paleozoic, (5) Late Paleozoic, (6) Mesozoic, (7) Late Mesozoic-Early Cenozoic; (5) major biotic events in the Phanerozoic (Alekseev, 1989, 1998).

Relics of the continental crust of Early Archean age (3.5–4.0 billion years) are known in a range of regions of the Earth—in south-western Greenland, in the south and east of the Canadian Shield, in the Sino-Korean crystalline core area, in Western Australia, and elsewhere. Rocks of this age are characterized by a specific composition and belong to the tonalite-trondhjemite-granodiorite association, for which is characteristic the predominance of sodium over potassium in their composition. Frequently occurring as well in the Early Archean complexes are basaltoids and ferruginous quartzites.

In the *Middle Archean* (3.5–3.0 billion years), a wide distribution is seen for granite-greenstone districts (Southern Africa, Western Australia, Ukraine, and elsewhere), and in the *Late Archean* (3.0–2.5 billion years) their quantity sharply grows (Canada-Greenland and Baltic shields, the Priazovskii block of the Ukrainian Shield, significant areas of the Siberian, Sino-Korean, and Indian cratons, eocratons of the basements of the South American, African, Western Australian cratons, etc.).

Appearing in the Late Archean, besides the widely distributed granite-greenstone districts, are granulite-gneiss belts, which continue to develop in the Proterozoic (Limpopo in south-eastern Africa, the Western gneiss belt of Australia, the White Sea belt of the Baltic Shield, etc.).

The first manifestations of manganese and manganese-containing rocks are known already in the Early Archean. They are present in the metamorphic complexes of many Early Precambrian crystalline shields and separate blocks, notably in Greenland—rocks of the Isua complex with an age of 3800 million years; Australia—rocks of the Yilgarn and Pilbara blocks with an age of 3000–3700 million years; India—rocks of the group of iron ores with an age of 3200–2950 million years and rocks of the Chitradurga group of the Dharwar supergroup with an age of more than 2600 million years; Brazil—rocks of the Rio das Velhas series with an age of 2800–2600 million years; Canada and the United States—deposits of the Michipicoten-Vermilion series with age 2730–2655 million years; and elsewhere.

Rocks of the Isua complex (Greenland) are represented by schists, phyllites, and tuffs that overlap ferruginous quartzites and include rare interbeds of volcanogenic rocks of basic composition. Their accumulation transpired in the shallow-water environment of the stable shelf; the rocks are weakly metamorphosed. In the composition of the rocks and ores are present pyrolusite, cryptomelane, manganite, and braunite.

Rocks of the Chitradurga group and the Dharwar supergroup (India) likewise were formed in shallow-water environments of the Archean intracratonic basin. Under conditions similar to these were accumulated rocks of the Khondalite complex (Western Ghats, India)—garnet-sillimanite-graphite granulites, limy granulites, garnet-containing quartzites, and charnockites. In the composition of the ores (deposits of Kodur, Garbham) are present braunite, hollandite, jacobsite, hausmannite, and vredenburgite and more rarely spessartine and rhodonite.

Deposits of the Rio das Velhas series of Minas Gerais state in Brazil are represented by granitic granulites, limy granulites, garnet-containing quartzites, and charnockites. The primary manganese-containing minerals (Morro da Mina deposit and others) are rhodochrosite, manganocalcite, tephroite, spessartine, rhodonite, and pyroxmangite. At the outset the rock represented terrigenous-carbonate deposits with an impurity of volcanogenic material formed under eugeosynclinal conditions. Subsequent processes of metamorphism contingent upon the quantity and composition of silicate impurity led to the formation of various metamorphic silicates of manganese (Dorr II, 1972).

Examples of smaller deposits associated with rocks of the Late Archean can be found in Zaba (Russia) and Khoshchevatskoe (Ukraine). The former is situated on the western shore of Baikal. The manganese rocks are represented by manganese limestones composing interbeds within a 100-m member of white and gray, layered, occasionally blocky crystalline limestones (marbles) of the lacustrine suite of the Ol'khon series of the Upper Archean, which is enclosed within the rock mass with interbeds and lenses of quartzite, gneisses of various composition (biotite, amphibolite, pyroxene, etc.), and amphibolites and porphyrites (Varentsov and Rakhmanov, 1974).

Manganese rocks of this deposit are associated genetically with marine geosynclinal carbonate facies of the Archean and represent initial sedimentary-diagenetic, poor-carbonate ores. Under conditions of regional metamorphism, they were recrystallized with the formation of crystalline-grained structures and mineral ore associations: manganese calcite—manganocalcite, more rarely manganese-containing silicate minerals of the garnet, amphibole, and pyroxene groups.

The Khoshchevatskoe deposit (Odessa Oblast [region], Ukraine) is associated with the weathering crusts of the rock mass of marbles with interbeds of arkosic quartzites, which are found in the composition of the complex of metamorphic rocks of the Archean (biotite and graphite gneisses, quartzites, marbles, quartz-magnesite and quartz-garnet rocks, granites, pegmatites, etc.) (Betekhtin, 1946). From the surface, it is represented by a thick zone of oxidation (up to 40 m) in the form of discrete lenses and nests. The ore mass is composed of friable ochreous-clayey rocks with a great quantity of concretions of pyrolusite and psilomelane, manganiferous brown iron ore, and hematite. The ores are classified as ferromanganese.

Economically significant concentrations of manganese are known as well in the metamorphic silicites of the Kharpenskaia series of the Anabar Shield (Varentsov et al., 1984).

It follows to note one important particularity: the frequent association of manganese-containing rocks of Archean age with carbon-containing formations. For example, in India there is wide distribution in metamorphic rocks (garnet-sillimanite-graphite granulites, calcite granulites, garnet-containing quartzites, charnockites, schists, phyllites, and tuffs) of oxides and silicates of manganese (deposits of Kodur, Garbham, and the Khondalite group; the Iron Ore group). In Brazil (Morro da Mina and other deposits) widely developed in the rocks of the Rio das Velhas series, represented by graphite phyllites, mica schists, metamorphosed conglomerates, amphibolites, and ferruginous quartzites, manganese carbonates, silicates, and oxides.

Stratiform deposits of manganese have been discovered in Madagascar (graphite system, group Ampanihy); among the graphite leptites and ambibole-pyroxene rocks have been established lenses of carbonate and silicate manganese ores. In the Khaptchan series of the Anabar Shield, seams of graphite schists and graphite quartzites are interbedded with basic crystalline schists and rhodonite-spessartine rocks. Manganese-containing rocks are known in the Lapponian carbon-bearing schists in northern Finland, as well as in a range of other basins of sedimentation of Late Archean age.

In this way, it can be supposed that manganese accumulation in basins of Archean age is widely scattered. As a rule, manganese was accumulated in the carbonate and terrigenous-carbonate rocks, and more rarely volcanogenic(hydrothermal)-sedimentary rocks, in close association with organic matter (C_{org}). Their initial contents of manganese were insignificant. Subsequent processes of metamorphism led to the formation of carbon-containing (graphite) metamorphic rocks and a range of manganese and manganese-containing minerals: manganocalcite, rhodochrosite, braunite, hollandite, jacobsite, hausmannite, rhodonite, pyroxmangite, spessartine, and others, as well as manganese-containing minerals of various classes and groups (pyroxenes—in the diopside-salite-ferrosalite-hedenbergite series, the diopside-aegerine series, and other series; amphiboles—in the cummingtonite-hübnerite series, the tremolite-actinolite-ferroactinolite series, and other series, joesmithite, kozulite, the epidote group, olivine, and garnets, as well as a range of sulfides, phosphates, borates, and vanadates).

Today, it is impossible to estimate the scale and accordingly the quantity of the accumulated metal in the water bodies of the Archean manganese-ore period. Nevertheless, there is no reason to doubt that this period in the metallogenic aspect is one of the principal periods. Although the initial manganese-containing metamorphic rocks—garnet-sillimanite-graphite granulites, limy granulites, garnet-containing quartzites, charnockites, schists, phyllites, ferruginous quartzites, carbonate rocks, etc.—are characterized by insignificant contents of manganese (up to 10%), the total manganese content of these rocks in many cases is high, which can lead in the conditions of the equatorial tropics to the formation in the weathering crusts of economically significant deposits of rich oxide ores.

The products of the breakdown of manganese-containing rocks of Archean age in the majority of cases serve as the sources of manganese for manganese-containing sediments of water bodies of subsequent periods, and above all of the Proterozoic, which contain the primary reserves of manganese of the Earth's upper crust. This permits the inference of initially significant resources of manganese in initial manganese-containing rocks of Archean age.

In the tectonic regard, the boundary of the Archean and Early Proterozoic that occurs on the level of approximately 2.5 billion years is one of the most important boundaries in the history of the Earth's geological development. It is proposed (Khain, 2001) that in the Late Archean (2.8–2.5 billion years), widely manifested deformations of compression, regional metamorphism, and granitization led to the formation of the first supercontinent—Pangaea-0. (A hypothetical continent or supercontinent of this time, according to the interpretations of Rogers (1996), is called Ur or Monogaea according to O.G. Sorokhtin and collaborators (Sorokhtin and Ushakov, 1991; Sorokhtin et al., 1999, 2001; Sorokhtin, 2004).)

Contemporary knowledge of the early stages of the establishment of the Earth's crust does not provide a clear understanding of the distribution of basins of sedimentation with manganese-ore specializations nor of the contours of consolidated blocks of this period of the Earth's development. Evidently, a simple convergence (merging) of consolidated blocks of the Earth's crust of Archean age within their contemporary boundaries on the initial protocontinent (Sorokhtin et al., 2001; Sorokhtin, 2004) is unable to shed light on their paleogeography in the early Archean.

Nevertheless, the present position of known manganese deposits and ores in the formations of the Middle and Late Archean within the borders of Pangaea can be seen in Fig. 4.2.

4.2 PROTEROZOIC METALLOGENIC PERIOD (2500–550 MILLION YEARS)

In terms of geochronology, the Proterozoic is the period of longest duration in the history of the Earth's development and the predominant period in the plan of accumulation of manganese ores. Associated with this period are the principal manganese resources contained in rocks of the landmass. They constitute more than 70% of those known today. The accumulation of manganese over the course of this metallogenic period occurred with varying intensity. It can be delineated into two metallogenic epochs: the Early-Middle Proterozoic and the Late Proterozoic.

4.2.1 EARLY-MIDDLE PROTEROZOIC EPOCH (2500–1000 MILLION YEARS)

This epoch in the metallogenic regard is the most important; in the temporal regard, it corresponds to the period of the existence of the hypothetical paleo(super)continents Pangaea-0 and Pangaea-1 (Khain, 2001), Monogaea, or the hypothetical supercontinent Paleopangaea. Manganese accumulation in this period was associated with the same areas (possibly basins), which allows us to combine the

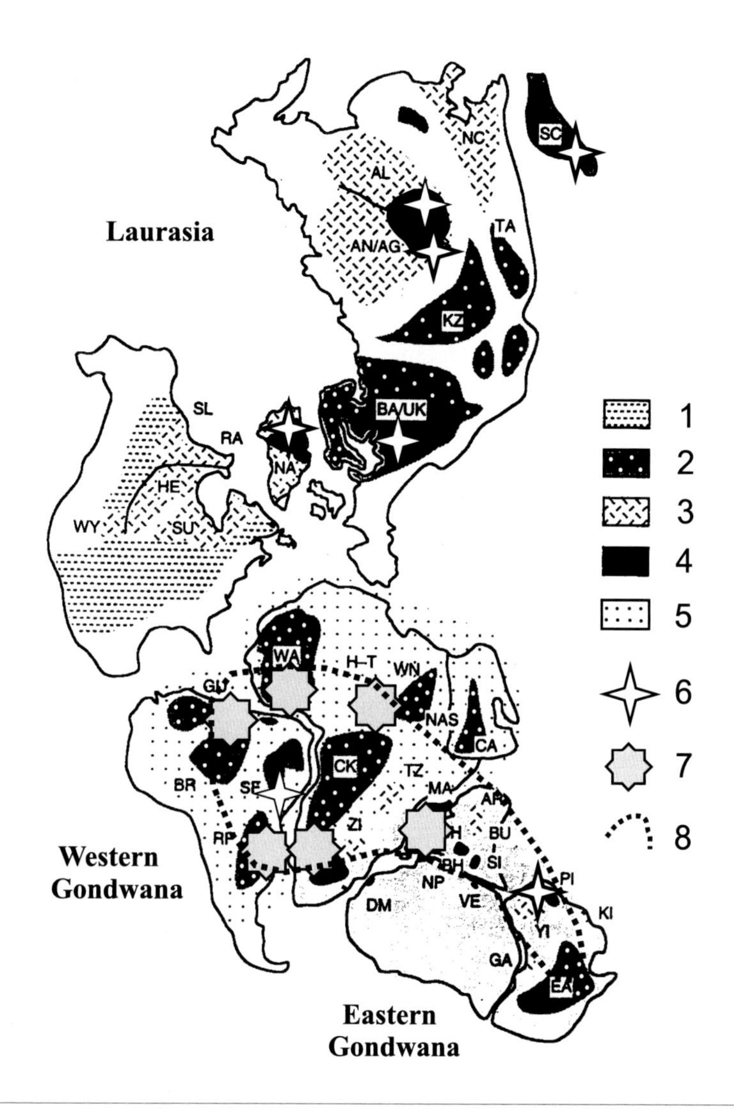

FIG. 4.2

Development of manganese rocks and ores in the Archean and Early Proterozoic in Pangea 0.3 Ga ago. Based on Rogers, J.J., 1996. A history of continents in the past three billion years. J. Geol. 104 (1) 91–107. (1–4) Age of rocks in continental blocks, Ga: (1) 1–2, (2) 1.5–2, (3) 2.5, (4) 3 or more; (5) boundary of Western Gondwana; (6) field of Archean manganese rocks and ores; (7) positions of Early Proterozoic manganese deposits; (8) field of Early Proterozoic sedimentation with manganese-ore specialization. Latin letter designations: *AL*, Aldan; *AR*, Aravalli; *AN/AC*, Anabar/Angara; *BA/UK*, Baltica/Ukraine; *BH*, Bhandara (Bastar); *BR*, Brazil (Guapore); *BU*, Bundelhand; *CA*, Central Arabia; *CK*, Congo/Kasai; *DH*, Dharwar (Western and Eastern); *DM*, Western Dronning Maud Land; *EA*, Eastern Australia; *GA*, Gavler; *GU*, Guayana; *HE*, Herne; *NT*, terranes including the Archean blocks of North Africa; *KA*, Kaapvaal; *KI*, Kimberley; *KZ*, Kazakhstan; *MA*, Madagascar; *NA*, North Atlantic (including Nain, Greenland, and Levisian); *NAS*, Pan African crust of the Nubian-Arabian Shield; *NC*, North Chinese (Sino-Korean); *NP*, Napier; *PI*, Pilbara; *RA*, Rae; *RP*, Rio de la Plata; *SC*, South China (Yangtze); *SF*, San Francisco (including Salvador); *SI*, Singhbhum; *SL*, Slave; *SU*, Superior; *TA*, Tarim; *TZ*, Tanzania; *VE*, Vestfold; *WA*, West Africa; *WN*, Western Nile; *YI*, Yilgarn; *ZI*, Zimbabwe.

Early and Middle Proterozoic into a single metallogenic epoch. However, the intensity and degree of manganese concentration in the initial sediments varies between the Early and Middle Proterozoic.

More important in the metallogenic aspect is the *Early Proterozoic phase (2500–1600 million years, phase 1)* (see Fig. 4.1), particularly its first half, during which time was accumulated the principal industrial potential of manganese (greater than half of its stocks and reserves). These are the deposits of Africa (Republic of South Africa, Gabon, Ghana, Congo), Brazil, Australia, India, Sweden, and other countries (see Fig. 4.2).

In this phase, most significant in terms of manganese reserves is without question the Kalahari manganese-ore basin of South Africa, within which is concentrated more than half of all resources of known deposits on the Earth's landmass (Laznica, 1992; Beukes et al., 1995; Gutzmer and Beukes, 1995; Astrup and Tsikos, 1998). Over the course of the geological history of the development of the lithosphere, manganese rocks of this basin to a significant degree were evidently destroyed (eroded, covered under tectonic nappes), and at present only its eastern part remains. The manganese ores here are represented by manganese (braunite) lutites, which in the vertical section are underlain and overlain by banded ferruginous (hematite) siliceous shales. Their formation occurred in the environment of a shallow-water basin situated on the western plunge of the consolidated Archean Kaapvaal block and was characterized by high bioproductivity (cyanobacterial mats) and a tranquil tectonic regime.

At present, manganese ores in other regions of the world that are analogous to the Kalahari Desert deposit are unknown. It can be supposed that this was a singular and unique event in the history of the formation of the lithosphere of the paleocontinent Atlantica (more precisely, its component part: Western Gondwana) (see Figs. 4.2 and 4.3).

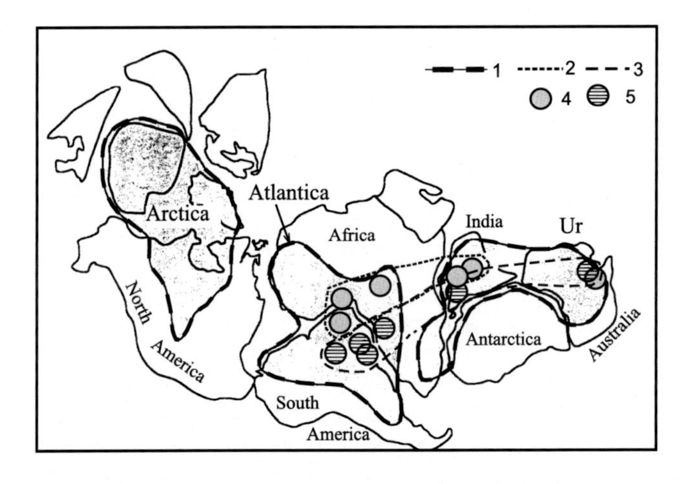

FIG. 4.3

Manganese rock terranes during the existence of Atlantica. Modified after Rogers, J.J., 1996. A history of continents in the past three billion years. J. Geol. 104 (1) 91–107. (1) Boundaries of paleocontinents; (2) domains of manganese deposits developed after gondites and ampelites; (3) boundaries of manganese deposits associated with the ferruginous-siliceous rocks; (4) major manganese deposits developed after gondites and ampelites; (5) major manganese deposits associated with the ferruginous-siliceous rocks.

Another type of basins of sedimentation of the primary manganese rocks of this time consists of intracratonic shallow-water basins with high bioproductivity, in which transpired the accumulation of carbonaceous shales—ampelites—containing rhodochrosite. In their weathering crusts were formed medium and large deposits of manganese (eg, the Moanda deposit in Gabon) (Weber, 1970).

Significant reserves of manganese are spatially and evidently genetically associated with rocks of iron-ore formations of the BIF (Banded Iron Formation) and their metamorphic analogues—ferruginous quartzites. The largest deposits of this type are known in Australia (Hamersley basin, Pilbara massif), Brazil (Mato Grosso state, Brazilian Shield), India, China, and other countries.

The accumulation of manganese occurred in conjunction with that of iron, but in quantity manganese was much lower relative to iron. The richest deposits of manganese (as well as of iron) are associated with the weathering crusts on these rocks. For example, significant quantities of manganese are present within the zones of hematite-goethite mineralization developed on the ferruginous quartzites of the iron-ore BIF-formation of Marra Mamba of Hamersley basin, Pilbara province (Western Australia). Here, within the boundaries of the Christmas Creek, Cloud Break, and Mount Lewin deposits, present in the weathering crusts are blocks of ferromanganese ores with high manganese content (the Mn/Fe ratios are not high: 2–3). The general reserves of iron ores within the boundaries of the indicated deposits constitute about 2.5 billion tons, which allows us to estimate the resources of ferromanganese ores at a few hundred million tons.

With rare exception, Early Proterozoic basins of sedimentation with ferromanganese specialization were characterized by a tranquil tectonic regime, shallow-water environment, and situated in districts of peneplains—confirmation of the practically total absence of clastic volcanic material in the rocks.

Industrially important deposits of high-quality manganese oxides are situated within the boundaries of the manganese-ore province of Pilbara (Western Australia), situated 400 km to the north-west of Port Hedland and stretching 30 km in the meridional and 10 km in the lateral direction. The largest deposits of this province are Ripon Hill, Woodie-Woodie, Ant Hill, Hi Tech, Ant Hill Sovereign, Enachedong Creek, Sandy Hill, Mount Cooke, Mount Sydney, Skull Springs, Bee Hill, Balfour Downs, Mount Nicholas, Roy Hill, etc. The deposits are connected, as a rule, with ancient and modern weathering crusts of initial manganese-containing Lower Proterozoic rocks of the Carawine Dolomite formation: carbonates, schists, and quartz-hematite and other complexes. Ore material is represented by manganese oxides.

Likewise deserving attention is the series of Lower Proterozoic manifestations of ferromanganese ores of the Bootu Creek deposit in the Northern Territory, Australia. It is situated 110 km to the north of the populated locality of Tennant Creek and 10 km east of Stuart Highway. The ores are confined to the lower part of the Bootu Formation of the Lower Proterozoic group of Tomkinson Creek, which constitutes the core of the Bootu syncline. A manganese-rich horizon stretches for 24 km around the core of the syncline and contains a series of fine occurrences of manganese oxides in the form of lenses and veins with rich iron content. The ore seam plunges toward the center of the syncline to a depth of over 100 m.

Associated as well with the Early Proterozoic stage are the deposits of India: the Aravalli supergroup—an age of approximately 2000 million years; the Sausar group—an age of approximately 2000 million years; and the Gangpur group—an age of 2000–1700 million years. These deposits are associated with metamorphosed terrigenous-sedimentary rocks and formed in a shallow-water (shelf) environment.

With metasedimentary rocks of the Lukosi complex (Democratic Republic of the Congo) with an age older than 1845 million years are associated shallow-water manganized stromatolite carbonates interlaid with manganese-containing silicate rocks and graphite schists. In Botswana, manganized stromatolites corresponding in age with deposits of the Transvaal supergroup have also been described.

A series of economically and genetically important deposits is represented by the karst formations of manganese oxides of the Postmasburg-Sishen region (Republic of South Africa) included in the Transvaal supergroup.

Significant quantities of manganese in the rocks of the Lower Proterozoic are also contained in poor manganese-containing rocks of the metamorphic complexes of crystalline shields, which are represented predominantly by manganese-containing garnets (spessartine) and carbonate-containing (rhodochrosite) carbonaceous shales. One characteristic example is the large deposit of oxide manganese ores developed in the metamorphic rocks of the Serra do Navio Formation in Amapá state (Brazil). The rocks of the formation are represented principally by quartz-biotite-garnet gneisses and crystalline schists with microcline, oligoclase, graphite, rhodochrosite, and rhodonite. Also occurring are lenses of marbles. The age of the metasediments is estimated at no fewer than 1770–10 million years (Herz and Banerjee, 1973).

The initial rocks of this deposit, in the opinion of certain authors (Dorr II, 1972; Scarpelli, 1970), represent shallow-water sediments formed in the unstable shelf zone or lagoon (a range of cycles of transgressions and regressions has been identified). Of industrial interest are the oxide ores of the hypergenesis zone (manganese content reaches 46% and higher; the initial reserves of rich manganese ores constitute more than 120 million tons). Manganese mineral resources in the initial rocks of the basin of sedimentation have not been evaluated.

There is no reason to doubt that significant reserves of manganese (evidently a few billion tons) are to be found in manganese-containing rocks of the Guiana and Brazilian shields (the territory of Brazil and of contiguous states); due to their inadequate state of geological study, an evaluation at present is not possible.

Of great practical value are deposits of manganese and ferromanganese ores associated with rock masses of carbonaceous terrigenous-carbonate rocks. In this regard, attention is due above all to the deposits confined to the areas of distribution of rocks of the Franceville series (Gabon). Reserves of manganese and ferromanganese ores here (within the modern boundaries, not subject to erosion of the plateau) constitute in total no fewer than 250–300 million tons.

One of the largest deposits of carbonate manganese ores in rocks of Early Proterozoic age and bearing important economic value in the global manganese-ore industry is the Nsuta deposit (Ghana, West Africa).

The primary manganese ores (carbonates) and manganese-containing rocks (gondites, phyllites) in this deposit are contained in the sequences of the Lower Proterozoic deposits of the Birimian series (supergroup); within its boundaries the manganese-containing horizon (thickness of up to 50–60 m, rarely greater) occupies an intermediary position—it is underlain and overlain by greenstone rocks (Lower and Upper greenstones).

A series of deposits of Early Proterozoic age is associated as well with the metamorphosed leptite formation of Svecofennian age (1900–1700 million years) of central Sweden (Långban, Harstigen, and other deposits).

In an embryonic state, the manganese process is manifested in Lower Proterozoic deposits of the Kursk Magnetic Anomaly (KMA) (Gurvich and Abramova, 1984). Here in the Tim suite of the Oskol series, represented by carbonaceous shales, quartzites, carbonate, ambibole-, and garnet-containing rocks, effusives, and tuffs of basic composition, spessartine-containing rocks are widely distributed. The most manganese-enriched are quartz-spessartine carbon-containing gneisses, in which has been established from 9.42% to 24.23% MnO and from 4.8% to 9.42% $C_{elementary}$. Also present is manganese sulfide (alabandite).

In this way, a brief survey of the most commercially and scientifically important manganese deposits associated with the rocks of the Lower Proterozoic illustrates the spatially extensive area occupied by these deposits—namely, the central part of Western Gondwana and the western part of Eastern Gondwana (see Fig. 4.2). It can be supposed that at some point, this was a single district of sedimentation of manganese and manganese-containing rocks; its lengthy existence (approximately 900 million years) featured a complex history of geological development characterized in different parts by different conditions of sedimentation and, correspondingly, by different types of sediments; these include carbonaceous clayey-carbonate deposits—ampelites (deposit Moanda, Gabon), braunite lutites (Kalahari deposits, Republic of South Africa), ferruginous cherts (deposits of Brazil), and the like.

Certain regularities can generally be noted at present in the confinement of the principal types of manganese rocks to given parts of the examined manganese district. For example, characteristic for the northern part of this district (the deposits of northern Brazil, Ghana, Gabon, and others) are (carbonate)-terrigenous carbonaceous associations (gondites, ampelites), whereas in its southern part manganese rocks are associated predominantly with banded ferro-siliceous rocks (BIF, see Fig. 4.3).

Also noted are districts where manganese rocks are enclosed in sequences, the composition of which contains significant quantities of volcanogenic-sedimentary rocks (deposits of Ghana and India). This can serve as evidence that the area of manganese-ore sedimentation in the Early-Middle Proterozoic Epoch represents a series of sedimentary basins characterized by various geological development and various types of sediment accumulation.

Middle Proterozoic manganese-ore metallogenic phase (1600–1000 million years, phase 2) (see Fig. 4.1) is likewise prominent in the economic regard. Areas of manganese accumulation for this phase were inherited from the preceding Early Proterozoic phase; on the whole, these are identified within the same basins of sedimentation within the boundaries of the erstwhile paleocontinent Pangaea-1 (according to Khain, 2001), to all appearances, however, they are significantly reduced in area.

It follows to note that during this metallogenic phase, as previously, were accumulated mainly manganese-containing rocks in which concentrations of manganese rocks did not reach ore concentrations. Industrially significant concentrations of manganese (reaching the size of deposits) were accumulated at a later geological time, predominantly within the weathering crusts on manganese-bearing rocks.

Significant resources of manganese in the Middle (as in the Early) Proterozoic are associated with rocks of the BIF. Ores of deposits of this age are classified predominantly as ferromanganese.

An example of this type of deposits are the manganese-ore deposits of several Brazilian states, especially those of the complex of rocks of the Minas series in Minas Gerais state. These rocks were formed in the conditions of the platform miogeosyncline and have a thickness of approximately 3500 m. The manganese content in rocks of the ferruginous formations of this series is very low. However, in the zones of weathering under hypergenic conditions occur manganese-ore bodies with manganese content of 30–48% constituting reserves up to 5 million tons. The extent of the separate ore bodies commonly constitutes a few hundred meters, but in certain cases may reach 5 km; zones with manganese mineralization can extend at a distance up to 10 km.

Also in close association with ferruginous formations are the manganese deposits of the Urandi region, in the south of Bahia state (Brazil).

Another type of manganese deposits in Brazil is closely associated with zones featuring the hypergenic breakdown of initial manganese-containing phyllites, where the accumulation of manganese oxides occurs (in the regions of Nazaré and Jacobina, Bahia state) (Dorr II, 1972).

Likewise confined to rocks of Middle Proterozoic age are a series of deposits in China (Editorial Committee of Mineral Deposits of China, 1995). Chief among them are the group of deposits (Wafangzi, Taipinggou, Zhongxincun, and others) confined to the middle part of the Tieling terrigenous-carbonate formation (Yanliao subsidence zone). The largest is the Wafangzi deposit, occupying an area of approximately 80 km^2, where have been identified three manganese-ore horizons containing 11 ore deposits (4 of them industrial); these are embedded in metaclastites and dolomites. Ores are represented by three types: initial (diagenetic)sedimentary—manganite-rhodochrosite; metamorphosed—bixbyite-braunite, mangano-silicate, and pyroxmangite-rhodochrosite; and hypergenic—oxide-hydroxide (pyrolusite and psilomelane-vernadite-goethite-hydrogoethite).

In the Middle Proterozoic deposits of China limited distribution is found as well for boron-manganese deposits such as Dongshuichang (Jixian type, Yanliao metallogenic province, north-eastern China). They are confined to the volcanogenic-sedimentary deposits of Gaoyuzhuan formation (volcanites, clayey dolomites, shales) and are represented by lensoid ore bodies. The primary ore minerals are rhodochrosite and chambersite ($Mn_3B_7O_{13}Cl$).

A series of deposits of Middle Proterozoic age is associated as well with the volcanogenic-sedimentary rocks of the Birimian complex in West Africa (Ghana, Côte d'Ivoire, Burkina Faso, eastern Liberia, and Guinea).

It should be noted that in the future, for a range of deposits of Middle Proterozoic age in many countries, with more precise measurements the radiological age of the host (or initial) rocks could turn out to be older—that is, Early Proterozoic. But this does not in principle render a notable influence on our understanding of manganese ore-genesis in the Early and Middle Proterozoic, insofar the basins of sedimentation in many cases were part of a single district.

4.2.2 LATE PROTEROZOIC (NEOPROTEROZOIC) EPOCH (1000–550 MILLION YEARS)

This epoch of manganese ore-genesis is associated, as is the preceding epoch, with complexes of rocks that vary by composition and origin (see Fig. 4.1). Primary manganese ores are poor and frequently associated with iron; of greatest practical interest, as for the above-examined Middle Proterozoic ores, are the hypergenic ores of the weathering crusts.

Basins of sedimentation of manganese rocks of this age were situated within the boundaries of the supercontinent Rodinia (Fig. 4.4).

Serving as a typical example of the deposits associated with rocks of ferro-siliceous formations are the deposits of Mato Grosso state, Brazil—the Morro do Urucum region, with an age of approximately 800–600 million years, possibly older (Walde et al., 1981; Urban et al., 1992; Klein and Ladeira, 2004). This region is classified as ferromanganese ore and is one of the world's largest, with iron-ore reserves of 36 billion tons (with Fe content from 50% to 67%) and manganese-ore reserves of 608 million tons (with Mn content from 25% to 49%).

Manganese ores are closely associated as well with rocks of the Santa Cruz Formation of the Jacadigo group, in the composition of which have been identified two rock masses: the lower mass, represented by fine-grained detrital sediments—iron-containing sandstones (arkoses), interlaid with jaspilites; and the upper mass, composed of banded (layered) hematite jaspilites with manganese-ore

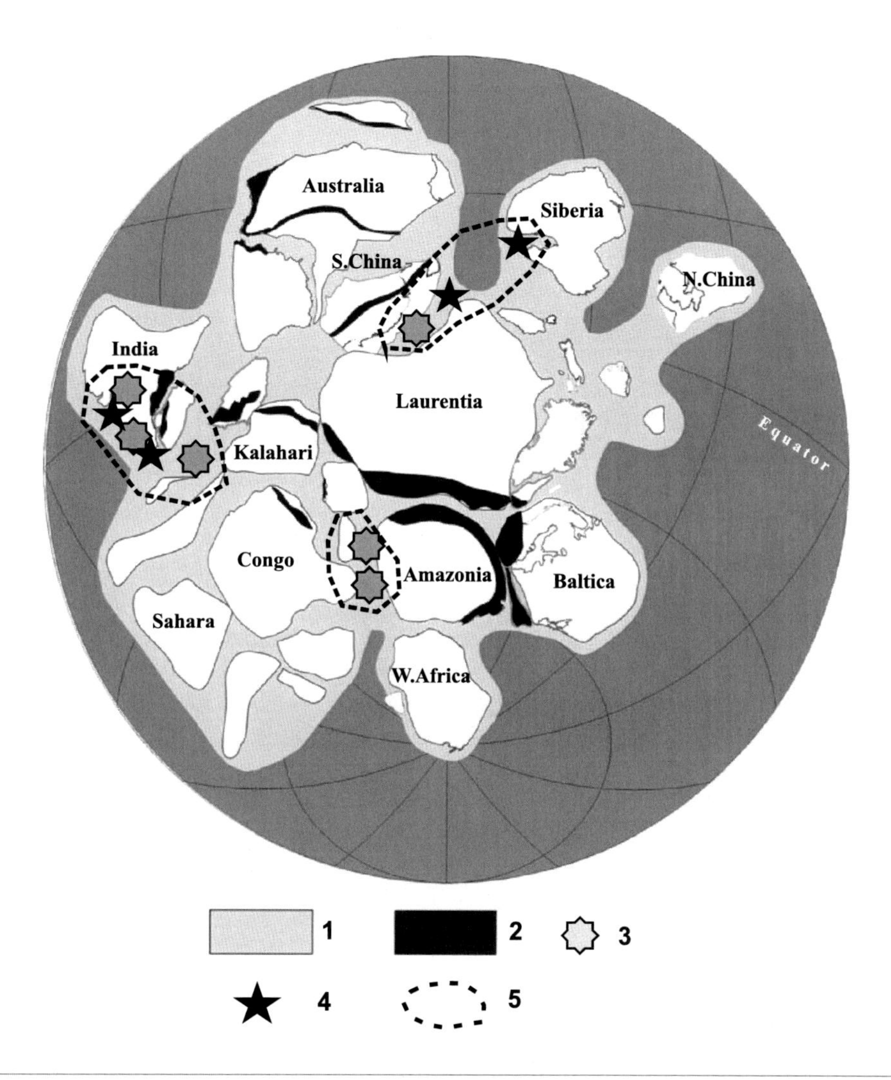

FIG. 4.4

Locations of manganese rocks and ores in the Late Proterozoic supercontinent Rodinia. Outlines of paleocontinents are shown as of 900 Ma ago, according to Bogdanova et al. (2009). (1) Inferred shelf margins; (2) major collisional orogens of the Rodinia breakup period; (3, 4) basins with manganese rock formation in the Middle and Late Proterozoic, respectively; (5) boundary of the major manganese-ore zone in the Middle and Late Proterozoic.

horizons and including horizons of detrital ores of various sizes. The lithological and mineralogical particularities serve as evidence for the continental conditions of their formation; this does not preclude a glacial origin for the horizons of detrital ores.

Of economic significance are the Early Sinian manganese deposits of south-eastern China (Hunan, Guizhou, Sichuan, and Hubei provinces) associated with black shales. Ores of these deposits belong to poor-carbonate ores of Xiangtan type, in the composition of which predominate rhodochrosite and kutnohorite associated with sulfides of iron. Ore seams are represented by various types of ores and manganese rocks—gray carbonate manganese and black anthraxolite oolitic carbonate ores as well as black manganese-containing dolomites.

China's Late Sinian manganese deposits are associated with the Doushantuo phosphate-manganese formation developed in the north of the country and represented by a productive seam of black shales and clayey dolomites with a thickness of 27 m. Ores likewise are represented by carbonates (the Gaoyang and Yangjiaba deposits).

Late Proterozoic deposits of manganese oxides are known in India as well; these are confined to the transgressive sandstone-carbonate-shale series of the Penganga Beds Formation. The ore horizon is enclosed within a stratum of limestones.

In association with the volcanogenic-sedimentary rocks of continental type are the deposits of manganese in the Anti-Atlas region of Morocco (the Tiouine, Idikel, Migouden, and Oufront deposits).

Small manganese deposits are known in association with the iron-ore formation of the Damara Supergroup (Namibia).

Within the boundaries of the southern part of the Hingan-Bureja median massif (Dal'nii Vostok, Russian Far East) ferromanganese mineral occurrences are developed in the composition of the upper part of the Hingan complex, represented to varying degree by metamorphosed terrigenous-carbonate and volcanogenic-sedimentary rocks of Upper Proterozoic (Riphean)-Lower Paleozoic (Cambrian) age. These occurrences form an ore field in the form of a narrow band with a width of 6–8 km, extending in the submeridional direction for approximately 60 km (Chebotarev, 1958).

The ore-bearing seam here is represented by three horizons—subore siliceous-argillaceous, carbonaceous shales, dolomite sandstones, breccia; ore horizons (manganese ore and iron ore); and supraore carbonaceous-argillaceous and argillaceous shales with lenses of limestones and dolomites.

Among the ores, the most widely distributed are silicon-braunite-hematite and silicon-hematite-hausmannite-rhodochrosite. In the contact zone of metamorphism appear silicate varieties of magnetite-spessartine-rhodonite, magnetite-spessartine-tephroite, and other compositions. Characteristic for the ores are a thin rhythmic banding (interlaying of interbeds of ore and jasperoid facies), micro-granularity, and close intergrowth of ore and non-ore minerals. Ore minerals are represented by rhodochrosite, oligonite, braunite, and hausmannite and contain spessartine, bustamite, tephroite, and piemontite.

Of the same age as the Hinggan ferromanganese ores are those of hydrothermal-sedimentary genesis of the Uda basin (Dal'nii Vostok, Russian Far East). These are deposited in a complex of volcanogenic-sedimentary rocks in association with banded ferruginous-siliceous horizons.

Substantial quantities of manganese also characterize the deposits of the Upper Precambrian Yenisei Range (Western Siberia), in which have been identified five manganese-bearing stratigraphic levels confined to three manganese formations: siliceous-terrigenous-carbonate, carbonaceous-siliceous-carbonate, and tuffaceous-siliceous-carbonate (deposits and ore-occurrences: Bol'she-Gremiachinskoe, Glushikhinskoe, Kiiskoe, Vylomskoe, and others). Manganese mineralization is represented by minerals of the isomorphic range of manganocalcite-rhodochrosite; the presence of manganoankerite has been established. In the zone of hypergenesis (weathering crust of Cretaceous-Paleogene age) on these minerals are developed manganese oxides constituting industrially viable sedimentary-infiltration ores (Golovko et al., 1982).

In Kazakhstan (Yerementau region) in association with the volcanogenic-sedimentary rocks of Vendian-Lower Cambrian age is the deposit of Kumdykol'. Here predominate manganous ferruginous ores of the type of ferruginous quartzites; ferromanganese ores are present in insignificant quantity. The primary ore minerals are hematite, magnetite, braunite, rhodonite.

Ferromanganese lensoid ore bodies are known as well in western and northern Pribalkhash (Kazakhstan), where they are associated with Vendian-Cambrian carbonaceous-siliceous and volcanogenic-shale rocks, jasperoids, and hematite-magnetite quartzites. The principal ore minerals are tephroite, magnetite, rhodonite.

Thus, an evaluation of the general regularities of manganese rock formation over the span of the Proterozoic zone will note the tendency of the manganese accumulation process of this time to be limited in the majority of cases to basins of sedimentation with iron-silicon specialization. Moreover, these basins (or parts of them) in certain cases remained in existence for a considerably long geological period of time, beginning in the Early Proterozoic and persisting over the course of practically the entire eon. Characteristic examples of such basins are the Lower, Middle, and evidently Upper Proterozoic iron-ore basins of Brazil and West Africa, certain of which survived up to the end of the Late Proterozoic (Mato Grosso, Brazil).

It follows to note as well the appearance by the end of the Proterozoic of new areas with silicon-iron and carbonate-terrigenous manganese sediment accumulation; for example, the area including the deposits of Western Siberia (Yenisei Range), Kazakhstan, and Transbaikal.

4.3 PHANEROZOIC METALLOGENIC PERIOD (ZONE)

This period of accumulation of manganese ores was likewise marked by the formation of a range of economically important deposits. Manganese ore-formation occurred in practically all parts of the continent Gondwana during its existence, and subsequently in its "continent-fragments," in the environment of a stable platform as well as in mio- and eugeosynclinal areas. Several metallogenic epochs and stages that led to the formation of a range of deposits of various scales (including such giant deposits as Nikopol') can be identified: the Lower-Middle Paleozoic, the Late Paleozoic, and the Mesozoic-Cenozoic.

4.3.1 EARLY-MIDDLE PALEOZOIC EPOCH (CM-D)

Connected with the Early-Middle Paleozoic Epoch (see Fig. 4.1, phase 4) is the formation of a range of small Lower Paleozoic deposits in Russia, Brazil, China, Kazakhstan, Canada, Israel, and other countries. From the outset of this epoch, the conditions of sedimentation by comparison with those of the Proterozoic on the whole were substantially changed: manganese accumulation occurs principally in areas with active volcanic activity and is closely associated with volcanogenic-sedimentary rocks. The sources of manganese were predominantly hydrothermal solutions discharged in submarine environments. The manganese deposits (initial manganese protolith) were formed by hydrothermal-sedimentary means, frequently in association with iron (Priishim'e, Karatau, Kazakhstan; Salair and Arga Ridge, Western Siberia; and elsewhere). Serving as the largest and most economically important for Russia is the Parnok ferromanganese deposit (Shishkin and Gerasimov, 1995).

In the Early Paleozoic in China were formed also manganese- and phosphorus-bearing organic-rich clay-carbonate deposits (Tiantaishan ore occurrence of Early Cambrian Tananpo Formation, Shaanxi Province); manganese—black shale (Taojiang deposit of Ordovician Modaoxi Formation, Hunan Province); and cobalt-manganese black shale (Jiaodingshan deposit of Ordovician, Sichuan Province).

A large quantity of minor deposits of manganese is associated with volcanogenic effusive and volcanogenic-sedimentary deposits of the Middle Paleozoic (S-D). These include especially a range of minor deposits of the eastern slope of the southern Urals that participate in the structure of the Magnitogorsk synclinorium. At the same time were formed the terrigenous-, more rarely volcanogenic-sedimentary rocks of Pai-Khoi, including a range of ore-occurrences and in rare cases minor deposits of manganese (Yudovich and Ketris, 1981; Platonov et al., 1992).

At this time, within the boundaries of the north-east of Russia and the Urals (the eastern part of the Middle Paleozoic continent of Laurentia) was deposited the Pai-Khoi-Ural manganese-bearing area (Fig. 4.5); within its boundaries existed the manganese-ore basins of Pai-Khoi-Polar-Urals (the Pai-Khoi group of ore-occurrences, the ore-occurrences of the upper reaches of the Sob' River basin, and the Parnok deposit) and the southern Urals (the Primagnitogorsk group of deposits).

A series of analogous occurrences (approximately 30) and deposits (Kyzyltal, Karamola, and others) have been uncovered as well among jasperoids and siliceous shales of the Upper Devonian-Lower Carboniferous of Dzhungar Alatau. The ores are braunite-hausmannite and from the surface are oxidized (psilomelane, pyrolusite, limonite, etc.). A series of small ore-occurrences is also known within the boundaries of Mugodzhar and Altai.

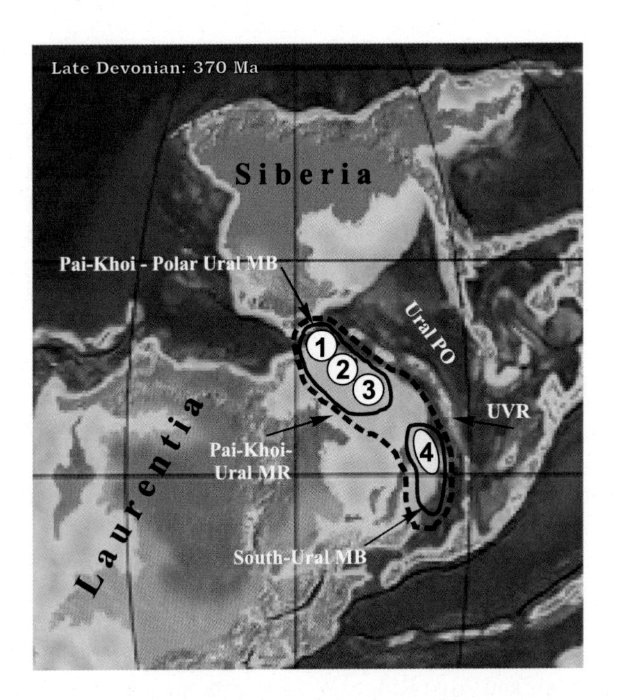

FIG. 4.5

Location of manganiferous basins and manganese deposits in PayKhoy Ural manganese region (Eastern Laurentia) in Middle Paleozoic (outlines of paleocontinents are based on http://jan.ucc.nau.edu.com, http://www.scotese.climate.com). Solid contours are PayKhoy-Polar Ural and South Ural manganese basins; deposits and occurrences: (1) PayKhoy group; (2) headwaters of Sob River; (3) Parnok; (4) Magnitogorsk group. *MB*, manganese basin; *MR*, manganese region; *UVR*, Ural volcanic ridge; *PO*, paleo-ocean (inferred position).

The accumulation of relatively insignificant quantities of manganese rocks occurred as well in basins of sedimentation with a tranquil tectonic regime; these are associated with carbonaceous and carbonate facies: deposits of Spain (Las Cabesas), France (Cazals, Brachy), and Germany (Rhenish Slate Mountains).

In the paleographic regard, this stage corresponds to the existence of such paleocontinents as Laurasia and Eastern and Western Gondwana.

4.3.2 LATE PALEOZOIC EPOCH (C-P)

This metallogenic epoch (see Fig. 4.1, phase 5) was marked by a drastic increase in accumulation of poor manganese rocks and ores. One of the most extensive manganese-ore basins in the history of the geological development of the Earth's crust during the Paleozoic is the Pai-Khoi-Novozemel'skii basin, which in the Permian period stretched from the northern part of the western flank of the Urals in the east and from Pai-Khoi in the south, encompassing the island of Novaya Zemlya; to all appearances, it might have extended even farther to the north-west (Fig. 4.6) (Voiakovskii et al., 1984; Platonov et al., 1992). This manganese-ore basin is characterized by a wide distribution of black shale deposits (Fig. 4.7) (Klimat, 2004).

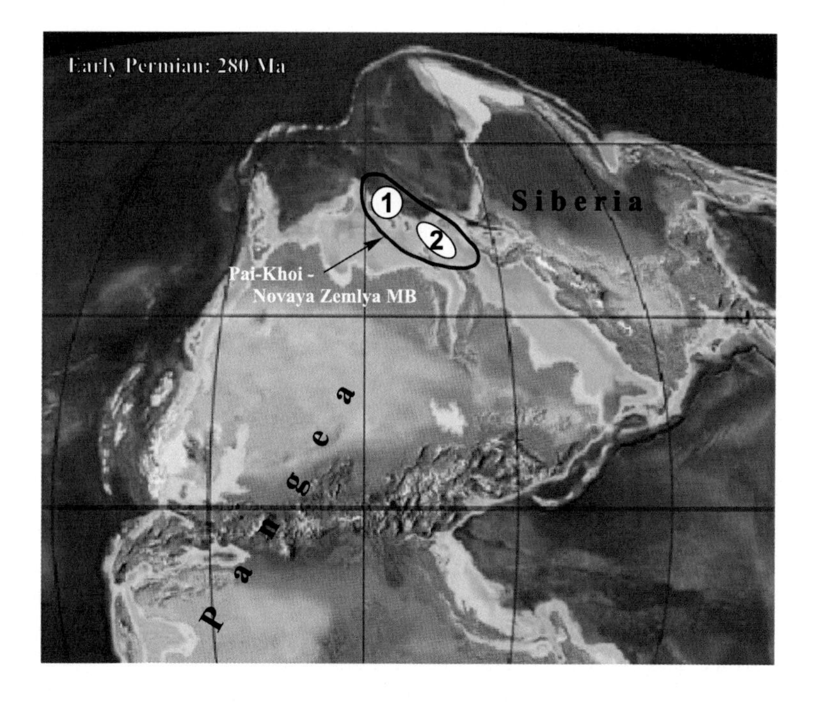

FIG. 4.6

Location of Novaya Zemlya-PayKhoy manganese basin (solid contour) in Early Permian (outlines of paleocontinents are based on http://jan.ucc.nau.edu.com, http://www.scotese.climate.com). (1) Deposits on Novaya Zemlya; (2) Deposits and occurrences in PayKhoy region. *MB*, manganese basin.

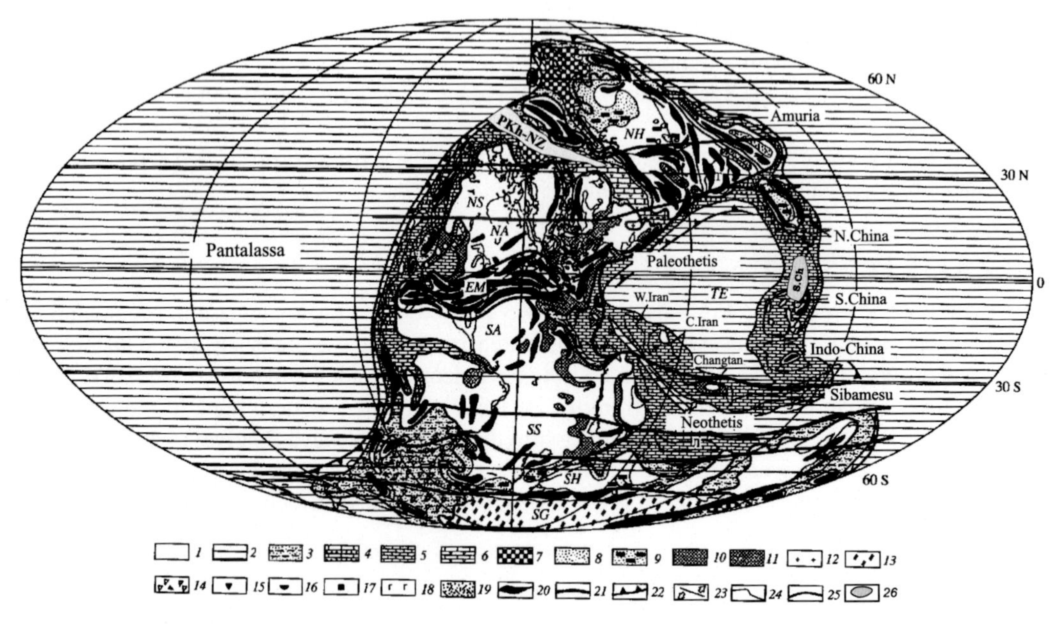

FIG. 4.7

Distribution of manganese-ore basins in the Late Paleozoic (C-P) (The outlines of paleocontinents and lithologic-paleogeographic map for Late Sarmatian-Early Artinskian time given to Climate (2004). (1) Land; (2) the oceans; (3) clastic shelf sea; (4) shelf clastic-carbonate sea; (5) carbonate platform; (6) evaporite-carbonate platforms; (7) black shale (anoxic) basins; (8) inland and coastal alluvial and alluvial-lacustrine basins in humid zones; (9) coal basins; (10) inland and coastal alluvial, aeolian and lacustrine basins of red sedimentation in arid zones; (11) inland and coastal alluvial, aeolian and lacustrine and sabkha basins of red gypsum-bearing sedimentation; (12) salt basins; (13) areas of development predominantly of continental glacial deposits; (14) mainly marine glacial deposits; (15) lateritic deposits, bauxite; (16) kaolin clay; (17) iron ores; (18) inland basalts; (19) volcanic-sedimentary deposits; (20) mountains; (21) border sedimentation-climatic zones; (*EM*, equatorial mountain; *NA*, northern arid, evaporite; *NS*, Northern semiarid; *NH*, Northern humid, carbonaceous; *SA*, southern arid, evaporite; *SS*, southern semiarid; *SH*, southern humid carbonaceous; *SG*, southern glacial; *TE*, tropical-equatorial-coal bauxite); (22) subduction zone; (23) spreading axis; (24) modern shorelines; (25) ancient shorelines; (26) outlines of manganese-ore basins (*PKh-NZ*, PayKhoy-Novaya Zemlya, *S.Ch*, South China).

Also associated with Lower Permian deposits is the Ulu-Teliak manganese deposit (Bashkortostan, Russia), where manganese mineralization is confined to the phyllite horizon of the Kungurian stage (P_1kg fl).

Manganese deposits of Permian age are also known in certain provinces of China (the Lohua-Loping platform deposits, Shanxi province, a range of deposits in the Middle-Guizhou-East Yunnan metallogenic province, and elsewhere) (Editorial Committee of Mineral Deposits of China, 1995).

Within the boundaries of the Middle-Guizhou-East Yunnan metallogenic province is situated a range deposits of manganese of industrial importance for China of Zunyi type (under development are Tongluojing, Fengjianwan, Gongqignhu, Tuanxi, Heshangchang, and Suanyegou). These are confined to the upper sequence of rocks of the Gufeng Formation (the "Bainitang" bed), represented by siliceous limestones with separate layers of manganese carbonates, as well (predominantly) toward the lower part of the sequence of Upper Permian deposits of the Longtan Formation. The latter is composed in the lower part by black shale or claystone with rhodochrosite and pyrite; in the middle by a horizon of rhodochrosite manganese ores with pyrite; and in the upper part by pyritized hydromica shales with rare inclusions of grains of rhodochrosite and manganese-iron ooliths and kaolin clays. Their formation most likely occurred in a shallow-water lagoon-bog environment.

Deposits of this age are also known among the volcanogenic-sedimentary deposits of Japan (Noda-Tamagawa, Iwate Prefecture; Kaso, Totigi Prefecture), United States (Calaveras Formation, Sierra Nevada, California; Dillon, Montana), Germany (Kellerwald, Harz), and others.

4.4 MESOZOIC-CENOZOIC MANGANESE EPOCH (T-PG)

This epoch is likewise of economic importance. Associated with the *Mesozoic metallogenic phase (T-K)* of manganese ore-genesis (see Fig. 4.1, phase 6) are deposits of Australia, Chile, Mexico, Morocco, Russia, Indonesia, China, Turkey, many countries of Europe (Hungary, Bulgaria, Italy, Romania), and others (Fig. 4.8). Many of these are classified as large deposits and occupy a prominent place in the global manganese-ore raw material base.

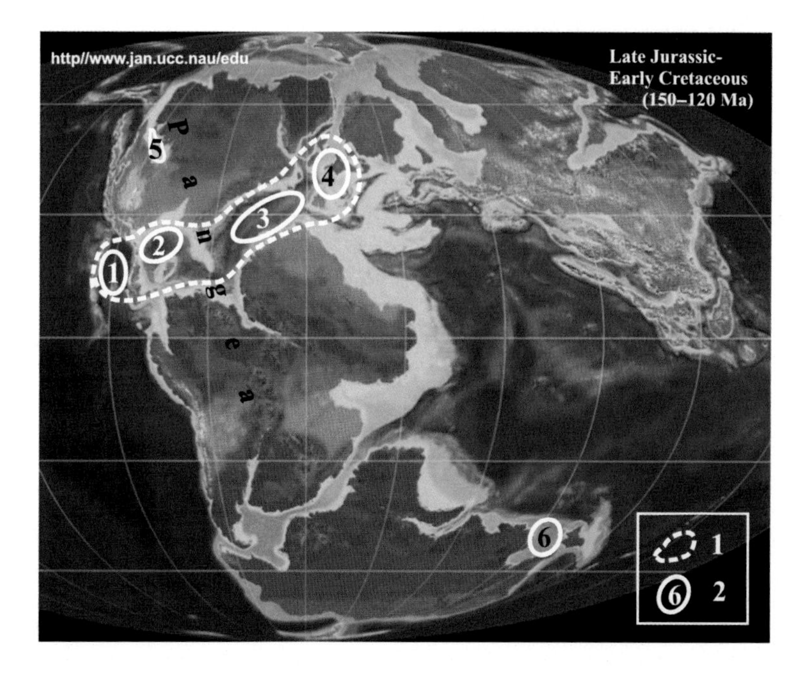

FIG. 4.8

Position of the main manganese-ore basins within the continent Pangaea after the beginning of its break-up (outlines of paleocontinents are based on http://www.jan.ucc.nau/edu). (1) Areas of manganese ore; (2) basins of accumulation of manganese ores and rocks: (1) Mexico, (2) California, (3) Morocco, (4) West European, (5) Yukon, (6) North-Australian.

Principal here for this period are sedimentary-diagenetic processes of accumulation of manganese ores (deposits of Morocco, Mexico, and Hungary), which are confined to carbonate and terrigenous sediments of intracontinental basins. At the same time were accumulated the manganese-containing carbonates of the Mullaman layers of the Carpentaria basin in the north of Australia (Fig. 4.9, basin 6).

Of note is this period's wide distribution of the processes of hydrothermal-sedimentary accumulation of manganese; for example, the large quantity (more than 400) minor deposits of California (see Fig. 4.9, basin 2) confined to the deep-water argillite-silicon deposits of the Franciscan Complex (the Ladd, Buckeye, Double-A Mine, and other deposits) (Hein and Koski, 1987). In terms of formation conditions, geological structure, and mineral composition they are similar to the Devonian deposits of the Magnitogorsk belt.

During the *Late Mesozoic-Cenozoic manganese-ore* period were formed the deposits of the Eastern Paratethys (see Fig. 4.1, phase 7). These include widely known deposits such as Nikopol and Bol'she-Tokmak in the southern Ukrainian basin, as well as a range of large deposits of Georgia (Chiatura, Kvirila) in the western Georgia basin, Bulgaria (Obrochishte and others) in the Varna depression, Kazakhstan (Mangyshlak) in the southern Mangyshlak trough, northern Caucasus (Laba), Turkey (Binkiliç) in the Thracian depression, and a range of minor deposits and occurrences in the Pannonian depression in Hungary and Slovakia; the total resources constitute no fewer than 2 billion tons (Nikopol'skii, 1964; Griznov and Selin, 1959; Chiaturskoe, 1964; Machabeli, 1986; Griaznov, 1960, 1967; Varentsov, 1963; Varentsov et al., 1997, 2004; Strakhov et al., 1967, 1968). During the same epoch were formed the deposits of the northern Urals manganese-ore basin (Yuzhno- and Novo-Berezovo, Ekaterininsk, Marsiat, Yurkinsk, Lozvinsk, Ivdel', Tynia, and others with total resources no fewer than 120 million tons) (Rabinovich, 1971; Kontar' et al., 1999). Their formation occurred, as in the Permian, by means of sedimentary-diagenetic processes, frequently in basins with free hydrogen sulfide in the water column (Fig. 4.10).

A drastic increase of manganese ore-formation within the boundaries of the Eastern Paratethys is dated to a change in the direction of tectonic movements—from the subsidence to the uplift of the bed of the paleobasin, which was accompanied by an ingression of the sea into the peripheral parts of the basin (ie, manganese-ore areas of Ukraine, Georgia, Kazakhstan, and other countries) and a simultaneous upwelling of manganese-rich, deep hydrosulfuric waters. In the shallow-water environments of the shelf, a separation of manganese in the form of hydroxides occurred at the chemocline (Stoliarov, 1993). The deposits are associated with the terrigenous sandstone-clayey facies of the epicontinental basins (Fig. 4.11).

It follows to note that manganese specialization of the Oligocene basin of the Eastern Paratethys evidently persisted up to the Neogene. According to the data of lithogeochemical sampling in the central regions of Russia, substantial reserves of manganese oxide ores can be enclosed in sandstone deposits. Thus, according to the results of the advance geochemical studies on the area of the northern part of the Verchnetsninskii district, G.S. Aver'ianov has identified seven anomalous geochemical fields and three point anomalies (Fig. 4.12) (Mineragenicheskii, 2008). The manganese-bearing deposits of the Sosnovskii suite of the Neogene (N_1^2) (Fig. 4.13) are represented by clays of various colors and facies of general thickness up to 86 m. The highest quantity of manganese is found in the upper part of the suite with a thickness of 4.3 m, in which has been identified a horizon with a thickness of 15 cm full of concretions of manganese oxides (Fig. 4.14). In terms of composition, the Sosnovskii suite is analogous to a sandstone-clayey glauconite of Nikopol' type.

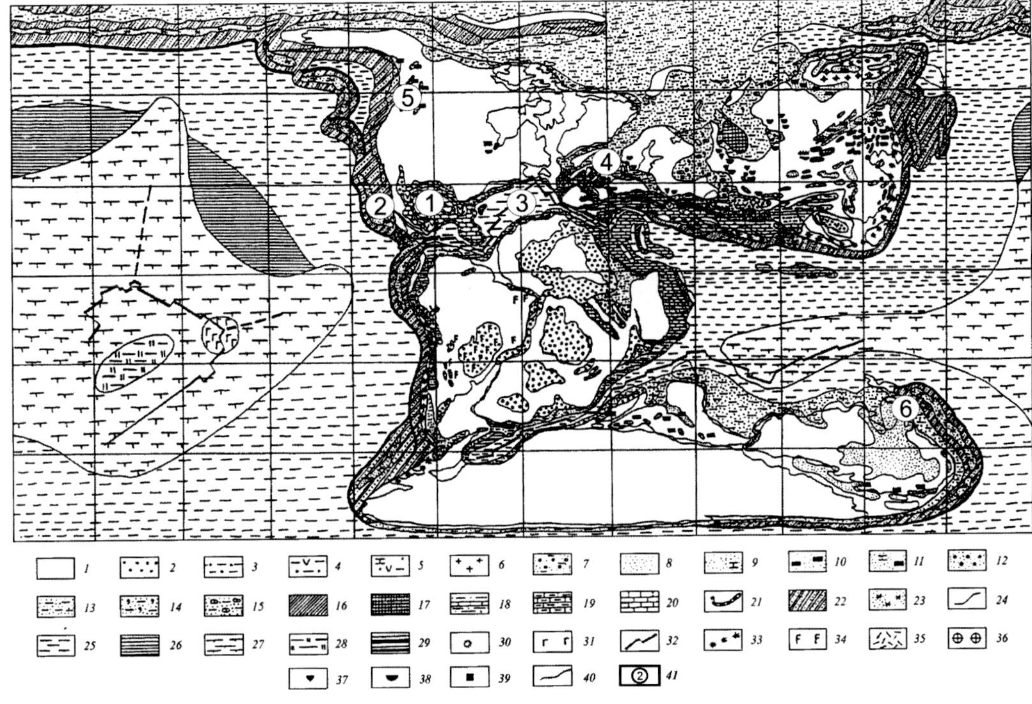

FIG. 4.9

Lithologic-paleogeographic map of the Berriasian century (Cretaceous) (by Climate, 2004). (1) Land; (2–6) deposits of alluvial and proluvial valleys, intermountain valleys, lakes, sabkhas, lagoons in arid zones: (2) red conglomerates, gravelites, sandstones, (3) red-colored and variegated sandstones, siltstones, clays, (4) sandstone, siltstone, clay plaster, (5) gypsum-bearing carbonate and terrigenous-carbonate, (6) salt-bearing; (7–12) deposits of alluvial and lacustrine-marsh plains, intermountain valleys, coastal plains, sometimes poured sea, and lagoons in humid areas: (7) gray-colored conglomerates, gravelites, sandstones, (8) gray-colored sandstone, siltstone, clay, (9) carbonate-terrigenous, (10) coal-terrigenous (inland), (11) carbonate-terrigenous coal-bearing, (12) red-colored terrigenous without carbonates; (13–21) deposits of shelf and epicontinental seas (13) sandstone, siltstone, clay, (14) glauconite-bearing, (15) phosphorite-bearing, (16) turbidites of slopes of the shelf and back-arc basins, (17) carbonaceous clay, clay-carbonate, carbonate-siliceous (black shales), (18) siltstone, clay, limestone, (19) argillaceous limestone, marl, (20) carbonate platform, (21) reefs; (22–25) deposits foot of the slopes of continents, island arcs, and peripheral zones of the oceans: (22) turbidites, (23) calc-alkaline and tholeiitic and terrigenous-volcanogenic complexes of island arcs, (24) deep-sea trenches, (25) hemipelagic clay, carbonate-clay, carbonate; (26–32) deposits of the central regions of the oceans: (26) pelagic clay (red clay), (27) pelagic calcareous and siliceous-carbonate, (28) pelagic siliceous, (29) carbonaceous clay, carbonate-clay, carbonate-chert, siliceous (black shales), (30) carbonate atolls, (31) alkaline and tholeiitic basalts intraplate, (32) mid-ocean ridges manifestations of tholeiitic basalts, (33) calc-alkaline igneous rocks marginal-plates volcano-plutonic associations, (34) intracontinental alkaline and tholeiitic basalts and bimodal association, (35) distal ash, (36) belts of granite massifs in the collision sutures, (37) bauxite, bauxite-bearing deposits, (38) kaolin clays, kaolin-bearing rocks, (39) iron ore, (40) boundary of lithological complexes and paleogeographic areas, (41) areas of manganese accumulation of rocks and ores: (1) Mexico, (2) California, (3) Morocco, (4) Western Europe, (5) Yukon, (6) North-Australia.

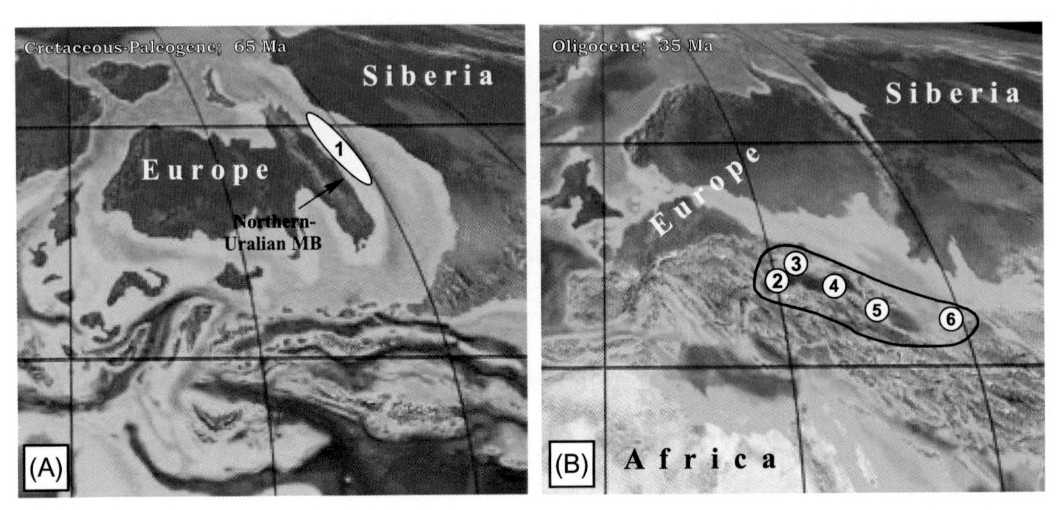

FIG. 4.10

Location of the main manganese-ore basins in Late Mesozoic-Early Cenozoic era in the Euro-Siberian region (configurations of paleocontinents are given on: http://www.jan.ucc.nau/edu). (A) Position of the North Ural manganese basin (1) in the Late Cretaceous-Early Paleogene. (B) Position in manganese ore deposits of manganese basins of the Eastern Paratethys (numerals in circles: (2) Varna, (3) Nikopol, (4) Chiatura and Kvirila, (5) Labinsk, (6) Mangyshlak.

The general predicted resources in the area of the northern part of the Tambov oblast' is estimated at no fewer than 50 million tons.

Modern manganese ore-formation is widely manifested on the floor of the global ocean as well as within the boundaries of the sea and lake water bodies of the continental landmass (marginal and intracontinental, fresh and salt-water). Thus, in certain evaluations, ferromanganese concretions on the bed of the Pacific, Indian, and Atlantic oceans contain approximately 2.5 trillion tons of manganese, which exceeds by 100 times the total reserves of all the calculated deposits on dry land. Moreover, it should be noted that with the formation and growth of new ferromanganese concretions the reserves of these ores increase by approximately 10 million tons annually.

Despite the enormous endowment of manganese to be found in oceanic ferromanganese concretions and crusts, which not infrequently form solid "armors," at present they lack industrial significance in the capacity of manganese-ore raw material. However, in a range of cases they are prospective for an entire complex of valuable components such as Co, Ni, Cu, and Pt.

In a range of intracontinental water bodies also occurs the formation of oxide (manganese and ferromanganese) and carbonate ores (in the Baltic Sea, the Gulf of Riga, Gulf of Finland, and other gulfs, and a range of deep-water depressions; certain lakes of Canada, Karelia, the Baltic Shield, the Urals, and others). These deposits are minor, characterized by non-grade ores and their interest at present is predominantly scientific. In Russia, however, in connection with an acute deficit of manganese-ore raw material, tests are being conducted on the industrial application of Baltic poor ferromanganese ores from the Gulf of Finland.

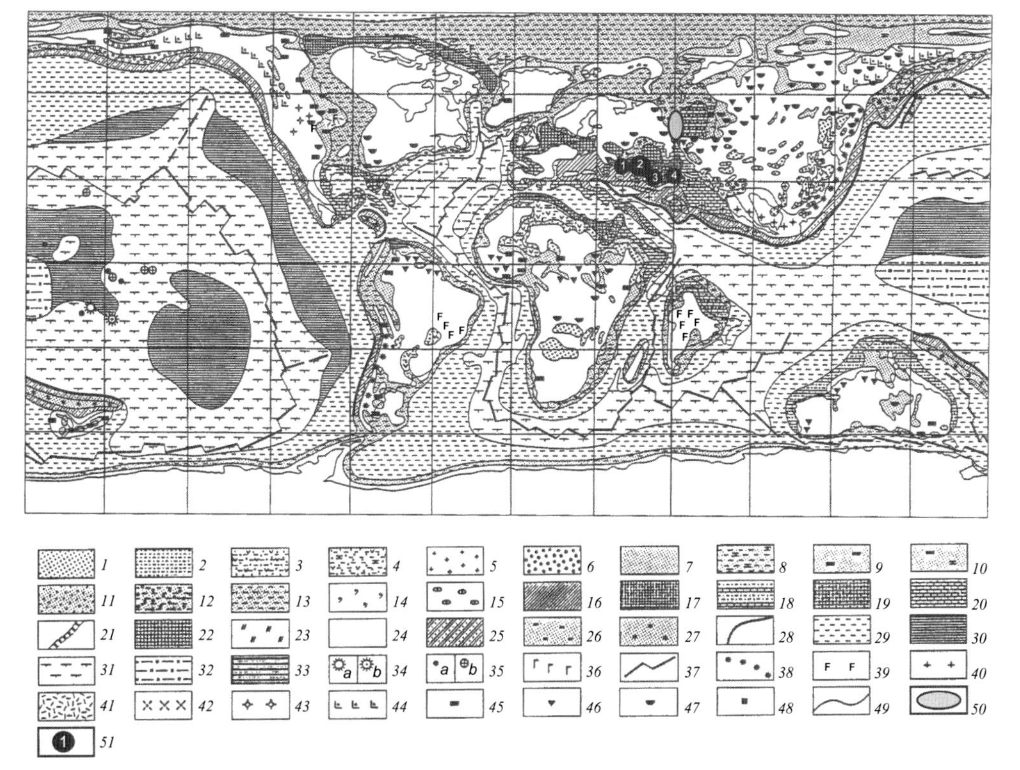

FIG. 4.11

Distribution of manganese-ore basins in the Mesozoic-Cenozoic (paleo outlines of the continents and sedimentation basins is to Maastricht for [Climate, 2004], with additions). (1–5) Deposits of alluvial and proluvial valleys, intermountain valleys, lakes, sabkhas, lagoons in arid zones: (1) red conglomerates, gravelites, sandstones, (2) red-colored and variegated sandstones, siltstones, clays, (3) sandstone, siltstone, anhydrite and gypsum-bearing clays, (4) gypsum-bearing carbonate and terrigenous-carbonate, (5) salt-bearing, (6–11) deposits of alluvial and lacustrine-marsh plains, intermountain valleys, coastal plains, sometimes poured sea, and lagoons in humid areas: (6) gray-colored conglomerates, gravelites, sandstones, (7) gray-colored sandstone, siltstone, clay, (8) carbonate-terrigenous, (9) coal-terrigenous (inland), (10) carbonate-terrigenous coal-bearing, (11) red-colored terrigenous without carbonates; (12–23) deposits of shelf and epicontinental seas, (12) conglomerates, sandstones, siltstones, clays, (13) sandstones, siltstones, clays, (14) glauconite-bearing, (15) phosphorite-bearing, (16) turbidites of slopes of the shelf and back-arc basins, (17) carbonaceous clay, clay-carbonate, carbonate-siliceous (black shales), (18) siltstones, clays, limestones, (19) argillaceous limestones, marl, (20) shallow carbonates (carbonate platform), (21) reefs, (22) chalk, (23) diatomite, (24) land, (25–29) foot deposits of the slopes of continents, island arcs, and peripheral zones of the oceans: (25) turbidites, (26) terrigenous deposits on the elevations of peripheral areas of the oceans, (27) calc-alkaline and tholeiitic and terrigenous-volcanogenic complexes of island arcs, (28) deposits of deep-sea trenches, (29) hemipelagic clay, carbonate-clay, carbonate, (30–37) deposits of the central regions of the oceans: (30) pelagic clay and carbonate-clay (red clay), (31) pelagic calcareous and siliceous-carbonate, (32) pelagic siliceous, (33) carbonaceous clay, carbonate-clay, siliceous-carbonate, siliceous (black shales), (34) intraoceanic islands: carbonate atolls (a), volcanic—alkaline and bimodal association (b), (35) guyots: with phosphate and ferromanganese hardgrounds (a) covered by pelagic carbonate silts (b), (36) alkaline and tholeiitic basalts intraplate, (37) tholeiitic basalts of mid-ocean ridges, (38) calc-alkaline igneous rocks marginal-plates volcano-plutonic associations, (39) intracontinental alkaline and tholeiitic basalts and bimodal association, (40) fading volcanic belts, (41) distal ash, (42) fading plutonic belts, (43) granitoids in orogenic belts, (44) gabbros and basalts in orogenic belts, (45) coaly, (46) bauxite, (47) kaolin ore, (48) iron ores, (49) boundary of lithological complexes and paleogeographic areas, (50) position of Severouralsk manganese basin, (51) most important manganese deposits of the Eastern Paratethys (1) Varna, (2) Nikopol, (3) Chiatura and Kvirila, (4) Mangyshlak.

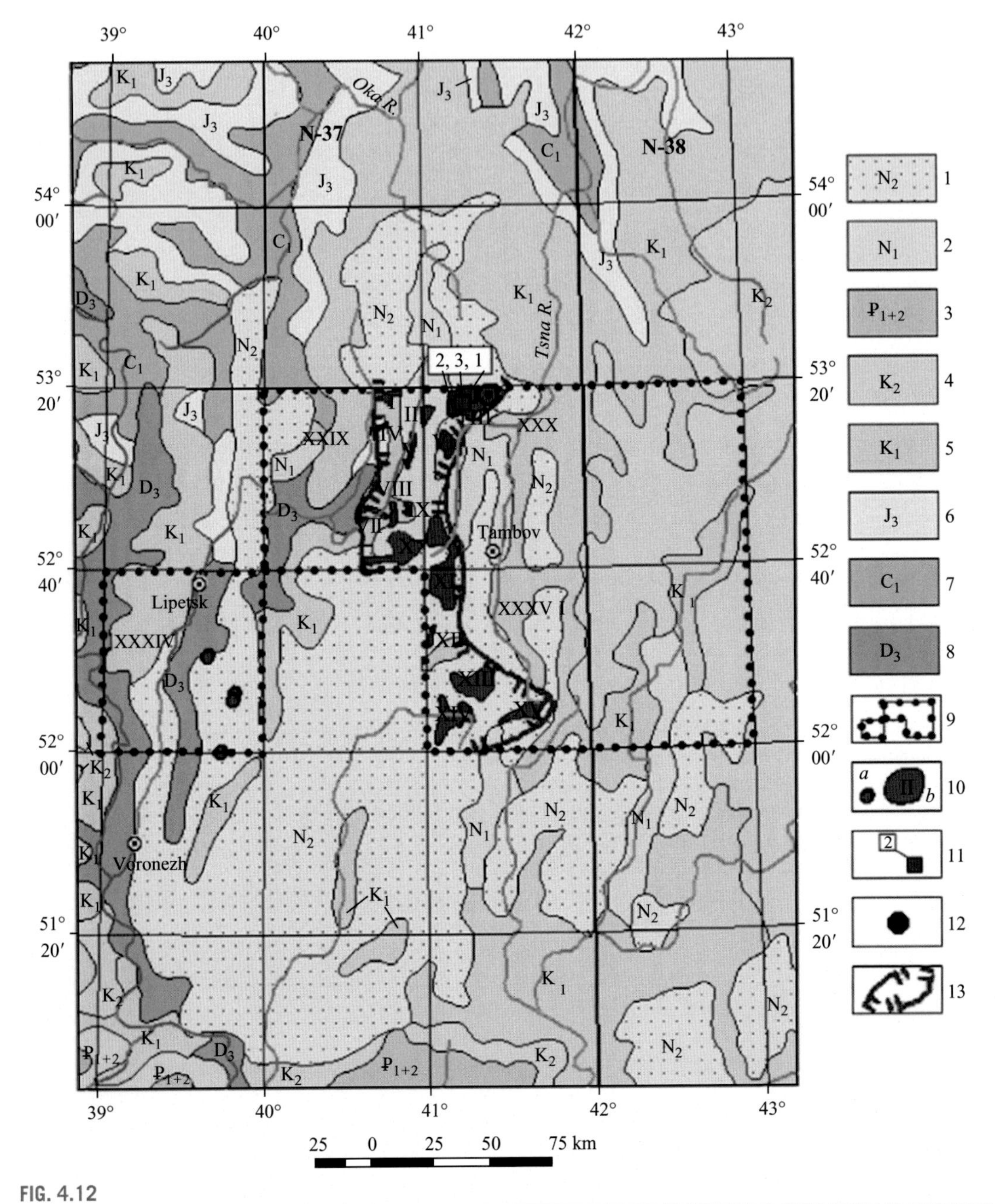

FIG. 4.12

Schematic structure of the Neogene manganiferous basin (Central Federal District) (by G.A. Averyanov). (1–8) Rocks: (1) sand, clay, and silt, (2) sand, glauconite sand, clay (with ferromanganese concretions in some places), and siltstone, (3) sand, clay, and sand with phosphorite inclusions, (4) quartz-glauconite sand, chalk, opoka, and phosphorite, (5) sand, sandstone, phosphorite, and clay, (6) clay, marl, sand, and phosphorite, (7) clay, sandstone, limestone, and coal, (8) dolomite, limestone, clay, and sandstone; (9) area of geochemical research; (10) Mn geochemical anomalies: (a) spot anomalies, (b) areal anomalies and their number; (11) manganese-ore occurrences: (1) Sosnovka, (2) outcrop 2, (3) outcrop 3; (12) boreholes showing clay with ferromanganese concretions; (13) Tambov-Lipetsk region with manganese potential.

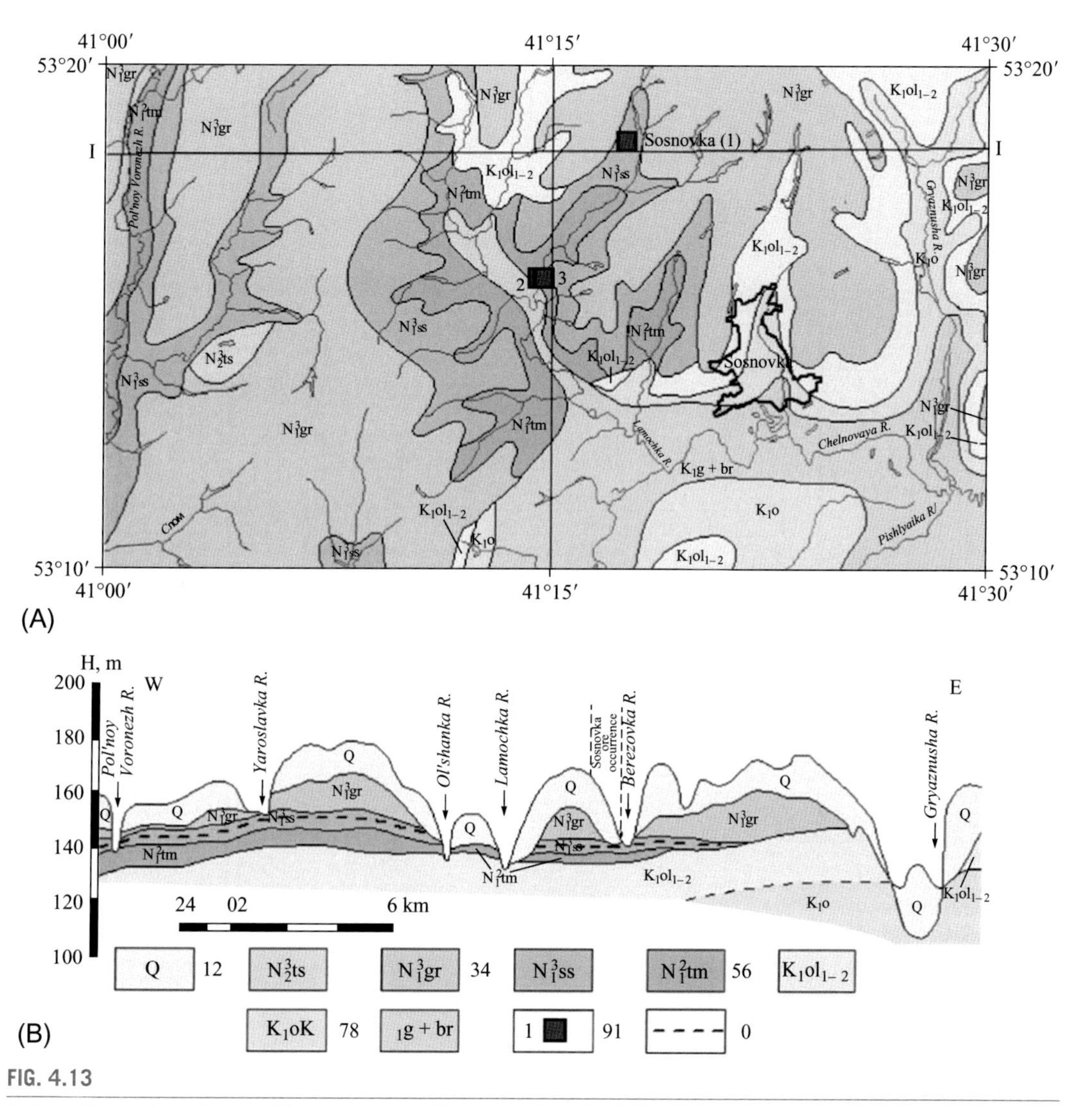

FIG. 4.13

Schematic geological map (A) and cross-section along line I-I′ (B) in the Sosnovka manganese-ore occurrence area (by G.A. Averyanov based on the materials of V.A. Golovko). (1) Quaternary sediments (clay, sand, and boulders); (2) Tikhaya Sosnovka Formation (sand and clay); (3) Gorelkin Formation (sand and clay); (4) Sosnovka Formation (variegated clay with coaly inclusions, ferromanganese grains, and concretions); (5) Tambov Formation (clay, sand, and sandstone); (6) Albian Stage (sand with phosphorite pebbles, clay, and silt); (7) Aptian Stage (sand with clay interbeds); (8) Hauterivian and Birimian stages (silt, clay, and sand); (9) manganese-ore occurrences: (1) Sosnovka, (2) outcrop 2, (3) outcrop 3; (10) inferred clay unit with manganese concretions.

Geological age	Geological column	Layer no.	Layer thickness, m	Total thickness, m	Brief description of rocks and manganese ores
QgIIdn	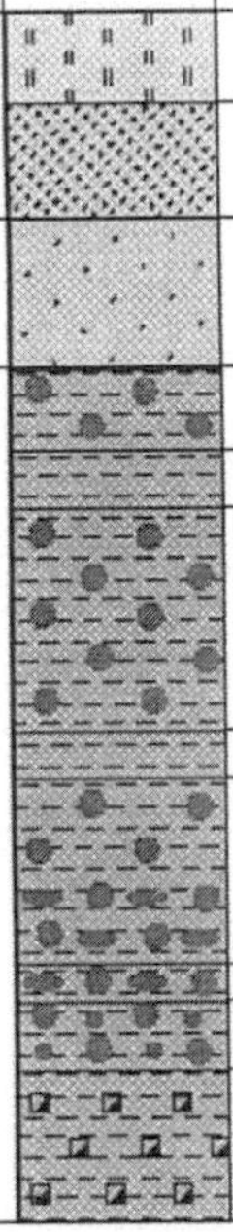	1	0.50	0.50	Soil layer
		2	0.75	1.25	Brown sand with granite boulders and cemented sand
N_1^3		3	0.90	2.15	Cross-wavy quartz sand with clay (locally coal) interlayers
N_1^3ss		4	0.60	2.75	Greenish gray clay with rare manganese oxide concretions (up to 11 cm), manganese hydroxide in microfissures
		5–6	0.25	3.00	Dark gray clay (ferruginated) with quartz grains
		7–9	1.44	4.44	Green clay (horizontal bedded) with frequent manganese oxide concretions
		10	0.20	4.64	Dark gray clay with manganized quartz grains
		11–12	1.20	5.84	Greenish gray clay with manganese ore concretions and spheruites, coaly inclusions, and quartz grains
		13	0.15	5.99	Layer with the highest concentration of manganese oxide concretions (mainly spherical, up to 0.15 m) sometimes merging into a continuous bed (concretions are amorphous with metallic luster in some places)
		14–15	0.46	6.45	Greenish gray clay with abundant small grains and concretions of manganese ores
		16–19	0.90	7.35	Dark gray to green clay with the fine-grained sand and gypsum crystals (up to 5 cm)

QgIIdn 1 N_1^3 2 N_1^3ss 3 4 5 6 7

FIG. 4.14

Sequence of manganese-bearing part of the Sosnovka Formation (by G.A. Averyanov based on the materials of Грищенко (1932)). (1) Quaternary sediments; (2) Gorelkin(?) Formation; (3) Sosnovka Formation; (4) gypsum crystals; (5) manganese concretions; (6) manganese spherulites; (7) manganese mineral grains.

4.5 THE EVOLUTION OF MANGANESE-ORE FORMATION IN THE EARTH'S HISTORY

The study of the geological structure, material composition, and conditions of formation of deposits of manganese ores along with the regularities of their distribution permits the conclusion that manganese accumulation in the rocks of the Earth's upper crust is closely associated with sedimentary, more rarely volcanogenic(hydrothermal)-sedimentary rocks. The scale and form of the burial of manganese in the history of the development of the Earth's continents has varied substantially and is contingent upon the regularities of the evolution of formation conditions, material composition, and subsequent processes of transformation of initial sediments in the ancient basins.

The elucidation of the regularities of the formation of manganese rocks and ores allows us to consider that the primary concentration of manganese reaching the size of deposits occurs in all stages of lithogenesis, beginning from the sedimentary-diagenetic stage (sedimentogenesis and early diagenesis), and continues in later stages of the transformation of sedimentary or volcanogenic-sedimentary rocks (various types of diagenesis and catagenesis, concretion-formation). A strong influence on the concentration of manganese under these conditions is imparted by the total initial bioproductivity of the paleo-water bodies—that is, by the content in the rock of organic matter and the products of its decomposition.

Despite the fact that deposits of manganese of Archean age of significant size remain unknown today, it can nevertheless be supposed that in the Archean period (zone) occurred a wide-scale accumulation of manganese that was buried in scattered form in carbonate and terrigenous-carbonate rocks, more rarely volcanogenic (hydrothermal)-sedimentary rocks. The basins of sedimentation were evidently shallow-water and intracratonic, with intensive development of organic (microbial, prokaryotes) life. The presence of high concentrations of carbon matter in the sediments of Early Precambrian (including Archean) water bodies led to the formation of carbonate and carbonaceous manganese-bearing facies (Gurvich et al., 1982; Veizer, 1988) and engendered reduction conditions in the diagenesis zone, whereby mobile Mn^{2+} was fixed in carbonates and hydroxides.

It can be assumed that from the moment of the occurrence of the first carbonates and carbonaceous matter in crustal rocks of the Archean, the atmosphere and hydrosphere with particularities of exogenic processes and conditions of sedimentation characteristic for that time were already in existence.

It follows to note that the time of the occurrence of the Earth's primary hydrosphere and atmosphere and their temperature have resulted in different evaluations by different authors. The most thorough examination of this question is found in the work of Sorokhtin et al. (1999). He has proposed that when the global ocean arose in the Middle Archean (approximately 3.4 billion years ago), the volume of water in the hydrosphere was small relative to today, and instead of a unitary ocean there existed only discontinuous shallow-water marine basins. The atmosphere of this time consisted predominantly of CO_2 (85–90%) and N_2 (10–15%), and the temperature of the oceanic water was high—up to 75–85°C.

According to Kazakova (1990), in terms of the character of crust-formation, the physical-chemical conditions on the planet's surface, and the specifics of the rock-formation processes, the early history of the Earth's development can be divided into two epochs: (1) the formation and primary consolidation of the crust (Early Archean) and (2) the development of geological processes upon the stabilized crust (Late Archean-Early Proterozoic). The geological boundary between these epochs is seen in the occurrence of the first greenstone belts—that is, approximately 3.5 billion years ago. From this boundary begins the development of exogenic processes and aqueous deposition of sediments.

The most representative information regarding the time of the occurrence of the first hydrosphere—that is, the appearance of aquatic basins of sedimentation, including those with carbonate sediment deposition—consists of isotope data, specifically data on the isotopic composition of sulfur, carbon, and oxygen. These are classified as cyclic elements; the separation of their isotopes occurs during exogenic processes with the active participation of C_{org}: the supply of matter into the basin of sedimentation—the separation (physical-chemical or biological) of isotopes—sedimentation. This leads to the formation of sedimentary rocks with characteristic isotopic signatures.

For carbon the principal cycle of separation of isotopes, as is well known, occurs during the process of dissolution of CO_2 from the atmosphere into the hydrosphere; that is, during the formation of a bicarbonate (HCO_3^-)-ion in the aquatic medium and the deposition of sediment (by chemical or biogenic means) in carbonate form. The difference in isotopic composition between initial carbon dioxide and the carbonate formed at 25°C is approximately 12–13‰ (Friedman and O'Neil, 1977). Available data on the composition of Archean carbonates indicate that beginning from ca. 3.7 to 3.5 Ga the system of the atmosphere-hydrosphere was established an isotopic equilibrium close to that of modern sediments (Nagy et al., 1974; Eichman and Schidlowski, 1975; Sidorenko et al., 1974; Schidlowski et al., 1975; Vinogradov and Kuleshov, 1976; Kuleshov, 1978, 1986a).

Important information on the existence of the early hydrosphere is derived from the isotopic composition of the sulfur of sulfates. The separation of these sulfur isotopes during the process of microbial sulfate-reduction turns out to have been established already by the end of the Archean. Prior to this time, according to V.I. Vinogradov, a "prebiological" stage took place in the Earth's development that was likely characterized by a "null" isotopic composition of sulfur ($\delta^{34}S = 0$); that is, by a composition particular to the juvenile matter of the Earth and to meteorites (Vinogradov, 1973, 1980).

Thus, it can be taken as established that sedimentary rocks deposited in the Precambrian were already well established in facies patterns typical of younger rocks by the time of the Middle to Late Archean transition. Sedimentary basins of this time featuring manganese specialization were shallow-water, marginal, or intracontinental, and characterized by high bioproductivity. However, the concentrations of manganese in the rock were low (up to 10%); specifically manganese ores (15% and higher) were not forming.

There is no doubting that the principal source of manganese (as well as iron) are deep rocks, chiefly those of basic or in some cases of ultrabasic composition. During rifting, rocks of these types emerge at the Earth's surface (as in ophiolites—Cypress and others) or are emplaced at shallow levels in the crust. In addition, heat flow is much higher in rifts, which stimulates circulation of hydrothermal fluids through these rocks. When they reach the surface, they are enriched in manganese and in iron.

The initial contents of manganese in the primary (ie, Archean) rocks (of sedimentary and volcanogenic-sedimentary genesis) were insignificant. Subsequent processes of metamorphism led to the formation of a range of manganese minerals: manganocalcite, rhodochrosite, braunite, hollandite, jacobsite, hausmannite, rhodonite, pyroxmangite, spessartine, etc., as well as manganese-containing minerals of various classes and groups (pyroxenes—members of the diopside-salite-ferrosalite-hedenbergite, diopside-aegerine, and other series; amphiboles—cummingtonite-hübnerite, tremolite-actinolite-ferroactinolite, and other series, joesmithite, kozulite, minerals of the epidote group, olivine, and garnets, as well as a range of sulfides, phosphates, borates, and vanadates).

Thus, an initially low content of manganese not reaching industrial concentrations is a characteristic particularity of manganese rocks of Archean age. Only under hypergenic conditions in the weathering crust on the initial manganese-containing and manganese rocks does there occur a secondary concentration of manganese featuring the formation of industrially viable deposits.

An evaluation today of the scale and correspondingly the quantity of the accumulated manganese in water bodies of the Archean manganese-ore epoch is impossible. Nevertheless, the products of the breakdown of manganese-containing rocks of Archean age in many cases undoubtedly served as a source of manganese in manganese-bearing sediments of the water bodies of subsequent epochs, especially those of the Proterozoic, which constitute the principal manganese reserves of the Earth's upper crust. This serves as evidence of a significant volume of initial manganese-containing rocks of Archean age with high contents (resources) of manganese.

Without question, the most wide scale and significant in terms of manganese resources is the Lower Proterozoic epoch. During this time, within the boundaries of the hypothetical supercontinent

Pangaea-0 (or the early stages of the erstwhile supercontinent Paleopangea, or Megagaea), a vast area has been identified containing within its boundaries manganese-ore basins that within the boundaries of the latest paleocontinent Atlantica occupied the central part of Western Gondwana, the western part of Eastern Gondwana, and the western part of Ur. The largest of these are spatially and in most cases genetically associated with rocks of iron-ore formations—BIF—and their metamorphic analogues—ferruginous quartzites.

In terms of manganese resources contained in rocks of the Lower Proterozoic, the South African basin is clearly distinguished. At present, we see only its eastern part—the Kalahari manganese-ore field. A review of the cross-section (Black Rock deposit) and the presence of a nappe overthrust onto the western part of the ore basin allow us to estimate a wider initial distribution of the basin of sedimentation toward the west compared with its modern position.

In the Lower Proterozoic basins of Australia (Hamersley, Pilbara massif) and Brazil (Mato Grosso state, Brazilian Shield), in association with the formation of ferruginous quartzites, is a range of ferromanganese deposits and ore-occurrences, the richest of which are confined to the weathering crust.

Another type of primary manganese rocks is poor rhodochrosite-containing carbonaceous shales—ampelites. In their weathering crusts were formed medium and large deposits of manganese (Moanda, Gabon; Rio do Navio, Brazil).

Wide distribution among Precambrian rocks is likewise found for manganese-containing metamorphic rocks—gondites, characterized by the presence of spessartine and quartz in the capacity of primary rock-forming minerals, and rhodonite, rhodochrosite, manganese-containing amphiboles, and others in the capacity of secondary rock-forming minerals. These rocks were formed as a result of the metamorphism of the initial manganese-containing carbonate-clayey sediments.

On the whole, basins of sedimentation of the Early Proterozoic with ferromanganese specialization were developed in areas with a tranquil tectonic regime and in shallow-water environments, and were situated in areas of peneplains, as evidenced by the practically total absence of clastic and volcanic material in these rocks. This same regime of sedimentation and character of accumulation of manganese-containing rocks persisted as well for the duration of the subsequent Middle Proterozoic manganese-ore epoch approximately within the boundaries of the distribution of the Early Proterozoic basins of sedimentation.

In the Neoproterozoic, already within the boundaries of the paleocontinent Rodinia, the character of manganese accumulation principally remained unchanged. The principal areas of sedimentation with manganese specialization were inherited from the Early and Middle Proterozoic basins in the central part of Ur (India). Additionally, new areas appeared in which were formed the sediments of future deposits of China, Siberia (Yenisei Ridge, southern Hingan massif, a range of minor deposits of the south of Siberia), and Kazakhstan.

With the breakup of the supercontinent Rodinia, the character of sediment accumulation and evidently the types of basins of sedimentation underwent fundamental change. In the Lower to Middle Paleozoic basins alongside widely distributed carbonaceous aleurite-clayey appeared manganese-containing jaspelite rocks of hydrothermal-sedimentary genesis. Basins of sedimentation in terms of scale and quantity of accumulated manganese (deposits of Kazakhstan, the Urals, China, Western Europe, and elsewhere) are significantly inferior to those of the Proterozoic. This time period lacked a wide distribution of vast shallow-water intracontinental and marginal basins. The geological development of the continents was characterized by high volcanic and tectonic activity. The shape of the paleocontinents (Western and Eastern Gondwana, Laurasia) as well as their relative arrangement underwent rapid change. All of this contributed to the brevity (in the geological sense) of the existence of basins of sedimentation and resulted in the insignificant quantity of manganese accumulated within them.

In the Carboniferous Period, the paleocontinent Pangaea on the whole was already formed. Within its boundaries during the early Permian in the north-east of the present-day Russian Platform was established a vast stagnant basin with a formation of carbonaceous (ie, rich in organic matter) carbonate-terrigenous deposits. Here was accumulated a colossal quantity (a few billion, possibly dozens of billions of tons) of poor-carbonate manganese ores (Novaya Zemlya, a range of ore-occurrences of the Komi Autonomous Republic, Bashkortostan). Evidently, an analogous but smaller basin existed at that time within the borders of China.

In the beginning of the Late Jurassic (during the Callovian) commenced the progressive breakup of the Pangaea supercontinent, which had been in existence for approximately 160 million years (from the end of the Devonian-beginning of the Carboniferous to the Upper Jurassic). Initially, Pangaea was divided into Laurasia and Gondwana; between them the central part of the future Atlantic Ocean began to form. In the Late Jurassic Eastern and Western Gondwana began to break off, and in the Middle Cretaceous (Albian) Western Gondwana was divided into South America and Africa. Approximately at the same time, or a bit earlier, India "split off" from Eastern Gondwana (Klimat, 2004).

To the moment of the breakup of Pangaea can be dated the beginning of the Mesozoic-Cenozoic manganese-ore epoch. Beginning with this stage in the development of the continental landmass, during the Mesozoic a significant quantity of basins of sedimentation was established; many of these featured intensive accumulation of manganese. In this way were formed deposits and ore-occurrences in Australia, Chile, Mexico, Morocco, Russia, Indonesia, Canada, China, Turkey, and many countries of Europe: Hungary, Bulgaria, Italy, Romania, and others (see Fig. 4.1, manganese-ore phase 6). A prominent place here is occupied by the manganese-ore basins of Australia (McArthur and Carpentaria Basins) and north-western Africa.

The final phase of the formation of the large and giant deposits of manganese is the Late Mesozoic-Cenozoic. During this time were formed the largest manganese deposits of the Paratethys province (Bol'she-Tokmak, Nikopol', Chiatura, Mangyshlak, Varna, and others with total resources no fewer than 2 billion tons). The manganese accumulation in the sediments was associated principally with the processes of diagenesis; hydrothermal-sedimentary and volcanic processes played a sharply subordinate role.

In the geological-historical aspect cannot be examined the formation of industrially important deposits of rich oxide manganese ores, such as the deposits of the Pilbara massif, Western Australia (Woodie-Woodie, etc.) as well as evidently a range of deposits of Postmasburg province (Republic of South Africa), the formation of which in many cases is associated with the processes of karst-formation. These processes are widely manifested as well in the Ulutelyak deposit (Bashkortostan, Russia) (Mikhailov, 1993, 2001).

To this same group can be classified a range of deposits that were formed during the postsedimentational stage in the geological life of the sediment. These include the known deposit of carbonate manganese ores in Kuznetskii Alatau—Usinskoe, which was formed during the process of manganese metasomatism in the initial organogenic-detrital limestones (detrital riftogenic facies) (Kuleshov and Bych, 2002), as well as the numerous occurrences of concretional manganese carbonates (eg, the occurrences and minor deposits of Pai-Khoi), formed in the early stages (post-early-diagenetic, catagenic) of lithogenesis (Kuleshov and Beliaiev, 1991).

Modern manganese ore-formation is manifested insignificantly. Besides oceanic ferromanganese concretions, which as a rule are scattered and comprise insignificant specific reserves (by comparison with deposits of the landmass) and occupy a vast field of the deep-water portions of the oceans, in a range of intracontinental water bodies occurs the formation of oxide (manganese and ferromanganese) and carbonate ores (in the Baltic Sea, the Gulf of Riga and other gulfs, and a range of deep-water depressions; the lakes of Karelia, the Scandinavian Peninsula, the Urals, and elsewhere). These deposits at present are of purely scientific interest.

In terms of the study of the processes of formation of rich manganese oxide ores, the zone of hypergenesis in a humid tropical climate environment undoubtedly merits attention.

It follows to note that the formation of manganese rocks and ores on many geological boundaries—for example, the boundary of the Devonian and Carboniferous, the Permian and Triassic, the Cretaceous and the Paleogene, and other stratigraphic levels—is closely connected with global climatic and tectonic restructurings (the breakup of the Gondwana continent, periods of glaciation and aridification) as well as biological events (mass die-offs of organisms) (see Chapter 5). These phenomena determined the quantity of organic matter in the water bodies and the entry of ore matter into the sedimentary basin via hydrothermal or continental runoff. The most prominent of these in the historical-geological regard on the whole are dated to the phases of manganese ore-genesis (Cretaceous-Paleogene, Triassic-Jurassic, Carboniferous-Permian, etc.) (Fig. 4.1).

CHAPTER

5

THE ROLE OF THE BIOSPHERE IN MANGANESE-ORE FORMATION IN THE GEOLOGICAL HISTORY OF THE EARTH

The material presented in this monograph permits the conclusion that the concentration of manganese in rocks of sedimentary and volcanogenic-sedimentary basins for the entire duration of the history of the formation of the Earth's crust is closely connected with the organic matter of the biosphere.

The primary concentration of manganese in sedimentary rocks, on the scale of ore occurrences and deposits, is associated in many cases with carbonates (manganese-containing and manganese). Essential to their formation is the participation of the carbon of oxidized organic matter. The mechanism of the participation of organic carbon in this process has been explicated in great detail in the example of early diagenetic carbonates of the Oligocene manganese-ore deposits of the Eastern Paratethys Basin (Strakhov et al., 1968).

The results of studies of the isotopic composition of carbon in manganese carbonates in known global deposits (Ukraine, Georgia, Kazakhstan, Mexico, Hungary, China, Republic of South Africa, and elsewhere) (Okita et al., 1988; Kuleshov and Beliaiev, 1999; Kuleshov, 2001; Kuleshov, 2003) fully confirm the carbonates' postsedimentational (sedimentary-diagenetic and catagenetic) nature. Noted in practically all Mn-carbonates, including ancient deposits of the land mass as well as sediments of modern water bodies (ocean, sea, and lake), is the participation of oxidized carbon of organic matter (Coleman et al., 1982; Kuleshov, 1999). The variation in the share (or degree) of participation by oxidized C_{org} in their formation is contingent upon several factors, chief among them are the quantity of organic matter in the sediment and the degree of activity of its microbial oxidation processes. In certain cases, the magnitudes of the isotopic composition of the carbon ($\delta^{13}C$) in manganese carbonates are characterized by values close (frequently analogous) to those of carbon C_{org} (−25‰ to −20‰), and in a range of cases by yet lower values particular to the carbon of hydrocarbons of the petroleum series and the carbon of methane (−50‰ to −35‰) (Kuleshov and Brusnitsyn, 2005).

The formation of authigenic manganese carbonates during diagenesis occurs in modern water bodies with the active participation of oxidized carbon of organic matter. The details of this process have been closely studied in examples of marine (Shterenberg, 1971; Shterenberg et al., 1968; Varentsov,

Isotope Geochemistry. http://dx.doi.org/10.1016/B978-0-12-803165-0.00005-7

1975) and lacustrine (Sokolova-Dubinina and Deriugina, 1967a,b; Shterenberg et al., 1966, 1970) sediments. In the isotopic regard, we have studied this process in the examples of the sediments of Lake Punnus-Iarvi (Kuleshov and Shterenberg, 1988) and the Baltic Sea (Kuleshov and Rozanov, 1998).

It follows to note that based on the data on the isotopic composition of carbon, the participation of carbon dioxide of oxidized carbon of organic origin (C_{org}) is the determining factor in the process of the concentration of manganese in authigenic carbonate oxides in modern water bodies as well as in the geological past. Of great significance to this process is the bioproductivity of the paleowater bodies.

Beginning with the deposits of the early Precambrian, the spatial, and in many cases, genetic association of manganese deposits with carbonaceous formations is readily observed. Carbon-containing rocks, including carbonaceous shales, are present in the vertical sequence of rocks of many Phanerozoic manganese deposits. This has enabled a range of researchers to propose models of the formation of manganese ores associated with carbonaceous facies of many basins of sedimentation over the history of the formation of the Earth's crust (Gurvich, 1980; Force and Cannon, 1988; Polgari et al., 2009).

Manganese carbonates in deposits can be enclosed in various forms, from concretional formations and secondary cementation to products of metasomatism in the substratum of the initial rocks, including various organic remains and mineralized cyanobacterial mats (biosomatism). Microbial structures in manganese ores enclosed in the sedimentary and volcanogenic-sedimentary rock masses are noted practically everywhere (conclusively beginning from the Early Proterozoic) (Issledovanie, 2012). The processes of biosomatism contributed to the formation of many known manganese deposits, evidently including the ores of Early Proterozoic supergiant deposits such as the Kalahari manganese-ore field (Republic of South Africa) (Kuleshov, 2012).

The formation of the most industrially valuable deposits of oxide-manganese ores occurs predominantly in the zone of hypergenesis, due to the breakdown of primary manganese and manganese-containing rocks of various genesis—primary sedimentary-diagenetic rocks and those of subsequent processes of transformation (late diagenesis, catagenesis), as well as secondary rocks in varying degree metamorphosed or metamorphic. These processes occur in the zone of hypergenesis with the active participation of protozoans (bacteria, algae, etc.) and are most intensively and widely manifested in areas with humid climate, particularly in subequatorial regions.

The Earth's exogenic system is driven mostly by the energy from the Sun. This energy is to a large extent filtered through the activities of animals, plants, and especially of microorganisms. In this way, the role of the biosphere as a system is a determining factor in the genesis of many types of ore deposits, and this is especially true for manganese. At surface temperatures, many key chemical reactions of manganese are impossibly slow and bacterial catalysis dominates manganese behavior. This is manifest at different stages of the formation of manganese rocks and ores: (a) in the participation in the composition of manganese carbonates by oxidized carbon of organic matter—that is, by CO_2 formed during processes of microbial oxidation of C_{org} (sulfate-reduction and oxidation by the oxygen of manganese oxides); (b) in the mineralization by manganese carbonates of organogenic (microbial, etc.) remains; and (c) in processes of microbial oxidation of manganese rocks and ores in the zone of hypergenesis.

The accumulation of manganese in the sedimentary and volcanogenic-sedimentary rocks in the history of the formation of the Earth's lithosphere, as shown above (see Fig. 4.1), occurred unevenly. Distinct periods of manganese metallogenesis coincide in many cases with biotic events. It can be accepted as established that sedimentary rocks from the earlier Precambrian sequences appeared already on the boundary of the Middle and Late Archean. The paleobasins of that time with manganese specialization were shallow water, marginal, or intracontinental, characterized by high bioproductivity. The earliest prokaryotes (Fig. 5.1) as well as stromatolite buildups (Fig. 5.2) are known already from the Early Archean. Initial contents of manganese in the primary rocks of Archean age were insignificant (up to 10%) but did not reach industrial concentrations. Subsequent processes of metamorphism and metasomatism (with the participation of carbon of oxidized organic matter) led to the formation of manganese and manganese-containing minerals and rocks. However, only in the weathering crust in hypergenic conditions with active participation by microorganisms is subsequent (secondary, tertiary, etc.) concentration of manganese found to occur on the scale of deposits with formations of high-quality oxide-manganese ores (Fig. 5.2).

Most wide-scale and significant in manganese resources is unquestionably the Early-Middle Proterozoic epoch of manganese accumulation, above all phase 1 (see Fig. 4.1). At this time existed the most significant area in terms of metallogenic potential of manganese—the South African basin (Kalahari), where today resources constitute no less than 5–6 billion tons. Manganese ores here are represented by braunite lutites evidently with an initial carbonate composition; these represent the product of biosomatism in the primary cyanobacterial mats (see Fig. 3.98).

A widespread type of nonoxidized manganese rocks of this age, besides gondites and quartz-hematite banded shales, are poor rhodochrosite-containing carbonaceous shales—ampelites. In their weathering crust form medium and large deposits of manganese (Moanda, Gabon; Rio do Navio, Brazil; and elsewhere).

Thus, the basins of sedimentation in the Proterozoic on the whole were evidently shallow water and characterized by abundant organic (microbial) life.

In the Lower-Middle Paleozoic basins, alongside carbonaceous silty-clayey rocks, wide distribution is found for manganese-containing jaspelite rocks of hydrothermal-sedimentary genesis. The basins of sedimentation in terms of scale and quantity of accumulated manganese (deposits of Kazakhstan, Urals, China, western Europe, and elsewhere) are significantly inferior to those of the Proterozoic. The Earth's geological development was characterized by intensive volcanic and tectonic activity. The outlines of the paleocontinents (Western and Eastern Gondwana, Laurasia) and their position relative to each other underwent relatively rapid change. All of this led to the brief (in the geological sense) duration of existence for the basins of sedimentation and, as a result, to the insignificant accumulation of manganese within them.

In this metallogenic period were formed detrital Riphean facies of deposits of Jiaodingshan and Dawashan (Sichuan Province, China) and Usa (Kuznetskii Alatau), in which were developed processes of the metasomatic replacement of the initial rocks by manganese carbonate with the participation of the carbon of oxidized organic matter (see Figs. 3.122 and 3.130).

In the Carboniferous the paleocontinent Pangaea was on the whole already formed. Within its boundaries in the Early Permian, in the north-east of the West European platform was formed a vast, stagnant basin with an accumulation of carbonaceous, carbonate-terrigenous deposits rich in organic matter. Here was accumulated a colossal quantity (a few billion to dozens of billions of tons) of poor carbonate manganese ores Novaya Zemlya, a range of ore occurrences of the Komi Autonomous Republic, (Bashkortostan). In formation of Mn-carbonates participated carbon dioxide, formed as a result of the oxidation of the carbon of organic matter. Evidently, an analogous but smaller basin existed at this time within the borders of China.

In the beginning of the Upper Jurassic began the progressive breakup of the supercontinent Pangaea. With its breakup is evidently associated the beginning of the Mesozoic-Cenozoic manganese-ore epoch.

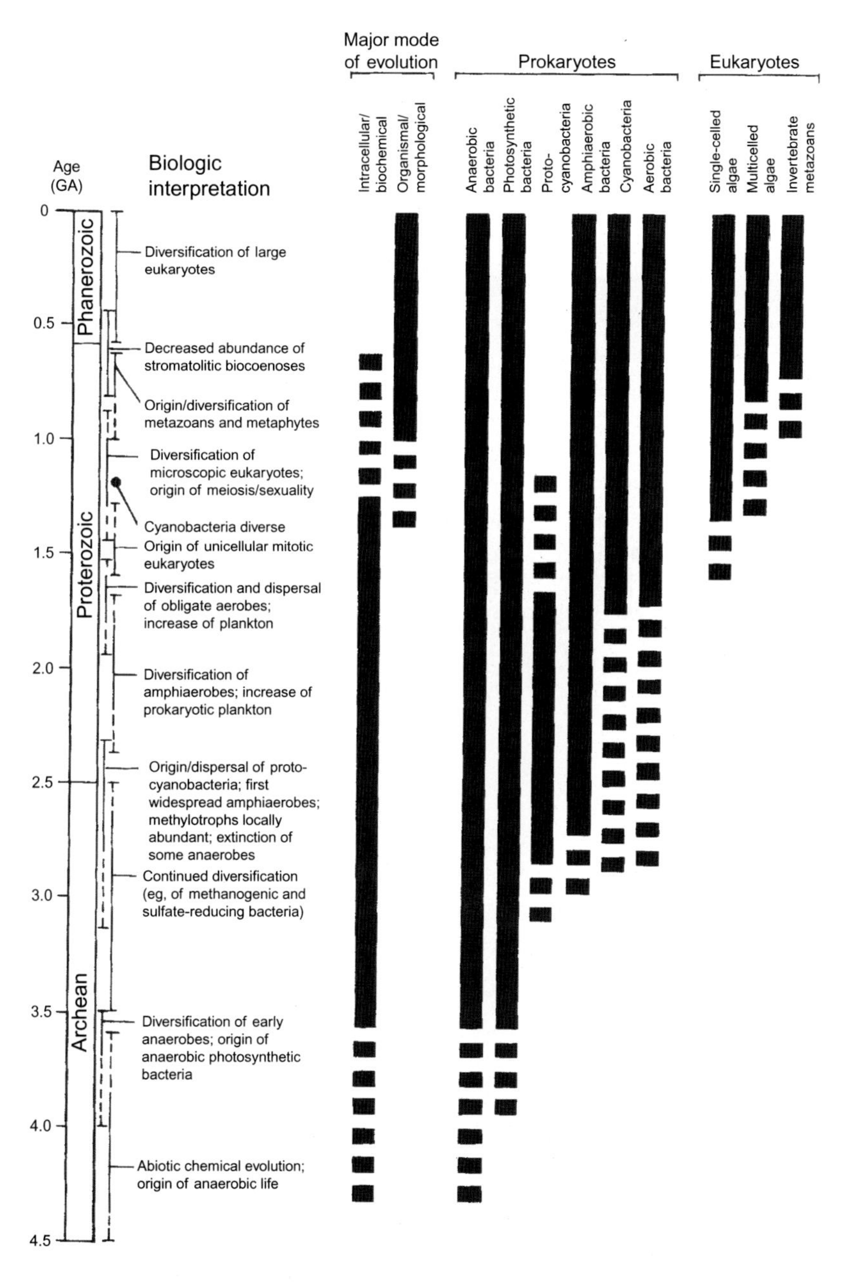

FIG. 5.1

Evolution of protozoo in the Earth history (Earth's..., 1983).

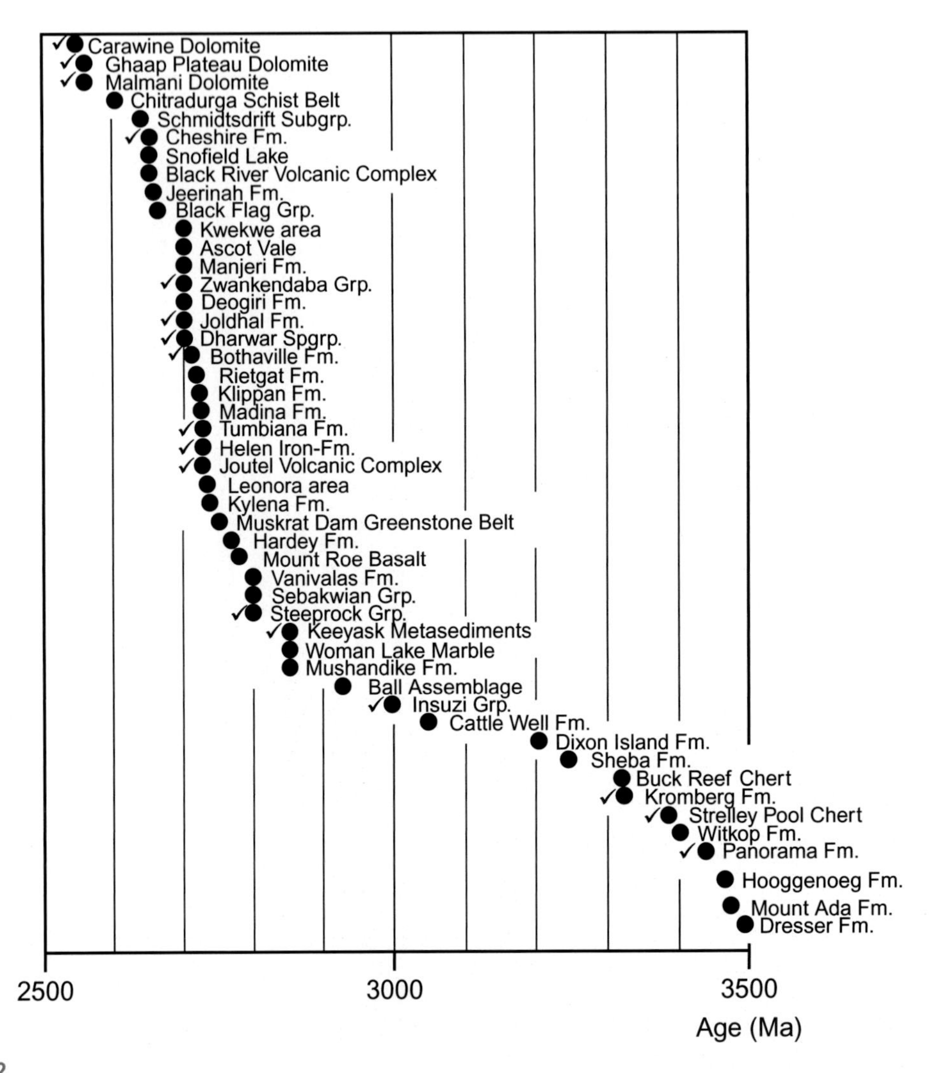

FIG. 5.2

Archean geological units containing stromatolites (Schopf, 2006).

Beginning from this stage in the development of the continental landmass was formed a significant quantity of basins of sedimentation with the accumulation of carbonaceous formations; many of these featured manganese accumulated (deposits and ore occurrences of Australia, Chile, Mexico, Morocco, Russia, Indonesia, Canada, China, Turkey, and many countries of Europe—Hungary, Bulgaria, Italy, Romania, and others) (Fig. 4.1, manganese-ore phase 6). Of note are the manganese-ore basins of Australia (McArthur and Carpentaria Basins) and of north-western Africa. Many of these deposits are associated with carbonaceous formations.

The final phase (Fig. 4.1, phase 7) of accumulation of large and giant deposits of manganese is the Late Mesozoic-Cenozoic, in which were formed the largest manganese-ore provinces of Paratethys

(Bolshoy-Tokmak, Nikopol, Chiatura, Mangyshlak, Obrochishte, etc. with general resources no less than 2 billion tons). The formation of manganese concentrations in the sediments is connected principally with the processes of early diagenesis featuring active participation of oxidized organic carbon. Hydrothermal-sedimentary and volcanic processes in the accumulation of manganiferous sediments played a sharply subordinate role.

In a range of modern marginal and intracontinental seas and lakes occur the formation of oxide (manganese and ferromanganese) and carbonate rocks and ores (in the Baltic Sea, a range of deep-water depressions, the Gulf of Riga, and other gulfs; the lakes of Karelia, the Scandinavian Peninsula, the Urals, and elsewhere). These manganese formations likewise form with the active participation of microorganisms and microbially oxidized organic carbon.

It follows to note an important circumstance: the fact that the accumulation of manganese rocks and ores on many geological boundaries, for example those of the Devonian and Carboniferous, Permian and Triassic, Cretaceous and Paleogene, and other stratigraphic levels, is closely connected with global climatic and tectonic restructurings (the breakup of the Gondwana continent, periods of glaciation and aridification), and biotic events (mass extinction of organisms, Alekseev, 1989, 1998). These phenomena determined the quantity of organic matter in the paleowater bodies, as well as the entry of ore matter into the sedimentary basin with hydrothermal or continental runoff. The most significant of these events in the historical-geological regard coincide on the whole with the identified phases of manganese ore-genesis in the history of the Earth's development (see Fig. 4.1).

Conclusion

In our opinion, the studied examples of the formation of manganese-ore deposits of various genetic types can serve as models in the interpretation of geological data for other, already known deposits and in surveying for new manganese-ore objects.

Primary manganese ores from the overwhelming majority of global deposits are represented by carbonates and oxides (more rarely, hydroxides) of manganese, which were formed in sedimentary-rock basins and are closely connected paragenetically (in many cases also genetically) with sedimentary and volcanogenic-sedimentary rocks. The mineral form of deposited compounds of manganese is determined by the prevailing physical-chemical conditions (Eh, pH, etc.) of various zones of the basin of sedimentation or of the stages of lithification of the initial sediment (lithogenesis).

Manganese oxides are formed in a medium with high oxidation potential (Eh) predominantly in the shore zones of shallow-water or on the bottom of the basins with submarine discharge of hydrothermals, whereas manganese carbonates are formed principally in reductive conditions within the sediment itself during various stages of lithogenesis.

In nature, specifically sedimentary manganese carbonates—that is, those that would have precipitated from the mass of lake or marine water—do not exist. All manganese carbonates are authigenic: their formation occurs within the sediment in conditions of early diagenesis as a result of the redistribution of manganese already contained in the sediment itself (in the form of oxides) or later during catagenesis as a result of the supply into the already lithified sediment by new portions of manganese.

A general regularity of the formation of manganese carbonates under conditions of early diagenesis is the necessary participation in this process by microbially oxidized carbon of organic matter; the formation of manganese carbonates of post-early-diagenetic stages (phases) of lithogenesis features participation by isotopically light carbon, which was formed during the metamorphism (destruction) of organic matter under conditions of the submersion of sedimentary strata into the depth (including during various phases of oil and gas generation).

The concentration of manganese up to the size of deposits occurs primarily in exogenous conditions of the Earth's crust—on the borders of the division between the atmosphere and the rock (the zone of hypergenesis) and in the sedimentary-diagenetic (early diagenesis) stages, and to a lesser degree in subsequent stages of lithification of the initial sediment and the transformation of sedimentary or volcanogenic-sedimentary rock (diagenesis of submersion, concretion formation). An important factor in the concentration of manganese is contributed by the presence of organic matter in the sediment (rock), resulting from the bioproductivity of the paleo-water body. Oxidized organic matter (carbon dioxide) facilitates the concentration of manganese up to ore concentrations.

The distribution of deposits of manganese in time (geological) and space (within the boundaries of the paleo-continents and the modern continents) can be taken as a proof that manganese accumulation occurred in sedimentary-rock basins with various regimes of sedimentation and in various paleotectonic and paleoclimatic environments, coincided with sequences of sedimentary (terrigenous-, chemogenic-, and biogenic-) and volcanogenic-sedimentary rocks, and formed during various stages of sediment and lithogenesis.

During the early stages of the establishment of the Earth's crust (in the Archean Eon), manganese was accumulated in rocks predominantly in scattered form; significant concentrations of manganese in rocks of this age have not been discovered. Industrially viable accumulations of manganese ores are represented by weathering crusts on manganese-containing rocks and poor manganese ores (metamorphosed).

This applies likewise to deposits of the Proterozoic manganese-ore period, when principally manganese-containing rocks (carbonaceous clayey, siliceous-ferruginous shales, and the like) were accumulated; these served as parent rocks for many large deposits of weathering crusts. The source of manganese was predominantly the products of the destruction of Archean rocks.

Standing apart in this set is the giant deposit of the Kalahari manganese-ore field (Republic of South Africa), the genesis of which has not been explained in its entirety. The available data allow us to propose its post-sedimentational (biometasomatic) origin.

In the Phanerozoic manganese period, the conditions of manganese accumulation changed substantially. Forming in shallow-water shore zones of many water bodies were deposits of primary oxide-manganese, and in sediments of more deepwater districts of the basin deposits primary carbonate manganese ores were formed. In districts with active manifestations of volcanogenic and hydrothermal (subaqueous) activity, predominantly ferromanganese deposits were formed in many basins.

The most industrially valuable manganese deposits today are represented by deposits of ancient and modern weathering crusts (the zone of hypergenesis), above all those of the near-equatorial districts of the countries of South America and Africa, as well as Australia and India. For this reason, these districts are unarguably the most prospective for the presence of new deposits of quality ores (eg, the Brazilian and Guiana shields of South America, the Precambrian rocks of western central Africa, and elsewhere).

Manganese "caps" of weathering crusts are developed also in many deposits in regions with a humid climate (Urals, Siberia, etc.); however, in terms of the scale of development of zones with rich contents of manganese, these are substantially inferior to those of the near-equatorial deposits.

Significant resources of manganese are contained in ferromanganese concretions of modern oceanic sediments. Their extraction under current circumstances remains nonviable.

Glossary of Terms Less Familiar to an English Audience

Definitions based on *Glossary and Index of Geology*, American Geological Institute, Washington, DC, 1972

Ampelite carbonaceous shale or black shale. This usage was at one time common in French language publications
Catagenesis (or katagenesis) later diagenetic alteration of a sedimentary rock
Elision a disturbance in the continuity of a sedimentary section; **Elisional Basin**—one with rapid clastic deposition that outruns normal compactional dewatering. Combined with clay dehydration, this compactional disequilibrium produces overpressuring. The eventual release of these deep waters along faults or sand aquifers causes migration of **Elisional Fluids** which then may produce **Elisional Catagenesis.** (These concepts were developed by V.N. Kholodov. See for example Kholodov, V.N., 2013. Elisional processes and salt tectonics: communication 1. Catagenetic transformations in saliferous sequences. Lithol. Miner. Resour. 48 (4), 267–284. Kholodov, V.N., 2012. Elision systems of the Dnieper-Donets aulacogen: communication 2. Catagenetic processes in the Donets and Pripyat depressions and some metallogenic features of the aulacogen. Lithol. Miner. Resour. 47 (1), 48–69.)
Gryphon a cone-shaped structure, as from a mud
Hypergenesis in Russian usage, surficial alteration of sedimentary rocks; when applied to mineral deposits similar to **Supergene**.
Hypogene said of a mineral deposit formed by ascending solutions; plutonic
Jaspillite banded ferruginous chert with more than 25% Fe. Compare **jasperoid**, a silicified limestone or dolostone
Ocellar igneous texture in which the phenocrysts are made up of aggregates of smaller crystals surrounding a larger euhedral crystal
Oligonite manganese-rich siderite
Opoka (Опока) lacks an exact English equivalent that we are aware of. Fine-grained rock composed dominantly of opal. "Opal claystone" would be a possible translation, but implies more clay than is generally present. **Diatomite** and its weathered equivalent **Tripoli** would be subcategories of Opoka.
Phthanite a silicified shale (typically 95–98%—quartz + chalcedony, up to 5%—organic matter)
Phytogenetic a rock produced directly or indirectly by the actions of plants, especially algae
Proluvial said of sediment dumped at the base of a slope
Stadial in English refers to a deposit formed during a glacial **stade**. In Russian usage, following Strakhov, **Stadial analysis** is used to refer to the identification of later (**Catagenic**) events in order to reconstruct the original sedimentogenic features of the rock. (eg, Yapaskurt, O.V., 2008. The stadial analysis of sedimentary process. Lithol. Miner. Resour. 43 (4), 326–337.) Somewhat analogous to **Geohistory Analysis** in the petroleum industry.
Syneclise used in the Russian literature to describe a broad depression in the continental platform. Dimensions are a few hundred to a thousand kilometers in diameter and 3–5 km in thickness, similar to a **Sag Basin** in American usage.
Taxitic said of a volcanic rock that has the texture and appearance of a sedimentary rock.

REFERENCES

Авалиани Г.А. Марганцевые месторождения Грузии (геология, минералогия, генезис). М.: Наука, 1982. 170 с.

Авалиани Г.А., Штеренберг Л.Е., Долидзе Д.П. и др. Некоторые особенности формирования Чиатурского месторождения. Тр. Груз. политехнич. ин-та. 1965. № 4. С. 111–122.

Алексеев А.С. Глобальные биотические кризисы и массовые вымирания в фанерозойской истории Земли. Биотические события на основных рубежах фанерозоя. М.: Московский государственный университет, 1989. С. 22–47.

Алексеев А.С. Массовые вымирания в фанерозое. Дисс. в форме научн. доклада на соиск. ученой степени доктора. геол.-мин. наук. М.: МГУ. 1998. 76 с.

Алексиев Б. Пирокластични седиментни скали с олигоценска възраст от Варненско. Изв. Геол. Бълг. АН. 1959. Кн. 7. С. 101–117.

Андрусов Н.И. Избранные труды. М.: Изд-во АН СССР, 1961. Т. 1. 687 с.

Атлас литолого-палеогеографических карт СССР. М.: Недра, 1967. Т. IV.

Басков Е.А. Значение и основные задачи палеогидрогеологического анализа при металлогенических построениях. Методика палеогидрогеологических исследований. Ашхабад, 1970. С. 31–37.

Басков Е.А., Пустовалова Г.И. Металлоносность термальных вод Кавказа. Металлоносность вод складчатых областей СССР. Л.: Труды ВСЕГЕИ, 1979. Т. 260. С. 30–46.

Беленькая И.Ю. Влияние углеводородных газов на аутигенное минераллобразование в осадках холодных сипов. Вестник МГУ. Серия: Геология. 2003. № 3. С. 15–21.

Беляев А.А., Семенов Г.Ф., Суханов Н.В. Изотопный состав кислорода и углерода карбонатных пород сланцевой зоны Пай-Хоя. Геология и полезные ископаемые Европейского Северо-Востока СССР. Тр. Института геологии Коми НЦ УрО АН СССР. 1983. Вып. 44. Сыктывкар: 1983. С. 53–54.

Беляев А.А., Кулешов В.Н. Изотопный состав и происхождение карбонатных марганцевых руд Карской зоны Пай-Хоя. Литогенез и геохимия осадочных формаций Тимано-Уральского региона. Тр. Ин-та геологии Коми НЦ УрО РАН. Вып. 84. Сыктывкар, 1994. С. 71–84.

Бердичевская М.Е., Рахманов В.П. О перспективах выявления марганцевых руд на полуострове Мангышлак. Полезные ископаемые в осадочных толщах. М.: Наука, 1981. С. 157–168.

Беляев А.А., Юдович Э.Я. Конкреционные комплексы палеозойских отложений Лемвинской фациально-структурной зоны на Урале и Пай-Хое. Конкреции и конкреционный анализ нефтегазоносных формаций. (Тезисы.докл. Всесоюзн. научн. конференции 11–13 окт. 1983г.). Тюмень: Тюменская правда, 1983. С.30-31.

Бетехтин А.Г. Промышленные марганцевые руды СССР. М.: АН СССР, 1946. 315 с.

Билибин Ю.А. Металлогенические провинции и металлогенические эпохи. М.: Госгеолтехиздат, 1955. 86 с.

Билибина Т.В., Кратц К.О., Лаверов Н.П. и др. Металлогения докембрия и металлогенические эпохи. Проблемы металлогении докембрия. Л.: 1978. С. 3–29.

Блажчишин А.И. Минеральный состав донных осадков. Геология Балтийского моря. Вильнюс. Мокслас, 1976. С. 221–254.

Богданова С.В. Пиаревский С.А. Ли Ч.Х. Образование и распад Родинии (по результатам МПГК 440). Стратиграфия. Геологическая корреляция. 2009. Т. 17. № 3. С. 29–45.

Борщевский Ю.А., Борисова С.Л., Попова Н.К. Новый метод выделения кислорода и углерода из карбонатов и карбонатно-силикатных пород для изотопного анализа. Тез. докл. V Всесоюзн. симпоз. по стаб. изотопам в геохимии. М.: ГЕОХИ, 1974. С. 207–209.

Брусницын А.И., Старикова Е.В., Жуков И.Г. Марганцевое месторождение Кызыл-Таш (Южный Урал, Россия): девонский прототип низкотемпературных гидротермальных построек современного океана. Геология рудных месторождений. 2000. Т. 42. № 3. С. 231–247.

Брусницын А.И., Жуков И.Г. Южно-Файзулинское марганцевое месторождение (Южный Урал): геологическое строение, петрография, процессы формирования. Литология и полезные ископаемые. 2005. № 1. С. 35–55.

Брусницын А.И., Жуков И.Г., Кулешов В.Н. Биккуловское марганцевое месторождение (Южный Урал): геологическое строение, состав металлоносных отложений, модель формирования. Литология и полезные ископаемые. 2009. № 6. С. 613–636.

Брусницын А.И., Жуков И.Г. Марганценосные отложения магнитогорского палеовулканического пояса (Южный Урал): строение залежей, состав, генезис. Литосфера. 2010. № 2. С. 77–90.

Брусницын А.И., Кулешов В.Н. Геохимия рудоносных отложений Парнокского железо-марганцевого месторождения (Полярный Урал). Металлогения древних и современных океанов – 2011. Рудоносность осадочно-вулканогенных и гипербазитовых комплексов. Миасс: ИМин УрО РАН, 2011. С. 97–104.

Буадзе В.И. Рудоносные системы Кавказа и проблемы его металлогении. Геология, прогнозирование и технологическая оценка месторождений полезных ископаемых Кавказа. М.: Недра, 1991. С. 5–39.

Булейшвили Д.А. Оценка перспектив нефтеносности третичных отложений восточной Грузии. Геология нефти. 1957. № 10. С. 8–16.

Булейшвили Д.А. Глубинные недра межгорной впадины Грузии как основной резерв запасов нефти и газа. Труды ВНИГНИ. 1972. Вып. 120. С. 211–229.

Бутузова Г.Ю. К познанию цеолитов гейландитовой группы. Цеолит из палеогеновых отложений юга СССР. Литология и полезные ископаемые. 1964. № 4. С. 66–79.

Быч А.Ф., Миртов Ю.В., Тарасова С.М. Марганцевое оруденение в рифейских отлождениях Кузнецкого Алатау. Новые данные по геологии и полезным ископаемым Западной Сибири. Томский гос. ун-т. Томск: 1975. Вып.10. С. 18–22.

Варенцов И.М. Некоторые вопросы геохимии Усинского марганцевого месторождения (Кузнецкий Алатау). ДАН СССР. 1961. Т. 138. № 5. С. 1175–1178.

Варенцов И.М. О геохимии Усинского марганцевого месторождения в Кузнецком Алатау. Осадочные руды железа и марганца. Тр. ГИН АН СССР. Вып. 70. М.: АН СССР, 1962а. С. 28–64.

Варенцов И.М. О главнейших марганценосных формациях. Осадочные руды железа и марганца. Тр. ГИН АН СССР. М.: АН СССР, 1962б. Вып. 70. С. 119–173.

Варенцов И.М. К геохимии олигоцена Южно-Украинского марганцеворудного бассейна. Тр. ГИН АН СССР. 1963. Вып. 97. С. 72–164.

Варенцов И.М. К познанию условий образования Никопольского и других месторождений Южно-Украинского марганцево-рудного бассейна. Литология и полезные ископаемые. 1964. № 1. С. 25–39.

Варенцов И.М. Геохимические аспекты формирования железо-марганцевых руд в современных шельфовых морях. Проблемы литологии и геохимии осадочных пород и руд. М.: Наука, 1975. С. 150–165.

Варенцов И.М. Месторождение марганца Грут Эйландт, Австралия. Геология и геохимия марганца. М.: Наука, 1982. С. 66–83.

Варенцов И.М., Рахманов В.П. Месторождения марганца. Рудные месторождения СССР. М.: Недра, 1974. Т. 1. С. 109–167.

Варенцов И.М., Базилевская Е.С., Белова И.В. и др. Особенности распределения Ni, Co, Cu, V, Cr в рудах и вмещающих отложениях Южно-Украинского марганцево-рудного бассейна. Марганцевые месторождения СССР. М.: Наука, 1967. С. 179–198.

Варенцов И.М., Блажчишин А.И., Соколова Г.В. Региональные вариации минерального состава железомарганцевых конкреций, корок Балтийского моря. Конкреции и конкреционный анализ. Харьков, Харьковский гос. ун-т, 1973. С. 157–159.

Варенцов И.М., Блажчишин А.И., Соколова Г.В. Региональные вариации минерального состава железомарганцевых конкреций и корок, Балтийское море. Конкреции и конкреционный анализ. М.: Наука, 1977. С. 180–187.

Варенцов И.М., Головин Д.И. Марганцевое месторождение Грут-Эйланд, Северная Австралия: K-Ar-возраст криптомелановых руд и аспекты генезиса. ДАН СССР. 1987. Т. 294. № 1. С. 203–207.

Варенцов И.М., Вейнмарн А.Б., Рожнов А.А. и др. Геохимическая модель формирования марганцевых руд фаменского рифтогенного бассейна Казахстана (главные компоненты, редкие земли, рассеянные элементы). Литология и полезные ископаемые. 1993. №3. С. 56–79.

Варенцов И.М., Столяров А.С., Ивлева Е.И. и др. К геохимической модели формирования раннеолигоценовых Mn-руд Восточного Паратетиса: Никопольское и другие месторождения Южно-Украинского бассейна. Геология рудных месторождений. 1997. Т. 39. № 1. С. 49–67.

Варенцов И.М., Музылев Н.Г., Николаев В.Г. и др. Процессы формирования крупнейших марганцеворудных накоплений в олигоценовых бассейнах Паратетиса. Статья 1. Стратиграфия и тектоническая позиция. Бюлл. МОИП. Отд. Геол. 2004. Т. 79. Вып. 6. С. 28–38.

Васильев Е.К., Васильева Н.П. Рентгенографический определитель минералов. Новосибирск: Наука, 1980. 141 с.

Виноградов В.И. Роль осадочного цикла в геохимии изотопов серы. М.: Наука, 1980. 192 с.

Виноградов В.И. Историзм в геохимии в свете данных по изотопному составу серы. Очерки отдельных элементов. М.: Наука, 1973. С. 244–274.

Виноградов В.И., Кулешов В.Н. Изотопный состав углерода и кислорода в архейских карбонатах Алданского щита. Тезисы докл. VI Всесоюзн. симпоз. по стабильным изотопам в геохимии. М.: ГЕОХИ, 1976. С. 23–24.

Волков И.И. Геохимия серы в осадках океана. М.: Наука. 1984. 170 с.

ВойткевичГ.В., Лебедько Г.И. Полезные ископаемые и металлогения докембрия. М.: Недра, 1975. 228. с.

Вояковский С.К., Ильин В.Ф., Павлов Л.Г. и др. Новая Земля – новая марганценосная провинция. Марганцевое рудообразование на территории СССР. М.: Наука, 1984. С. 174–177.

Вулканогенно-осадочные и гидротермальные марганцевые месторождения (Центральный Казахстан, Малый Кавказ и Енисейский кряж). М.: Наука, 1985. 196 с.

Гавашели А.В. Марганцевые руды Чиатуро-Сачхерского бассейна. Тбилиси: Сабчота Сакартвело, 1969. 87 с.

Гаврилов А.А. Марганценосный вулканогенно-осадочный комплекс ордовика Южного Урала и Северных Мугоджар. М.: Наука, 1967. 115 с.

Гаврилов А.А. Эксгаляционно-осадочное рудонакопление марганца. М.: Недра, 1972. 215 с.

Галимов Э.М., Гирин Ю.П. Изменение изотопного состава углерода в процессе образования карбонатных конкреций. Геохимия. 1968. № 2. С. 228–233.

Галимов Э.М., Мазур В.М. Связь изотопного состава углерода сидеритов с фациальной характеристикой отложений и условиями существования фауны (на примере верхнеюрких и нижнемеловых пород Западной Сибири). Известия вузов. Геология и разведка. 1972. № 10. С. 26–32.

Гальянов А.В., Яковлев В.Л. Сырьевая база промышленного комплекса черной металлургии. Екатеринбург: УрО РАН. 2006. 297 с.

Гамкрелидзе И.П. Основные черты тектонического строения и развития Кавказа. Проблемы нефтегазоносности Кавказа. М.: Наука, 1988. С. 10–19.

Гаррелс Р.М., Крайст Ч.Л. Растворы, минералы, равновесия. М.: Мир, 1968. 368 с.

Геологический словарь. М.: Недра, 1973a. Т. 1. 486 с.

Геологический словарь. М.: Недра, 1973b. Т. 2. 456 с.

Геология Балтийского моря. Ред. Гуделис В.К., Волков Е.М. Вильнюс: Мокслас, 1976. 384 с.

Геология и генезис докембрийских железисто-кремнистых и марганцевых формаций мира. Киев: Наукова думка, 1972. 388 с.

Геология и геохимия марганца. М.: Наука, 1982а. 265 с.
Герасимов Н.Н. Геологическое строение и генезис Парнокского железо-марганцевого месторождения (Полярный Урал). Автореф. дисс. канд. геол.-мин. наук. М.: МГУ, 2000. 24 с.
Гецева Р.В. Контроль химического обогащения карбонатных марганцевых руд Усинского месторождения минералогическими методами. Советская геология. 1947. Сб. 27. С. 96–100.
Гогишвили В.Г., Гогишвили Т.Ш., Зулиашвили Т.Г. и др. Высококремнистые цеолиты Закавказья. Вопросы геологии и технологии полезных ископаемых Кавказа. Тбилиси: Сабчота Сакартвело, 1979. С. 115–121.
Гогишвили В.Г., Хамхадзе Н.И., Гуниава В.И. Марганцеворудный пояс Закавказья. Новые данные по марганцевым месторождениям СССР. М.: Наука, 1980. С. 117–126.
Гогишвили В.Г., Хамхадзе Н.И., Гуниава В.И. Генетические типы кремнисто-марганцевой минерализации Закавказья. Геология и геохимия марганца. М.: Наука, 1982. С. 140–147.
Головко В.А. К вопросу о теории осадочного рудогенеза марганца. Литология и полезные ископаемые. 1985. № 5. С. 121–128.
Головко В.А., Мстиславский М.М., Наседкина В.Х. и др. Марганценосность докембрия Енисейского кряжа. Геология и геохимия марганца. М.: Наука, 1982. С. 94–104.
Голота В.В. Подготовительная стадия осадочного марганцеворудного процесса. Уфа: Башнефтехим, 2000. 56 с.
Голуб Д.Л., Сидоров Ю.С. Строение поверхности докембрийского фундамента Балтийского моря (по данным магнитных съёмок э/с «Заря»). Океанология. 1971. № 11. Вып. 2. С. 239–244.
Горная энциклопедия. М.: Советская энциклопедия, 1989. Т. 4. 623 с.
Горшкова Т.И. Химико-минералогические исследования осадков Баренцова и Белого морей. Тр. гос. Океанологического института. М.: 1931. Т. 1. Вып. 2–3. С. 83–127.
Горшкова Т.И. Марганец в донных отложениях северных морей. Марганцевые месторождения СССР. М.: Наука, 1967. С. 117–134.
Грибов Е.М. Алабандин из Улутелякского марганцевого месторождения (Башкирское Приуралье). Известия вузов. Геология и разведка. 1972а. № 8. С. 34–36.
Грибов Е.М. Улутелякское марганцевое месторождение (Башкирское Приуралье). Геология рудных месторождений. 1972б. № 6. С. 95–101.
Грибов Е.М. Условия образования нижнепермских марганценосных известняков Улутелякского месторождения (Башкирское Приуралье): Автореф. канд. дисс. геол.-мин. наук. М., 1974. 28 с.
Грибов Е.М. Вещественный состав рудовмещающей толщи Улутелякского месторождения. Металлогения осадочных и осадочно-метаморфических толщ. М.: Наука, 1978. С. 107–110.
Грибов Е.М. Марганец в отложениях регрессивной серии (Улутелякское месторождение). Геология и геохимия марганца. М.: Наука, 1982. С. 137–139.
Гриненко В.А., Гриненко Л.Н. Геохимия изотопов серы. М.: Наука, 1974. 274 с.
Гриненко В.А., Кроуз Х.Р., Рабинович А.Л. Изотопный состав кислорода речных сульфатов. Тез. докл. XI Всес. симпоз. по стаб. изотопам в геохимии. М.: ГЕОХИ, 1986. С.110-111.
Грязнов В.И. Минералогия никопольских марганцевых руд в связи с ролью диагенеза в рудном минералообразовании. Вопросы минералогии осадочных образований. Львов, 1956. Кн. 3–4. С. 212–226.
Грязнов В.И. Материалы к геохимии и промышленной оценке Больше-Токмакского марганцевого месторождения. Вопросы геологии и минералогии осадочных формаций Украинской ССР. Научн. Зап. Днепропетровского. ун-та. 1960. № 59. С. 3–32.
Грязнов В.И. Генезис мараганцевых руд Никопольского бассейна и методика прогноза морских осадочных месторождений марганца. Марганцевые месторождения СССР. М.: Наука, 1967. С. 135–146.
Грязнов В.И., Селин Ю.И. Основные черты геологии Больше-Токмакского марганцевого месторождения (УССР). Геология рудных месторождений. 1959. № 1. С. 35–55.

Гуделис В.К. История развития Балтийского моря. Геология Балтийского моря. Вильнюс: Мокслас, 1976. С. 95–116.

Гурвич Е.М. Углеродистые марганценосные формации. Геология рудных месторождений. 1980. Т. 22. № 2. С. 76–84.

Гурвич Е.М., Абрамова И.А. Некоторые особенности марганцевой минерализации в тимской свите Курской магнитной аномалии. Марганцевое рудообразование на территории СССР. М.: Наука, 1984. С. 64–73.

Гурвич Е.М., Грибов Е.М., Рахманов В.П. Углеродсодержащая марганценосная формация докембрия. Геология и геохимия марганца. М.: Наука, 1982. С. 47–59.

Данилов И.А., Юдович Я.Э. Первая находка осадочных марагнцевых руд в Северном Предтиманье. Рудообразование на Тимане и Севере Урала. Тр. Ин-та геологии Коми филиала АН СССР. Сыктывкар, 1981. Вып. 34. С. 94–99.

Данилов И.А., Пивень В.А., Гриб В.П., Алексеев Н.Е. Новые данные о рудоносновти юрских отложений не западном склоне Северного Тимана. Геология и геохимия севера европейской части СССР. Межведомственный сборник научных трудов Министерства просвящения РСФСР, Московский госуд. пед. институт им. Ленина. М., 1983. С. 145–156.

Дворов В.И., Соколова Е.А. О генезисе Мангышлакского марганцевого месторождения. Геология рудных месторождений. 1986. Т. 28. № 1. С. 98–100.

Дворов В.И., Соколова Е.А. Геолого-геохимические предпосылки формирования марганцевых руд Мангышлакского месторождения. Сообщение 2. Литология и полезные ископаемые. 1987. № 4. С. 28–47.

Дегенс Э.Т. Геохимия осадочных образований. М.: Мир, 1967. 299 с.

Дегенс Э.Т. Распределение устойчивых изотопов в карбонатах. Карбонатные породы: физико-химисеская характеристика и методы исследования. М.: Мир, 1971. Т. 2. С. 141–153.

Дзоценидзе Г.С. О генезисе Чиатурского месторождения марганца. Литология и полезные ископаемые. 1965. № 1. С. 3–17.

Дзоценидзе Г.С. Геологические условия формирования марганцевых месторождений Чиатуры и Квирильской депрессии. Новые данные по марганцевым месторождениям СССР. М.: Наука, 1980. С. 62–69.

Додин А.Л. Усинское месторождение марганцевых руд. Советская геология. 1947. Сб. 27. С. 55–63.

Долидзе Д.П., Мачабели Г.А., Табагари В.И. и др. Литогенезис олигоценовых мараганценосных отложений Квирильской депрессии и направление дальнейших поисково-разведочных работ. Новые данные по марганцевым месторождениям СССР. М.: Наука, 1980. С. 75–86.

Дорр Дж.В.Н., II. Железистые и связанные с ними марганцевистые формации Бразилии. Геология и генезис докембрийских железисто-кремнистых и марганцевых формаций мира. Киев: Наукова Думка, 1972. С. 103–111.

Дриц В.А., Петрова В.В., Горшков А.И. и др. Марганцевые минералы Fe-Mn-микроконкреций в осадках центральной части Тихого океана и их постседиментационные преобразования. Литология и полезные ископаемые. 1985. № 5. С. 17–39.

Дружинин И.П. Фациальный контроль марганцевого оруденения Мангышлака и циклический тип строения разреза. Новые данные по марганцевым месторождениям СССР. М.: Наука, 1980. С. 191–199.

Дружинин И.П. Фации нижнемайкопских марганценосных отложений Мангышлака. Полезные ископаемые в осадочных толщах. М.: Наука, 1981а. С. 183–204.

Дружинин И.П. Фации марганцевоносных отложений куюлусской свиты (Мангышлак). Известия вузов. Геология и разведка. 1981б. № 12. С. 30–33.

Дружинин И.П. Постумная стадия глубинного разлома, гидротермальный процесс и новый тип седиментационной ловушки при формировании морских марганцевых руд Мангышлака. ДАН СССР. 1986. Т. 289. № 5. С. 1194–1198.

Елисеев А.И., Юдович. Я.Э., Беляев А.А. и др. Осадочные формации Пай-Хоя и перспективы их рудоносности. Научные рекомендации – народному хозяйству. Коми филиал АН СССР, 1984. Вып. 48. 50 с.

Елисеев А.И., Юдович. Я.Э., Беляев А.А. и др. Осадочные формации Лемвинской зоны Урала и перспективы их рудоносности. Научные доклады. Коми филиал АН СССР, 1986. Вып. 151. 28 с.

Емельянов Е.М., Пустельников О.С. Количество взвешенных форм элементов (Сорг, SiO2, Fe, Al, Ti, Ni, Cu, Co) в водах Балтийского моря. Геохимия. 1975. № 7. С. 1049–1063.

Емельянов Е.М., Волков И.И., Розанов А.Г и др. Процессы восстановительного диагенеза в осадках впадин. Геохимия осадочного процесса в Балтийском море. М.: Наука, 1986. С. 131–155.

Есиков А.Д. Изотопная гидрология геотермальных систем. М.: Наука, 1989. 208 с.

Жабин А.В., Сиротин В.И. К вопросу о происхождении железистых кварцитов курской серии КМА. ДАН РАН. 2009. Т. 427. № 1. С. 64–66.

Жариков В.Г. Программа опытной разработки и изучения руд Улутелякского месторождения. Тр. первой научно-технической конференции, Екатеринбург, 12–14 мая 1999 г. Состояние марганцеворудной базы России и вопросы обеспечения промышленности марганцем. Екатеринбург: Екатеринбургская ассоциация малого бизнеса, 2000. С. 137–141.

Железные и марганцевые руды Дальнего Востока. Архипов/Г.И., Кулиш Е.А., Кулиш Л.И., Меркурьев К.М., Фрумкин И.М./Владивосток: ДВНЦ АН СССР, 1985. 296 с.

Железомарганцевые конкреции Мирового океана – новый вид минерального сырья. Тр. ВНИИ геол. и минерал. ресурсов Миров. Океана. 1984а. Т. 192. С. 7–10.

Железорудные месторождения докембрия Украины и их прогнозная оценка./Белевцев Я.Н., Епатько Ю.М., Веригин М.И. и др./Киев: Наукова думка, 1981. 232 с.

Жуков И.Г. Положение девонских марганцевоносных отложений в структурах Магнитогорской палеоостроводужной системы. Металлогения древних и современных океанов - 2002. Формирование и освоение месторождений в офиолитовых зонах. Миасс: ИМин УрО РАН, 2002. С. 148–153.

Жуков И.Е. Мизенс Л.И., Сапельников В.П. О находке бентосной фауны на низкотемпературном палеогидротермальном поле Южно-Файзулинского марганцевого месторождения (Южный Урал). Металлогения древних и современных океанов - 98. Руды и генезис месторождений. Миасс: ИМин УрО РАН, 1998. С. 111–115.

Жуков И.Г., Леонова Л.В. Бентосная фауна из придонной гидротермальной постройки Файзулинского низкотемпературного палеогидртермального поля (Южный Урал). Металлогения древних и современных океанов - 99. Рудоносность гидротермальных систем. Миасс: ИМин УрО РАН, 1999. С. 74–79.

Жукова Г.А., Киселев В.М. Оценка перспектив рационального использования железомарганцевых конкреций в качестве высокосортного марганцевого сырья. М.: ЦНИГРИ, 1990. 44 с. (Деп. В ВИНИТИ 08.08.90, № 4537-В90).

Жуховицкая А.Л., Генералова В.А., Жолнерович В.А. Марганец в осадках современных озер Белоруссии. ДАН БССР. 1986. Т. XXX. № 9. С. 839–842.

Завалишин А.А. Почвообразование и генезис болотных и озерных руд в таежной зоне севера европейской части СССР. Научн. бюлл. Ленинградского ун-та, 1951. № 27.

Зайков В.В. Вулканизм и сульфидные холмы палеоокеанических окраин (на примере колчеданоносных зон Урала и Сибири). М.: Наука, 2006. 429 с.

Зайков В.В., Масленников В.В., Зайкова Е.В. и др. Рудно-формационный анализ колчеданных месторождений Уральского палеоокеана. Миасс: Мин УрО РАН, 2002. 214 с.

Зайкова Е.В., Зайков В.В. Признаки придонного гидротермального происхождения месторождения железисто-кремнистых построек Магнитогорско-Мугоджарской островодужной системы Урала. Металлогения древних и современных океанов – 2003. Формирование и освоение месторождений в островодужных системах. Миасс: ИМин УрО РАН, 2003. С. 208–215.

Зайкова Е.В., Зайков В.В. Кремнисто-железистые постройки на гидротермальных полях окраинно-океанических структур (обзор). Металлогения древних и современных океанов – 2005. Формирование месторождений на разновозрастных океанических окраинах. Миасс: ИМин УрО РАН, 2005. Т. 1. С.91–97.

Зарицкий П.В. Конкреции угленосных отложений Донецкого бассейна. Харьков: Харьковский государственный университет, 1959. 240с.

Зарицкий П.В. Минералогия и геохимия диагенеза угленосных отложений. Харьков: Харьковский государственный университет, 1970. Ч. 1. 223с.

Зарицкий П.В., Македонов А.В. Конкрециеобразование и стадийность литогенеза. Геологический журнал. 1985. Т.45. № 6. С.101-105.

Зоненшайн Л.П., Кориневский В.Г., Казьмин В.Т. и др. Строение и развитие Южного Урала с точки зрения тектоники литосферных плит. История развития Уральского палеокеана. М.: ИО АН СССР, 1984. С. 6–56.

Зыкин Н.Н. Геологическое строение и генезис Парнокского железо-марганцевого месторождения (Полярный Урал). Вестник МГУ. Геология. 2002. № 2. С. 40–49.

Иванов М.Б., Леин А.Ю. Распространение микроорганизмов и их роль в процессах диагенетического минералообразования. Геохимия диагенеза осадков Тихого океана (Тихоокеанский профиль). М.: Наука, 1980. С. 117–137.

Иванов Л.А., Петровский А.Д. О новом возможном механизме формирования осадочных марганцевых руд на примере Улутелякского месторождения в Южном Приуралье. Закономерности строения осадочных толщ. Тез. докладов Третьего Уральского литологического совещания, Екатеринбург, 15–16 сент. 1998 г. Екатеринбург: УГГГА, 1998. С. 110–112.

Иванова В.П., Касатов Б.К., Красавина Т.Н. и др. Термический анализ минералов и горных пород. Л.: Недра, 1974. 400 с.

Икошвили Д.В. К литологии олигоценовых отложений Чиатурского марганцевого месторождения. Тр. КИМС. 1971. Вып. 9/11. С. 189–193.

Исследование марганцевой и железомарганцевой минерализации в разных природных обстановках методами сканирующей электронной микроскопии. Науч. ред. Г.Н.Батурин. М.: Эслан, 2012. 472 с.

Калинин В.В. Марганцевые и железомарганцевые месторождения восточного склона Южного Урала. Марганцевые месторождения складчатых областейСССР. М.: Наука, 1978. С. 55–90.

Карцев А.А., Вагин С.Б., Басков Е.А. Палеогидрогеология. М.: Недра, 1969. 150 с.

Кекелия С.А., Амбокадзе А.Н., Гудушаури М.П. К вопросу об источнике рудного вещества ртутных месторождений Кавказа. Материалы по полезным ископаемым Кавказа. Тбилиси: Ганатлеба, 1979. С. 163–181.

Киссин И.Г., Пахомов С.И. О возможности генерации углекислоты в недрах при умеренно высоких температурах. ДАН СССР. 1967. Т.174. № 2. С.134-137.

Климат в эпохи крупных биосферных перестроек. М.: Наука. 2004. 299 с.

Князев Г.И., Шевченко Е.С. Зональность и генезис марганцевых руд Никопольского бассейна. Литолого-фациальный анализ осадочных рудоносных формаций (Al-Fe-Mn). Л.: ВСЕГЕИ, 1986. С. 79–81.

Кодина Л.А., Галимов Э.М. Формирование изотопного состава углерода органического вещества "гумусового" и "сапропелевого" типов в морских отложениях. Геохимия. 1984. № 11.С. 1742–1756.

Кожевников К.Е., Рабинович С.Д. Североуральский марганцеворудный бассейн. Использование бедных марганцевых руд Северного Урала. Тр. ин-та металлургии Уральского филиала АН СССР. Свердловск, 1961. Вып. 7. С. 23–51.

Контарь Е.С., Либарова Л.Е. Металлогения меди, цинка, свинца на Урале. Екатеринбург: Уралгеолком, 1997, 233 с.

Контарь Е.С., Савельева К.П., Сурганов А.В. и др. Марганцевые месторождения Урала. Ред. К.К. Золоев, Е.С. Контарь. Екатеринбург, 1999. 120 с.

Коротеев В.А., Сазонов В.Н. Геодинамика, рудогенез, прогноз (на примере Урала). Екатеринбург: ИГГ УрО РАН, 2005. 258 с.

Кравцов В.А. Формы миграции Fe, Mn, Zn, Cd, Pb Ca в сероводородных водах Готландской впадины Балтийского моря. Геохимия. 1998. № 4. С. 404–415.

Краснова Н.И., Петров Т.Г. Генезис минеральных индивидов и агрегатов. СПб.: Невский курьер, 1997. 228 с.

Краткий справочник по геохимии. М.: Недра, 1970. 280 с.

Кротов Б.П. Выделение гидроокислов железа и марганца в озерах. ДАН СССР. 1950а. Т. 71. № 3. С. 533–536.

Кротов Б.П. Типы пресных озер и образующиеся в нем руды. ДАН СССР. 1950б. Т. 70. № 5. С. 907–910.

Кулешов В.Н. Изотопный состав углерода и кислорода карбонатов и графита метаморфических пород Алданского щита. Тез. докл. «Карбонатное осадконакопление и проблема эвапоритов в докембрии». Ростов-на-Дону: РГУ, 1978. С. 100–101.

Кулешов В.Н. Изотопный состав и происхождение глубинных карбонатов. М.: Наука, 1986а. 126 с.

Кулешов В.Н. К вопросу о генезисе марганцевых карбонатных руд Никопольского и Больше-Токмакского месторождений (по данным изотопного состава углерода и кислорода). Тез. докл. XI Всесоюзн. симпоз. по геохимии изотопов. М.: ГЕОХИ, 1986б. С. 210–212.

Кулешов В.Н. Карбонаты марганца в современных осадках: геохимия изотопов ($\delta^{13}C$, $\delta^{18}O$) и происхождение. Литология и полезные ископаемые. 1999. № 5. С. 483–502.

Кулешов В.Н. Эволюция изотопных углекислотно-водных систем в литогенезе. Сообщение 1. Седиментогенез и диагенез. Литология и полезные ископаемые. 2001а. № 5. С. 491–508.

Кулешов В.Н. Эволюция изотопных углекислотно-водных систем в литогенезе. Сообщение 2. Катагенез. Литология и полезные ископаемые. 2001б. № 6. С. 610–630.

Кулешов В.Н. Геохимия изотопов (13С, 18О) и происхождение карбонатных марганцевых руд. Геология, методика поисков разведки и оценки месторождений твердых полезных ископаемых. М. 2001в. 56 с.

Кулешов В.Н. Месторождения марганца. Сообщение 1. Генетические модели марганцевого рудогенеза. Литология и полезные ископаемые. 2011а. № 5. С. 527–550.

Кулешов В.Н. Месторождения марганца. Сообщение 2. Главнейшие эпохи и фазы накопления марганца в истории Земли. Литология и полезные ископаемые. 2011б. № 6. С. 612–634.

Кулешов В.Н. Месторождение-супергигант – марганцеворудное поле Калахари (северный Кейп, ЮАР): геохимия изотопов ($\delta^{13}C$; $\delta^{18}O$) и генезис. Литология и полезные ископаемые. 2012. № 3. С. 245–263.

Кулешов В.Н., Беляев А.А. Происхождение карбонатных марганецсодержащих конкреций в палеозойских отложениях Пай-Хоя (по данным изотопного состава углерода и кислорода). Литология и полезные ископаемые. 1999. № 3. С.252-261.

Кулешов В.Н., Брусницын А.И. О новом механизме формирования карбонатных марганцевых руд (по данным 13С и 18О для Южно-Файзулинского месторождения, Южный Урал). Докл. РАН. 2004. Т. 395. № 5. С. 661–666.

Кулешов В.Н., Брусницын А.И. Изотопный состав (^{13}C, ^{18}O) и происхождение карбонатов из марганцевых месторождений Южного Урала. Литология и полезные ископаемые. 2005. № 4. С. 416–429.

Кулешов В.Н., Быч А.Ф. Изотопный состав ($\delta^{13}C$ и $\delta^{18}O$) и происхождение карбонатных марганцевых руд Усинского месторождения (Кузнецкий Алатау). Литология и полезные ископаемые. 2002. № 4. С. 381–396.

Кулешов В.Н. Марганцевые породы и руды: геохимия изотопов, генезис, эволюция рудогенеза. М.: Научный мир, 2013. 540 с.

Кулешов В.Н., Гаврилов Ю.О. Изотопный состав ($\delta^{13}C$, $\delta^{18}O$) и происхождение карбонатных конкреций Восточного Предкавказья. Литология и полезные ископаемые. 2001. № 2. С. 187–190.

Кулешов В.Н., Домбровская Ж.В. Изотопный состав и условия образования Никопольских карбонатных марганцевых руд. Изотопная геохимия процесса рудообразования. М.: Наука, 1988. С. 233–258.

Кулешов В.Н., Домбровская Ж.В. Изотопный состав и происхождение карбонатных марганцевых руд Мангышлакского месторождения. Литология и полезные ископаемые. 1990. № 2. С. 50–62.

Кулешов В.Н., Домбровская Ж.В. К вопросу о генезисе карбонатных марганцевых руд Мангышлакского месторождения (по данным изотопного состава углерода и кислорода). Литология и полезные ископаемые. 1993. № 2. С. 34–43.

Кулешов В.Н., Домбровская Ж.В. Марганцевые месторождения Грузии. Сообщение 1. Геологические особенности и изотопный состав карбонатных марганцевых руд Чиатурского и Квирильского месторождений. Литология и полезные ископаемые. 1997а. № 3. С. 286–306.

Кулешов В.Н., Домбровская Ж.В. Марганцевые месторождения Грузии. Сообщение 2. Происхождение марганцевых руд (на примере Чиатурского и Квирильского месторождений). Литология и полезные ископаемые. 1997б. № 4. С. 339–355.

Кулешов В.Н., Розанов А.Г. Изотопный состав и происхождение карбонатов марганца в глубоководных осадках Центральной Балтики. Литология и полезные ископаемые. 1998. № 4. С. 525-531.

Кулешов В.Н., Чистякова И.А. (13С и (18O железомарганцевых конкреций Онежского залива Белого моря. Литология и полезные ископаемые. 1989. № 2. С. 126–130.

Кулешов В.Н., Штеренберг Л.Е. Изотопный состав Fe-Mn-конкреций оз.Красного (Карельский перешеек). Известия АН СССР. Геология. 1988. № 10. С. 92–104.

Кулешов В.Н., Горностай Б.А., Чопоров Д.Я. и др. Изотопный состав и происхождение карбонатных марганцевых руд Безмошицкого проявления. Советская геология. 1991. № 10. С. 34–39.

Лаврухина А.К. Юркина Л.В. Аналитическая химия марганца. М.: Наука, 1974. 219 с.

Лаврушин В.Ю., Кулешов В.Н. Оолиты Каспийского моря (закономерности распространения, геохимические особенности, генезис). Литология и полезные ископаемые. 1999. № 6. С. 596–618.

Лалиев А.Г. Майкопская серия Грузии. М.: Недра, 1964. 308 с.

Левин Л.Е., Фельдман С.Л. Балтийское море. Нефть и нефтегазоносность окраинных и внутренних морей СССР. М.: Недра, 1970. С. 190–251.

Леин А.Ю., Логвиненко Н.В., Волков И.И. и др. Минеральный и изотопный состав диагенетических карбонатных минералов конкреций из восстановительных осадков Калифорнийского залива. ДАН СССР. 1975. Т. 224. № 2. 426–429.

Леин А.Ю., Логвиненко Н.В., Сулержицкий Л.Д и др. Об источнике углерода и возрасте диагенетических карбонатных конкреций Калифорнийского залива. Литология и полезные ископаемые. 1979. № 1. С. 23–29.

Леин А.Ю., Ваншейн Б.М., Кашпарова Е.В. и др. Биогеохимия анаэробного диагенеза и материально-изотопный баланс серы и углерода в осадках Балтийского моря. Геохимия осадочного процесса в Балтийском море. М.: Наука, 1986. С. 155–176.

Леин А.Ю., Гальченко В.Ф., Покровский Б.Г. и др. Морские карбонатные конкреции как результат процессов микробиального окисления газгидратного метена в охотском море. Геохимия. 1989. № 10. С. 1396–1406.

Леин А.Ю., Горшков А.И., Пименов Н.В. и др. Аутигенные карбонаты на метановых сипах в Норвежском море: минералогия, геохимия, генезис. Литология и полезные ископаемые. 2000а. № 4. С. 339–354.

Леин А.Ю., Гричук Д.В., Гурвич Е.Г., Богданов Ю.А. Новый тип гидротермальных растворов, обогащенных водородом и метаном, в рифтовой зоне Срединно-Атлантического хребта. Доклады РАН, 2000б. Т. 375. № 3. С. 380–383.

Лемешев А.В., Вишев В.С., Дьяконова А.Г. и др. Новые данные о строении Парнокского железо-марганцевого месторождения (Приполярный Урал). Геология и минеральные ресурсы Европейского северо-востока России. Материалы XV геологического съезда республики Коми. Т. III. Сыктывкар: Геопринт, 2009. С. 216–218.

Лисицын А.П. Гидротермальные системы мирового океана – поставка эндогенного вещества. Гидротермальные системы и осадочные формации срединно-океанических хребтов Атлантики. Под ред. А.П.Лисицына, М.: Наука, 1993. С. 147–246.

Логвиненко Н.В. Состав и генезис карбонатов железа и марганца в осадочных образованиях. Литология и полезные ископаемые. 1972. № 1. С. 86–96.

Логвиненко Н.В., Волков И.И., Соколова Е.Г. Родохрозит в глубоководных осадках Тихого океана. ДАН СССР. Т. 203. 1972. № 1. С. 204–207.

Мавыев Н.Ч., Надиров Б.Р., Шарапов В.К. Катагенез и коллекторские свойства нижнемеловых песчано-алевритовых пород Мургабского осадочного бассейна. Осадочные бассейны и их нефтегазоносность. М.: Наука, 1983. С. 298–301.

Магадеев Б.Д., Грешилов К.В., Антонов К.В. и др. Минерально-сырьевая база Республики Башкортостан. Концепция её освоения. Горный журнал. 1997. № 5–6. С. 3–10.

Макушин А.А. О марганце в нижнепермской галогенной формации Башкирского Приуралья. ДАН СССР. 1970. Т. 191. № 6. С. 1381–1384.

Макушин А.А. О генезисе марганцевого оруденения среди отложений нижнепермской галогенной формации Башкирии. ДАН СССР. 1972. Т. 206. № 3. С. 683–686.

Макушин А.А. Закономерности распределения марганца в отложениях улутелякской свиты (Башкирское Приуралье). Литология и полезные ископаемые. 1975. № 2. С. 70–80.

Марганцево-железисто-кремнистая формация Дальнего Востока СССР. Е.А. Кулиш, Л.И. Кулиш, К.М. Меркурьев и др. М.: Наука, 1981. 208 с.

Марганцевое рудообразование на территории СССР. М.: Наука, 1984b. 280 с.

Марганцевые и железорудные концентрации джаильминской мульды. М.: Наука, 1966. 143 с.

Марганцевые месторождения Урала. Ред. К.К. Золоев, Е.К Контарь. Екатеринбург, 1999. 120 с.

Марганцевые месторождения складчатых областей. М.: Наука, 1978. 245 с.

Марганцевые месторождения в осадочных и вулканогенно-осадочных отложениях. Тез. докладов III Всесоюзн. совещания по марганцевым рудам СССР, 25–27 мая 1982 г., Каражал. М.: ОПЛОП ВИЭМС, 1982b. 115 с.

Мартмаа Т.А., Пиррус Р.О., Пуннинг Я.-М.К. Формирование карбонатных озерных отложений. Тез. докладов IX Всес. симпоз. по стаб. изотопам в геохимии. 1982 г. М.: ГЕОХИ АН СССР, 1982. С. 399–400.

Махарадзе А.И. Об источниках и путях привноса Mn, Si, Fe и P в нижнеолигоценовых отложениях Западной Грузии. ДАН СССР. 1972. Т. 202. № 4. С. 929–931.

Махарадзе А.И. Кремнисто-цеолитовые породы майкопской серии Грузии и условия их формирования. Материалы по полезным ископаемым Кавказа. Тбилиси, 1979. С. 207–218.

Махарадзе А.И., Чхеидзе Р.Г. Литология олигоценовых отложений Квирильской депресии и о генезисе связанных с ними полезных ископаемых. Тр. КИМС. 1971. Вып. 9(2).

Махнач А.А. Катагенез и подземные воды. Минск: Наука и техника, 1989. 335 с.

Махнач А.А. К вопросу об объеме и содержании стадии катагенеза. Литология и полезные ископаемые. 1992. № 4. С. 138–142.

Мачабели Г.А. Специфика седименто- и диагенеза олигоценовых марганценосных отложений Грузии. Условия образования рудных месторождений. М.: Наука, 1986. Т. 2. С. 839–849.

Мачабели Г.А., Хамхадзе Н.И. О литологических особенностях олигоценовых марганценосных отложений Кавказа и источнике марганца. Вопросы геологии и технологии полезных ископаемых Кавказа. Тбилиси: Сабчота Сакартвело, 1979. С. 33–39.

Медноколчеданные месторождения Урала: геологические условия размещения. Под ред. В.И.Смирнова. Свердловск: УНЦ АН СССР, 1985. 288 с.

Мерабишвили М.С., Чхеидзе Р.Г., Доленджтшвили Ц.Г. и др. Высококремнистые цеолиты Закавказья и возможные пути их миграции. Вопросы геологии и технологии полезных ископаемых Кавказа. Тбилиси: Сабчота Сакартвело, 1979. С. 221–230.

Мизенс Г.А. Рудопроявления марганца в нижнепермских отложениях Среднего Урала. Литология и полезные ископаемые. 1979. № 5. С. 155–156.

Минерагенический потенциал недр России. Вып. 1. Восточноевропейско-Баренцевская мегапровинция. М.: Геокарт, Геос, 2008. С. 338–343.

Минеральные месторождения Китая. Сост. В.П. Федорчук. М.: Геоинформмарк, 1999. 279 с.

Миртов Ю.В., Тарасова С.М. Марганценосность Древних отложений Алтае-Саянской складчатой области. Новые данные по марганцевым месторождениям СССР. М.: Наука, 1980. С. 211–215.

Миртов Ю.В., Тарасова С.М., Быч А.Ф. К стратиграфии позднего докембрия и раннего кембрия Усинско-Бельсинского района Кузнецкого Алатау. Геология. Томск: Томский государственный университет, 1973. С. 35–48.

Миртова С.М. О генезисе карбонатных руд Усинского месторождения марганца. Минералогия и петрография пород и руд главнейших рудных районов Сибири. Новосибирск: Наука, Сибирское отд., 1983. С. 138–143.

Михайлов Б.М. Прогнозная оценка России на марганцевые руды. Марганцевые руды. Тез. докладов Междуведомственного совещания «Актуальные проблемы образования, прогнозирования и поисков марганцевых руд». СПб.: 1992. С.69-72.

Михайлов Б.М. Марганцевые руды России. Литология и полезные ископаемые. 1993. № 4. С. 23–33.

Михайлов Б.М. Актуальные проблемы прогнозирования марганцевых месторождений на Урале. Литология и полезные ископаемые. 2001. № 1. С. 3–15.

Михайлов Б.М., Рогов В.С. Геологические предпосылки прогнозирования марганцевых месторождений на Урале. Советская геология. 1985. № 8. С. 24–31.

Михайлов Д.А. Метасоматическое происхождение железистых кварцитов докембрия. Л.: Наука, 1983. 168 с.

Мстиславский М.М. Палеотектонические особенности локализации олигоценовых марганцеворудных месторождений юга СССР. ДАН СССР. 1981. Т. 260. №5. С. 1207–1212.

Мстиславский М.М. О рудоподводящем канале Чиатурского месторождения марганца. Геология рудных месторождений. 1984. Т. 26. С. 68–76.

Мстиславский М.М. Существуют ли в природе «классически осадочные» месторождения марганца Чиатурского типа?. Геология рудных месторождений. 1985. Т. 27. № 6. С. 3–16.

Мстиславский М.М. Принципы прогнозирования промышленных месторождений марганца. Руды и металлы. 1994. № 3–5. С. 29–38.

Мухин А.С., Ладыгин П.П. Новые данные по геолого-промышленной характеристике Усинского месторождения марганцевых руд. Вестник западно-сибирского геологического управления. 1957. № 2. С. 29–37.

Никопольский марганцеворудный бассейн. М.: Недра, 1964а. 525 с.

Овчинников Л.Н. Полезные ископаемые и металлогения Урала. М.: Геоинформатик, 1998. 413 с.

Осадочные формации Пай-Хоя и перспективы их рудоносности. (А.И. Алексеев, Я.Э. Юдович, А.А. Беляев, Г.Ф. Семенов). Научные рекомендации - народному хозяйству. Вып. 48. Сыктывкар: Коми филиал УрО АН СССР, 1984. 50с.

Павлов Д.И. Связь осадочных месторождений железа и марганца с нефтегазоносными бассейнами. Геология рудных месторождений. 1989. № 2. С. 80–91.

Павлов Д.И., Домбровская Ж.В. Осадочные месторождения марганца как результат восходящей разгрузки подземных вод нефтегазоносных бассейнов. Отечественная геология. 1993. № 8. С. 21–26.

Папава Д.Ю. Основные направления геолого-разведочных работ на нефть и газ в Грузии. Проблемы нефтегазоносности Кавказа. М.: Наука, 1988. С. 26–30.

Парагенезис металлов и нефти в осадочных толщах нефтегазоносных бассейнов. Ред. Д.И. Горжевский, Д.И. Павлов. М.: Недра, 1990. 269 с.

Перри Ю.К., Тан Ф.К. Вариации изотопного состава углерода в карбонатах железистой формации Бивабик. Геология и генезис докембрийских железисто-кремнистых и марганцевых формаций мира. Киев: Наукова думка, 1972. С. 314–319.

Петровский А.Д. Иванов Л.А. О новых находках гониатитов на Пай-Хое. Известия АН СССР. Сер. геол. 1987. № 12. С. 122–123.

Платонов Е.Г. Повышева Л.Г., Устрицкий В.И. О генезисе карбонатных марганцевых руд Пай-Хойско-Новоземельского региона России. Литология и полезные ископаемые. 1992. № 4. С. 76–89.

Поляков В.П., Криштал Н.Н., Ткаченко А.Е. Формирование изотопного состава углерода карбонатной системы озерных вод в гумидных и аридных климатических условиях. Тез. докл. IX Всес. симпоз. по стаб. изотопам в геохимии. М.: ГЕОХИ, 1982. С. 474–475.

Потконен М.И. Марганец России: состояние, перспективы освоения и развития минерально-сырьевой базы. Минеральное сырье. Серия геолого-экономическая. М.: ВИМС, 2001. 84 с.

Пустовалов Л.В. Генезис липецких и тульских железных руд. Тр. ВГРО. 1933. Вып. 285. 28 с.

Пучков В.Н. Палеоокеанические структуры Урала. Геотектоника, 1993, № 3. С.18−33.

Пушкина З.В. К геохимии Усинского марганцевого месторождения. ДАН СССР. 1960. Т. 135. № 1. С. 176–178.

Рабинович С.Д. Северо-Уральский марганцеворудный бассейн. М.: Недра, 1971. 262 с.

Радугин К.В. Ивановское месторождение марганцевых руд (Зап. Сибирь). Советская геология. 1941. № 3. С. 61–74.

Рахманов В.П., Ерощев-Шак В.А. Некоторые особенности эпигенетического минералообразования в марганцевых рудах Усинского месторождения. Металлогения осадочных и осадочно-метаморфических пород. М.: Наука. 1966. С. 90–96.

Рахманов В.П., Чайковский В.К. Генетические типы осадочных марганценосных формаций. Советская геология. 1972. № 6. С. 22–32.

Рахманов В.П., Грибов Е.М.Медведовская. Изотопные особенности карбонатных и окисных руд марганцевых месторождений. Известия вуов. Геология и разведка. 1994. № 4. С. 91–98.

Рогов В.С., Галицкая Э.И. Проявление карбонатных марганцевых руд в нижнепермских отложениях Пай-Хоя. Советская геология. 1983. № 6. С. 101–103.

Рогов В.С., Галицкая Э.И. Нижнепермские руды марганца северо-восточного склона Пай-Хоя. Марганцевое рудообразование на территории СССР. М.: Наука, 1984. С. 177–182.

Рогов В.С., Галицкая Э.И., Давыдов В.И. и др. Новые данные по стратиграфии марганценосных отложений перми и карбона Пай-Хоя. Советская геология. 1988. № 7. С. 51–61.

Рожнов А.А., Бузмаков Е.И. Середа В.Я. и др. Новые данные о геологическом строении железо-марганцевых месторождений Атасуйского района (Центральный Казахстан). Новые данные по марганцевым месторождениям СССР. М.: Наука, 1980. С. 158–170.

Розанов А.Г. Исследование железомарганцевых конкреций северной экваториальной части Тихого океана. 41-й рейс научно-исследовательского судна «Дмитрий Менделеев» 4 января - 23 апреля 1988 г. Океанология. 1989. Т. 29. Вып. 3. С. 522–525.

Рой С. Месторождения марганца. Пер. с англ. М.: Мир, 1986. 520 с.

Салихов Д.Н., Ковалев С.Г., Брусницын А.И., и др. Полезные ископаемые республики Башкортостан (марганцевые руды). Уфа: Изд-во «Экология», 2002. 243 с.

Салли А. Марганец. Пер. с англ. М.: Металлургиздат, 1959. 296 с.

Салуквадзе Н.Ш. Об основных геологических событиях на территории Грузии в палеогеновое время. Сообщение АН ГССР. 1990. Т. 137. № 2. С. 341–343.

Сапожников Д.Г. Об условиях образования марганцевых месторождений юга Русской платформы и Крымско-Кавказской геосинклинали. Геология рудных месторождений. 1967. № 1. С. 74–87.

Саркисян С.Г. О возрасте рудоносных слоев Улу-Телякского марганцевого месторождения Башкирии (по данным минералогического анализа). ДАН СССР, нов. сер., 1945. Т. 46. № 4. С. 172–173.

Свальнов В.Н., Кулешов В.Н. Кальциевый родохрозит в осадках Гватемальской котловины. Литология и полезные ископаемые. 1994. № 3. С. 20–35.

Селин Ю.И. Стратиграфия и моллюски олигоцена Больше-Токмакского марганцево-рудного района. М.: Недра, 1984. 240 с.

Семененко Н.П. Металлогения докембрия. Советская геология. 1962. № 2. С. 50–60.

Семенович Н.И. Лимнологические условия накопления железистых осадков в озерах. Тр. Лимнол. станции АН СССР на оз. Пуннус-Ярви. Вып.1. М.- Л.: 1958. 188 с.

Серавкин И.Б., Косарев А.М., Салихов Д.Н. и др. Вулканизм Южного Урала. М.: Наука, 1992. 195 с.

Сидоренко А.В., Борщевский Ю.А. Изотопные особенности древнейших карбонатных отложений докембрия. Тез. докл. V Всесоюзн. симпоз. по геохимии стабильных изотопов. М.: ГЕОХИ, 1974. С. 2–4.

Силаев В.И. Механизмы и закономерности эпигенетического марганцевого минералообразования. Екатеринбург: УрО РАН, 2008. 386 с.

Скорнякова Н.С., Успенская Т.Ю. Железо-марганцевые конкреции южной части Гватемальской котловины. ДАН СССР. 1989. Т. 309. С. 958–962.

Скорнякова Н.С., Успенская Т.Ю., Мурдмаа И.О. Железо-марганцевые конкреции Гватемальской котловины. Литология и полезные ископаемые. 1996. № 6. С. 648–652.

Соколова Е.А. Распределение марганца и фосфора в различных типах пород усинской свиты (нижний кембрий Кузнецкого Алатау). ДАН СССР. 1960. Т. 135. № 3. С. 717–719.

Соколова Е.А. Положение Усинского марганцевого месторождения в отложениях нижнего кембрия хребта Кузнецкий Алатау. Известия АН СССР. Сер. геол. 1961. № 2. С. 20–34.

Соколова Е.А. Марганценосность вулканогенно-осадочных формаций. Тр. ГИН АН СССР. Вып. 360. М.: Наука, 1982. 195 с.

Соколова Е.А., Дворов В.И. Геолого-геохимические предпосылки формирования марганцевых руд Мангышлакского месторождения. Сообщение 1. Литология и полезные ископаемые. 1987. № 1. С. 60–79.

Соколова Е.А., Домбровская Ж.В., Тропп Е.Б. Особенности строения и формирования рудных залежей Мангышлакского марганцевого месторождения в Казахстане. Марганцевое рудообразование на территории СССР. М.: Наука, 1984. С. 242–249.

Соколова Е.И. Физико-химическое исследование железорудного озера Пуннус-Ярви. Очерки по металлогении осадочных пород. М.: АН СССР, 1962a. 216 с.

Соколова Е.И. Физико-химическое исследование осадочных железных и марганцевых руд и вмещающих их пород. М.: АН СССР, 1962b. 216 с.

Соколова-Дубинина Г.А., Дерюгина З.П. Роль микроорганизмов в образовании родохрозита в оз. Пуннус-Ярви. Микробиология. 1967а. Т. 36. Вып. 3. С. 536–542.

Соколова-Дубинина Г.А., Дерюгина З.П. Роль микробиологических факторов в образовании марганцевых руд в Карельских озёрах. Сообщ. 2. Образование марганцевых руд в оз.Пуннус-Ярви. Микробиология. 1967б. Т. 36. Вып. 5. С. 1066–1076.

Сорохтин О.Г. Глобальная эволюция Земли. В кн.: Современные проблемы геологии. Отв. ред.: Ю.О.Гаврилов, М.Д.Хуторской (Тр. Геологического института РАН, вып. 565). М.: Наука, 2004. С.203-222.

Сорохтин О.Г., Ушаков С.А. Глобальная эволюция Земли. М.: Московский государственный университет, 1991. 446 с.

Сорохтин О.Г., Ушаков С.А., Сорохтин Н.О. Глобальная эволюция и металлогения раннего докембрия. Отечественная геология. 1999. № 5. С. 56–63.

Сорохтин О.Г., Старостин В.И., Сорохтин Н.О. Эволюция Земли и происхождение полезных ископаемых. Известия секции наук о Земле РАЕН. 2001. Вып. 6. С. 5–25.

Старикова Е.В. Поведение железа и марганца в гидротермально-осадочном процессе: анализ природных и расчетных данных. Металлогения древних и современных океанов = 2001. История месторождений и эволюция рудообразования. Миасс: ИМин УрО РАН, 2001. С. 71–77.

Старикова Е.В., Брусницын А.И., Жуков И.Г. Палеогидротермальная постройка марганцевого месторождения Кызыл-Таш, Южный Урал. СПб.: Наука, 2004. 230 с.

Столяров А.С. Новые данные по стратиграфии олигоценовых отложений Южного Мангышлака. Бюлл. ОНТИ. М.: Госгеолтехиздат, 1958. № 3. С. 8–10.

Столяров А.С. О генезисе крупнейших фанерозойских осадочных концентраций марганца и прогнозе их генетических аналогов в России. Отечественная геология. 1993. № 5. С. 28–33.

Столяров А.С., Ивлева Е.И., Халезов А.Б. и др. Марганец России. Состояние, перспективы освоения и развития минерально-сырьевой базы. Минеральное сырье. М.: ВИМС, 2009. № 20. 147 с.

Страхов Н.М. Общая схема осадкообразования в современных морях и озерах малой минерализации. Образование осадков в современных водоемах. М.: АН СССР, 1954. Ч. 3. С. 375–387.

Страхов Н.М. К познанию диагенеза. Вопросы минералогии осадочных образований. Львов, 1956. Кн. 3–4. С. 7–26.

Страхов Н.М. Типы литогенеза и их эволюция в истории Земли. М.: Госгеолтехиздат, 1963. 535 с.

Страхов Н.М. О проблемах и некоторых итогах изучения геохимии палеогенового марганцеворудного бассейна юга СССР. Литология и полезные ископаемые. 1964. № 1. С. 3–10.

Страхов Н.М. Типы накопления марганца в современных водоёмах и их значение для познания марганцево-рудного процесса. Литология и полезные ископаемые. 1965. № 4. С. 18–49.

Страхов Н.М. Условия образования конкреционных железомарганцевых руд в современных водоёмах. Литология и полезные ископаемые. 1976. № 1. С. 3–19.

Страхов Н.М., Штеренберг Л.Е. К вопросу о генетическом типе Чиатурского месторождения. Литология и полезные ископаемые. 1965. № 1. С. 18–30.

Страхов Н.М., Варенцов И.М., Калиненко В.В. и др. К познанию механизма марганцеворудного процесса (на примере олигоценовых руд СССР). Марганцевые месторождения СССР. М.: Наука, 1967. С. 34–56.

Страхов Н.М., Штеренберг Л.Е., Калиненко В.В. и др. Геохимия осадочного марганцеворудного процесса. Тр. ГИН АН СССР. М.: Наука, 1968. Вып. 185. 495 с.

Табагари Д.В. Распределение и вещественный состав генетических типов марганцевых руд на Чиатурском месторождении. Новые данные по марганцевым месторождениям СССР. М.: Наука, 1980. С. 86–93.

Табагари Д.В. Некоторые особенности строения и формирования рудных залежей Чиатурского месторождения. Марганцевое рудообразование на территории СССР. М.: Наука, 1984. С. 109–216.

Тарасова С.М., Миртов Ю.В. Вещественный состав и условия образования марганценосной карбонатной формации рифея-нижнего кембрия в Кузнецком Алатау. Тез докл. к семинару «Литолого-фациальный анализ осадочных рудоносных формаций. Al-Fe-Mn». Л.: ВСЕГЕИ, 1971. С. 72–73.

Тарасова С.М., Миртов Ю.В., Быч А.Ф. Типы разрезов алданского яруса нижнего кембрия Кузнецкого Алатау в связи с их рудоносностью. Новые данные по геологии и полезным ископаемым Западной Сибири. Томск: Томский университет, 1973. Вып. 8. С. 31–37.

Тарасова С.М., Миртов Ю.В., Быч А.Ф. Новые типы разрезов усть-кундатской свиты в Кузнецком Алатау. Геология и металлогения протерозой-кембрийских отложений западной части Алтае-Саянской складчатой области. Тез. докл. научно-техн. конф.

Твалчрелидзе Г.А О главных металлогенических эпохах Земли. Геология рудных месторождений. 1970. Т. 12. № 1. С. 22–36.

Термальный эпигенез палеозойских отложений Пай-Хоя./А.А.Беляев, А.А.Иевлев, В.В.Юдин, Н.С.Овнатанова. – Сыктывкар: Коми научн. центр УрО АН СССР. 1989. Вып. 244. 24с.

Тихомирова Е.С. Палеогеография и геохимия нижнеолигоценовых марганцевых отложений Мангышлака. Литология и полезные ископаемые. 1964. № 1. С. 75–92.

Тихомирова Е.С., Черкасова Е.В. О распределении малых элементов в рудах Мангышлакского месторождения марганца. Марганцевые месторождения СССР. М.: Наука, 1967. С. 258–273.

Торохов П.В., Таран Ю.А., Сагалевич А., М. и др. Изотопный состав метана, углекислого газа и карбонатов термальных выходов подводного вулкана Пийпа (Берингово море). Геохимия, 1991. Т. 318. № 3. Стр. 728–732.

Трубецкой К.Н., Чантурия В.А., Воробьев А.В. и др. Марганец (Минерально-сырьевая база СНГ. Добыча и обогащение руд). Ред. акад. К.Н.Трубецкой. М.: Академия горных наук, 1999. 271 с.

Туманишвили Д.В. Литология и условия формирования олигоценовых марганцевых отложений Квирильской депрессии (Западная Грузия). Автореф. дисс. канд. геол.-мин. наук. Ростов-на-Дону: РГУ, 1989. 25 с.

Условия образования рудных месторождений. Тр. симпозума МАГРМ, Тбилиси 6–12 сент. М.: Наука, 1986. Т. 1–2. 862 с.

Федорчук В.П. Рудные корки – новый вид океанического минерального сырья. Разведка и охрана недр. 1988. № 1. С. 63–64.

Ферронский В.И., Поляков В.А. Изотопия гидросферы. М.: Наука, 1983. 280 с.

Формозова Л.Н. Формационные типы железных руд докембрия и их эволюция. М.: Недра, 1973. 172 с.

Фролов В.Т. Литология. М.: МГУ, 1990. Т.2. 429 с.

Фролов В.Т. Методическое руководство к лабораторным занятиям по петрографии осадочных пород. М.: МГУ, 1964. 310 с.

Хабаков А.В. Улутеляк – новое месторождение окисленных карбонатных марганцевых руд в пермских отложениях Башкирии (Западный склон Урала). Известия АН СССР. Сер. Геол. 1944. № 4. С. 70–85.

Хайн В.Е. Тектоника континентов и океанов. М.: Научный мир, 2001. 606 с.

Хамхадзе Н.И. О связи кремне- и рудообразования в марганцевых месторождениях Грузии. Вулканизм и литогенез. Тбилиси: Менциебера, 1981. С. 141–146.

Хамхадзе Н.И. Тектоно-гидротермальная активизация областей накопления кремнисто-марганцевых отложений Грузии в олигоценовое время. Условия образования рудных месторождений. М.: Наука, 1986. С. 834–838.

Хамхадзе Н.И., Туманишвили Г.П. Палеотектонические условия локализации марганцевых руд Квирильской депрессии. Марганцевое рудообразование на территории СССР. М.: Наука, 1984. С. 227–235.

Херасков Н.П. Геология и генезис восточно-башкирских марганцевых месторождений. Вопросы литологии и срратиграфии СССР. Памяти акад. А.Д. Архангельского. М.: АН СССР, 1951. С. 328–348.

Хёфс Й. Геохимия стабильных изотопов. М.: Мир, 1983. 201 с.

Хмелевский В.А. Геологическое строение и генетический тип Бурштынского месторождения марганца. Геологический сборник Львовского геологич. общ-ва. 1968. № 11. С. 108–115.

Хмелевский А.А., Янчук Э.А. Бурштынское (Предкарпатье) и Улутелякское (Приуралье) марганцевые месторождения аридного типа. Бюллетень МОИП. Отд. Геол. 1990. Т. 65. Вып. 1. С. 89–97.

Ходак Ю.В., Рахманов В.П., Ерощев-Шак В.А. Месторождение марганца Кузнецкого Алатау. М.: Наука, 1966. 104 с.

Ходак В.А. Вулканогенно-осадочный тип девонского марганцевого оруденения на Южном Урале. Полезные ископаемые в осадочных толщах. М.: Наука, 1973. С.156–175.

Холодов В.Н. Новое в познании катагенеза. Сообщение 1. Литология и полезные ископаемые. 1982. № 3. С. 3–21.

Холодов В.Н. Геохимия осадочного процесса. Тр. ГИН РАН. Вып. 574. Ред. Ю.Г. Леонов. М.: ГЕОС, 2006. 608 с.

Цветков Л.И., Вальяшихина Е.П., Пилоян Г.О. Дифференциальный термический анализ карбонатных минералов. М.: Наука, 1964. 168 с.

Чайковский В.К., Рахманов В.П., Ходак Ю.А. Принципы составления прогнозно-металлогенических карт марганценосных формаций. М.: Недра, 1972. 48 с.

Чебан С.Г., Гареев С.И. Объяснительная записка по геолого-экономической оценке прогнозных ресурсов. Марганцевые руды, марганцовистые известняки. Уфа: Башкиргеология, 1997. 85 с.

Чеботарев М.В. Геологическое строение Южно-Хинганского марганцевого месторождения и вещественный состав его руд. Советская геология. 1958. № 8. С. 114–136.

Чиатурское месторождение марганца. М.: Недра, 1964b. 244 с.

Чистякова А.И. Минеральный состав железомарганцевых стяжений Онежского залива Белого моря. Тез. докл. 2-ой Всесоюзн. конф. «Проблемы четвертичной палеоэкологии и палеогеографии северных морей». Апатиты, 1987. С. 110–111.

Чумаков И.С., Головин Д.И., Ганзей С.С. К геохронологии мэотического яруса (верхний миоцен) Восточного Паратетиса. ДАН. 1996. Т.347. № 3. С 372–373.

Чухров Ф.В., Горшков А.И., Дриц В.А Гипергенные окислы марганца. М.: Наука, 1989. 208 с.

Чухров Ф.В. Рудные месторождения Джезказгано-Улутаувского района в Казахстане. М.-Л.: АН СССР, 1940. 121 с.

Шарков А.А. Минерально-сырьевая база марганца России и проблемы ее использования. Разведка и охрана недр. 2000. № 11. С. 15–19.

Шатский Н.С. О марганценосных формациях и о металлогении марганца. Ст. 1. Вулканогенно-осадочные марганценосные формации. Известия АН СССР. Сер. Геол. 1954. № 4. С. 3–38.

Шишкин М.А. Тектоника юга Лемвинской зоны (Полярный Урал). Геотектоника. 1989. № 3. С. 86–95.

Шишкин М.А., Герасимов Н.Н. Парнокское железо-марганцевое месторождение (Полярный Урал). Геология рудных месторождений. 1995. № 5. С. 445–456.

Школьник Э.Л., Жегалло Е.А., Богатырев Б.А. и др. Биоморфные структуры в бокситах (по результатам электронно-микроскопического изучения). М: Ислан, 2004. 184 с.

Шнюков Е.Ф. О геологических условиях образования марганцевых оолитов в рудах Южно-Украинского марганцевого бассейна. Геология рудных месторождений. 1962. № 5. С. 77–83.

Шнюков Е.Ф., Савченко В.А., Григорьев А.В. О процессах оолитообразования в озерах северо-запада СССР (на примере оз. Красного на Карельском перешейке). Материалы по лимнологии, петрографии и геохимии осадочных пород и руд. Киев: Наукова Думка, 1976. Вып. 4. С. 113–120.

Штанчаева З.М. Геохимия некоторых тяжелых металлов в рассолах газонефтяных месторождений Северного Дагестана. Тр. Ин-та геологии Дагестанского филиала АН СССР. 1984. № 31. С. 12–25.

Штеренберг Л.Е. Очерк геохимии северо-уральских марганцевых месторождений. Геохимия осадочных месторождений марганца. Тр. ГИН АН СССР. Вып. 97. М.: АН СССР, 1963. С. 9–71.

Штеренберг Л.Е. О некоторых сторонах формирования железо-марганцевых конкреций Рижского залива. ДАН СССР. 1971. Т. 201. № 2. С. 457–460.

Штеренберг Л.Е. Осадконакопление и диагенез в озерах северной гумидной зоны (на примере Европейской части СССР). Ч. 1–2. Дисс. докт. геол.-мин. наук. Москва: ГИН АН СССР, 1979.

Штеренберг Л.Е. К вопросу о генезисе Чиатурского месторождения. Геология рудных месторождений. 1985. Т. 27. № 1. С. 91–101.

Штеренберг Л.Е., Базилевская Е.С., Чигирева Т.А. Карбонаты марганца и железа в донных отложениях озера Пуннус-Ярви. ДАН СССР. 1966. Т. 170. № 3. С. 691–694.

Штеренберг Л.Е., Горшкова Т.И., Нактинас В.М. Карбонаты марганца в железомарганцевых конкрециях Рижского залива. Литология и полезные ископаемые. 1968. № 4. С. 63–69.

Штеренберг Л.Е., Стравинская Е.А., Уранова О.В. Основные процессы, контролирующие рудообразование в озерах северной лесной зоны (на примере озера Пуннус-Ярви). Литология и полезные ископаемые. 1970. № 1. С. 27–42.

Штеренберг Л.Е., Зверев В.П., Лаврушин В.Ю. и др. Карбонаты марганца в осадках Центрально-Американского жёлоба (поднятие Эль-Гардо). Известия АН СССР. 1992. Сер. Геол. № 9. С. 94–103.

Эдилашвили В.Я., Леквинадзе Р.Д., Гогиберидзе В.В. О влиянии тектоники на марганценакопление в Грузии. Советская геология. 1973а. № 4. С. 106–114.

Эдилашвили В.Я., Леквинадзе Р.Д., Гогиберидзе В.В. и др. О геологических условиях марганценакопления в Грузии. Мат-лы Кавказ. ин-та минер. сырья. Сер. Геол. 1973б. Вып. 10(12). С. 135–142.

Эдилашвили В.Я., Леквинадзе Р.Д., Гогиберидзе В.В. и др. Геологическое строение района марганцевых месторождений Грузии и вопросы их перспективности. Новые данные по марганцевым месторождениям СССР. М.: Наука, 1980. С. 69–74.

Эмсли Дж. Элементы. М.: Мир, 1993. 256 с.

Эренбург Б.Г. О непрерывности изоморфного ряда $CaCO_3$-$MnCO_3$. Журнал неорганической химии. 1959. Т. 4. № 8. С. 1899–1902.

Эффендиева М.А. Рельеф кристаллического фундамента под акваторией Балтийского моря по магнитометрическим данным. Советская геология. 1967. № 4. С. 88–94.

Юдович Э.Я. Конкрецоиды. Конкреции и конкреционный анализ нефтегазовых формаций. Тез. докл. Всесоюзн. конф. 1–13 окт. 1983 г. Тюмень: Тюменская правда, 1983. С. 36–37.

Юдович Я.Э., Кетрис М.П. Марганцевые карбонатные руды на Пай-Хое. ДАН СССР. 1981. Т. 257. № 4. С. 988–991.

Юдович Я.Э., Суханов Н.В. Изотопно-аномальные карбонаты в черносланцевых толщах Пай-Хоя. ДАН СССР. 1984. Т. 275, № 2. С.445-449.

Юдович Э.Я., Беляев А.А., Кетрис М.П. Геохимия, минералогия и рудогенез марганца в черносланцевых формациях Пай-Хоя. Рудообразование на Тимане и севере Урала. Тр. Ин-та геологии Коми фил. АН СССР. Сыктывкар, 1981. Вып. 34. С.54-72.

Юдович Я.Э., Беляев А.А., Рогов В.С. Верхнепалеозойский уровень марганценосности Уральской складчатой системы. ДАН СССР. 1987. Т. 292. № 4. С. 952–956.

Юдович Э.Я., Беляев А.А., Кетрис М.П. Геохимия и рудогенез черных сланцев Пай-Хоя. СПб.: Наука, 1998. 365 с.

Яншин А.Л. Палеоген Мангышлака. Бюлл. МОИП. Нов сер. Отд. Геол. 1950. Т. XXV. Вып. 4. С. 3–42.

Archer, A.A., 1985. Metal resources in manganese (or polymetallic) nodules. Proc. Indian Natl. Sci. Acad., A. Phys. Sci. 51 (3), 630–637.

Ashley, P.M., 1989. Geochemistry and mineralogy of tephroite bearing rocks from the Hoskins manganese mine, New South Wales, Australia. Neues Jb. Mineral. Abh. 161 (1), 85–111.

Asikainen, C.A., Wehrle, S.F., 2007. Accretion of ferromanganese nodules that form pavement in the Second Connecticut Lake, New Hampshire. Proc. Natl. Acad. Sci. U. S. A. 104, 17579–17581.

Astrup, J., Tsikos, H., 1998. Manganese. In: Wilson, M.J., Anhaeusser, C.R. (Eds.), Mineral Resources of Southern Africa: Handbook. Council for Geosciences, Pretoria, pp. 450–460.

Backer, R.H., Clayton, R.N., 1976. Oxygen isotope study of a Precambrian banded iron-formation, Hamersley Range, Western Australia. Geochim. Cosmochim. Acta 40 (10), 1153–1165.

Bau, M., Romer, R.L., Lüders, V., Beukes, N.J., 1999. Pb, O, and C isotopes in silicified Mooidraai dolomite (Transvaal Supergroup, South Africa): implications for the composition of Paleoproterozoic seawater and "dating" the increase of oxygen in the Precambrian atmosphere. Earth Planet. Sci. Lett. 174 (1), 43–57.

Beckholmen, M., Tirén, S.A., 2009. The geological history of the Baltic Sea a review of the literature and investigation tools. SSM Report number: 2009:21, Swedish Radiation Safety Authority (SSM). 118 pp.

Beckholmen, M., Tirén, S.A., 2010a. Displacement along extensive deformation zones at the two SKB sites: Forsmark and Laxemar. SSM Report number: 2010:43, Swedish Radiation Safety Authority (SSM). 89 pp.

Beckholmen, M., Tirén, S.A., 2010b. Displacement along extensive deformation zones at the two SKB sites: Forsmark and Laxemar. Swedish Radiation Safety Authority.

Beukes, N.J., 1983. Palaeoenvironmental setting of iron formations in the depositional basin of the Transvaal Supergroup, South Africa. In: Trendall, A.F., Morris, R.C. (Eds.), Iron Formations: Facts and Problems. Elsevier, Amsterdam, pp. 131–209.

Beukes, N.J., Gutzmer, J., 1996. A volcanic-exhalative origin for the world's largest (Kalahari) manganese field. A discussion of the paper by D.H. Cornell and S.S. Schutte. Mineral. Deposita 31 (2), 242–245.

Beukes, N.J., Klein, C., 1990. Geochemistry and sedimentology of a facies transition—from microbanded to granular iron-formation—in the early Proterozoic Transvaal Supergroup, South Africa. Precambrian Res. 47 (1–2), 99–139.

Beukes, N.J., Smit, C.A., 1987. New evidence for thrust faulting in Griqualand West, South Africa: implications for stratigraphy and age of red beds. Trans. Geol. Soc. S. Afr. 90 (4), 378–394.

Beukes, N.J., Klein, C., Kaufman, A.J., Hayes, J.M., 1990. Carbonate petrography, kerogen distribution, and carbon and oxygen isotope variations in an early Proterozoic transition from limestone to iron-formation deposition, Transvaal Supergroup, South Africa. Econ. Geol. 85 (4), 663–690.

Beukes, N.J., Burger, A.M., Gutzmer, J., 1995. Fault-controlled hydrothermal alteration of Palaeoproterozoic manganese ore in Wessels Mine, Kalahari manganese field. S. Afr. J. Geol. 98 (4), 430–451.

Biro, L., 2009. 3D modeling and mathematical-statistical study of the manganese deposits, Urkut (W-Hungary). In: Manganese in the Twenty-First Century, Short Course, 5–9 September. Veszprem, 2009, pp. 68–78. Abstract volume.

Boardman, L.G., 1964. Further geological data on the Postmasburg and Kuruman manganese ore deposit, Northern Cape Province. In: Haughton, S.H. (Ed.), The Geology of Some Ore Deposits of Southern Africa, vol. 2. Geological Society of South Africa, Johannesburg, pp. 415–440.

Bolton, B.R., Frakes, L.A., Cook, J.N., 1988. Petrography and origin of inversely graded manganese pisolite from Groote Eylandt, Australia. Ore Geol. Rev. 4 (1–2), 47–69.

Bonatti, E., Zerbi, M., Kay, R., Rydell, H., 1976. Metalliferous deposits from the Apennine ophiolites: Mesozoic equivalents of modern deposits from oceanic spreading centers. Geol. Soc. Am. Bull. 87, 83–94.

Bostrom, K., Kunzendorf, H., 1986. Marine hard mineral resources. In: Kunzendorf, H. (Ed.), Marine Mineral Exploration. Elsevier, Amsterdam, pp. 21–53.

Bottcher, M.E., 1993. Die Experimentelle Untersuchung Lagerstatten-Relevanter Metall-Anreicherungsreaktionen aus Wasserigen Losungen Unter Besonderer Berucksichtigun der Bildung von von Rodochrosit ($MnCO_3$) (Unpubl., Ph.D. thesis). Georg-August-Universitet zu Gettingen. 237 s.

Böttcher, M.E., Huckriede, H., 1997. First occurrence and stable isotope composition of authigenic γ-MnS in the central Gotland Deep (Baltic Sea). Mar. Geol. 137 (3), 201–205.

Bottinga, Y., 1969. Calculated fractionation factors for carbon and hydrogen isotope exchange in the system calcite-carbon dioxide-graphite-methane-hydrogen-water-vapour. Geochim. Cosmochim. Acta 35 (1), 49–64.

Botz, R., Stoffers, P., Faber, E., Tietze, K., 1988. Isotope geochemistry of carbonate sediments from lake Kivu (East-Central Africa). Chem. Geol. 69 (3), 299–308.

Brusnitsyn, A.I., Zhukov, I.G., 2012. Manganese deposits of the Devonian Magnitogorsk palaeovolcanic belt (Southern Urals, Russia). Ore Geol. Rev. 47 (1), 42–48.

Burger, A.M., 1994. Fault-Controlled Hydrothermal Alteration of Palaeoproterozoic Manganese Ore in Wessels Mine, Kalahari Manganese Field (Dissertation). Rand Afrikaans University (Unpubl.).

Burke, W.H., Denison, R.E., Hetherington, E.A., et al., 1982. Variation of seawater $^{87}Sr/^{86}Sr$ through Phanerozoic time. Geology 10 (10), 516–519.

Cairncross, B., Beukes, N.J., 2013. The Kalahari Manganese Field. Random House Struik, Cape Town.

Cairncross, B., Beukes, N.J., Gutzmer, J., 1997. The Manganese Adventure: The South African Manganese Fields. Associated Ore & Metal Corporation Limited, Johannesburg.

Calvert, S.E., Price, N.B., 1977. Shallow water, continental margin and lacustrine nodules: distribution and geochemistry. In: Glasby, G.P. (Ed.), Marine Manganese Deposits. Elsevier, Amsterdam, pp. 45–86.

Cameron, H.M., 1984. Marine mineral resources. In: Indigo Raw Materials Industrial Proceedings Conference, 22–23 November 1983, London, pp. 57–61.

Cannon, W.F., Force, E.R., 1983. Potential for high-grade shallow-marine manganese deposits in North America. In: Shanks, W.C. III (Eds.), Unconventional Mineral Deposits. American Institute of Mining, Metallurgical, and Petroleum Engineers, Society of Mining Engineers, New York, pp. 175–190.

Carothers, W.W., Kharaka, Y.K., 1980. Stable carbon isotopes of HCO_3^- in oil-field waters—implications for the origin of CO_2. Geochim. Cosmochim. Acta 44 (2), 323–332.

Chetty, D., Gutzmer, J., 2012. REE redistribution during hydrothermal alteration of ores of the Kalahari manganese deposit. Ore Geol. Rev. 47, 126–135.

Claypool, G.E., Holger, W.T., Kaplan, J.R., et al., 1980. The age curves of sulfur and oxygen isotopes in marine sulfate and their mutual interpretation. Chem. Geol. 28 (3–4), 199–260.

Coleman, M., Fleet, A., Dobson, P., 1982. Preliminary studies of manganese-rich carbonate nodules from leg 68, suite 503, eastern equatorial Pacific. Initial Reports DSDP, vol. 68. U.S. Govt. Printing office. pp. 481–489.

Commeau, R.F., Clark, A., Johnson, C., et al., 1984. Ferromanganese crust resources in the Pacific and Atlantic oceans. In: Oceans'84: Conference Record, 10–12 September 1984, Washington, DC, vol. 1. IEEE, New York, pp. 421–430.

Compston, W., 1960. The carbon isotopic compositions of certain marine invertebrates and coals from the Australian Permian. Geochim. Cosmochim. Acta 18 (1), 1–22.

Corliss, J.B., Lyle, M., Dymond, J., Crane, K., 1978. The chemistry of hydrothermal mounds near the Galapagos Rift. Earth Planet. Sci. Lett. 40 (1), 12–24.

Cornell, D.H., Shutte, S.S., 1995. A volcanic-exhalative origin for the world's largest (Kalahari) manganese field. Mineral. Deposita 30 (3), 146–151.

Cornell, D.H., Schutte, S.S., Eglington, B.L., 1996. The Ongeluk basaltic andesite formation in Griqualand West, Southern Africa: submarine alteration in a 2222 Ma Proterozoic sea. Precambrian Res. 79 (1), 101–124.

Cornell, D.H., Armstrong, R.A., Walraven, F., 1998. Geochronology of the Proterozoic Hartley Basalt Formation, South Africa: constraints on the Kheis tectogenesis and the Kaapvaal Craton's earliest Wilson cycle. J. Afr. Earth Sci. 26 (1), 5–27.

Crerar, D.A., Cormick, R.K., Barnes, H.L., 1980. Geochemistry of manganese: an overview. In: Varentsov, I.M., Grasselly, G. (Eds.), Geology and Geochemistry of Manganese, vol. 1. Schweizerbart'sche, Stuttgart, pp. 293–334.

Crerar, D.A., Namson, J., Chyi, M.S., et al., 1982. Manganiferous cherts of the Franciscan assemblage: 1. General geology, ancient and modern analogues, and implications for hydrothermal convection at oceanic spreading centers. Econ. Geol. 77 (3), 519–540.

de Villiers, J. (Ed.), 1960. The Manganese Deposits of the Union of South Africa. Government Printer, Pretoria. 280 pp.

De Villiers, J.E., 1983. The manganese deposits of Griqualand West, South Africa: some mineralogical aspects. Econ. Geol. 78, 1108–1118.

Deconynck, A., Leprêtre, R., Missenard, Y., Bernard, A., Barbarand, J., Saint-Bezar, B., Saddiqi, O., Ruffet, G., Yans, J., 2014. Genesis of manganese ore deposits of the Imini District (Morocco). In: Académie Royale des sciences d'Outre-Mer 2014, Brussels, Belgium.

Du, Q., Yi, H., Hui, B., Li, S., Xia, G., Yang, W., Wu, X., 2013. Recognition, genesis and evolution of manganese ore deposits in southeastern China. Ore Geol. Rev. 55, 99–109.

Editorial Committee of Mineral Deposits of China (Eds.), 1995. Mineral Deposits of China, vol. 3. Geological Publishing House, Beijing 188 pp.

Efremova, T.V., Pal'shin, N.I., 2003. Formation of vertical thermal structure in lakes in northwestern Russia and Finland. Water Resour. 30 (6), 640–649.

Eichman, R., Schidlowski, M., 1975. Isotopic fractionation between coexisting organic carbon—carbonate pairs in Precambrian sediments. Geochim. Cosmochim. Acta 39 (5), 585–595.

Emelyanov, E.M., Pilipsnuk, M., Volostnykh, B.V., Khandros, G.S., Shaidurov, Y.O., 1982. Fe and Mn forms in sediments in the geochemical profile of the Baltic Sea. Baltica (Vilnius) 7, 153–171.

Epstein, S., Graf, D.F., Degens, E.T., 1964. Oxygen isotope studies on the origin of dolomites. In: Isotopic and Cosmic Chemistry. North-Holland, Amsterdam, pp. 169–180.

Fan, D., Yang, P., 1999. Introduction to and classification of manganese deposits of China. Ore Geol. Rev. 5 (1), 1–13.

Fan, D., Ye, J., Yin, L., et al., 1999. Microbial process in the formation of the Sinian Gaoyan manganese carbonate ore, Sichuan Province, China. Ore Geol. Rev. 5 (1), 79–93.

Fellerer, R., 1986. Manganknollen, mineralische Rohstoffe aus dem Meer—metallreserven der Zukunft. Aufschluss 37 (1), 1–18.

Ferenczi, P.A., 2001. Iron ore, manganese and bauxite deposits of the Northern Territory. Northern Territory Geological Survey, Report 13, Darwin. 113 pp. (Fig. 53).

Ferronsky, V.I., Brezgunov, V.S., Romanov, V.V., et al., 1996. Isotope studies of water dynamics. Implications of the rise Caspian Sea level. Isotopes in Water Resources Management, vol. 1. International Atomic Energy Agency, Vienna, pp. 129–140.

Fonselius, S., Valderrama, J., 2003. One hundred years of hydrographic measurements in the Baltic Sea. J. Sea Res. 49 (4), 229–241.

Force, E.R., Cannon, W.F., 1988. Depositional model for shallow-marine manganese deposits around black shale basins. Econ. Geol. 83 (1), 93–117.

Frakes, L.A., Bolton, B.R., 1984. Origin of manganese giants: sea level change and anoxic-oxic history. Geology 12 (1), 83–86.

Friedman, J., O'Neil, Y.R., 1977. Compilation of Stable Isotope Fractionation Factors of Geochemical Interest. Government Printing Office, Washington, DC. 110 pp. (US Geol. Surv. Prof. Pap.; N 440-KK).

Gautier, D.L., Pratt, L.M., 1986. Carbon, oxygen and sulfur isotopic trends in carbonate concretions from the Upper Cretaceous Sharon Springs Member of the Pierre Shale, Colorado. Terra Cognita 6 (2), 108–132.

Gieskes, J.M., Lawrence, J.R., Perry, E.A., et al., 1987. Chemistry of interstitial waters and sediments in the Norwegian-Greenland sea, Deep Sea Drilling Project Leg 38. Chem. Geol. 63 (1), 143–155.

Glasby, G.P., 1988. Manganese deposition through geological time: dominance of the post-Eocene deep-sea environment. Ore Geol. Rev. 4 (1–2), 135–144.

Gradstein, F.M., Ogg, J.G., Schmitz, M.D., Ogg, G.M. (Eds.), 2012 The Geologic Time Scale. 2012. Elsevier, Amsterdam. 1144 pp.

Green, D., Eng, C., Daly, M.C., 1982. Manganese mineralization in Zambia. Trans. Inst. Min. Metall. B: Appl. Earth Sci. 91, B33–B41.

Grenne, T., Slack, J.F., 2003. Bedded jaspers of the Ordovician Lokken ophiolite, Norway: seafloor deposition and diagenetic maturation of hydrothermal plume-derived silica-iron gels. Mineral. Deposita 38, 625–639.

Gutzmer, J., Beukes, N.J., 1995. Fault controlled metasomatic alteration of early Proterozoic sedimentary manganese ores in the Kalahari manganese field, South Africa. Econ. Geol. 90, 823–844.

Gutzmer, J., Beukes, N.J., 1996. Mineral paragenesis of the Kalahari manganese field, South Africa. Ore Geol. Rev. 11 (3), 405–428.

Gutzmer, J., Beukes, N.J., 1997. Effects of mass transfer, compaction and secondary porosity on hydrothermal upgrading of Paleoproterozoic sedimentary manganese ore in the Kalahari manganese field, South Africa. Mineral. Deposita 32 (1–2), 250–256.

Gutzmer, J., Beukes, N.J., Yeh, H.-W., 1997. Fault controlled metasomatic alteration of Early Proterozoic sedimentary manganese ores at Mamatwan mine, Kalahari manganese field, South Africa. S. Afr. J. Geol. 100 (1), 53–71.

Gutzmer, J., Pack, A., Luders, V., Wilkinson, J.J., Beukes, N.J., van Niekerk, H.S., 2001. Formation of jasper and andradite during low-temperature hydrothermal seafloor metamorphism, Ongeluk formation, South Africa. Contrib. Mineral. Petrol. 142 (1), 27–42.

Hariya, Y., Tsutsumi, M., 1981. Hydrogen isotopic composition of MnO(OH) minerals from manganese oxide and massive sulfide (Kuroko) deposits in Japan. Contrib. Mineral. Petrol. 77, 256–261.

Hartman, M., 1964. Zur Geochemie von Manganese und Eisen in der Ostsee. Meyniana 14 (53), 3–21.

Hartway, J.C., Degens, E.T., 1969. Methane-derived marine carbonates of pleistocene age. Science 165 (3894), 690–692.

Hein, J.R., Bolton, B., 1992. Stable isotope composition of Nikopol and Chiatura manganese ores, USSR: implications for genesis of large sedimentary manganese deposits. In: 29th International Geological Congress, Kyoto, Japan. p. 209 (Abstracts, 1-3-47, 1691).

Hein, J.R., Koski, R.A., 1987. Bacterially mediated diagenetic origin for chert-hosted manganese deposits in the Franciscan Complex, California Coast Ranges. Geology 15 (8), 722–726.

Hein, J.R., O'Neil, J.R., Jones, M.G., 1979. Origin of authigenic carbonates in sediment from the deep Bering Sea. Sedimentology 26 (5), 681–705.

Hein, J.R., Fan, D., Ye, J., et al., 1999. Composition and origin of Early Cambrian Tiantaishan phosphorite-Mn carbonate ores, Shaanxi Province, China. Ore Geol. Rev. 15 (1), 95–134.

Herz, N., Banerjee, S., 1973. Amphibolites of the Lafaiete, Minas Gerais, and the Serra Do Navio Manganese Deposits Brazil. Econ. Geol. 68 (8), 1289–1296.

Herzig, P.M., Becker, K.P., Stoffers, P., Backer, H., Blum, N., 1988. Hydrothermal silica chimney field in the Galapagos Spreading Center at 86°W. Earth Planet. Sci. Lett. 89 (2), 261–272.

Hoefs, J., 1992. The stable isotope composition of sedimentary iron oxides with special reference to banded iron formations. In: Clauer, N., Chaudhury, S. (Eds.), Lecture Note in Earth Sciences and Sedimentary Records. Springer-Verlag, Amsterdam, pp. 199–212.

Honnorez, J., Von Herzen, R.P., Barrett, T.J., et al., 1983. Hydrothermal mounds and young ocean crust of the Galapagos: preliminary deep sea drilling results. Initial Reports of Deep Sea Drilling Project, vol.70. U.S. Govt. Printing office, Washington, DC, pp. 459–481.

Hovland, M., Talbot, M.R., Qvale, H., et al., 1987. Methane-related carbonate cements in pockmarks of the North Sea. J. Sediment. Petrol. 57 (5), 881–892.

Hren, M.T., Love, D.R., Tice, M.M., et al., 2006. Stable isotope and rare earth element evidence within the Archean Barberton greenstone belt, South Africa. Geochim. Cosmochim. Acta 70 (6), 1457–1470.

Hudson, I.D., 1977. Stable isotopes and limestone lithification. J. Geol. Soc. 133 (6), 637–660.

Huebner, J.S., Flohr, V.J.K., Grossman, J.N., 1992. Chemical fluxes and origin of manganese carbonate-oxide-silicate deposit in bedded chert. Chem. Geol. 100 (1), 93–118.

Hunter, D.R. (Ed.), 1981. Precambrian of the Southern Hemisphere. Elsevier, Amsterdam.

Irvin, H., Curtis, C.D., Coleman, M., 1977. Isotopic evidence for source of diagenetic carbonates formed during burial of organic rich sediments. Nature 269 (2), 209–213.

Jennings, M., 1986. The Middelplaats manganese ore deposit, Griqualand West. In: Anhaeusser, C.R., Maske, S. (Eds.), Mineral Deposits of Southern Africa, vol. 1. Geological Society of South Africa, Johannesburg, pp. 979–983.

Johnson, C.J., Clark, A.L., 1985. Potential of Pacific Ocean nodule, crust, and sulfide mineral deposits. Nat. Res. Forum 9 (3), 179–186. Blackwell Publishing Ltd.

Jonsson, A., et al., 2001. Whole-lake mineralization of allochthonous and autochthonous organic carbon in a large humic lake (Örträsket, N. Sweden). Limnol. Oceanogr. 46 (7), 1691–1700.

Klein, C., Beukes, N.J., 1989. Geochemistry and sedimentology of a facies transition from limestone to iron-formation deposition in the early Proterozoic Transvaal Supergroup, South Africa. Econ. Geol. 84 (7), 1733–1774.

Klein, C., Ladeira, E.A., 2004. Geochemistry and mineralogy of Neoproterozoic banded iron-formations and some selected, siliceous manganese formations from the Urucum district, Mato Grosso do Sul, Brazil. Econ. Geol. 99 (6), 1233–1244.

Kleinschrot, D., Klemd, R., Brochern, M., et al., 1991. Die Manganlagerstatte Nsuta, Ghana. In: Berichte zur Lagerstatten und Rohstoff-Forschung, Hannover. Bd. 6. 119 s.

Kleinschrot, D., Klemd, R., Bröcker, M., Okrusch, M., Schmidt, K., 1993. The Nsuta manganese deposit, Ghana: geological setting, ore-forming process and metamorphic evolution. Z. Angew. Geol. 39, 48–50.

Klemm, D.D., 2000. The formation of Palaeoproterozoic banded iron formations and their associated Fe and Mn deposits, with reference to the Griqualand West deposits, South Africa. J. Afr. Earth Sci. 30 (1), 1–24.

Kleyenstuber, A.S.E., 1984. The mineralogy of the manganese-bearing Hotazel formation of the Proterozoic Transvaal sequence in Griqualand West, South Africa. Trans. Geol. Soc. S. Afr. 87, 257–272.

Koepnick, R.B., Burke, W.H., Denison, R.E., et al., 1985. Construction of the seawater $^{87}Sr/^{86}Sr$ curve for the Cenozoic and Cretaceous: supporting data. Chem. Geol. Isot. Geosci. 58 (1–2), 55–81.

Kudelin, B.I., Zektser, I.S., Meskheteli, A.V., Brusilovsky, S.A., 1971. The problem of groundwater discharge into the seas. EOS Trans. Am. Geophys. Union 52 (10), 717–722.

Kuleshov, V.N., 2003. Isotopic composition ($\delta^{13}C$, $\delta^{18}O$) and origin of carbonate manganese ores in Early Oligocene deposits, Eastern Paratethys. Chem. Erde 63, 329–363.

Kunzmann, M., Gutzmer, J., Beukes, N.J., Halverson, G.P., 2014. Depositional environment and lithostratigraphy of the Paleoproterozoic Mooidraai Formation. Kalahari manganese field, South Africa. S. Afr. J. Geol. 117 (2), 173–192.

Lawrence, J.R., Gieskes, J.M., 1981. Constraints on water transport and alteration in the oceanic crust from the isotopic composition of pore water. J. Geophys. Res. 86 (B9), 7924–7934.

Lawrence, J.R., Gieskes, J.M., Broecker, W.S., 1975. Oxygen isotope and cation composition of DSDP pore waters and the alteration of Layer II basalts. Earth Planet. Sci. Lett. 27 (1), 1–10.

Laznica, P., 1992. Manganese deposits in the global lithogenetic system: quantitative approach. Ore Geol. Rev. 7 (4), 279–356.

Leclerc, J., Weber, F., 1980. Geology and genesis of the Moanda manganese deposits, Republic of Gabon. In: Varentsov, I.M., Grasselly, G. (Eds.), Manganese Deposits on Continents. Geology and Geochemistry of Manganese, vol. 2. Academia Kiado, Budapest, pp. 89–109.

Lepland, A., Stevens, R.L., 1998. Manganese authigenesis in the Landsort Deep, Baltic Sea. Mar. Geol. 151 (1), 1–25.

Li, Ch., Love, G.D., Lyons, T.W., Scott, C.T., Feng, L., Huang, J., Chang, H., Zhang, Q., Chu, X., 2012. Evidence for a redox stratified Cryogenian marine basin, Datangpo Formation, South China. Earth Planet. Sci. Lett. 331–332, 246–256.

Liu, T.-B., Maynard, J.B., Alten, J., 2006. Superheavy S isotopes from glacier-associated sediments of the Neoproterozoic of south of China: oceanic anoxia or sulfate limitation? Mem. Geol. Soc. Am. 198, 1–18.

Longstaff, F.I., Ayalon, A., 1986. Oxygen isotopes studies of diagenesis in clastic rocks from the Viking and Belly River Formations, Alberta. Terra Cognita 6 (2), 109.

Lynn, D.S., Bonnati, E., 1965. Mobility of manganese in the diagenesis of deep-sea sediments. Mar. Geol. 3 (6), 457–474.

Macaulay, C.I., Haszeldine, R.S., Fallick, A.E., 1993. Distribution, chemistry, isotopic composition and origin of diagenetic carbonates: Magnus Sandstone North Sea. J. Sediment. Res. 63 (1), 33–43.

Manheim, F.T., 1961. A geochemical profile in the Baltic Sea. Geochim. Cosmochim. Acta 25 (1), 52–70.

Marchig, V., Erzinger, J., Rosch, H., 1987. Sediments from a hydrothermal field in the central valley of the Galapagos rift spreading center. Mar. Geol. 76 (2), 243–251.

Martynova, M.V., 2013. Exchange of Mn compounds between bottom sediments and water: 1. Mn flux from water to the bed. Water Resour. 40 (6), 640–648.

Martynova, M.V., 2014. Exchange of manganese compounds between bottom sediments and water: 2. Manganese flux from bed into water (a brief review of studies). Water Resour. 41 (2), 178–187.

Maynard, J.B., 2003. Chapter 15: Manganiferous sediments, rocks and ores. In: MacKenzie, F.T., (Ed.), Sediments, Diagenesis, and Sedimentary Rocks. Treatise on Geochemistry, vol. 7. Elsevier, Amsterdam, pp. 289–308.

Maynard, J.B., 2010. The chemistry of manganese ores through time: a signal of increasing diversity of earth-surface environments. Econ. Geol. 105 (3), 535–552.

Maynard, J.B., 2013. The Kalahari manganese field—the adventure continues. Econ. Geol. 108 (8), 2021 (Book Review).

Maynard, J.B., 2014. Manganiferous sediments, rocks and ores. In: second ed., In: Holland, H.D., Turekian, K.K. (Eds.), Treatise on Geochemistry. vol. 9. Elsevier, Oxford, pp. 327–349.

McCrea, J.M., 1950. On the isotopic chemistry of carbonates and a paleotemperature scale. J. Chem. Phys. 18 (6), 849–857.

McDuff, R.E., Gieskes, J.M., Lawrence, J.R., 1978. Interstitial water studies, leg 42A. Initial Reports of the DSDP, vol. 42 Pt. 1. U.S. Government Printing Office, Washington, DC, pp. 561–568.

Meister, P., Bernasconi, S.M., Aiello, I.W., Vasconcelos, C., McKenzie, J.A., 2009. Depth and controls of Ca-rhodochrosite precipitation in bioturbated sediments of the Eastern Equatorial Pacific, ODP Leg 201, Site 1226 and DSDP Leg 68, Site 503. Sedimentology 56, 1552–1568.

Melcher, F., 1995. Genesis of chemical sediments in Birimian greenstone belts: evidence from gondites and related manganese-bearing ricks from northern Ghana. Mineral. Mag. 59 (2), 229–251.

Miyano, T., Beukes, N.J., 1987. Physicochemical environments for the formation of quartz-free manganese oxide ores from the Early Proterozoic Hotazel formation, Kalahari manganese field, South Africa. Econ. Geol. 82 (3), 706–718.

Morad, S., Al-Aasm, I.S., 1997. Conditions of rhodochrosite-nodule formation in Neogene-Pleistocene deep-sea sediments: evidence from O, C and Sr isotopes. Sediment. Geol. 114 (2), 295–304.

Mottl, M.J., Lawrence, J.R., Keigwin, L.D., 1983. Elemental and stable isotope composition of pore waters and carbonate sediments from deep sea drilling project sites 501/504 and 505. Initial Reports of DSDP. vol. 69. U.S. Government Printing Office, Washington, DC, pp. 461–463.

Mozley, P.S., Burns, S.J., 1993. Oxygen and carbon isotopic composition of marine carbonate concretions: an overview. J. Sediment. Petrol. 63 (1), 73–83.

Muke, A., Dzigbodi-Adjimah, K., Annor, A., 1999. Mineralogy, petrography, geochemistry and genesis of the Paleoproterozoic Birimian manganese-formation of Nsuta, Ghana. Mineral. Deposita 34 (3), 297–311.

Nagy, B., Kunen, S.M., Zumberge, J.E., et al., 1974. Carbon content and carbonate ^{13}C abundances in the early Precambrian Swaziland sediments of South Africa. Precambrian Res. 1 (1), 43–48.

Nel, C.J., Beukes, N.J., De Villiers, J.P.R., 1981. *The* Mamatwan manganese mine of the Kalahari manganese field. In: Anhaeusser, C.R., Maske, S. (Eds.), Mineral Deposits of Southern Africa, vol. 1. Geological Society of South Africa, Johannesburg, pp. 963–978.

Neumann, Th., Heizer, U., Leosson, M.A., Kersten, M., 2002. Early diagenetic process during Mn-carbonate formation: evidence from the isotopic composition of authigenic Ca-rhodochrosites of the Baltic Sea. Geochim. Cosmochim. Acta 66 (5), 867–879.

Ntiamoa-Agiakwa, Y., 1979. Relationship between gold and manganese mineralization in the Birimian of Ghana, West Africa. Geol. Mag. 116 (5), 342–345.

Nyame, F.K., 2008. Petrography and geochemistry of intraclastic manganese-carbonates from the 2.2 Ga Nsuta deposit of Ghana: significance for manganese sedimentation in the Palaeoproterozoic of West Africa. J. Afr. Earth Sci. 50 (1), 133–147.

Nyame, F.K., Beukes, N.J., 2006. The genetic significance of carbon and oxygen isotopic variations in Mn-bearing carbonates from the Palaeo-Proterozoic (~2.2 GA) Nsuta deposit in the Birimian of Ghana. Carbonates Evaporites 21 (1), 21–32.

Nyame, F.K., Beukes, N.J., Kase, K., Yamamoto, M., 2002. Compositional variations in manganese carbonate micronodules from the lower proterozoic Nsuta deposit, Ghana: product of authigenic precipitation or post-formational diagenesis? Sediment. Geol. 154, 159–175.

Ohmoto, H., 1972. Systematic of sulfur and carbon isotopes in hydrothermal ore deposits. Econ. Geol. 67 (5), 551–578.

Okita, P.M., Shanks, W.C.I.I.I., 1988. $\delta^{13}C$ and $\delta^{34}S$ trends in sedimentary manganese deposits, Molango (Mexico) and Taojiang (China): evidence for mineralization in a closed system. In: International Association of Sedimentologists, Symposium on Sedimentology Related to Mineral Deposits, 30 July–4 August, Beijing, China, pp. 188–189 (Abstracts).

Okita, P.M., Shanks III, W.C., 1992. Origin of stratiform sediment-hosted manganese carbonate ore deposits: examples from Molango, Mexico, and Taojiang, China. Chem. Geol. 99 (1–3), 139–164.

Okita, P.M., Maynard, J.B., Spikers, E.C., et al., 1988. Isotopic evidence for organic matter oxidation by manganese reduction in the formation of stratiform manganese carbonate ore. Geochim. Cosmochim. Acta 52, 2679–2685.

Ostwald, J., 1980. Aspects of mineralogy, petrology and genesis of the Groote Eylandt manganese deposits. In: Varentsov, I.M., Grasslley, G. (Eds.), Manganese Deposits of Continents. Geology and Geochemistry of Manganese, vol. 2. Academia Kiado, Budapest, pp. 149–181.

Ostwald, J., 1988. Mineralogy of the Groote Eylandt manganese oxides: a review. Ore Geol. Rev. 4 (1), 3–45.

Ozturk, H., Hein, J.R., 1997. Mineralogy and stable isotopes of black shale-hosted manganese ores, southwestern Taurides, Turkey. Econ. Geol. 92, 733–744.

Palmer, M.R., Elderfield, H., 1985. Sr isotope evolution of sea waters over the past 75 Myr. Nature 314, 526–528.

Pedersen, T.F., Price, N.B., 1982. The geochemistry of manganese carbonate in Panama Basin sediments. Geochim. Cosmochim. Acta 46 (1), 59–68.

Perry, E.A., Gieskes, J., Lawrence, J.R., 1976. Mg, Ca and O^{18}/O^{16} exchange in the sediment-pore water system, Hole 149, DSDP. Geochim. Cosmochim. Acta 40 (3), 413–423.

Pierre, C., Rouchi, J.-M., 1986. Oxygen and sulfur isotopes in anhydrites from Givetian and Visean evaporites of northern France and Belgium. Chem. Geol. Isot. Geosci. 58, 245–252.

Polgari, M., Okita, P.M., Hein, J.R., 1991. Stable isotope evidence for the origin of the Urkut manganese ore deposit, Hungary. J. Sediment. Petrol. 61 (2), 384–393.

Polgari, M., Vigh, T., Szabo-Drubina, M., et al., 2009. Characterization and genetic aspects of the Urkut manganese deposit, the role of microbes in manganese accumulations in low temperature aquatic systems. In: Manganese in the Twenty-First century, Short Course, 5–9 September 2009, Veszprem, pp. 46–67. Abstract volume.

Polgári, M., Hein, J.R., Tóth, A.L., Pál-Molnár, E., Vigh, T., Bíró, L., Fintor, K., 2012. Microbial action formed Jurassic Mn-carbonate ore deposit in only a few hundred years (Úrkút, Hungary). Geology 40 (10), 903–906.

Post, J.E., 1999. Manganese oxide minerals: crystal structures and economic and environmental significance. Proc. Natl. Acad. Sci. 96 (7), 3447–3454.

Pracejus, B., Bolton, B., Frakes, L.A., 1988. Nature and development of supergene manganese deposits, Groote Eyland, Northern territory, Australia. Ore Geol. Rev. 4 (1–2), 71–98.

Preston, P.C.C.R., 2001. Physical and chemical characterization of the manganese ore bed at the Mamatwan mine, Kalahari manganese field. Dissertation (unpubl.). Johannesburg. 101 p. http://152.106.6.200:8080/dspace/bitstream/10210/1967/1/PaulaPreston.pdf.

Rarsotti, A.F., 1984. Copper, cobalt, nickel and manganese availability from land-based endowments. A perspective. Nat. Res. Forum 8 (3), 267–278.

Reeve, W.H., 1963. The geology and mineral resources of Northern Rhodesia. Geological Survey Bulletin (Zambia), vol. 3. 213 pp.

Reyna, J.G. (Ed.), 1956. Simposio Sobre Yacimientmos de Manganese: XX Congreso Geologico International, vols. 1–5. Mexico.

Robbins, J.A., Callender, E., 1975. Diagenesis manganese in Lake Michigan sediments. Am. J. Sci. 275 (5), 512–533.

Rogers, J.J., 1996. A history of continents in the past three billion years. J. Geol. 104 (1), 91–107.

Rosenbaum, J., Sheppard, S.M.F., 1986. An isotopic study of sediments, dolomites and ankerites at high temperatures. Geochim. Cosmochim. Acta 50 (5), 1147–1150.

Roy, S., 1969. Classification of manganese deposits. Acta Mineral. Petrogr. Szeged. 19, 67–83.

Roy, S., 1981. Manganese Deposits. Academic Press, London. 458 pp.

Sass, E., Bein, A., Almogi-Labin, A., 1991. Oxygen-isotope composition of diagenetic calcite in organic-rich rocks: Evidence for ^{18}O depletion in marine anaerobic pore water. Geology 19 (8), 839–842.

Savin, S.M., Epstein, S., 1970. The oxygen and hydrogen isotope geochemistry of clay minerals. Geochim. Cosmochim. Acta 34 (1), 25–42.

Scarpelli, W., 1973. The Serra do Navio manganese deposit (Brazil). In: Genesis of Precambrian Iron and Manganese Deposits. Proc. Kiev Symp. 1970. Earth sciences, vol. 9. UNESCO, Paris, pp. 217–227.

Schidlowski, M., Eichmann, R., Junge, C.E., 1975. Precambrian sedimentary carbonates: carbon and oxygen isotope geochemistry and implications for the terrestrial oxygen budget. Precambrian Res. 2 (1), 1–69.

Schidlowski, M., Appel, P.W.U., Eichman, R., 1979. Carbon isotope geochemistry of the $3.7x10^9$- yr-old Isua carbon and oxygen cycles. Geochim. Cosmochim. Acta 43 (2), 189–199.

Schopf, J.W. (Ed.), 1983. Earth's earliest biosphere: its origin and evolution. Princeton University Press, Princeton, NJ. 543 pp.

Schopf, J.W., 2006. The first billion years: when did life emerge? Elements 2 (4), 229–233.

Service, H., Dunn, J.A., 1943. The geology of the Nsuta manganese deposits. Gold Coast Geological Survey, Memoir, vol. 5 F.J. Milner, London. 32 pp.

Slee, K.J., 1980. Geology and origin of the Groote Eyland manganese oxide deposits, Australia. In: Varentsov, I.M., Grasselly, G. (Eds.), Manganese Deposits on Continents. Geology and Geochemistry of Manganese, vol. 2. Academia Kiado, Budapest, pp. 125–148.

Söderberg, P., Flodén, T., 1992. Gas seepages, gas eruptions and degassing structures in the seafloor along the Strömma tectonic lineament in the crystalline Stockholm Archipelago, east Sweden. Cont. Shelf Res. 12 (10), 1157–1171.

Sohlenius, G., Sternbeck, J., Andren, E., et al., 1996. Holocene history of the Baltic Sea a recorded in a sediment core from the Gotland Deep. Mar. Geol. 134 (3–4), 183–201.

Sorem R.K., 1986. Polymetallic resource estimates of east equatorial pacific manganese nodule deposits. Условия образования рудных месторождений. Тр.6 Симп. МАГРМ, Тбилиси, 6–12 сент., 1982. Т.2. М.: Наука. С. 718–722 (англ.).

Sorem, R.K., Cameron, E.N., 1960. Manganese oxides and associated minerals of the Nsuta manganese deposits, Ghana, West Africa. Econ. Geol. 55 (2), 278–310.

Staley, G.H.S., 1986. The diagenesis of the cretaceous Cardium and Viking Formations, Alberta Basin, Canada. Terra Cognita 6 (2), 107–108.

Suess, E., 1979. Mineral phases formed in anoxic sediments by microbial decomposition of organic matter. Geochim. Cosmochim. Acta 43 (3), 339–353.

Tang, S., Liu, T., 1999. Origin of the early Sinian Minle manganese deposit, Hunan Province, China. Ore Geol. Rev. 15 (1), 71–78.

Thienhaus, R., 1967. Montangeologische Probleme lateritischer Manganerz-Lagerstatten. Mineral. Deposita 2 (4), 253–270.

Thode, H.C., Shima, V., Rees, C.E., et al., 1965. Carbon-13 isotope effects in systems containing carbon dioxide, bicarbonate, carbonate and metal ions. Can. J. Chem. 43 (3), 582–595.

Timoleon, N., Sylvestre, G., Nono, G.D., Paul, N.J., 2015. Mineralogy and geochemistry of Neoproterozoic siliceous manganese formations from Ntui–Betamba (Cameroon Pan-African Fold Belt): implications for mineral exploration. Int. J. Earth Sci. 104 (5), 1123–1138.

Trendall, A.F., Morris, R.C. (Eds.), 1983. Iron Formations: Facts and Problems. Elsevier, Amsterdam.

Tsikos, H., Beukes, N.J., Moore, N.J., et al., 2003. Deposition, diagenesis, and secondary enrichment of metals in the Paleoproterozoic Hotazel iron formation, Kalahari manganese field, Southern Africa. Econ. Geol. 98, 1449–1462.

Urban, H., Stribrny, B., Lippolt, H.J., 1992. Iron and manganese deposits of the Urucum district, Mato Grosso do Sul, Brazil. Econ. Geol. 87, 1375–1392.

USGS mineral commodity summary—manganese. http://minerals.er.usgs.gov/minerals/pubs/commodity/manganese/index.html (accessed 04.04.16).

Varentsov, I.M., 1996. Manganese Ores of Supergene Zone: Geochemistry of Formation. Kluwer Academic Publishers, Dordrecht 302 pp.

Varentsov, I.M., 2002. Genesis of the Eastern Paratethys manganese ore giants: impact of events at the Eocene/Oligocene boundary. Ore Geol. Rev. 20 (1), 65–82.

Varentsov, I.M., Grasselly, G. (Eds.), 1980. Geology and Geochemistry of Manganese. vol. 2. Academiai Kiado, Budapest 513 pp.

Varentsov, I.M., Rakhmanov, V.P., Gurvich, E.M., et al., 1984. Genetic aspects of the formation of manganese deposits in the geological history of the Earth's crust. In: Metallogenesis and Mineral Ore Deposits. Proceedings of 27th International Geolological Congress. VNU Science Press, Utrecht, pp. 275–291.

Veizer, J., 1983. Trace elements and isotopes in sedimentary carbonates. In: Reeder, R.J. (Ed.), Carbonates: Mineralogy and Chemistry. Reviews in Mineralogy, vol. 11. Mineralogical Society of America, Washington, DC, pp. 265–299.

Veizer, Jan, 1988. The evolving exogenic cycle. In: Gregor, C.B., Garrels, R.M., Mackenzie, F.T., Maynard, J.B. (Eds.), Chemical Cycles in the Evolution of the Earth. John Wiley & Sons, New York, pp. 175–220.

Veizer, J., Compston, W., 1974. $^{87}Sr/^{86}Sr$ composition of seawater during the Phanerozoic. Geochim. Cosmochim. Acta 38, 1461–1484.

Veizer, J., Ala, D., Azmy, k., et al., 1991. $^{87}Sr/^{86}Sr$, $\delta^{13}C$ and $\delta^{18}O$ evolution of Phanerozoic seawater. Chem. Geol. 161 (1), 59–88.

Von Rad, U., Rosch, H., Berner, U., et al., 1996. Authigenic carbonates derived from oxidized methane vented from the Makram accretionary prism of Pakistan. Mar. Geol. 136 (1–2), 55–77.

Walde, D.H.G., Gierth, E., Leonardos, O.H., 1981. Stratigraphy and mineralogy of the manganese ores of Urucum, Mato Grosso, Brazil. Geol. Rundsch. 70 (3), 1077–1085.

Walls, R.A., Mountjoy, E.W., Fritz, P., 1979. Isotopic composition and diagenetic history of carbonate cements in Devonian Golden Spike reef, Alberta Canada. Geol. Soc. Am. Bull. 90 (10), 963–982. Pt. 1.

Walters, L.Y., Claypool, G.E., Choquette, P.W., 1972. Reaction rates and $\delta^{18}O$ variation for the carbonate-phosphoric acid preparation method. Geochim. Cosmochim. Acta 36 (2), 129–140.

Weber, F., 1973. Genesis and supergene evolution of the Precambrian sedimentary manganese deposit at Moanda (Gabon). In: Genesis of Precambrian Iron and Manganese deposits. Proc. Kiev Symp., 1970, Earth Sciences. vol. 9. UNESCO, Paris, pp. 307–322.

Webmineral: manganese-containing minerals. http://webmineral.com/chem/Chem-Mn.shtml#.VwP13BFViko (accessed 04.04.16).

Wedepohl, K.H., 1980. Potential sources for manganese oxide precipitation in the oceans. In: Varentsov, I.M., Grasselly, G. (Eds.), Geology and Geochemistry of Manganese, vol. 3. Publishing House of Hungarian Academy of Sciences, Budapest, pp. 13–22.

Winterhalter, B., 2001. On Sediment Patchiness at the BASYS coring site, Gotland Deep, Baltic Sea. Baltica 14, 18–23.

Xie, J., Sun, W., Du, J., Xu, W., Wu, L., Yang, X., Zhou, T., 2013. Geochemical studies on Permian manganese deposits in Guichi, eastern China: implications for their origin and formative environments. J. Asian Earth Sci. 74, 155–166.

Yapp, C.J., 1990. Oxygen isotopes in iron (III) oxides 2. Possible constraints on the depositional environment of a Precambrian quartz-hematite banded iron formation. Chem. Geol. 85 (3–4), 337–344.

Ye, L., Fan, D., Yang, P., 1988. Characteristics of manganese ore deposits in China. Ore Geol. Rev. 4 (1), 99–113.

Yeh, H.-W., Hein, J.R., Bolton, B.R., 1995. Origin of the Nsuta manganese carbonate proto-ore, Ghana: carbon- and oxygen-isotope evidence. J.Geol. Soc. China 38 (4), 397–407.

Yen, H.-N., Savin, S.M., 1976. The extent of oxygen isotope exchange between clay minerals and sea water. Geochim. Cosmochim. Acta 40 (7), 743–748.

Zeng, Y., Liu, T., 1999. Characteristics of the Devonian Xialei manganese deposit, Guangxi Zhuang Region, China. Ore Geol. Rev. 15 (1), 153–163.

Zhegallo, E.A., Rozanov, A.Yu., Ushatinskaya, G.T., et al., 2000. Atlas of Microorganisms From Ancient Phosphorites of Khubsugul (Mongolia). National Aeronautics and Space Administration, Marshall Space Flight Center, Hunstville, AL. 171 pp.

Zheng, S., Feng, X., Xu, X., et al., 1986. Stable isotopic studies on the Mn nodules. Terra Cognita 6 (2), 116. Letters from the organizers of JCOG VI.

Index

Note: Page numbers followed by *f* indicate figures and *t* indicate tables.

A

B

C

D

E

F

G

H

I

N

O

V

W

X

Z

Printed in the United States
By Bookmasters